Pedagogical Use of Color

The colors that you see in the illustrations of this text are used to impr clarity and understanding. Many figures with three-dimensional perspectives are airbrushed in various colors to make them look as realistic as possible.

Color has been used in various parts of the book to identify specific physical quantities. The following schemes have been adopted.

Chapters 1–10: Motion

Speed and Velocity

Acceleration

Force

Rotation

Linear momentum

Angular momentum

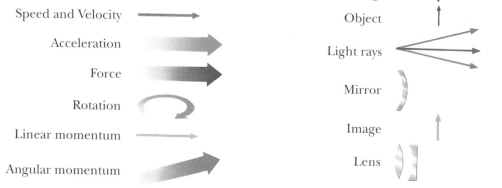

Chapters 17–19: Light and Optical Devices

Object

Light rays

Mirror

Image

Lens

Chapters 20–27: Electricity and Magnetism

Positive charge

Negative charge

Electric force and field

Magnetic force and field

Neutron

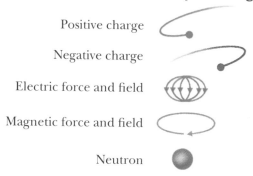

Passage of Time/Clock Icon Art

Some art shows the development of a phenomenon over time as a series of "snapshots." A clock icon indicates the passage of time in this art.

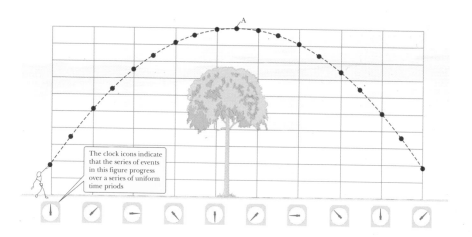

The clock icons indicate that the series of events in this figure progress over a series of uniform time priods

Physics

A Conceptual World View

7th EDITION

Larry D. Kirkpatrick | Gregory E. Francis

CENGAGE
Learning™

Australia • Brazil • Japan • Korea • Mexico • Singapore • Spain • United Kingdom • United States

Physics: A Conceptual World View

Executive Editors:
 Maureen Staudt
 Michael Stranz

Senior Project Development Manager:
 Linda deStefano

Marketing Specialist:
 Courtney Sheldon

Senior Production/Manufacturing Manager:
 Donna M. Brown

PreMedia Manager:
 Joel Brennecke

Sr. Rights Acquisition Account Manager:
 Todd Osborne

Cover Image:
Getty Images*

*Unless otherwise noted, all cover images used by Custom
Solutions, a part of Cengage Learning, have been supplied
courtesy of Getty Images with the exception of the Earthview
cover image, which has been supplied by the National
Aeronautics and Space Administration (NASA).

Physics, A Conceptual World View
Kirkpatrick | Francis

© 2011 Cengage Learning. All rights reserved.

For product information and technology assistance, contact us at
Cengage Learning Customer & Sales Support, 1-800-354-9706

For permission to use material from this text or product,
submit all requests online at **cengage.com/permissions**
Further permissions questions can be emailed to
permissionrequest@cengage.com

This book contains select works from existing Cengage Learning resources and
was produced by Cengage Learning Custom Solutions for collegiate use. As such,
those adopting and/or contributing to this work are responsible for editorial
content accuracy, continuity and completeness.

Compilation © 2011 Cengage Learning
ISBN-13: 978-1-133-36299-9

ISBN-10: 1-133-36299-0

Cengage Learning
5191 Natorp Boulevard
Mason, Ohio 45040
USA
Cengage Learning is a leading provider of customized learning solutions with
office locations around the globe, including Singapore, the United Kingdom,
Australia, Mexico, Brazil, and Japan. Locate your local office at:
international.cengage.com/region.

Cengage Learning products are represented in Canada by Nelson Education, Ltd.
For your lifelong learning solutions, visit **www.cengage.com/custom.**
Visit our corporate website at **www.cengage.com.**

Printed in the United States of America

Table of Contents

1

A World View

> ▶This photograph shows a cluster of galaxies. Each galaxy, a collection of billions of stars, appears small because it is very far from Earth. How far is it to the farthest galaxy that we may ever observe?
>
> (See page 13 for the answer to this question.)

Atlas Image courtesy of 2MASS/UMASS/IPAC-Caltech/NASA/NSF

A cluster of distant galaxies.

PHYSICS is the study of the material world. It is a search for patterns, or rules, for the behavior of objects in the universe. This search covers the entire range of material objects, from the smallest known particles—millions of millions of times smaller than a marble—to astronomical objects—millions of millions of times bigger than our Sun. The search also covers the entire span of time, from the primordial fireball to the ultimate fate of the universe. Within this vast realm of space and time, the searchers have one goal: to comprehend the course of events in the whole world—to create a world view.

First Grade

This course could be one of the most challenging experiences that you will ever have—except for first grade. But then you were too young to notice.

What happened in first grade?* Well, you learned to read, and that was really difficult. You first had to learn the names of all those weird little squiggles. You had to learn to tell a *b* from a *d* from a *p*. Even though they looked so much alike, you did it, and it even seemed like fun. Then you learned the sounds each letter represented, and that was not easy because the capitals looked different but made the same sound and some letters could have more than one sound.

Then one day your teacher put some letters on the board: first, the letter *C*, and you all knew it could make a "kuh" or "suh" sound; then the letter *A*, which had lots of possibilities; and finally a *T*, which luckily had only one sound. You tried out several combinations including "kuh-AAH-tuh." Then suddenly someone shouted out in triumph, "That isn't 'kuh-AAH-tuh'! It's a small furry animal with a long skinny tail that says 'meow.'" And your world was never the same again. When your car paused at an eight-sided red sign, you sounded out *stop* and understood how the drivers knew what to do. You saw the words *ice cream* on the front of a store and knew you wanted to go in.

If this book works, you will become aware of a whole world you never noticed before. You will never walk down a street, ride in a car, or look in a mirror without involuntarily seeing an extra dimension. Sometimes you will have to memorize what symbols mean—just as in first grade. Sometimes you will confuse things that seem as much alike as *b*, *d*, and *p* once did, until you suddenly see how different they are. And sometimes you will look at a combination of events and equations helplessly reciting "kuh-AAH-tuh" in total frustration. This has happened to all of us. But then the moment of insight will come, and you will see whole new images fitting together. You will see the *C-A-T* and will experience fully, and consciously, the exhilaration you felt in first grade.

So, welcome to one of the most challenging (and rewarding) courses you have ever taken in your life. If you work at it and let it happen, this experience will change your world view forever.

On Building a World View

The term *world view* has a fairly elastic meaning. When we think about world views, the interpretations can stretch from the philosophic to the poetic. In physics the world view is a shared set of ideas that represents the current explanations of how the material world operates. These include some rather common constructs, such as gravity and mass, as well as strange-sounding ones, such as quarks and black holes.

* Adapted from Barbara Wolff, "An Introduction to Physics—Find the CAT." *The Physics Teacher* 27 (1989): 427.

(a)

Figure 1-1 (a) Reproduction of a diagram from one of Albert Einstein's letters. (b) Our interpretation of this diagram.

(b)

Creative leap

Axiom

C_1 C_2 C_3 C_4 C_5 Consequences

The Real World

   The physics world view is a dynamic one. Ideas are constantly being proposed, debated, and tested against the material world. Some survive the scrutiny of the community of physicists; some don't. The inclusion of new ideas often forces the rejection of previously accepted ones. Some firmly accepted ideas in the world view are very difficult to discard; in the long run, however, experimentation wins out over personal biases. The model of the atom as a miniature solar system was reluctantly given up because the experimental facts just didn't support it; it was replaced by a mathematical model that's difficult to visualize.

   A few years before his death, Albert Einstein described the process of science to a lifelong friend, Maurice Solovine. Solovine was not a scientist but apparently enjoyed discussing science with his famous friend. In one of their last exchanges, Solovine wrote that he had trouble understanding a certain passage in one of Einstein's essays. The next week Einstein wrote back, carefully explaining his view of the process of science. Figure 1-1(a) is a reproduction of a diagram from this letter. Figure 1-1(b) shows our interpretation of the diagram.

   In his text Einstein explains the diagram. The lower horizontal line represents the real world. The curved line on the left signifies the creative leap a scientist makes in attempting to explain some phenomena. The leap is intuitive, and although it may be very insightful, it is not scientific. The scientific process begins, Einstein explains, when the scientist takes the idea, or axiom, and develops consequences based on it. These consequences are illustrated by a number of smaller circles connected to the axiom with lines. A very powerful axiom has a large number of consequences. The final task in the process of science is to test these consequences against the material world. In Einstein's

(a)

(b)

The study of physics began mostly in an attempt to understand the motions of planets, stars, and other bodies far from Earth. Shown are (a) Galileo's original telescope, which enabled him to discover the four largest moons orbiting Jupiter, and (b) a modern reflecting telescope.

drawing, these tests are vertical arrows returning to the world. If there is no match between the predicted consequences and the real world, the idea is scientifically worthless. If there is a match, there is hope that the idea has merit. With the publication of the idea, the community of scientists is brought into the process, and the original work is often modified.

Ideas that have simple physical models are easier to understand and accept. There is comfort in "picturing" electrons and protons as tiny balls, but those physicists who yearned for the electron to be a miniature billiard ball never got their wish. Cosmologist Sir Hermann Bondi, commenting on doing physics in realms beyond the range of direct human experience, said, "We should be surprised that the gas molecules behave so much like billiard balls and not surprised that electrons don't." This advice, though correct, may leave one feeling like Alice, in *Through the Looking Glass,* listening to the White Queen say, "Why, sometimes I've believed as many as six impossible things before breakfast."

Although some ideas in physics may appear to be contrary to common sense, they do in fact make sense. Normally, things that make sense—that don't violate your intuition, or common sense—are those that fit into your past experience. Common sense is a personal world view. Like the physics world view, your common sense is built on a large experimental base. The difference between what makes sense to you and what makes sense to a physicist is in part due to different ranges of experience. Whereas our observations are limited by the range of human sensations, the physicist has instruments that bring ultrahuman sensations into consideration. Our twinkling star is the physicist's window into the universe.

So, without common sense to guide them about whether to accept a new idea, how do physicists decide which ones to adopt? Acceptance is based on whether the idea works, how well it fits into the world view, and whether it is better than the old explanations. Although the most basic criteria for accepting an idea are that it agrees with the results of past experiments and successfully predicts the outcome of future experiments, acceptance is a human activity. Because it is a human activity, it has subjective aspects. The phrases *how well it fits* and *if it is better* imply opinions. Ideas have appeal, some more than others.

If an idea is very general, having many consequences in the Einsteinian sense, it can replace many separate ideas. It is regarded as more fundamental and thus more appealing. It is possible to construct a different explanation for each observation. For example, a scheme could be created to explain the disappearance of water from an open container; another, unrelated idea could be employed to account for the fluidity of water; and so on. An idea about the structure of liquids (not just water) that could be used to explain these phenomena and many others would be a highly valued replacement for the collection of separate ideas.

The simplicity of an idea also influences opinions about its worth. If more than one construct is proposed to explain the same phenomena and if they all predict the experimental results equally well, the most appealing idea is the simplest one. Although elaborate (Rube Goldberg) constructions are cute in cartoons, they hold very little value in the building of a physics world view.

Incredible as an idea may be initially, physicists seem to become more and more comfortable with it the longer it remains in their world view. Most physicists are comfortable with the relativistic notions of slowed-down time and warped space; when the ideas were first introduced, however, they caused quite a stir. As more and more experimental results support an idea, it gains stature and becomes a more established part of our beliefs. But even if an idea becomes very familiar and comfortable, it is still tentative. Experimental results can never prove an idea; they can only disprove it. If the predictions are borne out, the best that can be claimed is, "So far, so good."

© 1992 by Nick Downes; from Big Science.

"IN LAYMAN'S TERMS? I'M AFRAID I DON'T KNOW ANY LAYMAN'S TERMS."

Our goal in writing this book is to help you view the world differently. We describe the building of a physics world view and share with you some of the results. We start in areas that are familiar, where your common sense serves you well, and chart a course through less-traveled areas. Although there is no end to this journey, the book must stop. We hope it stops at a new place, a place you have never been before.

We also hope that you find joy in the process similar to that expressed by Isaac Newton:

> I do not know what I may appear to the world, but to myself I seem to have been only like a boy playing on the sea-shore, and diverting myself in now and then finding a smoother pebble or a prettier shell than ordinary, whilst the great ocean of truth lay all undiscovered before me.

Bode's Law

In 1776 Titius of Wittenberg developed a numerical rule that gave the relative sizes of the orbits of the planets. This rule was popularized by J. E. Bode and is now known as *Bode's law*. But is this rule a law of physics?

The rule can be developed through the following process: The first and second numbers in the first column of Table 1-1 are 0 and 3. All subsequent num-

Table 1-1	*Bode's Law*			
		Bode's Law (AU)	Modern Value (AU)	Planet
0	4	0.4	0.39	Mercury
3	7	0.7	0.72	Venus
6	10	1.0	1.00	Earth
12	16	1.6	1.52	Mars
24	28	2.8		
48	52	5.2	5.20	Jupiter
96	100	10.0	9.54	Saturn
192	196	19.6		
384	388	38.8		

bers are double the preceding one. The second column is obtained by adding 4 to each of the numbers in the first column. The third column is obtained by dividing the entry in the second column by 10. These are the numbers in Bode's law. Because all the radii of the orbits are measured relative to Earth's, the value for Earth (the "third rock from the Sun") must be 1. The radius of Earth's orbit, the mean distance between Earth and the Sun, is known as an *astronomical unit,* or AU.

The rule tells us that the radius of Mercury's orbit must be 0.4 times that of Earth's, or 0.4 astronomical unit. This was in agreement with the value known in 1776 and is close to the modern value of 0.39 astronomical unit. Likewise, the value of 0.7 astronomical unit for Venus's orbit agreed with the value at the time the law was developed and is close to the modern value of 0.72 astronomical unit. The modern values for the orbits of Mars and Saturn differ a bit from the rule, but the rule was consistent with the data known at the time the rule was proposed.

A scientific law must be testable—that is, it must make predictions that can be tested. Notice that there were no planets known in 1776 that matched the values in the fifth, eighth, or ninth rows. These rows can be viewed as predictions for the orbital radii of planets that had not yet been discovered. Uranus, discovered in 1781, has an orbit with a mean radius of 19.2 astronomical units, near the predicted value. So far, so good. The first of the asteroids, known as Ceres, was discovered in 1801, and its orbit has a mean radius of 2.8 astronomical units, in agreement with the prediction. Does this mean that Bode's law is a physical law of nature?

Neptune, discovered in 1846, has a mean orbital radius of 30.1 astronomical units; Pluto, discovered in 1930, has a mean orbital radius of 39.4 astronomical units. Clearly, Bode's law does not work for the two outer planets. Because it failed to predict these radii, Bode's law must be discarded.

However, Bode's law was never a candidate for a physical law. It never had any scientific basis; it was simply a way of remembering the values for the known radii. To this day, we know of no way of predicting the radii of the planetary orbits in the recently discovered planetary systems around nearby stars. Furthermore, we believe that each planetary system is unique.

We are not saying that numerical rules such as Bode's law have no role in science. As we will see numerous times in our study of physics, discovering patterns in nature may provide the first steps in developing physical laws.

Are You On the Bus?

Q: What are the three criteria required for an idea to become a physics law?

A: (1) It must account for the known data. (2) It must make predictions that can be tested. (3) It must have a scientific basis.

> Extended presentation available in the *Problem Solving* supplement

Measurements

If someone offered to sell you a bar of gold for $2000, you would immediately ask, "How large is the bar?" The size of the bar obviously determines whether it is a good buy. A similar problem existed in the early days of commerce. Even when standard units of measure were used, they were not the same from time to time and region to region. Later, several standardized systems of measurement were developed.

The two dominant systems are the *U.S. customary system* based on the foot, pound, and second, and the *metric system* based on the meter, kilogram, and second. Thomas Jefferson advocated that the United States adopt the metric system, but his advice was not taken. As a result, most people in the United

States do not use the metric system. It is used, however, by the scientific community and those who work on such things as cars. England and Canada have now officially changed over to the metric system. The United States is the only major country not to have made the change.

Having the entire world use a single system offers obvious advantages. It avoids the cumbersome task of converting from one system to another and aids in worldwide commerce. Disadvantages for countries that must change are the expense of converting machinery, signposts, and standards; the maintenance of dual systems for a time; and the abandonment of the familiar units for new ones for which people do not have intuitive feelings.

The metric system has advantages over the U.S. customary system and was chosen in 1960 by the General Conference on Weights and Measures. The official version is known as **Le Système International d'Unités** (International System of Units) and is abbreviated SI.

One problem of the U.S. customary system of measurement is illustrated by Table 1-2, in which we have listed a sample of units used to measure lengths. Some of these units are historic; others have been developed for use in specialized areas. Many of them may be unfamiliar. Few of us have any idea how long a fathom is, let alone how many inches there are in a fathom. Even among the more familiar units, converting from one unit to another is often a difficult task. For instance, to determine how many inches are in a mile, we must make the following computation

$$1 \text{ mile} = (5280 \text{ ft})\left[\frac{12 \text{ in.}}{1 \text{ ft}}\right] = 63,360 \text{ in.}$$

The metric system eliminates the confusion of having many different unfamiliar units and the difficulty of converting from one size unit to another. It does this by adopting a single standard unit for each basic measurement and a series of prefixes that make the unit larger or smaller by factors of 10. For instance, the

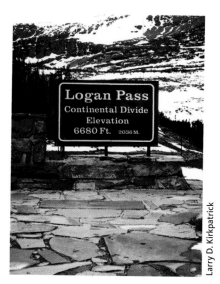

Some highway signs give the elevation in feet and meters—in this case 6680 feet and 2036 meters.

Larry D. Kirkpatrick

Table 1-2	*Partial List of Length Measures*
angstrom	micron
astronomical unit	mil
barleycorn	mile
cubit	nautical mile
fathom	pace
fermi	palm
foot	parsec
furlong	pica
hand	point
inch	rod
league	stadium
light-year	thumbnail
meter	yard

Are *You* On the Bus?

Q: The average adult human body contains about 45 miles of nerves. When you tell your sister that she is getting on every single inch of your nerves, how many inches of nerves is she getting on?

A: Each mile of nerves would be 63,360 inches of nerves. Forty-five miles of nerves would be $45 \times 63,360 = 2,851,200 \approx 2,900,000$ inches of nerves. That's a lot of nerve. For a discussion of significant figures, please see Section 1.1 of the *Problem Solving* supplement.

Table 1-3 *The Metric Prefixes*

Prefix	Symbol	Meaning	Power of Ten	Value
tera-	T	trillion	10^{12}	1,000,000,000,000
giga-	G	billion	10^{9}	1,000,000,000
mega-	M	million	10^{6}	1,000,000
kilo-	k	thousand	10^{3}	1,000
		one	10^{0}	1
centi-	c	hundredth	10^{-2}	0.01
milli-	m	thousandth	10^{-3}	0.001
micro-	μ	millionth	10^{-6}	0.000 001
nano-	n	billionth	10^{-9}	0.000 000 001
pico-	p	trillionth	10^{-12}	0.000 000 000 001
femto-	f	quadrillionth	10^{-15}	0.000 000 000 000 001

basic unit for measuring length is the **meter** (m), which is a little longer than a yard (yd). This unit is inconveniently small for measuring distances between cities, so road signs everywhere else in the world display *kilometers* (km) as a unit of length. The prefix **kilo-** means "one thousand" and indicates that the kilometer is equal to 1000 meters. The kilometer is about $\frac{5}{8}$ mile, so a distance of 50 miles equals about 80 kilometers. Speed limits appear as 100 kilometers per hour (km/hr) rather than 65 miles per hour (mph).

▶ *kilo-* = 1000

Smaller distances are measured in such units as the *centimeter* (cm). The prefix **centi-** means "one hundredth." It takes 100 centimeters to equal 1 meter. The other common prefixes are given in Table 1-3 along with their abbreviations and various forms of their numerical values. These units are pronounced with the accent on the prefix. Note also that the terms *billion* and *trillion* do not have the same meanings in all countries. In this textbook, 1 billion is 1,000,000,000, and 1 trillion is 1000 billion.

▶ *centi-* = $\frac{1}{100}$

WORKING IT OUT *Crawler Speed*

The crawler that carries the space shuttle to launch pad 39A moves very slowly. If it travels at 1 mph, what is its speed in furlongs per fortnight? One furlong is equal to ⅛ mile, and a fortnight is 2 weeks.

We often need to convert a measurement in one set of units into another set of units. There shouldn't be any confusion about whether to multiply or divide by the conversion factor if you follow a definite procedure each time. The procedure uses multiplication by 1. This works because 1 is the only number that does not change the value of the measurement. However, it can change the form of the measurement. The technique involves writing 1 as a fraction with its numerator equal to its denominator. For instance, we have learned that 1 fortnight is the same as 14 days. Thus,

$$\frac{14 \text{ days}}{1 \text{ fortnight}} = 1 \text{ or } \frac{1 \text{ fortnight}}{14 \text{ days}} = 1$$

We can now convert measurements in fortnights to the equivalent values in days (or vice versa) by multiplying by 1 in one of these forms. How do you know which form to use? Always choose the fraction that cancels out the old unit. This will automatically put the new unit in its proper place. We are now ready to convert miles per hour into furlongs per fortnight.

$$1\frac{\text{mile}}{\text{hr}}\left[\frac{8 \text{ furlongs}}{1 \text{ mile}}\right]\left[\frac{24 \text{ hr}}{1 \text{ day}}\right]\left[\frac{14 \text{ days}}{1 \text{ fortnight}}\right] = 2688\frac{\text{furlongs}}{\text{fortnight}}$$

Because all the prefixes are multiples of 10, conversions between units are done by moving the decimal point—that is, multiplying and dividing by 10s. For instance, because **milli-** means "one thousandth," we can convert from meters to *millimeters* (mm) by multiplying by 1000. There are 5670 millimeters in 5.67 meters.

◀ *milli-* $= \frac{1}{1000}$

In its purest form, the SI system allows no other units for length. However, other units that have grown up historically will remain in use for some time (and maybe forever). For instance, the terms *micron* and *fermi* continue to be used for *micrometer* and *femtometer,* respectively. A common unit of length on the astronomical scale is the *light-year,* the distance light travels in 1 year. Although it is not an SI unit, it is a naturally occurring unit of length on this scale and will continue to be used. On the other hand, the *angstrom* (10^{-10} meter) has been very popular, but is now being replaced by the *nanometer* (10^{-9} meter).

The metric system also differs from the U.S. customary system in that mass is considered the primary unit and weight (force) the secondary unit. In the U.S. customary system, the situation is reversed. (The distinction between mass and weight is explained in Chapter 3.) The basic unit in the U.S. customary system is the *pound* (lb), but the basic unit in the SI system is the **kilogram** (kg). [It may seem strange to use the kilogram rather than the gram (g) as the basic unit, but the gram was deemed to be too small for the basic unit.] The weight of 1 kilogram is 2.2 pounds. A U.S. nickel has a mass that is very close to 5 grams. The term *megagram* (1000 kilograms) is not often used in the metric system. This is known as a *metric ton* and has a weight equal to a *long ton* (2200 pounds).

Since the invention of the metric system in 1791, many unsuccessful attempts have been made to change the time system over to a decimal basis so that time units would also be multiples of 10. They have all failed. The SI system has the same units of time as the U.S. customary system.

Because the metric system is the primary system used in science, we should gain some familiarity with it. On the other hand, our principal goal is to estab-

Everyday Physics | *Monumental Metric Mistakes*

The Mars Climate Orbiter was supposed to go into orbit around Mars on September 23, 1999, acting as a weather satellite studying the chemical properties of the Martian atmosphere. Instead, the $125 million satellite flew too close to the planet and burned up. Later investigation found that the tragedy was the result of a simple miscommunication between engineers at Lockheed Martin and those at NASA. Lockheed Martin sent thrust data to NASA in units of pounds. NASA was expecting the data to be in metric units of newtons (N) and interpreted them as such. One pound is equal to 4.448 newtons. This mistake caused the orbiter to drift slowly off course each time it fired its booster rockets to stabilize itself during its 9-month, 420-million-mile journey. When the orbiter finally arrived at its destination, it was 60 miles off course, closer to the planet's surface than expected. The craft was most likely torn apart and burned by the Martian atmosphere that it had been sent to study. House Science Committee chairman F. James Sensenbrenner

Jr., sounding stunned, released a two-word statement after hearing the news about the miscommunication: "I'm speechless."

In a related story with a happier ending, a student intern working for an American oil company in London was given the job of checking the design of a new refinery that was being built. He noticed that the American engineers working for the company had calculated the length of pipe needed in units of feet. The London office was about to place an order for the pipe, assuming the numbers represented meters. He saved the company from ordering 48 kilometers of extra pipe. As a reward, they took the student to dinner at a fancy restaurant. (*New Scientist,* December 11, 1999, p. 55)

1. NASA engineers would describe the 60-mile error in metric units. By how many kilometers was the Mars Climate Orbiter off course when it reached the red planet?

2. How many feet of pipe were needed for the refinery?

lish connections between your commonsense world view and the physics world view. Learning the metric system at the same time as beginning your study of physics is complicated because you need to develop a feeling for the new units as well as for the scientific ideas. As a compromise, we will use the metric system predominantly, but we will give the approximate U. S. equivalents in parentheses when it is useful.

Sizes: Large and Small

Imagine taking a photograph of three children lying on a blanket in their backyard.* Assume that the camera is located directly above the children and that the scene captured on the film is 1 meter wide and 1 meter tall. One meter, a little more than 3 feet, is the scale of children. In fact, this is also the scale of adults because factors of 2, 3, or even 5 don't matter when we are talking about the approximate sizes of objects.

If we now move the camera 10 times as far away from the children, the film will capture a scene that is 10 meters wide by 10 meters tall. The new scene has 100 times the area and could include approximately 300 children. If we once again move the camera 10 times as far away, the new scene will be 100 meters on a side, about the length of a football or soccer field. At yet another 10 times as far away, our scene would be 1000 meters, or 1 kilometer (0.62 mile) on a side and could include the children's neighborhood.

As we continue increasing our distance from the children by additional factors of 10, the scenes captured by the camera will get larger and larger. In fact, they will become so large that it will be very difficult to keep track of the number of zeros in the length and width of the photograph. Therefore, we use **powers-of-ten notation,** which displays the number of zeros in these numbers. In mathematics the notation 10^2 means 10×10, which is equal to 100. Similarly, 10^3 means $10 \times 10 \times 10 = 1000$. The superscript is called an *exponent*

---

* Based on Philip Morrison, Phylis Morrison, and the Office of Charles and Ray Eames, *Powers of Ten* (Redding, Conn.: Scientific American Books, 1982). You can watch this video online at http://powersof10.com.

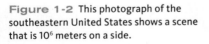

Figure 1-2 This photograph of the southeastern United States shows a scene that is 10^6 meters on a side.

Jacques Descloitres, MODIS Land Rapid Response Team, NASA/GSFC

Figure 1-3 Earth spills over the edges of the photograph at a scale of 10^7 meters.

Figure 1-4 The Milky Way Galaxy has a scale of 10^{21} meters.

and is equal to the number of 10s that are multiplied together. The exponent is also equal to the numbers of zeros in these numbers. (Note that $10^0 = 1$.)

Positive values for exponents indicate that the numbers are large. Negative exponents indicate small numbers. For instance, $10^{-1} = \frac{1}{10^1} = 0.1$ and $10^{-2} = \frac{1}{10} \times \frac{1}{10} = \frac{1}{10^2} = 0.01$. The minus sign indicates that the power of 10 is in the denominator; that is, it is to be divided into 1. The **order of magnitude** for a quantity is its value rounded off to the nearest power of ten.

At a scale of 10^6 meters (620 miles), our scene could encompass all but the largest states as shown in the Landsat photograph in Figure 1-2. At 10^7 meters, most of Earth would fit into our scene as shown in Figure 1-3. The Moon's orbit would be included at a scale of 10^9 meters, and the entire solar system would be included at a scale of 10^{13} meters. As we continue outward, the photographs look much the same: a field of stars of varying brightness. At a scale of 10^{21} meters, our scene shows the Milky Way Galaxy (Figure 1-4), a collection of billions of stars. Continuing outward, the Milky Way shrinks in size and many other galaxies enter the picture. At even larger distance, the galaxies look like stars. (See the chapter-opening photograph.) At a scale of 10^{26} meters, we approach the edge of the universe that is visible from Earth.

Let's now return to our children and imagine looking at smaller and smaller scales, each time decreasing the size of our photograph by a factor of 10. With a photograph that is $\frac{1}{10}$ meter on a side (a scale of 10^{-1} meter), our photograph could include a child's hand. At a scale of 10^{-2} meter = 1 centimeter [a little less than $\frac{1}{2}$ inch (in.)], we might only see a child's fingernail. The thickness of the fingernail is approximately 10^{-3} meter = 1 millimeter. The red blood cell shown in Figure 1-5 is approximately 10^{-5} meter across. Blood capillaries are only a little larger than the red blood cells. The smallest living cells are $\frac{1}{100}$ that size at a scale of 10^{-7} meter. The atoms making up the molecules in these cells have diameters on the order of 10^{-10} meter. On a scale smaller by a factor of approximately 10,000, we find the nuclei of these atoms; at a scale of 10^{-15} meter, we find the protons and neutrons that make up the nuclei of atoms. Even these protons and neutrons are made up of quarks, which we will study in Chapter 27.

Figure 1-5 A scanning electron micrograph of a red blood cell. Red blood cells have diameters of approximately 10^{-5} meter. Notice the white blood cell in the upper right corner.

The size of the visible universe is an incredible 10^{41} times the size of protons and neutrons. Physicists study the material world at both extremes of the size scale and everywhere in between.

WORKING IT OUT *Powers of Ten*

In this book we study objects and events that go far beyond the normal human scale of objects and events. When we look at phenomena on very large and very small scales, the sizes of the numbers quickly get out of hand. For instance, the approximate radii of the visible universe and a proton are as follows:

$$\text{radius of visible universe} = 140{,}000{,}000{,}000{,}000{,}000{,}000{,}000{,}000 \text{ m}$$

$$\text{radius of proton} = 0.000\,000\,000\,000\,001\,2 \text{ m}$$

These numbers are very difficult to read, write, and manipulate mathematically. It is even easy to make errors in counting the zeros unless they are grouped in threes as we have done.

Using powers-of-ten notation, the radius of the visible universe is written as 1.4×10^{26} m. This indicates that the number in front is to be multiplied by ten 26 times to get the actual number. Because multiplication by 10 moves the decimal point one position to the *right*, the superscript 26 indicates the total number of places the decimal point must be moved. By convention the number out in front is usually written so that it has a value between 1 and 10. You should check that moving the decimal point 26 places to the right in the number 1.4 gives the value in the first paragraph.

The radius of the proton is written as 1.2×10^{-15} m. The number in front must be divided by ten 15 times. Or equivalently, the number can be obtained by moving the decimal point 15 positions to the *left*.

Sometimes you may see a number written with only the power of 10. If the usual number preceding a power of 10 is missing, it is assumed to be 1; that is, $10^5 = 1 \times 10^5$. Similarly, if the exponent is missing, it is assumed to be zero; that is, $4 = 4 \times 10^0$.

The greatest power of using this notation comes when you have to multiply or divide very large or small numbers. For multiplication, multiply the two numbers in front and *add* the exponents. For example,

$$R_{universe} \times R_{proton} = (1.4 \times 10^{26}\ \text{m}) \times (1.2 \times 10^{-15}\ \text{m})$$
$$= (1.4 \times 1.2) \times 10^{26+(-15)}\ \text{m} \times \text{m}$$
$$= 1.7 \times 10^{11}\ \text{m}^2$$

This number represents an area that has half the width of a proton and a length that extends from Earth to the edge of the visible universe.

For *division*, you divide the two numbers in front and *subtract* the exponent in the denominator from that in the numerator. For example,

$$\frac{R_{universe}}{R_{proton}} = \frac{1.4 \times 10^{26}\ \text{m}}{1.2 \times 10^{-15}\ \text{m}} = \frac{1.4}{1.2} \times 10^{26-(-15)}\ \frac{\text{m}}{\text{m}}$$
$$= 1.2 \times 10^{41}$$

This number gives the relative size of the visible universe and the proton—that is, how many times larger the visible universe is than a proton.

Summary

This course could be one of the most challenging experiences that you've had since first grade. At the same time, it could be one of your most rewarding experiences, one that could change your view of the world around you.

The physics world view is a shared set of ideas that represents current explanations of how the material world operates. It is a dynamic view with new ideas being proposed, debated, and tested. For a new idea to be accepted, it must (1) agree with the existing data, (2) make predictions that can be tested, and (3) have a scientific basis.

The measurement system used in science (and most of the world) is the metric, or SI, system based on the meter, kilogram, and second. Larger and smaller units are obtained through the use of prefixes.

The sizes of objects studied in physics range from the entire universe (at a scale of 10^{26} meters) down to neutrons and protons (at a scale of 10^{-15} meter).

CHAPTER 1 *Revisited*

The farthest galaxies that we may ever observe lie near the edge of the visible Universe, at a distance of approximately 10^{26} meters from Earth.

Key Terms

centi- A prefix meaning "one hundredth." A centimeter is $\frac{1}{100}$ meter.

kilo- A prefix meaning "one thousand." A kilometer is 1000 meters.

kilogram The SI unit of mass. A kilogram of material weighs about 2.2 pounds on Earth.

Le Système International d'Unités The French name for the metric, or System International (SI), system of units.

meter The SI unit of length equal to 39.37 inches, or 1.094 yards.

milli- A prefix meaning "one thousandth." A millimeter is $\frac{1}{1000}$ meter.

order of magnitude The value of a quantity rounded off to the nearest power of ten.

powers-of-ten notation A method of writing numbers in which a number between 1 and 10 is multiplied by 10 raised to a power.

> Questions and exercises are paired so that most odd-numbered are followed by a similar even-numbered.
> Blue-numbered questions and exercises are answered in Appendix B.
> ⸙ indicates more challenging questions and exercises.
> WebAssign Many Conceptual Questions and Exercises for this chapter may be assigned online at WebAssign.

Conceptual Questions

1. Compare and contrast the physics world view with your own personal world view.

2. What is a physics world view?

3. Why is Bode's law giving the sizes of the orbits of the planets not considered a physical law?

4. Why should you be suspicious of a book titled *The Theory of Everything*?

5. What are the criteria for accepting a theory as a physical law?

6. Which of the criteria for a physical law are not satisfied by Bode's law, which gives the sizes of the orbits of the planets?

7. What role does the prestige of the scientist play in accepting a theory as a physical law?

8. Would you be more likely to accept a scientific theory proposed by a professor at a major university or one developed by the handyman down the street?

9. Which major countries (if any) have not adopted the metric system as the primary system of measurement?

10. What are the advantages and disadvantages of adopting the metric system?

11. What is the height of a typical person, in centimeters?

12. What is the length of a typical newborn baby, in centimeters?

13. What is the typical height of a bedroom ceiling, in meters?

14. What is the length of a full-sized bed, in meters?

15. What is the typical mass of a 6-foot-tall man, in kilograms?

16. What is the typical mass of a 5-foot, 6-in.-tall woman, in kilograms?

17. What is the order of magnitude for the distance across the United States?

18. What is the order of magnitude for the world's population?

Exercises

19. How many seconds are there in 1 day?

20. How many seconds are there in 1 year?

21. How long is a 100-m dash in yards if 1 m = 1.094 yd?

22. If 1 in. = 2.54 cm, how tall (in centimeters) is a 6-ft-tall basketball player?

23. How many inches are there in 1 m?

24. How many inches are there in 1 km?

25. Write each of the following numbers in powers-of-ten notation:
    a. 89,760 in.
    b. 0.000 000 000 000 707 g

26. Write each of the following numbers in powers-of-ten notation:
    a. 2,378,000,000 m
    b. 0.003 24 ft

27. Write each of the following numbers as ordinary numbers:
    a. 4.3×10^3 g
    b. 8.12×10^{-5} m

28. Write each of the following numbers as ordinary numbers:
    a. 5.782×10^6 s
    b. 6.9×10^{-3} ft

29. Complete the following computations:
    a. $(6.8 \times 10^{-3}) \times (2.3 \times 10^4)$
    b. $\dfrac{6.8 \times 10^5}{2.0 \times 10^{-3}}$

30. Complete the following computations:
    a. $(4.2 \times 10^8) \times (2.2 \times 10^4)$
    b. $\dfrac{4.4 \times 10^4}{1.4 \times 10^2}$

31. Approximately how many times larger is the orbit of Pluto than the orbit of the Moon about Earth?

32. Approximately how much larger is a child's fingernail than one of the protons in the fingernail?

Describing Motion

▶The blurred image of the train clearly shows that it's moving. To appear blurred in the photograph, the train had to be in different places during the time the shutter was open. If we know how fast the train is moving, can we determine how long it will take to reach its destination?

(See page 29 for the answer to this question.)

© Georgina Bowater

The blurring of the train's image shows that it is moving.

A PROPERTY common to everything in the universe is change. Some things are big, some are small; some are red, some have no color at all; some are rigid, some are fluid; but they all are changing. In fact, change is so important that the fundamental concept of time would be meaningless without it.

Change even occurs where seemingly there is none. Water evaporating, colors fading, flowers growing, and stars evolving are all examples of changes that are beyond our casual observations. Also beyond our sensations is the fact that these changes are a result of the motion of material, often at the submicroscopic level. Because change—and thus motion—is so pervasive, we begin our exploration of the ideas of physics with a study of motion.

Within our commonsense world view, we generally group all motions together, simply observing that an object is moving or that it is not moving. Actually, there is an extraordinary diversity of motion, ranging from the very simple to the extremely complicated. Fortunately, the complex motions—ones more common in our everyday experiences—can be understood as combinations of simpler ones. For example, Earth's motion is a combination of a daily rotation about its axis and an annual revolution around the Sun. Or, closer to home, the motion of a football can be treated as a combination of a vertical rise and fall, a horizontal movement, and a spinning about an axis.

We therefore begin our discussion of motion by trying to describe and understand the simplest kinds of motion. This will yield a conceptual framework within our world view from which even the most complicated motions, such as those associated with a hurricane or with a turbulent waterfall, can be understood.

The motion of a football is a combination of three simpler motions: a vertical rise and fall, a horizontal motion, and a spinning about an axis.

Havasu Creek on the Colorado River in Grand Canyon National Park.

Average Speed

Imagine driving home from school. For simplicity, assume that you can drive home in a straight line. Normally, you might describe this trip in terms of the time it takes. If pressed for a more detailed account, you would probably give the distance or the actual route taken, adding points of interest along the way. For our purposes we need to develop a more precise description of motion.

First, we note that your position continually changes as you drive. Second, we observe that it takes time to make the trip. These two fundamental notions—space and time—are at the core of our concept of motion. Furthermore, different positions along the trip can be matched with different times.

One relationship between space and time can be illustrated by answering the question, "How fast were you going?" Actually, there are two ways to answer this question; one way looks at the total trip, whereas the other considers the moment-by-moment details of the trip. For the total-trip description, we use the concept of an **average speed**, which is defined as the total distance traveled divided by the time it took to cover this distance.

$$\text{average speed} = \frac{\text{distance traveled}}{\text{time taken}} \blacktriangleright$$

We can write this relationship more efficiently by using symbols as abbreviations:

$$\bar{s} = \frac{d}{t}$$

where $\bar{s}$ is the average speed, d is the distance traveled, and t is the time taken for the trip. A bar is often used over a symbol to indicate its average value.

This ratio of distance over elapsed time gives the average rate at which the car's position changes. Speed is a quantitative measure of how rapidly the change takes place. The definition of average speed states a particular relationship between the concepts of space and time. If any two of the three quantities are known, the third is determined.

A humorous story about a small-time country farmer illustrates this relationship. The farmer was visited by his big-time cousin. Anxious to make a good impression, the host spent the morning showing his cousin around his small farm. At lunch the cousin could not resist the urge to brag that on his ranch he could get into his car in the early morning and drive until sunset and he would still be on his property. The country farmer thought for a moment and then said, "I had a car like that once."

To determine speed, we need a device for measuring distance, such as a ruler, and one for measuring time, such as a clock. Most highways have mile markers along the side of the road so that maintenance and law enforcement officials can accurately find certain locations. These mile markers and your wristwatch give you all the information you need to determine average speeds.

Assuming that we begin "thinking metric," speeds have units such as meters per second (m/s) or kilometers per hour (km/h). A person walks about $1\frac{1}{2}$ meters per second, and a car traveling at 70 miles per hour is going approximately 113 kilometers per hour.

Images of Speed

◀ Extended presentation available in the *Problem Solving* supplement

From the earliest cave drawings to modern time-lapse photography, it has been a part of human nature to try to represent our experiences. Artists, as well as scientists, have devised many ways of illustrating motion. A blurred painting or photograph such as that in Figure 2-1 is one way to "see" motion. One difference between the artist and the scientist is that the scientist uses the representations to analyze the motion.

A clever image of motion that also provides a way of measuring the speed of an object is the multiple-exposure photograph. These photographs are made in a totally dark room with a *stroboscope* (usually just called a *strobe*) and a camera with an open shutter. A strobe is a light source that flashes at a constant, controllable rate. The duration of each flash is very short (about 10 millionths of a second), producing a still image of the moving object.

If the strobe flashes 10 times per second, the resulting photograph will show the position of the object at time intervals of $\frac{1}{10}$ of a second. Thus, we can "freeze" the motion of the object into a sequence of individual events and use this representation to measure its average speed within each time interval.

As an example of measuring average speed, let's determine the average speed of the puck in Figure 2-2. The puck travels from a position near the

Mile markers and your wristwatch can be used to calculate your average speed.

Figure 2-1 The blurring of the background tells us that the race car is moving.

Figure 2-2 A strobe illustration of a moving puck shows its position at different times.

4-centimeter mark to one near the 76-centimeter mark, a total distance of 72 centimeters. Because there are seven images, there are six intervals and the total time taken is six times the time between flashes—that is, 0.6 second. Therefore, the average speed is

$$\bar{s} = \frac{d}{t} = \frac{72 \text{ cm}}{0.6 \text{ s}} = 120 \text{ cm/s}$$

We can also determine the average speed of the puck between each pair of adjacent flashes. Allowing for the uncertainties in reading the values of the positions of the puck, the average speed for each time interval is the same as the overall average. Therefore, the puck was traveling at a constant speed of 120 centimeters per second.

Suppose you live 40 miles from school and it takes you 2 hours to drive home. Your average speed during the trip is

$$\bar{s} = \frac{d}{t} = \frac{40 \text{ miles}}{2 \text{ h}} = 20 \frac{\text{miles}}{\text{h}}$$

This means that, on the average, you travel a distance of 20 miles during each hour of travel. This answer is read "20 miles per hour" and is often written as 20 miles/hour, or abbreviated as 20 mph. It is important to include the units with your answer. A speed of "20" does not make any sense. It could be 20 miles per hour or 20 inches per year, very different average speeds.

Are You On the Bus?

Q: What is the average speed of an airplane that flies 3000 miles in 6 hours?

A: Using our definition for average speed, we have

$$\bar{s} = \frac{d}{t} = \frac{3000 \text{ miles}}{6 \text{ hours}} = 500 \text{ mph}$$

WORKING IT OUT *Average Speed*

If you know the average speed, you can determine other information about the motion. For instance, you can obtain the time needed for a trip. Suppose you plan to drive a distance of 60 miles with the cruise control set at 50 mph. How long will the trip take?

Without consciously doing any calculation, you probably know that the answer is a little over 1 h. How do you get a more precise answer? You divide the distance traveled by the average speed. For our example we obtain

$$t = \frac{d}{\bar{s}} = \frac{60 \text{ miles}}{50 \text{ mph}} = 1.2 \text{ h}$$

◀ time taken $= \dfrac{\text{distance traveled}}{\text{average speed}}$

You can also calculate how far you could drive if you traveled with a specified average speed for a specified time. Suppose, for example, you plan to maintain an average speed of 50 mph on an upcoming trip. How far can you travel if you drive an 8-h day?

$$d = \bar{s}t = \left(50 \, \frac{\text{miles}}{\text{h}}\right)\left(8 \text{ h}\right) = 400 \text{ miles}$$

◀ distance traveled
= average speed × time taken

Therefore, you would expect to drive 400 miles each day.

Actually, you probably weren't moving at 20 mph during much of your trip. At times you may have been stopped at traffic lights; at other times you may have traveled at 50 mph. The use of average speed disregards the details of the trip. Despite this, the concept of average speed is a useful notion.

Instantaneous Speed

The notion of average speed is limited in most cases. Even something as simple as your trip home from school is a much richer motion than our concept of average speed indicates. For example, it doesn't distinguish the parts of your trip when you were stopped waiting for a traffic light to change from those parts when you were exceeding the speed limit. The simple question, "How fast were you going as you passed Third and Vine?" is not answered by knowing the average speed.

To answer the question, "How fast were you going at a specific point or at a specific time?" we need to consider a new concept known as the *instantaneous speed*. This more complete description of motion tells us how fast you were traveling at any instant during your trip. Because this is the function of your car's speedometer (Figure 2-3), the idea is not new to you, although its precise definition may be new.

Actually, the definitions of average and instantaneous speeds are quite similar. They differ only in the size of the time interval involved. If we want to know how fast you are going at a given instant, we must study the motion during a very small time interval that contains that instant. The **instantaneous speed** is equal to the average speed over a time interval that is very, very small.

As a first approximation in measuring the instantaneous speed, we could measure how far your car traveled during $\frac{1}{10}$ of a second and calculate the average speed for this time interval. With precise equipment we could determine the average speeds during time intervals of $\frac{1}{100}$ of a second, $\frac{1}{1000}$ of a second, or an even smaller interval. How small an interval do we need? For practical purposes, we need a time interval that is small enough that the average speed doesn't change very much if we use an even smaller time interval. It is the

Figure 2-3 A speedometer tells you the car's instantaneous speed.

Did the bullet have a speed at the instant this picture was taken? If the bullet had a speedometer, would it read zero when this picture was taken or some other value?

◀ instantaneous speed is the average speed over a very small time interval

Everyday Physics | *Fastest and Slowest*

The fastest speed in the universe is the speed of light, 300 million meters each second (186,000 miles per second), and the slowest speed is of course zero. Between those extremes is a vast range of speeds. With the exception of a few very small subatomic particles that have been catapulted through huge electrical voltages or released in nuclear reactions, most things move at speeds close to the slow end of the range.

The fastest large objects are planets moving at speeds up to 107,000 mph. Our own Earth orbits the Sun at 67,000 mph. (Even this admittedly high speed is 1/10,000 the speed of light.) Closer to home, but still in space, the *Apollo* spacecraft returned to Earth traveling at 24,800 mph, and the space shuttles orbit Earth at 17,500 mph.

Our people-carrying machines have a wide range of speeds, from supersonic airplanes with a record speed of 2193 mph (the Lockheed SR-71A Blackbird) to moving stairways that approximate fast-walking speeds of 4 mph. In between we have record speeds set by passenger planes (the Concorde) at 1450 mph, jet-powered cars at 1228 mph, race cars at 410 mph, magnetically levitated trains at 343 mph, motorcycles at 191 mph (special motorcycles have traveled as fast as 323 mph), and bicycles at 167 mph (while drafting behind a vehicle).

When we give up our machines, we slow down considerably. The fastest recorded human speed is Usain Bolt's 100-meter dash in 9.69 seconds, or about 23.0 mph. The fastest time recorded by a female sprinter is 10.49 seconds (21.3 mph) by Florence Griffith Joyner. (In the time it took either sprinter to run the race, a beam of light could go to the Moon, bounce off a mirror, and return to Earth with 7 seconds to spare!) As the distances get longer, human speeds slow: The 1-mile record is held by Hicham El Guerrouj in a time of 3 minutes and 43.13 seconds, which corresponds to a little more than 16 mph. The record pace for a marathon is a little more than 12 mph.

Other animals range from slow (three-toed sloths that creep at 0.07 mph, giant turtles that lumber along at 0.23 mph, and sea otters that swim at 6 mph) to the very fast (killer whales and sailfish, which swim at 35 and 68 mph, respectively, and cheetahs, reported to run up to 68 mph). In 1973 Secretariat set the record for the Kentucky Derby by running $1\frac{1}{4}$ miles in 1 minute and 59.2 seconds, for an average speed of almost 38 mph. The streamlined peregrine falcon diving for its prey has been clocked at 217 mph.

Nobody knows the slowest-moving object. A good candidate for a natural motion is a continent drifting at 1 centimeter per year, or 0.7 billionth of a mile per hour. A few years ago, a machine was built for testing stress corrosion that moves at a million millionths of a millimeter per minute, or 37 billion billionths of a mile per hour. At this rate it would take about 2 billion years to move 1 meter!

1. The distance between LAX airport in Los Angeles and JFK airport in New York is about 2500 miles. How long would this flight take in the Lockheed SR-71A Blackbird?

2. Our nearest star is the Sun. The next closest star is Alpha Centauri, a distance of 4.365 light-years away. A light-year is the distance that light travels in one year. What is the distance to Alpha Centauri, measured in kilometers?

© AP Photo/The Canadian Press, Ryan Remiorz

Usain Bolt broke the men's 100-meter record at the 2008 Olympics in Beijing.

instantaneous speed rather than the average speed that plays an important role in the analysis of nearly all realistic motions.

Speed with Direction

We have made a lot of progress in attempting to accurately represent motion. However, as we develop the rules for explaining (and thus predicting) the behavior of objects in the next chapter, we will need to go further. Objects do more than speed up and slow down. They can also change direction, sometimes keeping the same speed, but at other times changing both their speed and their direction. Either the average speed or the instantaneous speed tells us how fast an object is moving, but neither tells us the direction of motion. If we are discussing a vacation trip, direction doesn't seem important; you obviously know in which direction you're going. However, we are trying to develop rules of motion for all situations, and the direction is as important as the speed. You can get a sense for this by remembering situations in which there is an abrupt change in direction; for example, maybe a car you were riding in swerved sharply. The squeal of the tires and your own body's reaction are clues that new factors are involved when an object changes direction.

In the physics world view, we combine speed and direction into a single concept called **velocity**. When we talk of an object's instantaneous velocity, we give the instantaneous speed (for example, 15 mph) and just add the direction (north, to the left, or 30 degrees above the horizontal). The speed is known as the **magnitude** of the velocity; it gives its size. We use the symbol v to represent the magnitude of the instantaneous velocity.

◀ velocity equals speed with a direction

Quantities that have both a size and a direction are called **vectors**. Vectors do not obey the normal rules of arithmetic. We will study the rules for combining vector quantities in Chapter 3. For now, it is only important to realize that the direction of the motion can be as important as the speed.

There is another important difference between average speed and average velocity besides direction. The average speed is defined as the distance traveled divided by the time taken, whereas the **average velocity** is defined as the **displacement** divided by the time taken. Displacement is a vector quantity; its magnitude is the straight-line distance between the initial and final locations of the object, and its direction is from the initial location to the final location.

The magnitude of the displacement is the same as the distance traveled for motion along a straight line in a single direction. The magnitudes of the displacement and the distance traveled differ when the motion retraces part of its straight-line path or takes place in more than one dimension. For instance, assume that you travel 10 kilometers due west along a straight stretch of road, turn around, and travel 5 kilometers due east along the same road. You have traveled a distance of 15 kilometers, but your displacement is 5 kilometers west; that is, your ending location is 5 kilometers to the west of your starting location. If the trip takes 1 hour, your average speed is 15 kilometers per hour, and your average velocity is 5 kilometers per hour west.

Q: A car travels due north a distance of 50 kilometers, turns around, and returns to the starting place along the same route. What distance did the car travel, and what was its displacement?

A: The car travels a distance of 100 kilometers, but it has a displacement of zero because it returns to its starting place.

Are You On the Bus?

The magnitude of the average velocity of an object is the change in position divided by the time taken to make the change:

$$\overline{v} = \frac{\Delta x}{\Delta t}$$

◀ average velocity
$$= \frac{\text{change in position}}{\text{time taken}}$$

We have used x to represent the position of the object. The symbol Δx is called "delta ex." The delta symbol Δ is used to represent a change in a quantity. Thus, Δx represents the change in position—the displacement—and must not be thought of as the product of Δ and x. To calculate the displacement, we subtract the position of the object at the beginning of the time interval from its position at the end. For example, if a car travels from milepost 120 to milepost

180, the displacement is 180 miles − 120 miles = 60 miles. Notice that we have also written the time taken as Δt to indicate that it is an interval of time rather than an instant of time.

For the rest of this chapter, we will deal with objects traveling in only one dimension. They may be going left and right, east and west, or up and down, but not turning corners. Notice that by doing this we eliminate motions as simple as a home run. The payoff is that we learn to manipulate the new concepts before tackling the many realistic but more-difficult situations.

Acceleration

Because the velocities of many things are not constant, we need a way to describe how velocity changes. We now define a new concept, called *acceleration*, which describes the rate at which velocity changes. The magnitude of the **average acceleration** of an object is the change in its velocity divided by the time it takes to make that change:

average acceleration ▶
$$= \frac{\text{change in velocity}}{\text{time taken}}$$

$$\overline{a} = \frac{\Delta v}{\Delta t}$$

As we did with speed, we can speak of either the average acceleration or the instantaneous acceleration, depending on the size of the time interval.

The units of acceleration are a bit more complicated than those of speed and velocity. Remember that the units of velocity are distance divided by time: for example, miles per hour or meters per second. Because acceleration is the change in velocity divided by the time interval, its units are (distance per time) per time—for example, (kilometers per hour) per second or (meters per second) per second (abbreviated m/s/s or m/s^2).

The concept of acceleration is probably familiar to you. We talk about one car having "better acceleration" than another. This usually means that it can obtain a high speed in a shorter time. For instance, a Dodge Grand Caravan can accelerate from 0 to 60 miles per hour in 11.3 seconds; a Ford Taurus requires 8.7 seconds and a Chevrolet Corvette requires only 4.8 seconds.

Are You On the Bus?

Q: Which car has the largest average acceleration?

A: The Corvette has the largest average acceleration because it reaches 60 mph in the shortest time interval.

Another way of becoming more familiar with acceleration is by experiencing it. For example, when an elevator begins to move up (or down) rapidly, the sensation you get in your stomach is due to the elevator (and you) quickly changing speed. Astronauts feel this when the space shuttle blasts off from its launch pad. Exciting examples of the same effect can be achieved on a roller coaster. In fact, amusement parks can be thought of as places where people pay money to experience the effects of acceleration.

In contrast, you don't feel motion when you're traveling in a straight line at a constant speed—that is, motion with zero acceleration. The motion you do feel when riding in a car on a straight highway is due to small vibrations of the car. (These vibrations are tiny changes in direction or small accelerations of the car caused by bumps in the road.)

If you are standing on the sidewalk watching a moving object, how can you tell whether the object is accelerating? One way is to take a strobe photograph of its motion. Figure 2-4 is a drawing of two such pictures. Which of the two corresponds to the car accelerating?

If you answered car (b), you have a qualitative understanding of acceleration. Car (a) travels the same distance during each time interval and therefore is traveling at a constant speed. Car (b) travels farther during each successive time interval; it is accelerating.

Even if the car were slowing down, it would be accelerating. (We don't usually use the word *deceleration* in physics because the word *acceleration* includes slowing down as well as speeding up.) In this case the distances traveled during successive time intervals would be shorter. Acceleration refers to any change in speed or direction—that is, to any change in velocity.

Acceleration is a vector quantity. When the acceleration is in the same direction as the velocity, the speed of the object is increasing. When the acceleration and the velocity point in opposite directions, the object is slowing down. The idea of an acceleration having a direction might seem a little abstract and, perhaps, unnecessary. A car's velocity obviously has a direction and probably seems easier to comprehend. However, as we continue our study of acceleration, we will see many examples in which the direction of the acceleration has physical consequences. The discussion of accelerations due only to a change in the direction of the velocity appears in Chapter 4.

Robin Smith/Stone/Getty

Amusement parks sell the thrill of acceleration.

(a)

(b)

Figure 2-4 Strobe drawings of two cars. Which car is accelerating and which is traveling at a constant speed?

Are You On the Bus?

Q: You see two cars side by side as they exit a tunnel. The red car has a speed of 40 meters per second and an acceleration of 20 (meters per second) per second. The blue car has a speed of 20 meters per second and an acceleration of 40 (meters per second) per second. At the instant they leave the tunnel, which car is passing the other?

A: The car with the larger instantaneous speed will travel farther down the road in the next small interval of time. Therefore, the red car is passing the blue car as they exit the tunnel. The blue car will have the greater change on its speedometer in the next small interval of time and will eventually overtake and pass the red car.

Q: What is an example from everyday life of something that is slowing with an acceleration vector pointing upward?

A: If this "something" is slowing, its velocity vector must be pointing in the opposite direction of its acceleration vector, or downward. Therefore, we are looking for something that is moving toward the ground and slowing down. This could be a diver, right after she hits the water, or a parachutist right after the chute is opened. Try to think of another example.

WORKING IT OUT *Acceleration*

Consider a car traveling along a straight highway at 40 mph that speeds up to 60 mph during a time interval of 20 s. What is the car's average acceleration?

Using the symbols v_i and v_f to represent the initial and final velocities, we have

$$\bar{a} = \frac{\Delta v}{\Delta t} = \frac{v_f - v_i}{t_f - t_i} = \frac{60 \text{ mph } - 40 \text{ mph}}{20 \text{ s}} = \frac{20 \text{ mph}}{20 \text{ s}} = 1 \text{ mph/s}$$

The car accelerates at 1 mph/s; that is, during each second, its speed increases by 1 mph.

If, on the other hand, the car made this change in velocity in 10 s, our new calculation would yield an average acceleration of 2 mph/s. These calculations illustrate that acceleration is more than just a change in velocity; it is a measure of the rate at which the velocity changes.

A First Look at Falling Objects

With these few ideas, we can now look at a common motion: a ball falling near Earth's surface. How does the ball fall? Does it fall faster and faster until it hits the ground? Or does it reach a certain speed and then remain at that speed for the duration of its fall? Does the rate at which it falls depend on its weight? For example, would a cannonball fall faster than a feather?

Questions about this rather simple motion have fascinated scientists since at least the time of Aristotle (4th century BC). They turned out to be quite difficult to answer. In fact, modern answers to these questions were not given until early in the 17th century.

Until that time, the accepted answers were those attributed to Aristotle. Every motion required a mover (object) and a goal toward which the object moved. An object falling from a height to its natural resting place on an immobile Earth—its goal—would travel with a speed determined by its weight divided by the resistance of the medium through which it traveled. Heavier objects would naturally travel faster than light ones. Much heavier objects would presumably fall even more rapidly. A cannonball would fall more slowly in molasses than in air. A feather would flutter down in air, but would float in molasses. In Aristotle's view, all change was motion. Falling bodies were just one example of change—a change of place. Changes of temperature, color, or texture were other examples of motion in Aristotle's world view.

The advantage of Aristotle's system was that he dealt with concrete, observable situations that we encounter every day. This advantage led to serious and rewarding discussions over the next 1500 years. Great scientists such as Galileo carefully analyzed Aristotle's ideas, which included a prediction that a 10-pound rock should fall significantly faster than a 1-pound rock.

You can perform an equivalent experiment to test the theory. Hold a heavy book (a physics text is quite appropriate) and a piece of paper at equal heights above the floor and drop them simultaneously. Which falls faster? Now repeat

CAN THAT BE RIGHT?

the experiment, but this time wad the paper into a tight ball. How do the results differ?

In the first case, the book fell much faster than the paper. This is in qualitative agreement with Aristotle's claim. But the second case certainly disagrees with Aristotle. Using the Aristotelian rule to predict the fall of the paper and book leads to a conflict with reality. If the book is significantly heavier than the paper, according to Aristotle, the book should drop at a speed significantly faster than the paper. This means that the book would hit the floor well before the paper! Clearly, Aristotle was wrong.

Our experiment with the book and paper might lead you to believe that in the absence of any resistance, as, for example, in a vacuum, two objects would fall side by side, independent of their weights. This means that a cannonball and a feather would fall together in a vacuum. This was the opinion of Galileo Galilei, an Italian physicist of the 17th century.

Galileo | *Immoderate Genius*

On February 15, 1564, Galileo Galilei was born in Pisa, Italy, into the family of a Florentine cloth merchant. As a boy he was schooled in Latin, Greek, and the humanities at the local monastery until his family relocated to Florence, where his father assumed the boy's education in mathematics and music. The young Galileo mastered the lute while learning advanced mathematics, physics, and astronomy.

At age 17, Galileo returned to the town of his birth to study medicine. Before completing his training, however, Galileo withdrew from medical school because of conflicts with his mentors caused by his curiosities in the sciences. At age 25 Galileo enlisted the aid of his father's friends in academia and received an appointment as professor of

© mediacolor's/Alamy

Galileo Galilei

mathematics at the University of Pisa. Other prestigious appointments followed as Galileo engaged in a variety of scientific pursuits.

Galileo studied time, motion, floating bodies, the nature of heat, and the construction of telescopes and microscopes. In 1610 he achieved fame for improving on previous telescope designs; his design allowed magnification of the heavens up to 32 times. Galileo observed features of the Moon, sunspots, and some planets, including Venus in its various phases. He discovered four of Jupiter's moons and showed that the Milky Way consists of an enormous number of stars.

Galileo's many observations supported the Copernican Sun-centered (heliocentric) theory of the solar system. In this belief, Galileo directly contradicted the Roman Catholic Church dogma of the day, which held that Earth was the stationary center of the universe; such a model as proposed by Aristotle was supported by biblical references and was held sacred. To deny this view was considered heresy, a crime that carried the highest penalty of being burned at the stake. In 1615 the Church began its investigation of Galileo's beliefs; in 1633, after publishing *Dialogue Concerning the Two Chief World Systems,* he was taken to Rome, charged with heresy, and sentenced to life imprisonment. Later, the sentence was commuted to house arrest at his villa at Arcetri.

In his last years, Galileo reflected on his life in a manuscript entitled *Discourses and Mathematical Demonstrations Concerning Two New Sciences,* which was smuggled out of Italy and published in Holland in 1638. The two new sciences were dynamics and the strength of materials. After completion of this work, Galileo became blind and died while still under house arrest in 1642.

Sources: S. J. Broderick, *Galileo: The Man, His Work, His Misfortunes* (New York: Harper & Row, 1964); AIP Niels Bohr Library; Stillman Drake, *Galileo at Work: His Scientific Biography* (Chicago: University of Chicago Press, 1978).

Legend has it that Galileo dropped balls from this leaning tower in Pisa while developing his ideas about free fall.

A hammer and feather dropped on the Moon hit the ground at the same time because there is no air.

Galileo is often called the founder of modern science, as much because of his style of building a physics world view as because of his particular contributions. His style was characterized by a strong desire to verify his theories with measurements; that is, he performed experiments to check his ideas. His goal was simple: to find rules of nature—often in the form of equations—that expressed the results of his investigations.

His work led to some new ideas about motion. He developed the concepts and mathematical language necessary to describe motion. For example, he invented the concept of acceleration.

Although Galileo was unable to directly test his ideas about free fall, he did suggest the *thought* experiment illustrated in Figure 2-5. Imagine dropping three identical objects simultaneously from the same height. The Aristotelians would agree that the three would fall side by side. Now imagine repeating the experiment but with two of the objects close to each other. Nothing significant has changed, so there will again be a three-way tie. Finally, consider the situation in which the two are touching. Because there was a tie before, there will be no dragging of one on the other, and again there will be a tie. But if the two are touching, they can be considered a single object that is twice as big. Consequently, big and small objects fall at the same rate!

A common present-day test of this idea is to drop two objects, such as a coin and a feather (a cannonball is somewhat impractical), inside a plastic tube from which the air has been removed. In a vacuum the race always ends in a tie. Astronauts conducted an ultramodern demonstration of this on the Moon. Because there is no atmosphere on the Moon, a hammer and a feather fell at the same rate.

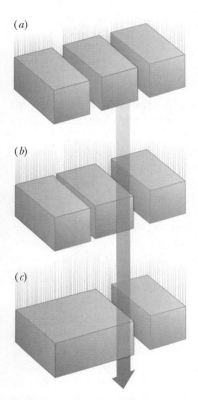

(*a*)

(*b*)

(*c*)

Figure 2-5 Galileo's thought experiment. All objects fall at the same rate in a vacuum.

Free Fall: Making a Rule of Nature

If we can find a rule for the motion of an object in free fall, the rule will be equally valid for all objects, heavy or light. Although we will obtain a rule that strictly speaking is valid only in a vacuum, it will be useful in many other situations—whenever the effects of air resistance can be ignored.

The quantitative measurement of a free-falling object was impossible with the technology of Galileo's time. Things simply happened too fast. To get around this, he studied a motion very similar to free fall, that of a ball starting from rest and rolling down a ramp. As the ramp is inclined more and more, the average speed of the ball increases. When the ramp is vertical, the ball falls freely. Galileo hoped (correctly) that he would be able to ignore the complications of rolling and be able to deduce the rule for free fall before the ramp became too steep and the ball's motion too fast to measure.

Experimenting at a single ramp angle, Galileo discovered that the ball traveled with a constant acceleration. After establishing this rule at a single angle, he increased the tilt of the ramp and started over again. At steeper angles he found what you might expect intuitively: the ball traveled down the ramp in less time. The exciting discovery, however, was that at each new ramp angle he discovered the *same* rule: the acceleration in each case was constant. This relationship held *independently* of the ramp angle, a fact that was crucial to the success of his proposal to extrapolate these results to the free-fall situation. He correctly concluded that the ball would obey the same relationship if the ramp was vertical.

Using modern techniques we can take a strobe photograph (Figure 2-6) of a falling ball and "see" its motion. The ball is clearly not moving at a constant speed. We can tell this by noting that the distances between the images are continually increasing. As Galileo showed, the ball falls with a constant acceleration.

Constant acceleration means that the speed changes by the same amount during each second. If, for example, we find that the ball's speed changed by a certain amount during 1 second at the beginning of the flight, then it would change by the same amount during any other 1 second of its flight.

We now know that for the case of the vertical ramp (free fall), the acceleration is about 9.8 (meters per second) per second (32 [feet per second] per second). This value is known as the *acceleration due to gravity* and varies slightly from place to place on Earth's surface. Students at Montana State University in Bozeman have determined that the value of the acceleration due to gravity in the basement of the old physics building is 9.800 97 (meters per second) per second. For convenience in calculations we will usually round off this value to 10 (meters per second) per second.

At any time during the fall, if we know its speed and its acceleration, we can calculate how fast the ball will be moving 1 second later. Assume that the ball is traveling 40 meters per second with an acceleration of 10 (meters per second) per second. This acceleration means that the speed will change by 10 meters per second during each second. Because the ball is speeding up, 1 second later the ball will be traveling with a speed of 50 meters per second. One second after that, the speed will be 60 meters per second, and so on.

As a final example of free fall, consider a thrill-seeking skydiver who jumps from a plane and decides not to pull the parachute cord until 30 seconds have elapsed. Our diver accelerates at a rate of 10 meters per second each second.

Figure 2-6 This strobe photograph of a falling ball shows that the ball has an acceleration.

WORKING IT OUT   *Should You Jump?*

You are standing at the top of a waterfall looking down at a deep pool of water below. Your friends think it is safe to jump, but you are worried that you may be too high for comfort. You pick up a rock and drop it into the pool. You count "one-one thousand, two-one thousand, three-one thousand," and find that the rock takes about 3 s to fall.

How fast would you be going just before you hit the water?

Any falling object, a rock or a person, speeds up by 10 m/s for every second that goes by (if we neglect air resistance). If you drop with an initial speed of zero, you would be traveling 10 m/s after the first second, 20 m/s after the second second, and 30 m/s (nearly 70 mph!) right before you hit the water. We suggest that you point your toes.

How tall is the waterfall?

The rock hit the water going 30 m/s, but clearly the rock did not travel this fast for the entire fall. The rock's initial speed was zero and increased uniformly to 30 m/s. The rock's average speed during the fall was 15 m/s (halfway between zero and 30 m/s), and the rock had this average speed for 3 s. That means that, on average, the rock fell 15 m every second for 3 s, for a total of 45 m.

FLAWED REASONING

A cart starts from rest and rolls down a 12-meter ramp in 3 seconds. The instantaneous speed of the cart is measured at the bottom of the ramp to be 8 meters per second. Shannon is not surprised by this measurement of final speed. She correctly reasons that the average speed of the cart during the trip is 4 meters per second (12 meters divided by 3 seconds) and that the average speed must be halfway between the slowest speed of zero and the final speed of 8 meters per second. She then brags to her friend Christian, "Because we know that the average speed happens halfway between the slowest speed and the fastest speed, we can predict that the cart will be going 4 meters per second (the average speed) when it is exactly halfway down the track." What is wrong with Shannon's reasoning, and where will the cart actually be on the track when its instantaneous speed is equal to the average speed of 4 meters per second?

ANSWER The acceleration on the ramp will be constant in time, meaning the cart will speed up by the same amount each second of travel. The cart will be going 4 meters per second halfway *in time*, not halfway down the track. The entire trip took 3 seconds, so the cart's instantaneous speed will be equal to its average speed exactly 1.5 seconds after it is released. The cart is moving much slower during the first 1.5 seconds of the trip and faster during the last 1.5 seconds of the trip. If the cart had a speedometer, it would read the average speed of 4 meters per second when the car is less than halfway down the track. Careful reasoning shows that it will actually be only a quarter of the way down the track.

Thus, at the end of $\frac{1}{2}$ minute (assuming our skydiver can resist pulling the cord for that long), the speed will be about 300 meters per second. That is about 675 mph! Actually, as we will see in Chapter 3, this description of free fall is quite inaccurate when the effect of air resistance becomes important.

Starting with an Initial Velocity

What happens if the object is already in motion when we start our observations? Suppose, for example, a ball is thrown vertically upward. Our experience tells us that it will slow, stop, and then fall. Examining strobe photographs of objects thrown vertically upward shows that the behavior of the rising object is symmetrical to that of the same object falling. The speed changes by 10 meters per second during each second. In fact, the strobe photograph in Figure 2-6 could just as well have been a photograph of a ball rising. (Of course, what goes up must come down; in taking the photograph, we would have to close the camera's shutter before the ball started back down.)

We can use the symmetry between motion vertically upward and downward in answering the following question: if you throw a ball vertically upward with an initial speed of 20 meters per second, how long will it take to reach its maximum height? Ignoring air resistance, we know that the ball slows down by 10 meters per second during each second. Therefore, at the end of 1 second, it will be going 10 meters per second. At the end of 2 seconds, it will have an instantaneous speed of zero. Therefore, the ball takes 2 seconds to reach the top of its path.

Air resistance prevents the skydivers from having a constant acceleration.

Are You On the Bus?

Q: If you could throw the ball with a vertical speed of 40 meters per second, how long will it take to reach its maximum height?

A: The ball will take 4 seconds to reach its maximum height.

A Subtle Point

Let's pause at the end of this chapter to emphasize the fact that Galileo used experiment and reasoning to discover a pattern in nature. He could discern the motion of objects subject only to the pull of Earth's gravity. Using simple mathematics and the rule that he formulated, you can calculate the outcome of future experiments. In a limited but very real way, you can predict the future. Predictions based on this rule are not the crystal-ball type popularized in science fiction, but they represent a very real accomplishment. The discovery of patterns and the creation of rules of nature are central in physicists' attempts to build a world view.

Summary

We began building a physics world view with the study of motion because motion is a dominant characteristic of the universe. We can obtain data about the motion of objects from strobe photographs. The average speed $\bar{s}$ of an object is the distance d it travels divided by the time t it takes to travel this distance, $\bar{s} = d/t$. The units for speed are distance divided by time, such as meters per second or kilometers per hour.

Instantaneous speed is equal to the average speed taken over a very small time interval. Speed in a given direction is known as velocity, a vector quantity.

Displacement is a vector quantity giving the straight-line distance and direction from an initial position to a final position. Average velocity is the change in position—displacement—divided by the time taken, $\bar{v} = \Delta x/\Delta t$.

Acceleration is the change in velocity divided by the time it takes to make the change, $\bar{a} = \Delta v/\Delta t$. Acceleration is a vector. The units for acceleration are equal to those of speed divided by time such as (meters per second) per second or (kilometers per hour) per second.

Galileo reasoned that all objects fall at the same rate in the absence of any air resistance. Furthermore, he discovered that these free-falling objects fall with a constant acceleration of about 10 (meters per second) per second.

CHAPTER 2 *Revisited*

We can determine how long it will take a train to reach its destination if we know its average speed and how far it has to go. The travel time is equal to the distance divided by the average speed.

Key Terms

average acceleration The change in velocity divided by the time it takes to make the change, $\bar{a} = \Delta x/\Delta t$. Average acceleration is measured in units such as (meters per second) per second. An acceleration can result from a change in speed, a change in direction, or both.

average speed The distance traveled divided by the time taken, $\bar{s} = d/t$. Average speed is measured in units such as meters per second or miles per hour.

average velocity The change in position—displacement—divided by the time taken, $\bar{v} = \Delta x/\Delta t$.

displacement A vector quantity giving the straight-line distance and direction from an initial position to a final position.

instantaneous speed The average speed for a very small time interval. The magnitude of the instantaneous velocity.

magnitude The size of a vector quantity. For example, speed is the magnitude of the velocity.

vector A quantity with a magnitude and a direction. Examples are displacement, velocity, and acceleration.

velocity A vector quantity that includes the speed and direction of an object.

Questions and exercises are paired so that most odd-numbered are followed by a similar even-numbered.

Blue-numbered questions and exercises are answered in Appendix B.

𝓈 indicates more challenging questions and exercises.

WebAssign Many Conceptual Questions and Exercises for this chapter may be assigned online at WebAssign.

Conceptual Questions

1. Describe the motion depicted in the following strobe drawing.

   ● ● ● ● ● ● ● ● ● ● ● ● ● ● ● ●●

𝓈2. Describe the motion of the pucks in the following strobe photographs. Assume the pucks move from left to right and do *not* retrace their paths.

© Cengage Learning/David Rogers

3. Where does the ball shown in the following strobe drawing have the slowest speed?

   ● ● ● ● ● ● ● ● ● ● ● ●

4. Where is the speed the fastest in the following strobe drawing?

   ●●●●●●●●● ● ● ● ● ● ● ●

5. Sketch a strobe drawing for the following description of a caterpillar moving along a straight branch. The caterpillar begins from rest and slowly accelerates to a constant speed. It then slows down to a slower constant speed. Finally it gets tired and stops for a rest.

6. A car is driving along a straight highway at a constant speed when it hits a mud puddle, slowing it down. After the puddle, the driver speeds up until he is going faster than before hitting the puddle and then sets the cruise control. Make a strobe drawing for this motion.

7. Draw a strobe photograph for a sprinter running the 100-meter dash. Represent the sprinter's motion from the firing of the starting gun until she stops after passing the finish line.

8. An ice climber falls from a frozen waterfall into a large snowdrift and gradually comes to rest. Draw a strobe diagram of the climber's motion from the moment he falls until he comes to rest.

9. Which (if either) has the greater average speed: a truck that travels from milepost 92 to milepost 100 in 10 minutes, or a car that travels from milepost 113 to milepost 120 in 10 minutes?

10. Which (if either) has the greater average speed: a car that travels from milepost 35 to milepost 40 in 5 minutes, or one that travels from milepost 68 to milepost 78 in 10 minutes?

11. You are driving down the road, with the cruise control set to 45 mph. You see a rabbit on the road, hit your brakes, and bring your car to rest. Is your average speed while braking greater than, equal to, or less than 45 mph?

12. In Aesop's fable of the tortoise and the hare, the "faster" hare loses the race to the slow and steady tortoise. During the race, which animal has the greater average speed?

13. Pat and Chris both travel from Los Angeles to New York along the same route. Pat rides a bicycle while Chris drives a fancy sports car. Unfortunately, Chris's car breaks down in Salt Lake City for more than a week, causing the two to arrive in New York at exactly the same time. Compare the average speeds of the two travelers.

14. A book falls off a shelf and lands on the floor. Which is greater, the book's average speed or its instantaneous speed right before it lands?

15. For the following strobe drawing, compare the instantaneous speeds at points C and D to the average speed for the time interval between C and D.

   A B   C   D   E
   ●●   ●   ●   ●

16. For the following strobe drawing, compare the instantaneous speeds at points C and D to the average speed for the time interval between C and D.

   A   B   C   D E
   ●   ●   ●   ●●

17. How might you estimate your speed if the speedometer in your car is broken?

18. Why is it *not* correct to say that time is more important than distance in determining speed?

19. A truck driver averages 92 kilometers per hour between 2 p.m. and 6 p.m. Can you determine the speed of the truck at 4 p.m.?

20. An ancient marathoner covered the first 20 miles of the race in 4 hours. Can you determine how fast he was running when he passed the 10-mile marker?

21. Which of the following can be used to measure an average speed: stopwatch, odometer, or speedometer? Which can be used to measure an instantaneous speed?

22. What are the units of the physical properties measured by a stopwatch, an odometer, and a speedometer?

23. What is the essential difference between speed and velocity?

24. If you are told that a car is traveling 65 mph east, are you being given the car's speed or its velocity?

25. In the following strobe drawings, which object (if either) has the greater acceleration?

26. The following strobe drawings represent the motions of two cars, a and b. During which interval of the motion of car a is the average speed of car a approximately equal to the average speed of car b?

A B   **C**   **D**   **E**
a. ● ●   ●   ●   ●
b. ●   ●   ●   ●   ●   ●

27. Which of the following (if any) could *not* be considered an "accelerator" in an automobile—gas pedal, brake pedal, or steering wheel?

28. In what sense can the brakes on your bicycle be considered an "accelerator"?

29. Assume that an airplane accelerates from 550 mph to 555 mph, a car accelerates from 60 mph to 67 mph, and a bicycle accelerates from 0 to 10 mph. If all three vehicles accomplish these changes in the same length of time, which one (if any) has the largest acceleration?

30. If an Acura Integra accelerates from 0 to 60 mph in 4 seconds and a Dodge Stealth accelerates from 20 mph to 75 mph in 4 seconds, which one has the larger acceleration?

31. A motorcycle travels down a straight highway with uniform speed of 35 mph. A sports car starts from rest and accelerates at 10 mph/s. Which will be moving faster after 3 seconds?

32. A Dodge Caravan has a speed of 50 mph and an acceleration of 2 (mph) per second. A Ford Taurus has a speed of 55 mph and an acceleration of 1 (mph) per second. Which car has the higher speed after 10 seconds have elapsed?

33. Carlos and Andrea are driving down the same road in the same direction, with Carlos behind Andrea. Carlos is slowing down and Andrea is speeding up, yet the distance between their cars is getting smaller. Give an example to show how this could happen.

34. Mary and Nathan are driving on a freeway in the same direction. At exactly noon, the cars are side by side. Mary is traveling at constant speed and Nathan is speeding up, yet Mary is passing Nathan. Explain how this could happen.

35. When we say that light objects and heavy objects fall at the same rate, what assumption(s) are we making?

36. Free fall near the surface of the Moon can be described as motion with a constant _____.

37. You are standing on a high cliff above the ocean. You drop a pebble, and it strikes the water 4 seconds later. Ignoring the effects of air resistance, how fast was the pebble traveling just before striking the water?

38. You throw a ball straight up in the air. The instant after leaving your hand, the ball's speed is 30 meters per second. Ignoring the effects of air resistance, predict how fast the ball will be traveling 2 seconds later.

39. What happens to the acceleration of a ball in free fall if the ball's mass is cut in half?

40. Two balls have the same size but are made from different materials: one from rubber and the other from steel. How do their accelerations compare after they are dropped?

41. You are bouncing on a trampoline while holding a bowling ball. As your feet leave the trampoline, you let go of the bowling ball. When you reach your maximum height, is the bowling ball above, beside, or below you?

42. You are bouncing on a trampoline while holding a bowling ball. As your feet leave the trampoline, you let go of the bowling ball. Do you rise to a higher, the same, or a lower height than if you had held on to the bowling ball?

43. A penny and a feather are placed inside a long cylinder, and the air is pumped out. When the cylinder is inverted, which hits the bottom first—the penny or the feather?

44. The Moon is a good place to study free fall because it has no atmosphere. An astronaut on the Moon simultaneously dropped a hammer and a feather from the same height. Which one hit the ground first?

45. How did the ideas of Galileo and Aristotle differ concerning the motion of a freely falling object?

46. A sheet of paper and a book fall at different rates unless the paper is wadded up into a ball, as shown in the following figure. How would Galileo and Aristotle account for this?

47. A student decides to test Aristotle's and Galileo's ideas about free fall by simultaneously dropping a 20-pound ball and a 1-pound ball from the top of a grain elevator.

The two balls have the same size and shape. What actually happens?

48. A table tennis ball and a golf ball have approximately the same size but very different masses. Which hits the ground first if you drop them simultaneously from a tall building? Do not neglect the effects of air resistance.

© Cengage Leaerning/George Semple

49. A table tennis ball and a marble are dropped side by side from the top of the biology building. Which ball has the greater acceleration? Do not ignore the effects of air resistance.

50. A table tennis ball and a marble are both thrown straight up in the air at the same initial speed. Which ball has the greater acceleration? Do not ignore the effects of air resistance.

51. A hard rubber ball is bounced on the floor. Compare the ball's acceleration on the way down to its acceleration on its way back up.

52. How (if at all) does the acceleration of a cylinder rolling up a ramp differ from that of one that is rolling down the ramp?

53. If we ignore air resistance, the acceleration of an object that is falling downward is constant. How do you suppose the acceleration would change if we do *not* ignore air resistance? Explain your reasoning.

54. If we do not neglect air resistance, during which of the first 5 seconds of free fall does a ball's speed change the most?

55. A rubber ball is thrown straight up into the air with an initial speed of 20 m/s. If we do not neglect air resistance as the ball moves upward, is the acceleration of the ball greater than, equal to, or less than the acceleration due to gravity?

56. As the ball in the previous questions returns to the ground, is the acceleration of the ball greater than, equal to, or less than the acceleration due to gravity?

Exercises

57. The top speed of the Blackbird is 2193 mph. Given that 1 mile = 1.61 km, what is this speed in km/h?

58. Top professional pitchers can throw fastballs at speeds of 100 mph. Given that 1 mph = 0.447 m/s, what is this speed in meters per second?

59. At exactly noon, you pass mile marker 50 in your car. At 2:30 p.m. you pull into a rest stop at mile marker 215. What was your average speed during this time?

60. To be eligible to enter the Boston Marathon, a race that covers a distance of 26.2 miles, a runner must be able to finish in less than 3 hr. What minimum average speed must be maintained to accomplish this?

61. In 1993 Sue Ellen Trapp broke the U.S. women's record for a 24-hr run by covering a distance of 145.3 miles. What was her average speed?

62. The 10,000-m run world record is 26 min. 17.53 s. What was the runner's average speed in m/s?

63. How far can a bus travel in 8 h at an average speed of 60 mph?

64. At an average speed of 10 m/s, how many kilometers can a cyclist travel in an 8-h day?

65. Starting at 9 a.m., you hike for 3 h at an average speed of 4 mph. You stop for lunch from noon until 2 a.m. What is your average speed over the interval from 9 a.m. to 2 a.m.?

66. Your plan was to be on the road by 9 a.m., but you did not leave the garage until 10 a.m. You then drove with the cruise control set at 75 mph until stopping at noon. What was your average speed over the time interval from 9 a.m. to noon?

67. If a cheetah runs at 25 m/s, how long will a cheetah take to run a 100-m dash? How does this compare with human times?

68. How many hours would be required to make a 4400-km trip across the United States if you average 80 km/hr?

69. If a runner can average 4 mph, can he complete a 100-mile ultramarathon in less than 24 h?

70. At an average speed of 125 mph, how long would a race car take to complete a 500-mile race?

71. If a Chevrolet Corvette can accelerate from 0 to 60 mph in 4.8 s, what is the car's average acceleration in mph/s?

72. If a Cessna 172 requires 20 s to reach its liftoff speed of 120 km/hr, what is its average acceleration?

73. A car speeds up from 40 mph to 70 mph to pass a truck. If this requires 6 s, what is the average acceleration of the car?

74. The world record for top fuel dragsters is 4.477 s to travel $\frac{1}{4}$ mile from a standing start. The dragster was traveling 332.75 mph at the end of the quarter mile. What was the dragster's average acceleration? What was its average speed?

75. A rock climber drops a piton. If the piton passes you with a speed of 7 m/s, how fast will the piton be traveling 2 s later?

76. A roofer drops a nail that hits the ground traveling at 23 m/s. How fast was the nail traveling 2 s before it hit the ground?

77. A child traveling 5 m/s on a sled passes her younger brother. If her average acceleration on the sledding hill is 2 m/s², how fast is she traveling when she passes her older brother 4 s later?

78. You throw a ball straight up at 30 m/s. How many seconds elapse before it is traveling downward at 10 m/s?

79. You accidentally drop a watch from the roof of a six-story building. While picking up the watch, you notice that it stopped 2 s after it was dropped. How tall is the building?

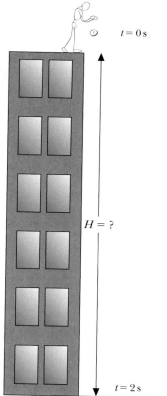

80. A rock is dropped into an abandoned mine, and a splash is heard 4 s later. Assuming that sound takes a negligible time to travel up the mineshaft, determine the depth of the shaft and how fast the rock was falling when it hit the water.

81. A ball is dropped from a height of 80 m. Construct a table showing the height of the ball and its speed at the end of each second until just before the ball hits the ground.

82. A ball is fired vertically upward at a speed of 30 m/s. Construct a table showing the height of the ball and its velocity at the end of each second until just before the ball hits the ground.

83. You decide to launch a ball vertically so that a friend located 45 m above you can catch it. What is the minimum launch speed you can use? How long after the ball is launched will your friend catch it?

84. A dummy is fired vertically upward from a cannon with a speed of 40 m/s. How long is the dummy in the air? What is the dummy's maximum height?

85. Hold a dollar bill so that it is vertical. Have your friend hold his thumb and index finger on each side of the middle of the bill. Tell him that he can keep the dollar if he can catch it when you let go. To catch the bill, he must be able to react within 0.13 s. If he just barely catches the bill, how fast must it be moving right before he stops it?

86. If your friend from the previous exercise holds his fingers near the bottom of the bill instead, he has 0.18 s to react. Use this information to find the length of a dollar bill.

3

Explaining Motion

▶ An object moving through a fluid must contend with resistance to its motion. For instance, bike racers streamline their bikes and clothing to minimize the air resistance. The effects of air resistance are very noticeable when skydivers open their parachutes. The acceleration is not constant; in fact, at some point the acceleration becomes zero. What happens to the speed of the falling object at this time?

(See page 54 for the answer to this question.)

Courtesy of U.S. Army Parachute Team, Golden Knights

Air resistance slows the descent of the U.S. Army Parachute Team, the Golden Knights.

WHAT is meant by explaining motion? Don't motions just happen? Philosophers could probably debate these questions for hours. In the physics world view, to explain something means to create a scheme, or model, that can predict the outcome of experiments. These experiments don't have to be elaborate; they can be as simple as throwing a baseball or looking at a rainbow.

Physicists try to create a set of ideas that explains how the world might work. Notice the word *might*; we have no proof that the ideas are correct or unique. Equally good (or better) schemes may yet be discovered.

Is it reasonable to expect that rules of nature exist? Motions appear to be reproducible; that is, if we start out with the same conditions and do the same thing to an object, we get the same resulting motion. The same motion occurs regardless of (1) when the experiment is done—the results on Monday match those on Tuesday—and (2) whether the experiment is done in San Francisco or Philadelphia. (Imagine the consequences if this were not true, and the "experiment" were throwing a baseball. Baseball teams would need a different pitcher for every day of the week and for every ballpark!) This reproducibility is a necessary condition for even attempting to search for a set of rules that nature obeys. Einstein reflected this idea when he commented that he believed that God was cunning but not malicious. His point was that although the rules of nature may be difficult to find, as we search for them we realize that they do not change.

An Early Explanation

Aristotle developed an explanation of motion that lasted for nearly 2000 years. Many of Aristotle's ideas seemed common sense—they were based on our most common experiences.

Aristotle believed that the world was composed of four elements: earth, water, air, and fire. These four were the building blocks of the material world. Each substance was a particular combination of these four elements. If this seems naive to you, consider our modern world view. We take chemical elements, each with its own special attributes, and combine them to form compounds that have quite different attributes. For example, we take hydrogen, a very explosive gas, and oxygen, the element required for combustion, and combine them to form water, which we use to fight fires!

Each Aristotelian element had its own natural place in the hierarchy of the universe. Earth, the heaviest, belonged to the lowest position. Water was next, then air and fire. Aristotle reasoned that if any of these were out of its hierarchical position, its natural motion would be to return. These natural motions occurred in straight lines, toward or away from the center of Earth.

If you try to test this part of Aristotle's world view, it works! Put some water and earth (soil) in a glass and wait. Watch as each element settles into its natural place (Figure 3-1).

For other than natural motions, Aristotle would probably challenge you to think of your own experiences. To move something, you have to make an effort. Even though we have developed machines to make the effort for us, we still agree that an effort must be made and that after the effort is stopped, the object comes to rest.

Although this seems reasonable, the Aristotelian explanation has problems. For example, objects don't stop immediately. An arrow continues to fly even after it loses contact with the bowstring. The Aristotelian explanation invokes an interaction between the arrow and the air. As the arrow moves through the air, it creates a partial vacuum behind it. The air, rushing in behind the arrow to fill the void, pushes on the arrow and causes the continued motion. This

Figure 3-1 In the Aristotelian world view, water rises while earth falls.

© Cengage Learning/Charles D. Winters

explanation, however, predicts that motion without an effort is impossible in a vacuum, while seeming to imply that in air it is perpetual.

The Beginnings of Our Modern Explanation

What is the motion of an object when there is nothing external trying to change its motion? One might guess, in agreement with Aristotle, that the only natural motion of an object is to return to Earth; otherwise, it has no motion—it remains at rest. Medieval thinkers agreed that objects have this tendency not to move.

Let's do a simple experiment. Give this book a brief push across a table or desk. Although the book starts in a straight line at some particular speed, it quickly slows and stops. It seems natural for an object to come to rest and remain at rest.

Remaining at rest is a natural state. However, there is another state of motion that is just as natural but not nearly as obvious. Suppose you were to repeat this book-pushing experiment on a surface covered with ice. The book would travel a much greater distance before coming to rest. Our explanation of the difference in these two results is that the ice is slicker than the desktop. Different surfaces interact with the book with different strengths. The book's interaction with the ice is less than that with the wood. Can you predict what would happen to the book if the surface were perfectly slick? The book would not slow down at all; it would continue in a straight line at a constant speed forever. Stated differently, when the interaction is reduced to zero, the book's motion is constant.

Thus, it seems that a natural motion is one in which the speed and direction are constant. Note that this statement covers the object whether it is "at rest" or "in motion." An interaction with an external agent is required to cause an object to change its velocity. If left alone, it would naturally continue in its initial direction with its initial speed.

Galileo reached this same conclusion by deducing the outcome of a thought experiment. He thought about the motion of a perfectly round ball placed on a tilted surface free of "all external and accidental obstacles." He noted—presumably from the same experiences we have all had—that a ball rolling down a slope speeds up (Figure 3-2[a]). Conversely, if the ball rolls up the slope, it naturally slows down. The ball experiences an interaction on the falling slope that speeds it up and an interaction on the rising slope that slows it down.

Next, Galileo asked himself, what would happen to the ball if it were placed on a level surface? Nothing. Because the surface does not slope, the ball would neither speed up nor slow down (Figure 3-2[b]). The ball would continue its motion forever.

It is important to remember that this is another of Galileo's thought experiments and not an account of an actual experiment. He assumed that there were no resistive interactions between the ball and the surface. There was no friction. By doing this he was able to strip motion of its earthly aspects and focus on its essential features.

Figure 3-2 Galileo's thought experiment. (a) On a tilted surface, a ball's speed changes. (b) On a level surface, a ball's speed and direction are constant.

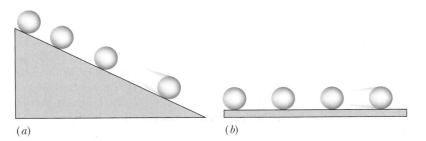

(a)   (b)

Galileo was the first to suggest that constant-speed, straight-line motion was just as natural as at-rest motion. This property of remaining at rest or continuing to move in a straight line at a constant speed is known as **inertia.**

The common use of the word *inertia* usually refers to an emotional state, one of feeling sluggish. Often when people say something has a lot of inertia, they are referring to their difficulty in getting it moving. (Sometimes they are talking about themselves.) As you build your physics world view, it is important to distinguish between the everyday uses of words and the usage within physics.

Are You On the Bus?

Q: What is the main difference between the everyday usage of the word *inertia* and its use in physics?

A: The physics usage also includes the idea that objects tend to keep moving.

You already have a vast set of experiences that are directly related to inertia as it is used here. If something is at rest, an interaction of some kind is required to get it moving. A magician uses the inertial property of cups and saucers when abruptly pulling a tablecloth from beneath them. If the tablecloth is smooth enough, the interaction with the cloth will be small, and the dishes will remain (nearly) at rest. Figure 3-3 shows before and after photographs of this trick.

Another example of how inertia can be fascinating is the circus strongman who boasts of his strength by asking someone to hit him on the head with a sledgehammer. The strongman always does this demonstration holding a big, heavy block on his head. The inertia of the block is large enough that the blow from the sledgehammer does little to move it. The happy strongman only has to be strong enough to hold up the block!

But there is more to inertia than getting things moving. If something is already moving, it is difficult to slow it down or speed it up. An example is drying your wet hands by shaking them. When you stop your hands abruptly, the water continues to move and leaves your hands. In a similar way, seat belts counteract your body's inertial tendency to continue forward at a constant speed when the car suddenly stops.

However, all objects do not have the same inertia. For example, imagine trying to stop a baseball and a cannonball, each of which is moving at 150 kilometers per hour (about the speed at which a major-league pitcher throws a baseball). The cannonball has more inertia and, as you can guess, requires a much larger effort to stop it. Conversely, if you were the pitcher trying to throw them, you would find it much harder to get the cannonball moving.

Although Galileo did not fully explain motion, he did take the first important step and, by doing so, radically changed the way we view the motion of objects. His work profoundly influenced Isaac Newton, the originator of our present-day rules of motion.

Newton's First Law

Isaac Newton, an Englishman, was born a few months after Galileo's death. Although he is probably best known for his work on gravitation, his most profound contribution to our modern world view is his three laws of motion. Like Galileo, Newton was interested in the interactions that occur while an object is in motion rather than in its final destination. He formulated Galileo's observations into what is now called **Newton's first law of motion.** It is also referred to as the **law of inertia.**

Figure 3-3 Impress your friends by pulling a tablecloth from under some dishes. To increase your chances of avoiding a disaster, you should do the following: Use a smooth, hemless cloth about the size of a pillowcase. Don't be timid; pull the cloth quickly in a downward direction across the straight edge of the table. Choose dishes that are stable (and cheap!).

Newton : *Diversified Brilliance*

On December 25, 1642, Isaac Newton was born in Woolsthorpe in Lincolnshire, England. Newton's father—a country gentleman in Lincolnshire—had died three months before he was born, leaving him to be reared by his mother. When he was 3 years old, his mother married a local pastor and moved a few miles away, leaving young Newton in the care of the housekeeper.

At age 14, Newton returned home from school at the request of his mother to help work on the family farm. He proved to be not much of a farmer, however, and spent most of his time reading. When he could be alone, Newton amused himself by building model windmills powered by mice, water clocks, sundials, and kites carrying fiery lanterns, which frightened the country folk. A local schoolmaster recognized Newton's abilities and helped him enter Trinity College at Cambridge at age 18, where he received his bachelor of arts degree four years later, in 1665.

Later that same year, the bubonic plague raged through the English countryside and, consequently, the university was closed. Newton returned to Woolsthorpe, and the next 18 months proved to be his most productive. It was during this interlude that Newton developed his theories and ideas about optics, celestial mechanics, calculus, the laws of motion, and his famous law of gravity. After the Great Plague, Newton returned to Cambridge, where he was appointed professor of mathematics at age 26. From here Newton went on to develop a reflecting telescope, one that used a mirror

© Bettman/CORBIS

Isaac Newton

to collect light instead of a lens, which was used in earlier models. Newton also published his most notable book—with the help of Edmund Halley—titled *Principia Mathematica Philosophiae Naturalis* (*Mathematical Principles of Natural Philosophy*) in 1687.

In 1701 Newton was appointed master of the mint, and in 1703 he was elected president of the Royal Society, a position he retained until his death. Newton's honors did not end there, however; in 1705 Queen Anne knighted him in recognition of his many accomplishments, forever changing his name to Sir Isaac Newton. This was the first knighthood given for scientific achievement.

Sir Isaac Newton's life was not all discoveries and honors. In fact, he spent a great deal of his later life quarreling with fellow scientists. Robert Hooke accused Newton of stealing some of his ideas about gravity and light. Newton also fought bitterly with Gottfried Leibniz, a German mathematician who claimed to have developed calculus first, and Christiaan Huygens, who worked independently on the wave theory of light. In 1727 Sir Isaac Newton became seriously ill, and on March 20 one of the greatest physicists of all time died. He was accorded a state funeral and interred in the nave of Westminster Abbey—a high and rare honor for a commoner.

Sources: Adapted from R. A. Serway and J. S. Faughn, *College Physics* (Philadelphia: Saunders, 1992); AIP Niels Bohr Library; F. E. Manuel, *A Portrait of Isaac Newton* (Cambridge, Mass.: Harvard University Press, 1968); Richard S. Westfall, *Never at Rest: A Biography of Isaac Newton* (Cambridge, England: Cambridge University Press, 1980).

Newton's first law ▶ **The velocity of an object remains constant unless an unbalanced force acts on the object.**

For the velocity of an object to remain constant, its speed and its direction must both remain constant. Note that this law applies to the special case of an object at rest: an object at rest remains at rest unless acted on by an unbalanced force.

The first law incorporates Galileo's idea of inertia and introduces a new concept, **force.** In the Newtonian world view, the book sliding across the table slows down and stops because there is a force (called friction) that opposes the motion. Similarly, a falling rock speeds up because there is a force (called gravity) continually changing its speed. In short, there is no acceleration unless there is a net, or unbalanced, force.

All of us have an intuitive understanding of forces; casually speaking, a force is a push or a pull. But it should be noted that the concept of force is a human construct. Because we have grown up with forces as a part of our personal world view, most of us feel quite comfortable with them. But we don't actually *see* forces. We see objects behave in a certain way, and we infer that a force is present. In fact, alternative world views have been developed that do

Stretch

Figure 3-4 The stretch of the spring is a measure of the applied force.

Figure 3-5 Equal forces acting in opposite directions cancel, and the cart does not accelerate.

not include the concept of force. This concept, however, has greatly aided the process of building a physics world view.

Although the concept of force includes much more than our intuitive ideas of push and pull, we use these thoughts as a beginning. A force can be defined in terms of the observed behavior of objects. For example, a force measurer (Figure 3-4) constructed with rubber bands or springs would allow us to quantify our observations of force by measuring the amount of stretch using some arbitrary scale. We will have more to say about these devices after we learn about Newton's second law.

Another important characteristic of forces is that they are directional, meaning that the direction of the force is as important as its size. Different results are produced by forces of the same size when applied in different directions. Imagine a skater coasting across the ice. A force in the direction of the original motion increases the skater's speed. A force applied in the opposite direction slows the skater. So we need to incorporate this difference into our understanding of motion. As you may guess from the discussion in Chapter 2, we will do this by treating forces as vectors.

Remember that Newton's first law refers to the *unbalanced* force. In many situations there is more than one force on an object. There is an unbalanced force only if the sum of the forces is *not* zero. When two forces of equal size act along a straight line but in opposite directions, they cancel each other. In this case the forces tend to stretch or compress the object, but the unbalanced, or *net*, force is zero. The "helpers" in Figure 3–5 could each be exerting a very large force, but if the forces are equal in size and opposite in direction, there is no unbalanced force on the cart.

The converse of this also holds. When we observe an object with no acceleration, we infer that there is no unbalanced force on that object. If you see a car moving at a constant speed on a level, straight highway, you infer that the frictional forces balance the driving forces. This is not to say that there are no forces acting on the car, because there are many. The crucial point is that the sum of all of these forces is zero; there is no unbalanced, or net, force acting on the car.

Q: What is the net force acting on an airplane in level flight flying at 500 mph due east?

A: Because the speed and direction are constant, there is no acceleration, and the net force must be zero.

Adding Vectors

Extended presentation available in the *Problem Solving* supplement

Mathematicians have developed rules for combining vector quantities such as displacements, velocities, accelerations, or forces. We can represent any vector by an arrow; its length represents the magnitude of the quantity, and its direction represents the direction of the quantity. To complete this representation, we assign a convenient scale to our drawing. For instance, we let 1 centimeter in the drawing in Figure 3–6 represent a distance of 20 meters on the ground. Then an arrow that is 4 centimeters long represents a displacement of 80 meters in the direction the arrow points.

Are You On the Bus?

Q: How many meters would a 10-centimeter arrow represent?

A: Because each centimeter represents 20 meters, a 10-centimeter arrow represents 200 meters.

Figure 3-6 A scale for displacement vectors. One centimeter on paper represents 20 meters on the ground.

In texts, vector quantities are represented by boldface symbols (such as **x**) and in handwritten materials with an arrow over the symbol (such as $\vec{x}$). The size, or magnitude, of the vector quantity is represented by an italic symbol. Therefore, a force is written as **F,** and its magnitude is written as *F*.

We can combine vectors using a graphical method and the scale shown in Figure 3-6. Let's assume that you walk a distance of 80 meters due north. This displacement is represented by an arrow 4 centimeters long pointing straight up the page. Then you continue walking due north for another 60 meters. This displacement is represented by an arrow 3 centimeters long pointing straight up the page. Your total displacement is 140 meters north. Notice in Figure 3-7(a) that we can represent this graphically by drawing the second arrow starting at the head of the first arrow, just as your second displacement started at the end of the first displacement. The sum of the two arrows is the arrow drawn from the tail of the first arrow to the head of the second. In this case, the arrow representing the sum is 7 centimeters long and points north, representing a displacement of 140 meters north.

Now let's say you walk 80 meters due north, turn around, and walk 60 meters due south along the same path. What is your displacement? You are 20 meters north of your starting place, so your displacement is 20 meters north. This is shown graphically in Figure 3-7(b).

The third time you walk 80 meters due north, turn to the right, and walk 60 meters due east. To find your displacement on the ground, you could put stakes at your beginning and ending locations, measure the distance between the stakes, and use a compass to get the direction from the beginning stake to the ending stake. Graphically, we draw a 4-centimeter arrow pointing up the page. We then draw a 3-centimeter arrow to the right, as shown in Figure 3-7(c).

Figure 3-7 (a) Two displacements in the same direction. (b) Two displacements in opposite directions. (c) Two displacements perpendicular to each other. In all cases the thicker arrow represents the vector sum of the two thinner arrows.

The displacement is the arrow drawn from the tail of the first arrow to the head of the second arrow, as shown. To get the displacement, we measure the length of the arrow. We get 5 centimeters, telling us that the distance is 100 m. We then use a protractor to measure the angle indicated, obtaining 37 degrees. Thus, we obtain a displacement of 100 meters at 37 degrees east of north.

This method of adding vectors is easily generalized to more than two vectors. After the first arrow is drawn, each succeeding arrow is drawn beginning at the head of the previous arrow. The arrow drawn from the tail of the first arrow to the head of the last arrow represents the vector sum. The order of the arrows does not affect the final answer.

When more than one force acts on an object, we can find the net force acting on the object by adding all of the forces using the method just described. Consider the three forces acting on the ball shown in Figure 3-8(a). To add these forces, we move F_2 (without changing its direction) and place its tail on the head of F_1, as shown in Figure 3-8(b). We then place the tail of F_3 on the head of F_2. The sum of these three forces is the arrow from the tail of F_1 to the head of F_3. The size of the force is determined by the scale used in the drawing, and the direction is determined with a protractor.

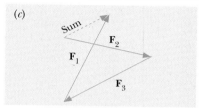

Figure 3-8 The three forces acting on the ball (a) can be added to find the net force (b). The order in which the forces are added (c) does not matter.

Newton's Second Law

Newton's first law tells us what happens when there is no net force acting on an object: the speed and direction don't change. If there is a net force, the object accelerates and thus changes its velocity. Newton's second law describes the relationship between a net force and the resulting acceleration.

Our development of the second law will present some simple experiments that illustrate this relationship before we state it formally. Assume that we have a collection of identical springs (the force measurers mentioned earlier), a collection of objects, and all the necessary equipment to measure accelerations. Further assume that if we stretch a spring by a fixed amount and maintain this stretch, the spring exerts a constant force.

Because the second law describes the net force, we need a situation in which the frictional forces are so small that they can be disregarded and any other forces are balanced so that the forces that we apply are the only ones affecting the acceleration. A horizontal air-hockey table is a good experimental surface. The hockey puck rides on a cushion of air, so it experiences very small frictional forces.

If we pull a hockey puck with a spring stretched by a certain amount and maintain the direction and amount of force even while the puck moves, we find that the puck experiences a constant acceleration in the direction of the force. After doing this many times with differing amounts of stretch, we conclude that a constant net force produces a constant acceleration. Furthermore, the direction of the acceleration is always in the same direction as the net force.

Let's now compare the results we obtained when pulling the puck with one spring with what happens when we pull the puck with two springs. When two springs are pulling side by side, as shown in Figure 3-9, the force is twice as

Figure 3-9 Two springs pulling side by side exert twice the force of one spring, as indicated by the scale readings.

large as that of the single spring. If we stretch each of the two springs by the same amount as before, we find that the two springs produce twice the acceleration. If we use three springs, they produce three times the acceleration, and so on. In general, we find that the acceleration of an object is **proportional** to the net force acting on it. This relationship will be part of the second law.

But this is not the entire story. Imagine pushing on a cannonball with the same force that is used on the hockey puck. Intuition tells you that the acceleration of the cannonball will be smaller. If pressed for a reason, you might respond, "Because there is more 'stuff' in a cannonball," or, "It weighs more." We will soon see that although the term *weight* is not technically correct, this intuition leads in the correct direction.

We build on this intuition by investigating how the acceleration of an object depends on the amount of matter in the object. The **mass** of an object is a measure of the amount of matter in the object. We assume that masses combine in the simplest possible way: the masses add. Therefore, the combined mass of two identical objects is twice the mass of one of them.

Again, we use a spring to pull on one of the hockey pucks and record its acceleration. We then look at the acceleration of two pucks (somehow tied together) pulled by a single spring. If the spring is stretched by the same amount as before, the acceleration is one-half the original. Likewise, one spring pulling on three pucks yields one-third the acceleration, and so on. Mass and acceleration are **inversely proportional,** where *inversely* indicates that the changes in the two values are opposite each other. If the mass is *increased* by a certain multiple, the acceleration produced by the force is *reduced* by the same multiple.

Notice that the more mass an object has, the more force it takes to produce a given acceleration. This means that the object has more inertia. We therefore take an object's mass to be the measure of the amount of inertia the object possesses.

Newton put the two preceding ideas together into one of the most important physical laws of nature ever proposed. This law states that the acceleration of an object is equal to the net force on the object divided by its inertial mass and can be written symbolically as

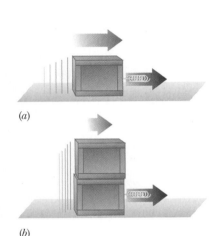

(a)

(b)

The acceleration of two masses pulled by identical springs is one-half as large as that for a single mass.

$$\mathbf{a} = \frac{\mathbf{F}_{net}}{m}$$

acceleration = $\dfrac{\text{net force}}{\text{mass}}$ ▶

where we have written the acceleration and the net force as vectors to emphasize that they always point in the same direction.

Any such mathematical equation can be rearranged using algebra. **Newton's second law of motion** is more commonly written as

net force = mass × acceleration ▶

$$\mathbf{F}_{net} = m\mathbf{a}$$

Newton's second law ▶

> The net force on an object is equal to its mass times its acceleration and points in the direction of the acceleration.

The second law describes a specific relationship between three quantities: net force, mass, and acceleration. Although we have a prescription for determining the numerical value of an acceleration, we have not yet done this for the other two quantities. We have a choice to make. We can choose a standard spring stretched by a specific amount as our definition of 1 unit of force, or we can take a certain amount of matter and define it as 1 unit of mass, or we could even choose the two units independently.

Historically, a certain amount of matter was chosen as a mass standard. It was assigned the value of 1 **kilogram** (kg). The mass of a liter of water (a little more than a quart) has a mass of 1 kilogram. The value of the unit force is then defined in terms of the observed acceleration of this standard mass. The force needed to accelerate a 1-kilogram mass at 1 (meter per second) per second is called 1 **newton** (N), in honor of Isaac Newton. The gravitational force on a very small apple is about 1 newton.

In the United States, a commonly used unit of force is the *pound* (lb). The unit of mass, a *slug,* is used so seldom that you may never have heard of it. One pound is the force required to accelerate a mass of 1 slug at 1 (foot per second) per second.

Once again we have found a pattern in nature. We can use Newton's second law to predict the motion of objects before we actually do the experiment.

WORKING IT OUT *The Second Law*

What is the net force needed to accelerate a 5-kg object at 3 m/s²? Applying the second law, we have

$$F_{net} = ma = (5 \text{ kg})(3 \text{ m/s}^2) = 15 \text{ kg} \cdot \text{m/s}^2 = 15 \text{ N}$$

In using any rule of nature, we must use a consistent set of units. The units are an integral part of the rules of nature. In the preceding case, when accelerations are measured in (meters per second) per second, the masses must be in kilograms, and the forces in newtons. The combination $\text{kg} \cdot \text{m/s}^2$ is equal to a newton.

Q: Suppose that in this situation you discovered that a 5-N force of friction is opposing the motion. How large is the applied force acting on the object?

A: The net force is the vector sum of the applied force and the frictional force. To obtain a net force of 15 N, the applied force must be 20 N. That is, 20 N in the forward direction plus 5 N in the backward direction gives a sum of 15 N in the forward direction.

We can use the second law to ask other questions. For example, what acceleration would be produced by a 2-N net force acting on the 5-kg object? Rearranging the second law and putting in the values of mass and force, we get

$$a = \frac{F_{net}}{m} = \frac{2 \text{ N}}{5 \text{ kg}} = \frac{2 \text{ kg} \cdot \text{m/s}^2}{5 \text{ kg}} = 0.4 \text{ m/s}^2$$

Q: A crate falls from a helicopter and lands on a very deep snowdrift. The snow slows the crate and eventually brings it to a stop. During the time that the crate is moving downward through the snow, is the magnitude of the upward force exerted on the crate by the snow greater than, equal to, or less than the magnitude of the gravitational force acting downward on the crate?

A: Because the crate is moving downward, its velocity is pointing down. Because the crate is losing speed, its acceleration must be pointing in the opposite direction—that is, up. The net force always points in the same direction as the acceleration. Therefore, the force acting upward on the crate must be larger than the force acting downward. Thus, the snow exerts the greater force.

Mass and Weight

Mass is often confused with weight. Part of the confusion lies in the fact that mass and weight are proportional to each other; doubling the value of one doubles the value of the other. In addition, the differences don't come up in our everyday experience. In the physics world view, however, the differences are profound and thus important to understanding motion.

We measure our **weight** by how much we can compress a calibrated spring, such as that in a bathroom scale. We compress the spring because Earth is attracting us; we are being pulled downward. Our weight depends on the strength of this gravitational attraction. If we were on the Moon, our weight would be less because the Moon's gravitational force on us would be less.

Our mass, however, is *not* dependent on our location in the universe. It is a constant property that depends only on how much there is of us. If we were far, far away from any planet or other celestial body, we would be weightless but not massless. And, because we are not massless, the force required to accelerate us is still given by Newton's second law.

The idea of weightlessness fascinates science fiction writers. Some of them, however, confuse the concepts of mass and weight. Contrary to some fictional accounts of weightlessness, the laws of motion still hold in these situations. For example, suppose that while far out in space, where all gravitational forces are negligible, you float across your spacecraft and collide with a wall. You won't just bounce off, feeling no pain. The wall provides a force to slow and reverse your motion. Newton's second law tells us that this force depends only on your mass and the acceleration you experience, both of which are the same as here on Earth. If the force can break bones on Earth, it can do the same in the spacecraft. Being weightless does not mean that you are massless. Similarly, imagine a huge truck in outer space "hanging" from a spring scale. Although the scale would read zero, if you tried to kick the truck, you would find that it resisted moving.

The marketplace is another place where mass and weight are often confused. We talk about buying "a pound of butter." A pound is a unit of weight and is determined by how much the butter stretches a spring in the scales. It is a measure of the gravitational attraction and varies slightly from place to place. The shopper doesn't really care about the weight of the butter but is interested in purchasing a certain amount of butter; the important thing is its mass. Stores using spring scales calibrate them with standard masses to compensate for the value of the local gravitational attraction. In the rest of the world, the units on spring scales are usually mass units to reflect the fact that you are buying a certain mass of the product.

This confusion between mass and weight will not be resolved by switching over to the metric system. Most people will probably still refer to the standard masses as "weights" and the process of determining the amount of butter as "weighing." What is really meant by saying "the butter weighs 1 kilogram" is that the amount of butter has a weight that is equal to the weight of 1 kilogram. Because that is quite a mouthful, people will probably refer to the weight of the butter as being 1 kilogram.

A 1-kilogram mass near Earth's surface has a weight of 9.8 newtons, or about 2.2 pounds. Therefore, a "pound" of butter has a weight of 4.5 newtons and a mass of a little less than $\frac{1}{2}$ kilogram. When the United States is fully converted to the metric system, butter will most likely be purchased by the $\frac{1}{2}$ kilogram, as is done presently in most of the world.

Although the distinctions between mass and weight are not important in the marketplace, the spacecraft example demonstrates that we have to be careful when discussing these concepts in physics.

Shoppers use supermarket scales to determine the masses of the produce.

Weight

The force causing a dropped object to accelerate toward Earth's surface is just the gravitational force. We often call this force the weight of the object.

In the idealized situation of no air resistance described in Chapter 2, we concluded that all objects near Earth's surface fall at a constant acceleration. Let's represent the acceleration due to gravity by the symbol **g,** where we've used a vector to indicate both the size and direction. If we replace the net force $\mathbf{F}_{net}$ by the weight **W** and the acceleration **a** by the acceleration due to gravity **g** in Newton's second law, we obtain

$$\mathbf{W} = m\mathbf{g}$$

◄ weight = mass × acceleration due to gravity

This is just a mathematical way of saying that the weight of an object is proportional to its mass and directed downward.

WORKING IT OUT *Weight*

As a numerical example, let's calculate the weight of a child with a mass of 25 kg:

$$W = mg = (25 \text{ kg})(10 \text{ m/s}^2) = 250 \text{ N}$$

Therefore, the child has a weight of 250 N (about 55 lb).

Q: What is the weight of a wrestler who has a mass of 120 kg?

A: 1200 N.

Are *You* On the Bus?

This process can be reversed to obtain the mass of a dog that has a weight of 150 N:

$$m = \frac{W}{g} = \frac{150 \text{ N}}{10 \text{ m/s}^2} = 15 \text{ kg}$$

Free-Body Diagrams

Imagine that you are pulling your little sister on a sled and that the sled is speeding up. There are many forces acting on the sled. The rope is exerting a tension on the sled, pulling it forward. Earth is pulling down on the sled with a gravitational force. The snow is pushing up on the sled with a force commonly called a *normal force.* (*Normal* means "perpendicular," and this force acts perpendicular to the surface between the sled and the snow.) Your sister is pushing down on the sled with a normal force and back with a friction force, and the snow is resisting your efforts with a frictional force that acts parallel to the surface of the snow.

Which of these forces do we use in Newton's second law to find the acceleration of the sled? Would it be the force of the rope? Would it be the largest of the forces? No. It is the net force. The net force is the vector sum of all the forces acting *on the sled.* It is important, therefore, to correctly identify *all* the forces acting on an object when analyzing its motion.

We identify the forces by drawing a *free-body diagram.* As the name suggests, we isolate, or free, the object in question (in this case the sled) from everything else. We represent that object by a dot. We then draw all the forces acting on the object with each tail starting on the dot. We label each vector to indicate what type of force it represents—**W** for a gravitational force, **N** for a normal force, **f** for a frictional force, and **T** for a tension force (a pull exerted by a string or a rope). Because every force is an interaction between two objects (things you can touch, taste, and smell), it is also useful to include two subscripts for each force label, one to indicate which object is exerting the force and the second to indicate which object is being acted on. For example, the tension force exerted by the rope on the sled would be labeled $T_{rope,sled}$. If you are not able to identify the object that is exerting a force, you should consider the possibility that the force does not exist.

Take a moment and draw a free-body diagram for the sled just described. When you are finished, compare your diagram to Figure 3-10. All the second subscripts should be "sled" because only forces acting on the sled appear on this free-body diagram.

Remember that your diagram should be consistent with the acceleration of the sled. We know that the sled is speeding up. This means that the sled's acceleration must point in the same direction as its velocity. If the sled is moving to the right and speeding up, the acceleration must also point to the right. If the acceleration is to the right, the sum of the forces acting to the right must be larger than the sum of the forces acting to the left. Because the sled is not accelerating upward or downward, the sum of the forces acting upward must balance the sum of the forces acting downward.

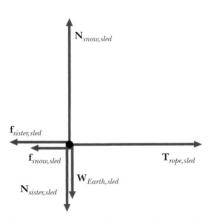

Figure 3-10 The free-body diagram for the sled.

WORKING IT OUT *Normal Force*

If the sister in this example has a mass of 31 kg and the sled has a mass of 28 kg, find the magnitude of the normal force exerted by the snow on the sled, $N_{snow,sled}$. The sled is speeding up to the right, so its acceleration and the net force that causes the acceleration must both point to the right. Because there is no acceleration in the vertical direction, all the force acting upward on the sled must balance all the force acting downward on the sled, or

$$N_{snow,sled} = N_{sister,sled} + W_{Earth,sled}$$

The weight of the sled is $W_{Earth,sled} = m_{sled}g = (28 \text{ kg})(10 \text{ m/s}^2) = 280$ N. The sister must be pushing down on the sled with a force equal to her weight: $N_{sister,sled} = W_{Earth,sister} = m_{sister}g = (31 \text{ kg})(10 \text{ m/s}^2) = 310$ N. We can now calculate the normal force exerted by the snow:

$$N_{snow,sled} = 310 \text{ N} + 280 \text{ N} = 590 \text{ N}$$

Drawing a free-body diagram should always be the first step in solving a problem involving Newton's second law. The time required to draw the diagram is seldom wasted, because most real problems are too complicated to correctly answer without first drawing a diagram. We are reminded of the story of the woodcutter sawing down a tree. A passing hiker asks, "How long have you been sawing down that tree?" and receives the reply, "Nearly three hours!" The hiker then asks, "Why is it taking so long?" and is told, "My saw is very dull." "Why don't you sharpen your saw?" asks the hiker. "I am too busy sawing down this tree."

Free Fall Revisited

Objects falling on Earth don't fall in a vacuum but through air, which offers a resistive force to the motion. Thus, in realistic situations, a falling object has two forces acting on it simultaneously: the weight acting downward and the air resistance acting upward. Among other factors, the force due to air resistance depends on the speed of the object. The greater the speed, the greater the air resistance. You can experience this by sticking your hand out a car window as the car's speed increases.

With these facts in mind, consider the downward motion of a falling rock. Initially, it falls at a low speed, and the air resistance is small. There is a net force downward that is equal to the weight of the rock minus the force of the air resistance (Figure 3-11[a]). Because there is a net force, the rock accelerates, thus increasing its speed. As the rock speeds up, however, its weight remains constant while the air resistance increases (Figure 3-11[b]). Thus, the net force and the acceleration decrease. The rock continues to speed up but at a decreasing rate. Eventually, the rock reaches a speed for which the air resistance equals the weight (Figure 3-11[c]). There is no longer a net force acting on the rock, and it stops accelerating—its speed remains constant. This maximum speed is called the **terminal speed** of the object.

The terminal speeds of different objects are not necessarily equal. Even if the shape and size of two objects are identical—thus having identical frictional forces—the objects have different terminal speeds if they have different masses. In all cases, though, the object continues to accelerate until the frictional force is the same size as its weight. The value of the terminal speed is determined by a combination of many factors: the size, shape, and weight of the object, as well as the properties of the medium. A BB has a much larger terminal speed than a feather, primarily because the feather's shape creates a resistive force that quickly becomes comparable to its weight.

Let's look again at the skydivers discussed in Chapter 2. Assuming no air resistance, we calculated that the skydivers would be falling at a speed of 1080 kilometers per hour (675 mph) after only 30 seconds. As skydivers know, the maximum speed they can obtain near sea level is a little over 300 kilometers per hour (190 mph). Obviously, air resistance is responsible for this difference. Skydivers also know that their terminal speed can be altered by changing their shape—falling feet-first or spread-eagle—because the larger the surface of the object facing into the wind, the greater the air resistance.

Galileo versus Aristotle

Recall that in Chapter 2 we discussed the views of Aristotle and Galileo on the subject of falling objects, and came down firmly on the side of Galileo. Now it seems that we are agreeing with Aristotle! If a falling object reaches a terminal speed whose value is determined by the object's weight and its interaction with the medium, wasn't Aristotle correct? It might seem so.

The motion of a bowling ball dropped from a great height shows that each man correctly described a part of the ball's motion. Initially, before the air

As the speed of a falling rock increases...

...the force of the air resistance increases until it equals the weight of the rock.

(a)   (b)   (c)

Figure 3-11 As the speed of a falling rock increases (a and b), the force of the air resistance increases until it equals (c) the weight of the rock.

Everyday Physics : *Terminal Speeds*

How fast something can move depends on the forces that retard its motion as well as those that propel it forward. The primary retarding forces are frictional forces, often due to the medium through which the object moves. A 100-meter race in 3 feet of water would produce times far slower than the current record of 9.72 seconds in air. But even in air, there is resistance to motion. A clean, waxed car has a measurable increase in gas mileage over a dirty car.

Downhill skiers gain speed by reducing their air resistance and the resistance of the skis with the snow.

© TempSport/CORBIS

When the retarding forces equal the propelling forces, there is no net force on the object, and the object stops accelerating; it reaches a constant speed known as the terminal speed. Minimizing the retarding forces increases the terminal speed.

Streamlining an object minimizes its air resistance. In 1980 Steve McKinney set the world unpowered land-speed record by paying a great deal of attention to minimizing air resistance and friction. He skied down a 40-degree slope at slightly more than 200 kilometers per hour (125 mph!). The current downhill skiing record is held by Philippe Goitschel at 251 kilometers per hour (156 mph). The woman's record is held by Karine Dubouchet at 242 kilometers per hour (151 mph). The peregrine falcon, already streamlined, dives for prey at speeds of up to 350 kilometers per hour.

The effect of minimizing air resistance was convincingly demonstrated by U.S. Air Force captain Joseph Kittinger when he jumped from a balloon at 31,330 meters and attained a speed of more than 1006 kilometers per hour (625 mph) after falling approximately 4000 meters. At this speed he nearly broke the sound barrier! He was then slowed down as the atmosphere became denser.

1. What criterion determines the terminal velocity of an object?

2. Why does a car get better gas mileage when it is clean?

---

resistance becomes significant, the ball exhibits constant acceleration, as hypothesized by Galileo. As the air resistance grows, the acceleration is no longer constant but decreases to zero. From that point on, the object travels at a constant speed, as described by Aristotle.

Each man described different extremes of the motion: Galileo, the extreme of negligible air resistance; Aristotle, the extreme of maximum air resistance. One might naively suggest that we determine how much of the motion is accelerated and how much is at a constant speed. We could then award a physics prize to the person whose explanation holds for the longest time. Aristotle would win; we only need to drop the object from higher and higher positions, making the constant-speed portion of the fall as large as desired.

But building a physics world view doesn't always progress by choosing on such a basis. Galileo has fared better in the eyes of science historians because his idealization stripped away the nonessentials of falling motion and thus uncovered the more fundamental behavior of motion. Galileo therefore paved the way for Newton's work, which explains the entire motion of a free-falling object (including Aristotle's observations). As long as all the forces acting on an object are known, the resulting acceleration can be calculated.

Friction

Newton's insight can be turned around; rather than predicting the motion from the forces, we can use an object's motion to tell us something about the forces acting on it. Imagine pushing horizontally on a large wooden crate (Figure 3-12[a]). At first you don't push hard enough to move the crate. If it doesn't move, there is no acceleration, and according to Newton's second law, there can be no net force on the crate. This means that there must be at least one other force canceling out the push. This other force is the force of friction exerted on the crate by the floor. As long as the crate does not move, the frictional force must be equal in size and opposite in direction to your applied force. This force is called **static friction,** to distinguish it from the frictional force that occurs when the crate moves.

This static frictional force seems a bit mysterious. Because it is equal to the force you exert, the frictional force is small if you push with a small force. But if you push with a large force, the frictional force is large (Figure 3-12[b]). It is a force that opposes the applied force and ceases to exist when the applied force is removed. The static frictional force can have any value from zero up to a maximum value determined by the surfaces and the weight of the crate. Notice that the behavior of the static frictional force is very similar to force exerted by a spring.

If your applied force exceeds the maximum static frictional force, the crate accelerates in the direction of your applied force. Although the crate is now sliding, there is still a frictional force (Figure 3-12[c]). The value of this **kinetic friction** is less than the maximum value of the static frictional force. Unlike air resistance, *kinetic friction has a constant value, independent of the speed of the object.*

It is important to understand the difference between static and kinetic friction when making an emergency stop in an automobile. Because you want to stop the car as quickly as possible, you want to have the maximum frictional force with the road. This occurs when the tires are rolling because the surface of the tire is not sliding along the surface of the road, and it is the larger static friction that is important. Therefore, you should not brake so hard that the tires skid. The same thing occurs when a car takes a corner too fast. Once the tires start to skid, the frictional force is reduced, making it difficult to recover from the skid. If you've ever been in one of these unfortunate situations, you may recall how fast the car slides once it starts to skid.

A significant advance in the automotive industry is based on the fact that static friction is greater than kinetic friction. Antilock brakes are a computer-controlled braking system that keeps the wheels from skidding, thus maximizing the frictional forces. Sensors monitor how fast the wheels are rotating and continuously feed the data to an onboard computer. The computer controls the braking by repeatedly applying and releasing pressure to the brake pads. Without antilock braking, a driver who jams on the brakes, hoping to avoid danger, causes the wheels to lock, which often results in a loss of control and an increase in the stopping distance.

Newton's Third Law

There is still one more Newtonian law to consider. Imagine that you are playing tennis and have just hit a ball. The racket exerts a force on the ball that causes it to accelerate. The high-speed photograph in Figure 3-13 shows that the strings of the tennis racket are pushed back at the same time the ball is flattened. The ball is squashed by the force of the racket *on the ball;* at the same time, the racket strings are stretched by the force of the ball *on the racket.* At the same time the racket is exerting a force on the ball, the ball is exerting an opposite force on the racket.

If you wish to pursue this point further, find a friend who will help with a simple experiment. Give your friend a shove. At the same time you are pushing, you will feel a force being exerted on you, regardless of whether your

(a)

(b)

(c)

Figure 3-12 The static frictional force is equal and opposite to the applied force if the crate does not accelerate. The applied force can be small (a) or large (b) as long as it doesn't cause the crate to move. (c) The kinetic frictional force has a constant value independent of the speed.

© Amoz Eckerson/Visuals Unlimited

Figure 3-13 A high-speed photograph illustrating Newton's third law. The ball exerts a force on the strings, and the strings exert an equal and opposite force on the ball.

friend pushes back. If you can, try this wearing ice skates or in-line skates; it will be even more dramatic.

Let's carry this one step further. Lean on a wall. Notice the force of the wall pushing back on you. If this force did not exist, you would fall.

These kinds of experiences may have led Newton to his third law of motion. He realized that there is no way to push something without being pushed yourself. For every force there is always an equal and opposite force. The two forces act on different objects, are the same size, and act in opposite directions. Formally, we state **Newton's third law of motion** as follows:

Newton's third law ▶

> **If an object exerts a force on a second object, the second object exerts an equal force back on the first object.**

Because Newton referred to these forces as *action* and *reaction*, they are often known as an *action–reaction pair.* However, because the two forces are equivalent, it doesn't matter which one is called the action and which the reaction. Another statement of the third law might be that for every action there is an equal and opposite reaction.

Forces always occur in pairs. In Newton's words, "If you press a stone with your finger, the finger is also pressed by the stone." These forces *never act on the same body.* When you press the stone with your finger, you exert a force *on the stone.* The reaction force is *acting on you.*

Consider a ball with a weight of 10 newtons falling freely toward Earth's surface. Ignoring air resistance, there is only one force acting on the ball; Earth's gravity is pulling it downward with a force of 10 newtons. What is the second force in the action–reaction pair? The first force is the force of Earth acting on the ball and can be labeled $\mathbf{W}_{Earth,ball}$ (where the subscript *Earth,ball* reminds us that this is the force of *Earth on ball*). The second force involves the objects in the reverse order and is written $\mathbf{W}_{ball,Earth}$ for *ball on Earth* (Figure 3-14). Therefore, Newton's third law tells us that the ball must be exerting an upward force on Earth of 10 newtons. Although your common sense may tell you that Earth must exert a larger force because it is so much larger, this is not

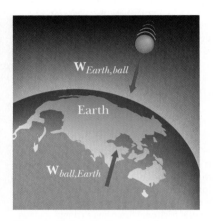

Figure 3-14 Earth exerts a force $\mathbf{W}_{Earth,ball}$ on the ball. According to Newton's third law, the ball exerts an equal and opposite force $\mathbf{W}_{ball,Earth}$ on Earth.

FLAWED REASONING

Let's reconsider the crate that fell from the helicopter into a deep snowdrift. Three students are discussing which force is bigger, the force exerted by the snow upward on the bottom of the crate or the force exerted downward by the bottom of the crate on the snow.

Jennifer: "The crate must be pushing down on the snow more than the snow is pushing up on the crate because the crate is still moving down through the snow."

Monica: "The snow must be pushing up on the crate harder than the crate is pushing down on the snow because the crate is slowing down."

Peter: "The two forces, crate on snow and snow on crate, are part of the same interaction. They must always be equal in magnitude and opposite in direction by Newton's third law."

With which student do you agree?

ANSWER Third-law forces always involve the same players. If A pushes on B, then B pushes back on A. When we refer to the two forces as crate on snow and snow on crate, it becomes obvious that these are third-law forces. They must always be equal in magnitude and opposite in direction.

true. No matter what the origin of the forces, Newton's third law tells us that the forces must be equal in size and opposite in direction.

But if this is true, why doesn't Earth accelerate toward the ball? It does, but if we put the values into Newton's second law, we find that Earth's mass is so large that its acceleration is minuscule. Earth does accelerate, but we don't notice it.

A very important point concerning third-law forces is that one of the forces acts on the ball while the other acts on Earth; one causes the ball to accelerate, and the other causes Earth to accelerate. Because the two forces act on different objects, the two forces cannot cancel; if these are the only two forces acting, both objects accelerate. Third-law forces *never* appear on the same free-body diagram.

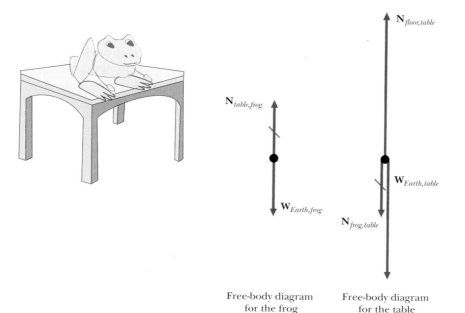

Free-body diagram for the frog

Free-body diagram for the table

Figure 3-15 Earth exerts a force $W_{Earth,frog}$ on the frog, which causes the frog to exert a force $N_{frog,table}$ on the table. By Newton's third law, the table exerts an equal and opposite force $N_{table,frog}$ on the frog. Although $W_{Earth,frog}$ and $N_{table,frog}$ are equal and opposite, they are not an action–reaction pair; they both act on the frog.

Free-body diagram for the frog

Free-body diagram for the table

Figure 3-16 If the frog accelerates upward, the two forces on the frog, $W_{Earth,frog}$ and $N_{table,frog}$, are no longer equal in magnitude, proving that they could not have been third-law companion forces.

Without the third law, paradoxical events would occur in the Newtonian world view. Consider, for example, a frog (really a prince) sitting on a table. Why doesn't the frog fall through the table? There is a gravitational force pulling the frog down. If this were the only force acting on the frog, according to Newton's second law, the frog would accelerate downward through the table. Because the frog is not accelerating, the net force on the frog must be zero. Thus, the question arises: What is the force that balances the downward gravitational force?

Newton's third law provides the answer. Earth attracts the frog with a force that we can label $\mathbf{W}_{Earth,frog}$, as shown in Figure 3-15. The frog pushes down on the table with a force that we label $\mathbf{N}_{frog,table}$. According to the third law, the table pushes upward on the frog with a force $\mathbf{N}_{table,frog}$ that is equal in size and opposite in direction to $\mathbf{N}_{frog,table}$. Therefore, there are two forces acting on the frog: the frog's weight $\mathbf{W}_{Earth,frog}$ and the upward force of the table $\mathbf{N}_{table,frog}$. Although these two forces are equal in size and act in opposite directions, they are not third-law companion forces; they both act on the frog. They are equal and opposite because the frog is not accelerating; therefore, by Newton's second law, the net force on the frog must be zero. If the frog sees a princess and suddenly jumps straight up in the air, his free-body diagram changes (see Figure 3-16). During the jump, he is accelerating upward and his free-body diagram must indicate an upward net force. The frog's weight cannot change (except through diet, exercise, or magic), so the normal force exerted by the table must increase. The frog jumps by pushing down on the table with an increased force, and by Newton's third law, the upward force by the table on the frog must also increase. Note that the third-law companion force associated with the frog's weight $\mathbf{W}_{Earth,frog}$ is the gravitational force exerted by the frog on the Earth $\mathbf{W}_{frog,Earth}$. This force is just one of the many forces appearing on Earth's free-body diagram.

Are You On the Bus?

Q: A branch exerts an upward force on an apple in a tree. What is the third-law companion to this force?

A: It is the downward force of the apple on the branch. Note that it is not the downward force of gravity on the apple. Although the gravitational force is equal and opposite to the upward force on the apple, both forces act on the apple and they cannot be action–reaction forces.

When you fire a rifle, it recoils. Why? As explained by the third law, when the rifle exerts a forward force on the bullet (by virtue of an explosion), the bullet simultaneously exerts an equal force on the rifle but in the backward direction. But why doesn't the rifle accelerate as much as the bullet? The force of the rifle on the bullet produces a large acceleration because the mass of the bullet is small. The force of the bullet on the rifle is the same size but produces a small acceleration because the mass of the rifle is large.

Even the common act of walking is possible only because of third-law forces. In walking you must have a force exerted on you in the direction of your accel-

Figure 3-17 As the man walks to the left, he exerts a force on the boat that causes the boat to move to the right.

eration. And yet the force you produce is clearly in the opposite direction. The solution to this apparent paradox lies in the third law. As you start to walk, you exert a force against the floor (down and backward); the floor therefore exerts a force back, causing you to go forward (and up a little). If there is any sand or loose earth where you walk, you can see that it has been pushed *back*. If you want a clearer demonstration of the fact that you push backward against the floor, try walking on a skateboard or in a rowboat (Figure 3-17). But be careful!

WORKING IT OUT *Putting It All Together*

Let us return to the example of pulling your sister on the sled. If the rope pulls on the sled with a force of 400 N, and the resulting acceleration is 5 m/s/s, find the magnitudes of the two friction forces acting on the sled. The first step whenever we are solving a force problem is to draw a free-body diagram for each object in the problem. Try drawing these diagrams and then check your answer with those in Figure 3-18. Note that tick marks have been used to indicate third-law companion forces.

The free-body diagram for your sister makes it clear that she is accelerating. She may not be moving relative to the sled, but the sled is accelerating and therefore so is she. Indeed, both she and the sled have acceleration of magnitude 5 m/s/s. We also see from her free-body diagram that the static friction force acting between her and the sled is the only horizontal force acting on her, so it is equal in magnitude to her net force. What net force is required to accelerate a 21-kg sister by 5 m/s/s? Newton's second law gives us $F_{net} = f_{sled,sister} = ma = (31\ kg)(5\ m/s/s) = 155\ N$.

We can now use Newton's third law to claim that the static friction force exerted by your sister on the sled must also be of magnitude 155 N. The net force required to accelerate a 28-kg sled by 5 m/s/s is 140 N. A net force of 140 N to the right means that the total force acting to the right must be 140 N greater than the total force acting to the left. The total force acting to the left must therefore be 400 N − 140 N = 260 N. The kinetic friction force exerted by the snow must have magnitude 105 N.

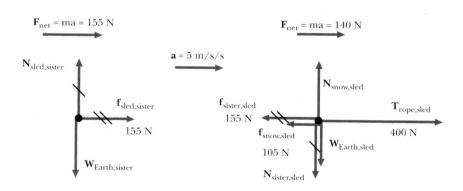

Free-body diagram for your sister   Free-body diagram for the sled

Figure 3-18 Free-body diagrams for your sister and the sled. Both diagrams must have more force to the right than to the left because both your sister and the sled are accelerating to the right.

Summary

According to Newton's first law of motion, an object at rest remains at rest, and an object in motion remains in motion with a constant velocity unless a net outside force acts on the object. The net force is determined by adding all the forces acting on an object according to the rules for combining vector quantities. The converse of Newton's first law is also true: if an object has a constant velocity (including the case of zero velocity), the unbalanced, or net, force acting on the object must be zero.

If there is a net force, the object accelerates with a value given by Newton's second law of motion, $\mathbf{a} = \mathbf{F}_{net}/m$. The direction of the acceleration is always the same as the net force. In the metric system, the unit of mass is the kilogram, and the unit of force is the newton.

Newton's second law of motion can be used to study frictional forces. Static friction can range from zero to a maximum value that depends on the force pushing the surfaces together and the nature of these surfaces. Kinetic friction has a constant value less than the maximum static value. A special kind of friction is air resistance, which varies with the speed of the object. As the air resistance acting on a falling object becomes equal to the object's weight, the acceleration goes to zero, and the object falls at its terminal speed.

The weight of an object close to Earth's surface is given by $\mathbf{W} = m\mathbf{g}$, where **g** is the acceleration due to gravity, about 10 (meters per second) per second downward. Weight, a force, should not be confused with mass.

There is no such thing as an isolated force. All forces occur in pairs that are equal in size and opposite in direction. As you stand on the floor, you exert a downward force *on the floor*. According to Newton's third law, the floor must exert an upward force *on you* of the same size. These two forces do not cancel, as they act on different objects—one on you and one on the floor. The other force acting on you is the force of gravity.

CHAPTER 3 | *Revisited*

The acceleration of an object in free fall becomes smaller and smaller because the force due to the air resistance increases as the speed of the object increases, becoming closer and closer in magnitude to the gravitational force pulling the object down. Therefore, the net force continually decreases. When the net, or total, force becomes zero, the acceleration also becomes zero and the speed assumes a constant value. This speed is known as the terminal speed.

Key Terms

force   A push or a pull. Force is measured (in newtons) by the acceleration it produces on a standard, isolated object, $\mathbf{F}_{net} = m\mathbf{a}$.

inertia   An object's resistance to a change in its velocity.

inversely proportional   A relationship in which two quantities have a constant product. If one quantity increases by a certain factor, the other decreases by the same factor.

kilogram   The standard international system (SI) unit of mass. A kilogram of material weighs about 2.2 pounds on Earth.

kinetic friction   The frictional force between two surfaces in relative motion. This force does not depend very much on the relative speed.

law of inertia   Newton's first law of motion.

mass   A measure of the quantity of matter in an object. The mass determines an object's inertia. Mass is measured in kilograms.

motion, Newton's first law of   The velocity of an object remains constant unless an unbalanced force acts on the object.

motion, Newton's second law of $\mathbf{F}_{net} = m\mathbf{a}$. The net force on an object is equal to its mass times its acceleration. The net force and the acceleration are vectors that always point in the same direction.

motion, Newton's third law of   If an object exerts a force on a second object, the second object exerts an equal force back on the first object.

newton   The SI unit of force. A net force of 1 newton accelerates a mass of 1 kilogram at a rate of 1 (meter per second) per second.

proportional   A relationship in which two quantities have a constant ratio. If one quantity increases by a certain factor, the other increases by the same factor.

static friction   The frictional force between two surfaces at rest relative to each other. This force is equal and opposite to the net applied force if the force is not large enough to make the object accelerate.

terminal speed   The speed obtained in free fall when the upward force of air resistance is equal to the downward force of gravity.

weight $W = mg$. The force of gravitational attraction of Earth for an object.

Questions and exercises are paired so that most odd-numbered are followed by a similar even-numbered.

Blue-numbered questions and exercises are answered in Appendix B.

🌶 indicates more challenging questions and exercises.

Web**Assign** Many Conceptual Questions and Exercises for this chapter may be assigned online at WebAssign.

Conceptual Questions

1. Assume you drop a bag of snacks while riding in an airplane that is flying due west at 800 kilometers per hour. Will the bag fall straight down, or will it angle toward the front or back of the airplane? Explain your reasoning.

2. The room you are sitting in is currently moving at about 400 meters per second as a result of Earth spinning about its axis. The walls of the room are attached to Earth, but if your keys fall out of your pocket, they are not. Why do the keys *not* appear to fly back toward the west wall?

3. Assume that you are pushing a car across a level parking lot. When you stop pushing, the car comes to a stop. Does this violate Newton's first law? Why?

4. If you give this book a shove so that it moves across a tabletop, it slows and comes to a stop. How can you reconcile this observation with Newton's first law?

5. How does the net force on the first subway car compare with that on the last car if the subway train has a constant velocity?

6. What can you say about the forces acting on a motorcycle that is traveling at a constant speed down a straight stretch of highway?

7. Why does a tassel hanging from the rearview mirror appear to swing forward as you apply the brakes?

8. When dogs finish swimming, they often shake themselves to dry off. What is the physics behind this?

9. Assume that you're not wearing your seat belt and the car stops suddenly. Why would your head hit the windshield?

10. Modern automobiles are required to have headrests to protect your neck during collisions. For what type of collision are these headrests most effective?

11. Why does a blacksmith use an anvil when hammering a horseshoe?

© Brakefield Photo/Brand X Pictures/Jupiterimages

12. You find that every time you pound a steak on your kitchen counter, the bottles fall out of the spice rack hanging on your wall. To solve the problem, you buy a large oak cutting board, which you place on the counter under the steak. Why does this help?

13. In everyday use, *inertia* means that something is hard to get moving. Is this the only meaning it has in physics? If not, what other meaning does it have?

14. How would you determine whether two objects have the same inertia?

15. When a number of different forces act on an object, is the net force necessarily in the same direction as one of the individual forces? Why?

16. You are analyzing a problem in which two forces act on an object. A 200-newton force pulls to the right and a 40-newton force pulls to the left. Your classmate asserts that the net force is 200 newtons because that is the dominant force that is acting. What is wrong with this assertion?

17. Forces of 40 newtons and 90 newtons act on an object. What are the minimum and maximum values for the sum of these two forces?

18. Two ropes are being used to pull a car out of a ditch. Each rope exerts a force of 700 newtons on the car. Is it possible for the sum of these two forces to have a magnitude of 1000 newtons? Explain your reasoning.

19. You apply a 75-newton force to pull a child's wagon across the floor at constant speed. If you increase your pull to 80 newtons, will the wagon speed up to some new constant speed or will it continue to speed up indefinitely? Explain your reasoning.

20. You push a crate full of books across the floor at a constant speed of 0.5 meter per second. You then remove some of the books and push exactly the same as you did before. How does the crate's motion differ, if at all?

21. If the net force on a boat is directed due east, what is the direction of the acceleration of the boat? Would your answer change if the boat had a velocity due north but the net force still acted to the east?

22. If the net force on a hot-air balloon is directed vertically upward, what is the direction of the acceleration of the balloon? What would be the direction of the acceleration

if the balloon were being blown westward (with the net force still acting vertically upward)?

23. You are riding an elevator from your tenth-floor apartment to the parking garage in the basement. As you approach the garage, the elevator begins to slow. What is the direction of the net force on you?

24. You are riding an elevator from the parking garage in the basement to the tenth floor of an apartment building. As you approach your floor, the elevator begins to slow. What is the direction of the net force on you?

25. If you double the net horizontal force applied to a wagon, what happens to the wagon's acceleration?

26. What happens to the acceleration of a rocket if the net force on it is cut in half?

27. A car can accelerate at 2 (meters per second) per second when towing an identical car. What will its acceleration be if the towrope breaks?

28. How does the net force on the first subway car compare with that on the last car if the subway train has a constant acceleration? Assume that the subway cars are identical.

29. When an astronaut walks on the Moon, is either her mass or her weight the same as on Earth? Explain.

30. If you buy a bag of pretzels labeled 0.1 kilogram, are you buying the pretzels by mass or by weight?

31. What happens to the weight of an object if you triple its mass?

32. What happens to the weight of an object if you take it from Earth to the Moon, where the acceleration due to gravity is one-sixth as large?

33. A skier is slowing down as she skis over level ground. Draw a free-body diagram for the skier.

34. A car on a level section of highway is speeding up to pass a truck. Draw a free-body diagram for the car.

35. Under what conditions will a golf ball and a table tennis ball that are dropped simultaneously from the same height reach the ground at the same time?

36. If a golf ball and a table tennis ball are simultaneously dropped from the same height, they do not reach the ground at the same time. How would Aristotle explain this? How would Galileo?

37. A marble dropped into a bottle of liquid soap quickly reaches a terminal speed. Draw a free-body diagram for the marble just before it hits the bottom of the bottle. What is the acceleration of the marble at this time?

38. Draw a free-body diagram for a parachutist who has reached terminal speed. What is his acceleration?

39. Sara is taking the high-speed elevator, which travels at a constant speed of 5 meters per second, to the 43rd floor of a high-rise building. Sam is stuck making the same trip in the freight elevator, which travels at a constant speed of only 1.5 meters per second. Compare the net forces on Sara and Sam.

40. Pat and Chris are pushing identical crates across a rough floor. Pat's crate is moving at a constant 1 meter per second while Chris's crate is moving at a constant 2 meters per second. Compare the net forces on the two crates.

41. A friend falsely claims, "Newton's first law doesn't work if there is any friction." How would you correct this claim?

42. One of your classmates falsely asserts, "Newton's second law only works when there are no frictional forces." How would you correct this assertion?

43. You are applying a 400-newton force to a freezer full of chocolate chip ice cream in an attempt to move it across the basement. It will not budge. Is the frictional force exerted by the floor on the freezer greater than, equal to, or less than 400 newtons?

44. You find that you must push with a force of 12 newtons to keep a textbook sliding at constant speed across your desk. With the book at rest, you apply a force of 13 newtons. Is it possible that the book will stay at rest? Explain.

45. What force is required to pull a dog in a wagon along a level sidewalk, as in the following figure, at a constant speed if the frictional force is 250 newtons?

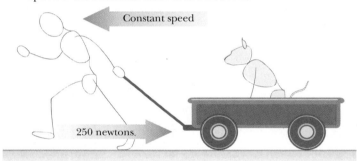

46. A skier is skiing down a steep slope, traveling at constant speed (that is, the skier has reached terminal velocity). What are the size and direction of the net force on the skier?

47. You are driving along the freeway at 75 mph when a bug splats on your windshield. Compare the force of the bug on the windshield to the force of the windshield on the bug.

48. You leap from a bridge with a bungee cord tied around your ankles. As you approach the river below, the bungee cord begins to stretch and you begin to slow down. Which is greater (if either), the force of the cord on your ankles to slow you or the force of your ankles on the cord to stretch it? Explain.

49. Is the force that the Sun exerts on Earth bigger, smaller, or the same size as the force that Earth exerts on the Sun? Explain your reasoning.

50. Earth exerts a force of 500,000 newtons on an orbiting communications satellite. Is the force that the satellite exerts on Earth greater than, equal to, or less than 500,000 newtons? Explain.

---

51. What is the net force on an apple that weighs 4 newtons when you hold it at rest?

52. Suppose you are holding an apple that weighs 4 newtons. What is the net force on the apple just after you drop it?

53. Why do the cannons aboard pirates' ships roll backward when they are fired?

54. Why does a tennis racket slow down when it hits a ball?

55. Describe the force or forces that allow you to walk across a room.

56. We often say that the engine supplies the forces that propel a car. This is an oversimplification. What are the forces that actually move the car?

57. A ball with a weight of 40 newtons is falling freely toward the surface of the Moon. What force does this ball exert on the Moon?

58. The following figure shows a ball hanging by a string from the ceiling. Identify the action–reaction pairs in this drawing. Argue that the magnitudes of all the forces are the same.

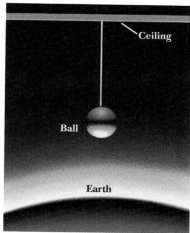

59. If the force exerted by a horse on a cart is equal and opposite to the force exerted by the cart on the horse, as required by Newton's third law, how does the horse manage to move the cart?

60. Gary reads about Newton's third law while sitting in a room with a single closed door. He reasons that if he applies a force to the door, there will be an equal and opposite force that will cancel his pull and he will never be able to escape. He moans, "Why did I ever take physics?" What is wrong with Gary's reasoning?

61. A soft-drink can sits at rest on a table. Which of Newton's laws explains why the upward force of the table acting on the can is equal and opposite to Earth's gravitational force pulling down on the can?

62. A book sits at rest on a table. Which force does Newton's third law tell us is equal and opposite to the gravitational force acting on the book?

Exercises

63. Find the size of the net force produced by a 6-N and an 8-N force in each of the following arrangements:
    a. The forces act in the same direction.
    b. The forces act in opposite directions.
    c. The forces act at right angles to each other.

64. Find the size of the net force produced by a 5-N and a 12-N force in each of the following arrangements:
    a. The forces act in the same direction.
    b. The forces act in opposite directions.
    c. The forces act at right angles to each other.

65. Two horizontal forces act on a wagon, 550 N forward and 300 N backward. What force is needed to produce a net force of zero?

66. Three forces act on an object. A 3-N force acts due west and a 4-N force acts due south. If the net force on the object is zero, what is the magnitude of the third force?

67. What is the acceleration of a 600-kg buffalo if the net force on the buffalo is 1800 N?

68. What is the acceleration of a 2000-kg car if the net force on the car is 4000 N?

69. A .30-06 bullet has a mass of 0.010 kg. If the average force on the bullet is 9000 N, what is the bullet's average acceleration?

70. The net horizontal force on a 60,000-kg railroad boxcar is 6000 N. What is the acceleration of the boxcar?

71. What net force is needed to accelerate a 60-kg ice skater at 2 m/s²?

72. If a sled with a mass of 20 kg is to accelerate at 4 m/s², what net force is needed?

73. If a 30-kg instrument has a weight of 50 N on the Moon, what is its acceleration when it is dropped?

74. A salesperson claims that a 1200-kg car has an average acceleration of 4 m/s^2 from a standing start to 100 km/h. What average net force is required to do this?

75. If a skydiver has a net force of 300 N and an acceleration of 4 m/s^2, what is the mass of the skydiver?

76. A child on roller skates undergoes an acceleration of 0.6 m/s^2 due to a horizontal net force of 24 N. What is the mass of the child?

77. A 0.5-kg ball has been thrown vertically upward. If we ignore the air resistance, what are the direction and size of each force acting on the ball while it is traveling upward?

78. A 1-kg ball is thrown straight up in the air. What is the net force acting on the ball when it reaches its maximum height? What is the ball's acceleration at this point?

79. Skip Parsec, intrepid space explorer, travels to a new planet and finds that he weighs only 320 N. If his mass is 80 kg, what is the acceleration due to gravity on this planet?

80. A fully equipped astronaut weighs 1500 N on the surface of Earth. If the astronaut has a weight of 555 N standing on the surface of Mars, what is the acceleration due to gravity on Mars?

81. A crate has a mass of 24 kg. What applied force is required to produce an acceleration of 3 m/s^2 if the frictional force is known to be 90 N?

82. A rope is used to pull a 10-kg block across the floor with an acceleration of 3 m/s^2. If the frictional force acting on the block is 50 N, what is the tension in the rope?

83. If a pull of 210 N accelerates a 40-kg child on ice skates at a rate of 5 m/s^2, what is the frictional force acting on the skates?

84. If you stand on a spring scale in your bathroom at home, it reads 600 N, which means your mass is 60 kg. If instead you stand on the scale while accelerating at 2 m/s^2 upward in an elevator, how many newtons would it read?

85. Terry and Chris pull hand over hand on opposite ends of a rope while standing on a frictionless frozen pond. Terry's mass is 75 kg and Chris's mass is 50 kg. If Terry's acceleration is 2 m/s^2, what is Chris's acceleration?

86. A mother of mass 50 kg and her daughter of mass 25 kg are ice-skating. They face each other, and the mother pushes on the daughter such that the daughter's acceleration is 2 m/s^2. What is the force exerted by the mother on the daughter? What is the force exerted by the daughter on the mother? What is the mother's acceleration?

Motions in Space

►Newton's first law states that an object naturally travels in a straight line. Yet in nearly every real-world example, objects execute much more complicated motions. Snowboarders fly through the air, and cars maneuver tight corners. What causes these motions? And how can we reconcile our subjective experience when we become part of a motion—say, as a passenger in a car rounding a corner—with an explanation based on our physics world view?

See page 73 for the answer to this question.

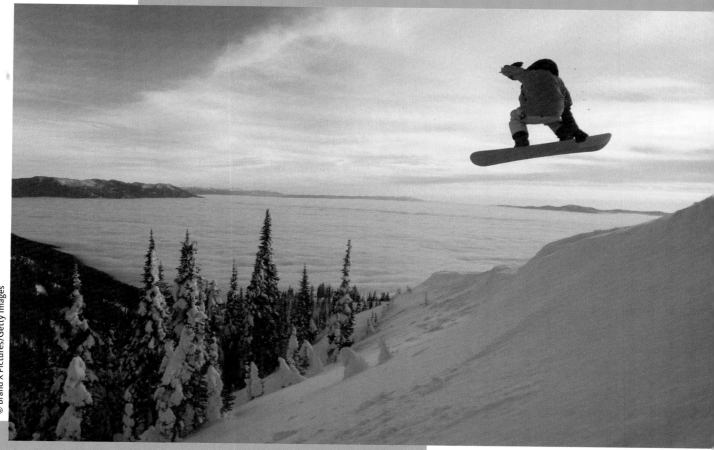

© Brand X Pictures/Getty Images

The motion of the snow boarder is a combination of a horizontal motion with constant speed and a vertical motion with constant acceleration.

THUS far, we have restricted our discussion to straight-line, or one-dimensional, motion. Most motions, however, take place in more than one dimension—most commonly in three-dimensional space.

Going from one dimension to two or three dimensions is less difficult than you might anticipate because all complex motions can be analyzed separately in each of the three dimensions. A knuckleball's motion can be thought of as the sum of three separate motions: up/down, left/right, and near/far (along the line between the pitcher and catcher). This separation means that we can apply the laws that were developed for one dimension to many common motions in space.

We will look at two-dimensional motion—that is, motion confined to a flat surface. Adding a second dimension allows us to study a variety of examples, from the path of a football to Earth's annual journey around the Sun.

Circular Motion

The motion of Earth around the Sun, the Moon around Earth, a race car around a circular track, and a ball swinging on the end of a string are examples of circular (or nearly circular) motion. Each of these occurs in a two-dimensional plane.

As we look at these motions, we start with the simplest situation, motion in which the speed of the object remains constant. In the one-dimensional cases we studied in the previous chapter, constant speed implied the absence of a net force. This is not the case in two dimensions. An object moving along a circular path at a constant speed must have a net force acting on it.

To see this, imagine whirling a ball on the end of a string in a circle above your head, as in Figure 4-1. Because you have to pull on the string to do this, the string must be exerting a force on the ball. And because a string can exert a force only along its length, this force must act toward the center of the circle. It has the special name **centripetal force**, which means *center-seeking* force. Could there also be a force pulling outward on the ball? What object could be exerting this force? According to Newton's first law, the ball will travel in a straight line unless acted on by a net force. The inward force of the string is required to pull the ball inward along the circular path. No outward force is required, and none exists.

It is important to distinguish between the adjectives centri*petal* (center-seeking) and centri*fugal* (center-fleeing). The word *centripetal* is rarely used in our everyday language. The word *centrifugal* is much more common and often mistakenly substituted for *centripetal*. The force we are discussing, the centripetal force, is directed toward the center of the circle. We discuss centrifugal effects in Chapter 9.

If you cut the string, the ball no longer moves in a circle but flies off in a straight line (if we ignore the force of gravity). Because of its inertia, the ball continues to travel in the direction of its velocity at the time the string was cut (Figure 4-2). In the absence of applied forces, the ball exhibits straight-line motion, as described by Newton's first law.

A car racing on an unbanked, circular track (again, at a constant speed) doesn't have a string anchoring it to the center of the track. The centripetal force, in this case, is provided by the friction between the track and the car's tires. A frictional force on the tires is directed *toward the center* of the track. An oil slick on the track could affect the frictional force, resulting in a potentially disastrous situation.

Let's reexamine Newton's first law of motion. It says that an object will go in a *straight line* unless acted on by a net force. This suggests that whenever an object changes direction, a net force must be acting on it. Newton's first

Figure 4-1 The tension force acting inward along the string keeps the ball in a circular path.

▶ a centripetal force is a center-seeking force

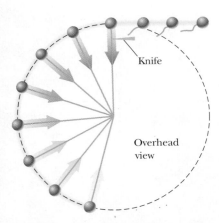

Figure 4-2 When the string is cut, the ball travels in a straight line tangent to the circular path.

law—the law of inertia—applies to direction as well as speed. Conversely, if the object is traveling in a straight line, then the net force, if it exists, can only be directed along the path of motion (if the object is speeding up), or against the path of motion (if the object is slowing down).

> **Q:** Assuming the frictional force goes to zero when the car hits the oil slick, describe its motion.
>
> **A:** The car would slide in a straight line in the direction it was traveling at the time it hit the oil slick. Because there is no net force acting on the car, the car maintains a constant velocity.

This discussion can be generalized to an even larger class of motions. Any slight change in direction can be thought of as traveling a brief interval of motion along an arc of a circle, as shown in Figure 4-3. This requires that a force act toward the inside of the turn. This is an explanation of why it is dangerous to drive your car on icy, curved roads—you need friction (traction) to change the direction of the car's motion as you turn the corners.

As mentioned in Chapter 3, the law of inertia applies to all objects in all situations. If a spacecraft were far from all massive bodies (where gravitational forces can be ignored), it would coast in a straight line. The only time NASA would need to turn on the spacecraft's engines would be to *change* its direction or speed.

Acceleration Revisited

> Extended presentation available in the *Problem Solving* supplement

We have established that an object moving in a circle with constant speed must have a net force acting on it. As a consequence, according to Newton's second law, the object must be accelerating. This acceleration is equal to the change in velocity divided by the time it takes to make the change. In this case the velocity changes direction without changing size. If this seems a little bizarre— that an object whose speed remains constant is accelerating—remember that a vector such as velocity can change in two ways.

Like the force vector, the velocity vector can be represented by an arrow, as shown in Figure 4-4. The length of the arrow in this case tells us the speed (for example, 1 centimeter = 5 meters per second), and the direction of the arrow is the direction of motion. The velocity vectors corresponding to circular motion at a constant speed are tangent to the circular path and have equal lengths. Although they have equal lengths, they are different vectors because their directions are different.

The change in velocity $\Delta\mathbf{v}$ is determined by subtracting the initial velocity vector from the final one. When we subtract one vector from another, we are finding the difference between the two vectors. If we are interested in the difference in height between two children, we would stand them up, back to back, and compare their heights. The comparison would not be fair if one of the children were standing on tiptoe. They both have to start at the same place, heel to heel. The same is true when we are comparing vectors to see how different they are (subtraction). Suppose we start with a vector describing the initial velocity of a car, $\mathbf{v}_i$, and find that one second later the velocity vector of the car, $\mathbf{v}_f$, was different (Figure 4-5[a]). This difference can be written as:

$$\Delta\mathbf{v} = \mathbf{v}_f - \mathbf{v}_i$$

We find this difference vector, $\Delta\mathbf{v}$, graphically by moving one of the vectors (being careful not to change its magnitude or direction) so that it is heel to

Figure 4-3 As the race car turns the corner, it momentarily moves along a portion of a circular path. The net force on the car is inward, toward the center of this "circle."

Figure 4-4 The velocity vector for circular motion with constant speed changes direction but not size (magnitude).

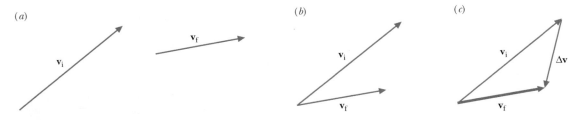

Figure 4-5 (a) Arrows v_i and v_f represent initial and final velocities of the car. (b) The initial and final velocity vectors are placed tail to tail to find their difference. (c) The initial velocity plus the change in velocity equals the final velocity.

heel with the other vector (Figure 4-5[b]). We can rewrite the above equation by adding $+\mathbf{v}_i$ to both sides:

$$\mathbf{v}_i + \Delta\mathbf{v} = \mathbf{v}_f$$

This equation indicates that the difference vector, $\Delta\mathbf{v}$, is the vector that must be added to the initial velocity vector to turn it into the final velocity vector, as shown in Figure 4-5(c).

The fact that the acceleration is defined as a vector ($\Delta\mathbf{v}$) divided by a number (the time elapsed) means that acceleration is also a vector. Its direction is the same as that of $\Delta\mathbf{v}$, and its magnitude is obtained by dividing the magnitude of $\Delta\mathbf{v}$ by the elapsed time:

acceleration = $\dfrac{\text{change in velocity}}{\text{time taken}}$ ▶

$$\mathbf{a} = \frac{\Delta\mathbf{v}}{\Delta t} \ \text{(where } \Delta t \text{ is small)}$$

Unlike the velocity vector, which points in the direction of motion, the direction of the acceleration vector is not always intuitive. The safest procedure is to check your intuition carefully by formally subtracting the velocity vectors to obtain the direction of the acceleration vector.

Let's check this definition for acceleration against the familiar case of straight-line motion, in which the object changes speed but not direction. Consider a marble rolling up a ramp, as shown in Figure 4-6(a). The initial velocity vector $\mathbf{v}_i$ and the final velocity vector $\mathbf{v}_f$ (for some time interval Δt) point in the same direction, but are not the same length. The velocity vector at the later time will have shorter length, as the marble is slowing down. The change in velocity $\Delta\mathbf{v}$ is obtained by placing $\mathbf{v}_i$ and $\mathbf{v}_f$ tail to tail (Figure 4-6[b]), and then finding the vector that must be added to the initial velocity to turn it into the

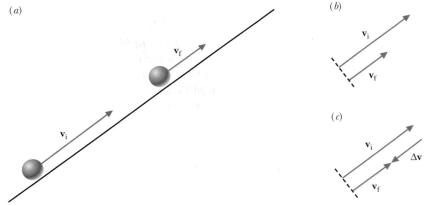

Figure 4-6 (a) A ball slows as it rolls up a ramp. (b) The initial and final velocity vectors are placed tail to tail to find their difference. (c) The initial velocity plus the change in velocity equals the final velocity.

final velocity (Figure 4-6[c]). This change-in-velocity vector points down the ramp, in the opposite direction as the velocity. This must also be the direction of the acceleration vector. In Chapter 2 we found that an object slows down when its acceleration vector points in the opposite direction as its velocity vector, and speeds up if those two vectors are pointing in the same direction. The same analysis of Figure 4-6 can be applied to show that the acceleration vector is also pointing down the ramp after the marble reaches its turnaround point and is rolling back down the ramp.

Q: Which way does the acceleration vector point at the turnaround point?

A: The marble's velocity at the turnaround point is zero, but its acceleration is not. The acceleration points in the same direction as the change-in-velocity vector for a time interval that includes the turnaround. Just before reaching the turnaround point, the marble's velocity vector points up the ramp. Just after the turnaround, the velocity points down the ramp. What vector must be added to a velocity up the ramp to change it into a velocity down the ramp? The change in velocity must point down the ramp, as does the acceleration vector. We can also answer this question using Newton's second law. The free-body diagram for the marble is the same at the turnaround point as it is while moving up the ramp. If the net force on the marble does not change at the turnaround point, then its acceleration cannot change. The marble's acceleration is constant (magnitude and direction) throughout its motion up and down the ramp.

Are You On the Bus?

The new physics student often greets this more complete definition of acceleration with disbelief and bewilderment. "How can you claim an acceleration when the speed stays constant?" is a typical reaction. One reason for the confusion is that the concept of acceleration has a more complex meaning in physics than in everyday language.

Although this difference in meaning may seem strange and perhaps even unnecessary, remember that our goal is to provide a complete description of motion. We must account for the observation that a change in *just* the direction (without a change in speed) requires a net force. Therefore, according to Newton's second law, there must be an acceleration.

FLAWED REASONING

At the county fair Billy finds himself pressed up against the wall of the Rotor, a circular room that spins about a vertical axis. When the room is spinning fast enough, the floor drops from under the people. Two students are arguing about the free-body diagram for Billy when he is at the location shown in Figure 4-7(a).

Isabel: "The gravitational force is pulling down on Billy, and the frictional force of the wall is keeping him from falling. The wall is exerting a normal force inward and the centrifugal force is acting outward, pinning Billy to the wall" (Figure 4-7[b]).

Caitlin: "I agree with you except for the centrifugal force. Billy would travel in a straight-line path if it weren't for the normal force and the inward centripetal force pushing him into a circular path" (Figure 4-7[c]).

Each student has made a critical error in reasoning. Identify each student's mistake, and draw the correct free-body diagram for Billy.

ANSWER Isabel thinks that Billy's free-body diagram should be balanced to show that he is not moving (and hence not accelerating) relative to the wall. Billy is, however, accelerating with respect to the ground. He is moving along a circle at constant speed, so his acceleration must be pointing toward the center of the circle. The free-body diagram should indicate a net force toward the right. There is no centrifugal force.

Caitlin correctly recognizes that Billy is accelerating toward the center of the circle, but believes that an extra centripetal force is needed to cause this acceleration. However, centripetal force is just a name for the *net* force in the special case of uniform circular motion. Only forces exerted by real objects should appear on free-body diagrams. Three forces act on Billy: the gravitational force, the frictional force, and the normal force. (See Figure 4-8.) The normal force is exerted by the wall and provides the centripetal force causing Billy to travel in a circle.

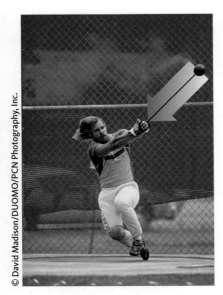

The hammer thrower provides the centripetal force required to make the hammer go in a circle.

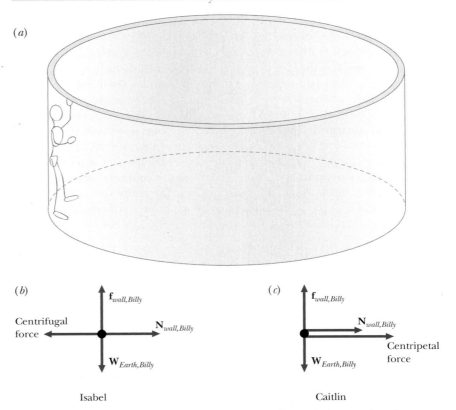

(a)

(b)

$\mathbf{f}_{wall, Billy}$

Centrifugal force

$\mathbf{N}_{wall, Billy}$

$\mathbf{W}_{Earth, Billy}$

Isabel

(c)

$\mathbf{f}_{wall, Billy}$

$\mathbf{N}_{wall, Billy}$

Centripetal force

$\mathbf{W}_{Earth, Billy}$

Caitlin

Figure 4-7 (a) Billy riding the Rotor. (b and c) *Incorrect* free-body diagrams for Billy.

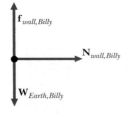

$\mathbf{f}_{wall, Billy}$

$\mathbf{N}_{wall, Billy}$

$\mathbf{W}_{Earth, Billy}$

Figure 4-8 The correct free-body diagram for Billy.

Acceleration in Circular Motion

In the case of circular motion at a constant speed, we know that the net force acts inward toward the center of the circle. We find the **centripetal acceleration** by subtracting two velocity vectors separated by a short time interval. This gives a change in velocity that points close to the center of the circle. As we shorten the time interval, the change in velocity points closer to the center. As the time interval becomes very small, we obtain the instantaneous-change-in-velocity vector, which points directly at the center. This agrees with Newton's second law because the instantaneous acceleration must always point in the same direction as the net force.

You can get a feeling for how the centripetal force depends on the parameters of the motion by twirling a ball on a string above your head and noticing how hard you have to pull on the string. To increase the speed of the ball while keeping the radius the same, you have to pull harder. This suggests that centripetal force increases as the speed of the circular motion increases. You can also lengthen the string while keeping the speed along the circular path the same. You will find that less pull on the string is required. This suggests that centripetal force varies inversely with the radius of the circular path.

WORKING IT OUT *Centripetal Acceleration*

A more detailed analysis of circular motion with a constant speed tells us that the centripetal acceleration a is equal to the object's speed v squared divided by the radius r of the circle:

$$a = \frac{v^2}{r}$$

◀ centripetal acceleration = $\dfrac{\text{speed squared}}{\text{radius}}$

As an example, we can calculate the centripetal acceleration of a 0.2-kg ball traveling in a circle with a radius of 1 m if it completes one revolution every second. Because this circle has a circumference of $2\pi r = 6.3$ m, the ball has a speed of 6.3 m/s:

$$a = \frac{v^2}{r} = \frac{(6.3 \text{ m/s})^2}{1 \text{ m}} = 40 \text{ m/s}^2$$

Applying $F_{net} = ma$ allows us to find the centripetal force required to produce this circular motion:

$$F_{net} = ma = (0.2 \text{ kg})(40 \text{ m/s}^2) = 8.0 \text{ N}$$

Q: If you double the speed of the ball traveling in a circle, what happens to the centripetal acceleration and the centripetal force?

A: Both the acceleration and the force quadruple.

Which bullet will hit the ground first if they are simultaneously released from the same height?

Projectile Motion

When something is thrown or launched near Earth's surface, it experiences a vertical gravitational force of constant magnitude. Motion under these conditions is called **projectile motion**. It occurs whenever an object is given some initial velocity and thereafter travels in a parabolic trajectory subject only to the force of gravity. Examples of projectile motion include cannonballs and a pass in football if we ignore the effects of air resistance (which we will do throughout the remainder of this chapter).

The study of projectile motion is simplified because the motion can be treated as two mutually independent, perpendicular motions, one horizontal and the other vertical. This reduces a complicated situation to two independent, one-dimensional motions that we already know how to handle.

Even though this separation of horizontal and vertical motion works, its consequences are often difficult to accept. For example, suppose a bullet is fired horizontally from a pistol and simultaneously another bullet is dropped from the same height. Which bullet hits the ground first?

A tempting but incorrect answer is that the dropped bullet hits first. The time it takes to reach the ground is determined by the vertical motion. Although the two bullets have different horizontal speeds, their vertical motions are identical *if we ignore the air resistance.* Therefore, they take the same time to reach the ground. Note that we do not claim that they travel the same distance or with the same speed; clearly the fired bullet has a much greater speed and consequently travels much farther by the time it hits the ground.

This result is convincingly demonstrated in the strobe photograph in Figure 4-9. The ball on the right was fired in a horizontal direction at the same time the ball on the left was dropped. The horizontal lines are included as a visual aid to

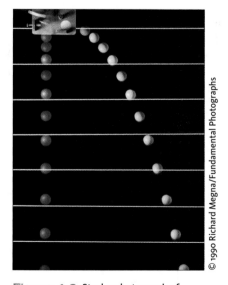

Figure 4-9 Strobe photograph of two balls. The ball on the right was fired horizontally; the other was dropped at the same instant.

Everyday Physics | *Banking Corners*

It takes a centripetal force to make a car go around a corner. On a flat parking lot or an unbanked curve, this force comes from the frictional interaction between the tires and the road. The size of the maximum possible frictional force is affected little by the type of tire and the type of surface but depends mostly on whether the surface is dry, wet, or icy. This maximum frictional force does not depend very much on the speed. However, as the speed of the car increases for a given turning radius, the actual centripetal force that friction must provide also increases. At some speed, the car is moving too fast, and the maximum possible frictional force is simply not large enough for the car to make it around the curve. Disaster may result!

There is another way to get a centripetal force. If the road is tilted, or banked, the demands on the frictional forces can be reduced. Let's first consider the case when there is no frictional force. As shown in the following figure, the car's weight acts vertically downward, pushing the car against the road. Because there is no frictional force, the road can only exert a force on the car that is perpendicular to the road. (If we assume that the car maintains the same elevation in the curve, the car does not accelerate in the vertical direction. Therefore, the vertical part of the force due to the road must just cancel out the car's weight.) As shown in the next figure, the sum of these two forces on the car—the force due to gravity and the force due to the road—is a horizontal force acting toward the inside of the curve. It is this horizontal force that provides the centripetal acceleration that makes the car go around the curve.

For a given bank to the curve, there is a well-determined centripetal force and a corresponding centripetal acceleration. It therefore requires a specific speed to execute circular motion with a radius that matches the curve. This speed is the design speed of the curve and is the speed that does not require any frictional forces between the tires and the road. This is the speed that you want to use if the road is icy. If your speed is higher, the circle your car follows will be larger than that of the roadway, and you will skid off the outside of the curve. If your speed is lower than the design speed, the centripetal force will cause your car's path to have too small a radius, and the car will slide off the road to the inside!

The relationship between the design speed and the banking angle is precise and can be calculated. To get a feeling for this, consider a curve with a radius of 100 meters. For a bank angle of 13 degrees, the design speed is found to be 15 meters per second (34 mph). Doubling the design speed to 30 meters per second requires a bank angle of 43 degrees, and tripling the design speed to 45 meters per second (100 mph) requires a bank angle of 65 degrees.

In actuality the centripetal force is provided by the banking of the curve *and* the force of friction. If the car is traveling faster than the design speed of the curve, the frictional force acts toward the inside of the curve to provide the extra centripetal force needed. However, if the car is traveling slower than the design speed, the frictional force acts outward, reducing the net centripetal force. Thus, friction means that you can safely execute corners at speeds above and below their design speeds. For instance, the corner with the 13-degree bank described in the previous paragraph could be taken at speeds up to 30 meters per second (67 mph).

The centripetal force experienced by a car going around a curve increases as the weight of the car increases. This is fortunate because the larger mass of the car requires a greater centripetal force to make the car go around the corner. Therefore, the design speed of the curve is the same for all cars. Can you imagine what the road signs would be like if this were not true?

1. The centripetal force causing a car to travel around an unbanked corner is provided by the frictional force between the car and the road. Is this *kinetic* friction or *static* friction? Explain the reasoning for your answer.

2. Explain how you would calculate the "design speed" for a banked corner.

The forces acting on a car going around a banked curve with no friction.

Banking corners allows cars to take curves at high speeds.

© DUOMO/CORBIS

help you see that the vertical race ends in a tie. This experiment provides evidence that vertical motion is unaffected by the presence of horizontal motion.

We can also show that the presence of vertical motion does not affect the horizontal motion of an object. Use a marking pen to make an *X* near the front of a skateboard. While standing at rest on the skateboard, drop a small beanbag such that it lands on the *X*. Now repeat the experiment while riding the skateboard at high constant speed on a level surface (Figure 4-10). If you drop the beanbag the same way you did while stopped, it still lands on the *X*. Galileo was the first to suggest this experiment. He claimed that a stone dropped from the top of the mast of a moving ship would land at the base of the mast.

Let's look at the motion of the ball thrown in Figure 4-11. Assuming that gravity is the only force acting on the ball, the acceleration of the ball must be in the vertical direction. Therefore, only the vertical speed changes; the horizontal speed remains constant throughout the flight.

How do we describe the vertical motion? The ball takes off with a certain initial vertical speed and has a constant downward acceleration throughout the flight. On the upward part of its flight, the downward acceleration slows the ball's vertical speed by 10 meters per second during each second. At some point the vertical speed becomes zero (point A in Figure 4-11). Then the ball starts its vertical descent, speeding up by 10 (meters per second) each second.

Although the vertical speed is zero at point A, the ball is still moving horizontally. The horizontal spacings between the images are equal, indicating that the horizontal speed is constant along the entire path. This curve also explains why basketball players such as Michael Jordan appear to hang in the air when they drive to the basket for a slam dunk. Note that near the top of the curve the ball travels very little in the vertical direction while it covers a much larger distance horizontally.

To further our understanding, we contrast the motion of a projectile on Earth with its corresponding motion on a mythical planet called Narang, a planet with **N**o **A**ir **R**esistance **A**nd **N**o **G**ravity. We fire a bullet from a horizontal rifle. On Narang the bullet travels in a straight, horizontal line at a constant

Figure 4-10 Modern version of Galileo's experiment. The beanbag hits the *X* when the skateboard is at rest or when it is moving with constant speed on a straight, horizontal road.

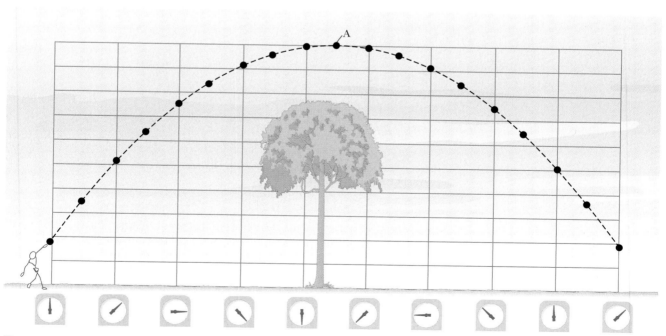

Figure 4-11 Strobe drawing of a thrown ball's motion. Note that the horizontal motion has a constant speed.

Michael Jordan appears to defy gravity as he slam-dunks the ball.

WORKING IT OUT *Projectile Motion*

An ugly giant rolls a bowling ball with a uniform speed of 30 m/s (approximately 60 mph!) across the top of his large desk. The ball rolls off the end of the desk and lands on the floor 120 m from the edge of the desk. How high is the desk?

The horizontal motion of the ball remains constant throughout the flight; every second the ball is in the air, it travels another 30 m in the horizontal direction. If the ball travels 120 m from the edge of the desk before it lands, it must have been in the air for 4 s:

$$t = \frac{d}{v} = \frac{120 \text{ m}}{30 \text{ m/s}} = 4 \text{ s}$$

The vertical motion of the ball is more complicated. It starts out with zero speed in the downward direction. Once the ball leaves the edge of the desk, it is in free fall and speeds up in the downward direction with an acceleration of 10 m/s². In 4 s the vertical speed changes from zero to

$$\Delta v = a\Delta t = (10 \text{ m/s}^2)(4 \text{ s}) = 40 \text{ m/s}$$

Which of these speeds, zero or 40 m/s, tells us how far the ball drops in 4 s? Neither. We must use the average speed of 20 m/s. The height of the desk is therefore

$$h = \bar{v}t = (20 \text{ m/s})(4 \text{ s}) = 80 \text{ m}$$

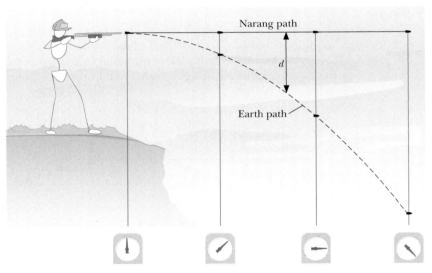

Figure 4-12 Paths of bullets fired horizontally on Narang and Earth. The distance *d* between the two paths is due to free fall.

velocity, as it experiences no force after it leaves the rifle barrel. On Earth the bullet's path differs because Earth's gravity affects its vertical motion. The horizontal motion of the bullet, however, is exactly the same on each planet, as shown in Figure 4-12.

Imagine now that we fire the rifle at an upward angle. On Narang the bullet again follows a straight line, but this time its path is inclined upward. On Earth the bullet continually falls downward from the straight-line path. At any time during the flight, the distance between the two paths is equal to the distance an object would fall from rest during the same elapsed time (Figure 4-13).

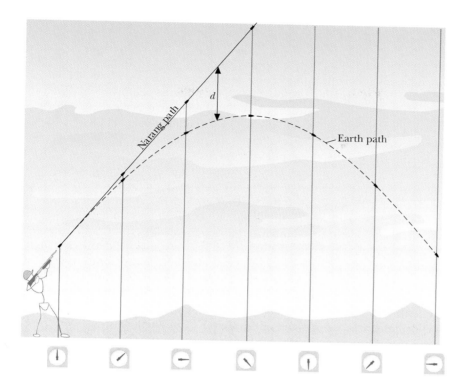

Figure 4-13 Paths of bullets fired at an upward angle on Narang and Earth. The distance *d* between the two paths is due to the presence of gravity.

Figure 4-14 Tranquilizing a gorilla in a tree. Where should you aim?

Q: You are hunting gorillas with a rifle that shoots tranquilizer darts. Suddenly you see a gorilla wearing a bright red button hanging from a limb (Figure 4-14). At the instant you pull the trigger, the gorilla lets go of the limb. Where should you have aimed to hit the gorilla's red button?

A: Directly at the button. The dart falls away from the line of sight at the same rate as the gorilla, and therefore the falling dart will hit the falling gorilla.

Figure 4-15 An unsuccessful launch results in projectile motion.

Launching an Apple into Orbit

Legend has it that Isaac Newton developed his ideas about gravity after seeing an apple fall and wondering if this was the same force that caused the Moon to orbit Earth. As a tribute to this legend, let's discuss the problem of launching an apple into orbit around Earth. We will of course neglect the effects of air resistance. (Incidentally, our discussion parallels the discussion in Newton's famous book *Principia;* the apple part is ours.)

To accomplish this thought experiment, we need a very high place to stand and a very, very strong arm. For each launch we will throw the apple horizontally. On our first attempt, we throw the apple with an ordinary speed and find that the apple follows a projectile path (Figure 4-15) like the ones we discussed in the previous section.

On our next attempt, imagine that we throw the apple much faster. The apple still falls to the ground, but the path is not exactly like the first one. The apple will travel a greater horizontal distance before it reaches the ground. If the apple travels far enough, Earth's curvature becomes important. The force of gravity points in slightly different directions at the beginning and end of the path.

Normally, we are not aware of the curvature of Earth's surface because Earth is so huge. Over large distances, however, this cannot be ignored. If we imagine that Earth is perfectly smooth—without hills and valleys—and construct a large horizontal plane at our location as shown in Figure 4-16, Earth's surface will be 5 meters below the plane at a distance of 8 kilometers away from where we stand. (This is about 16 feet at a distance of 5 miles.)

Imagine, then, that on our next attempt to launch the apple, we throw it with a speed of 8 kilometers per second. (This corresponds to 18,000 mph!) During the first second, the apple drops 5 meters, but so does the surface of Earth. Thus, the apple is still moving horizontally at the end of the first second. The motion during the next second is a repeat of that during the first. And so on. The apple is in orbit.

If air resistance is negligible, the speed of this apple remains constant. The only force is the force of gravity. It is a centripetal force acting perpendicular to the instantaneous velocity. By throwing the apple hard enough, we have changed the motion from projectile motion to circular motion. Unlike projectile motion, our apple will not come down even though it is continually falling. The illustration Newton used in his discussion of this is reproduced in Figure 4-17.

FLAWED REASONING

A newspaper report reads in part, "The space shuttle orbits Earth at an altitude of nearly 200 miles and is traveling at a speed of 18,000 mph. The shuttle remains in orbit because the gravitational force pulling it toward Earth is balanced by the centrifugal force (the force of inertia) that is pushing it away from Earth." **Explain why this newspaper should hire a new reporter.**

ANSWER All forces are exerted by one object on another object. Earth exerts the gravitational force on the shuttle. We have great difficulty, however, finding an object responsible for exerting a centrifugal, or outward, force on the shuttle. This is our first clue that such a force does not exist and, indeed, is not needed. Circular motion requires a net force acting *toward the center* of the circle, and the gravitational force provides this force. There is also no such force as a "force of inertia." Objects travel at constant velocity in the *absence* of a force, not *because* of a force.

Figure 4-16 A plane touching a spherical Earth is 5 meters above Earth's surface at a distance of 8 kilometers from the point of contact.

Rotational Motion

When something is thrown, it usually undergoes another motion. As it moves through space, it often spins or tumbles. We need to ask if it is legitimate to treat these motions as independent of each other.

The strobe photograph in Figure 4-18 shows the top view of a wrench sliding on a nearly frictionless surface. The wrench is doing two things: it is moving along a path, and it is rotating. Motion along a path is called **translational motion**. Notice that the spot marked with the white dot in Figure 4-18 is moving in a straight line and at a constant speed; it is moving at a constant velocity.

If we mentally shrink an object so that its entire mass is located at a certain point, the translational motion of this new, very compact object would be the same as the original object. Furthermore, if the object is allowed to rotate, it will rotate about this same point. This point is called the **center of mass**. The center of mass of the wrench in Figure 4-18 is marked by the white dot. There is a simple way of finding the center of mass of the wrench. If you place the wrench with the dot over your finger, the wrench will balance on your finger.

Although realistic motions are a combination of rotational and translational motions, the photograph shows that treating them as two separate motions can reduce the complexity. The presence of rotational motion does not affect the translational motion, and the presence of translational motion does not affect rotational motion. In other words, we can look at the rotational motion as if the object were not moving along a path and look at the translational motion as if the object were not rotating. We will take a detailed look at rotational motion in Chapter 8.

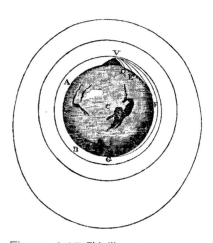

Figure 4-17 This illustration was used in Newton's *Principia* in the discussion of launching an object into orbit around Earth.

© 1990 Richard Megna/
Fundamental Photographs

Figure 4-18 Strobe photograph of a wrench sliding across a horizontal surface. Notice that the white dot representing the center of mass of the wrench moves with a constant velocity.

Everyday Physics | *Floating in Defiance of Gravity*

Many performers—long jumpers, basketball players, and ballet dancers, to name a few—would like to defy gravity and stay in the air longer. Regardless of their desires, however, the force of gravity is ever present, and the center of mass of the projectile—be it a pebble or a performing artist—follows the parabolic path discussed in this chapter. The path of the center of mass is determined by the initial conditions at the moment of launch. A different angle or different launch speed alters the projectile's path. However, once the object is launched, the center of mass follows the predetermined path.

The grand jeté in ballet seems to be a contradiction. In this popular ballet move, dancers execute a running leap across the stage, creating a seemingly floating motion that suspends them in the air longer than gravity should allow (Figure A).

There are two parts to this illusion. First, all objects spend most of the time of flight near the peak of the motion. By counting the images of the ball in Figure 4-11, it is easy to verify that the ball spends half of the time in the top one quarter of the vertical space. During this time, the object is moving mostly horizontally.

The second part of the illusion depends on the dancer's skill. Although the dancer's center of mass follows a parabolic path, a skillful dancer can change the position of the center of mass within her body during the flight. This allows the head and torso to stay at a nearly fixed height for a longer time. As illustrated in Figure B, the movements of the arms and legs raise and lower the location of the center of mass within the body. So, during the beginning and end of the jump, the dancer's arms and legs are pointing downward, keeping her center of mass low relative to her torso. During the middle of the flight, the dancer's arms are up and the legs are outstretched,

Figure A

© Stone/Getty Images

raising the center of mass relative to the rest of the dancer's body. The total effect is a flattening of the path of the head and torso.

1. What criteria determine the trajectory of a projectile's center of mass?

2. How does the ballet dancer change the location of her center of mass relative to her body?

Source: Kenneth Laws, *The Physics of Dance* (New York: Schirmer, 1984).

Figure B The dancer raises and lowers the center of mass within the body during the grand jeté.

Summary

Multidimensional motion can be divided into separate motions—translation in each of the three dimensions (up/down, left/right, and near/far). The laws for one-dimensional motion apply to each dimension separately.

An object moving along a circular path at a constant speed must have a net force acting on it. This centripetal force causes the circular motion; without it the object would fly off in a straight line, moving in the direction of its velocity at the time of release. Whenever there is any change in the velocity of an object, it experiences an acceleration and therefore must experience a net force. A net force with a constant magnitude acting perpendicular to the velocity produces circular motion with constant speed.

Projectile motion results from the constant, downward force of gravity. Again, these problems are simplified by the fact that the horizontal and vertical motions are independent. Assuming that air resistance is negligible, the only acceleration is in the vertical direction, the direction of the force of gravity. Therefore, the vertical motion is just that of free fall. The horizontal speed remains constant throughout the flight. For an extended object, there is a single point, the center of mass, that follows the projectile path.

Ignoring air resistance, a projectile launched just above Earth's surface with a horizontal speed of 8 kilometers per second will go into orbit around Earth. The projectile's speed and altitude remain constant, and it does not return to Earth even though it is continually falling.

CHAPTER 4 : *Revisited*

If the motion of an object changes direction, then the object is accelerating. There must be a net force acting on the object to cause this acceleration. A car turns a corner because the pavement exerts a force on its tires that provides the centripetal force toward the center of the curve. If you are a passenger in the car, your inertia keeps you moving forward in a straight line. The frictional forces between you and the car's seat are usually large enough to provide the force that makes you follow the same curve. You may get any additional needed force from the door. You feel that you've been pushed outward against the door, when in fact the door came to you and is pushing you into the curved path.

Key Terms

center of mass   The balance point of an object. This location has the same translational motion as the object would if it were shrunk to a point.

centripetal acceleration   The acceleration of an object toward the center of its circular path. For uniform circular motion, it has a magnitude v^2/r.

centripetal force   The net force required to keep an object moving in a circular path. This force is directed toward the center of the circular path. For uniform circular motion, the centripetal force has a magnitude mv^2/r.

projectile motion   A type of motion that occurs near Earth's surface when the only force acting on the object is that of gravity.

translational motion   Motion along a path.

> Questions and exercises are paired so that most odd-numbered are followed by a similar even-numbered.
>
> Blue-numbered questions and exercises are answered in Appendix B.
>
> indicates more challenging questions and exercises.
>
> WebAssign Many Conceptual Questions and Exercises for this chapter may be assigned online at WebAssign.

Conceptual Questions

Important: Ignore the effects of air resistance in the following questions and exercises.

1. A motorcycle drives through a vertical loop-the-loop at constant speed, as shown in the following figure. Draw arrows to show the directions of the instantaneous velocity and the net force on the motorcycle at points A, B, and C.

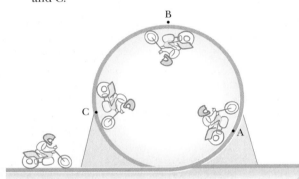

2. The following figure shows a racetrack with identical cars at points A, B, and C. The cars are moving clockwise at constant speeds. Draw arrows indicating the direction of the net force on each car and the instantaneous velocity of each car. In what direction would car A travel if there were an oil slick at its position? Why?

3. What is the force that causes a communications satellite to orbit Earth?

4. What is the force that allows a person on in-line skates to turn a corner? What happens if this force is not strong enough?

5. Consider the motorcycle in the figure for question 1 when it is at point B. In which directions do the velocity, acceleration, and net force point?

6. A child rides on a carousel. In which directions do the velocity, acceleration, and net force vectors point?

7. How do the velocity vectors differ on opposite sides of the path for uniform circular motion? How are they the same?

8. How do the acceleration vectors differ on opposite sides of the path for uniform circular motion? How are they the same?

9. An object executes circular motion with a constant speed whenever a net force acts perpendicular to the object's velocity. What happens to the speed if the net force is not perpendicular to the velocity?

10. An object is acted on by a force that is always perpendicular to the velocity. If the force continually increases in magnitude, does the speed change? Draw a sketch of the path of the object.

11. A car is initially traveling 30 mph due north. One minute later, the car is traveling 30 mph due west. What is the direction of the change-in-velocity vector during this minute?

12. A car is initially traveling 50 mph due east. One minute later, the car is traveling 50 mph due south. What is the direction of the change-in-velocity vector during this minute?

13. A water bug is skittering across the surface of a pond. In each case in the following figure, the bug's initial and final velocity vectors are shown for a time interval Δt. For each case, find the direction of the bug's average acceleration during this interval.

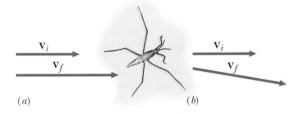

(a) (b)

14. A dog is running loose across an open field. In each case in the following figure, the dog's initial and final velocity vectors are shown for a time interval Δt. For each case, find the direction of the dog's average acceleration during this interval.

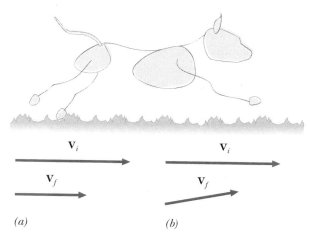

(a)   (b)

15. Each case in the following figure depicts an object's velocity vector and acceleration vector at an instant in time. State whether the object is (i) speeding up, slowing down, or maintaining the same speed and (ii) turning right, turning left, or moving in a straight line.

(a)   (b)

16. Each case in the following figure depicts an object's velocity vector and acceleration vector at an instant in time. State whether the object is (i) speeding up, slowing down, or maintaining the same speed and (ii) turning right, turning left, or moving in a straight line.

(a)   (b)

17. You are driving your car at 40 mph due north. Describe the general direction of your acceleration if you hit the brakes and turn right. Would the angle between your velocity vector and your acceleration vector be less than, equal to, or greater than 90 degrees?

18. You are driving your car at 40 mph due east. Describe the general direction of your acceleration if you hit the gas and turn left. Would the angle between your velocity vector and your acceleration vector be less than, equal to, or greater than 90 degrees?

19. Give examples from everyday life in which an object's acceleration is pointed downward and the object is (a) speeding up and (b) slowing down.

20. Give examples from everyday life in which an object's acceleration is pointed upward and the object is (a) speeding up and (b) slowing down.

21. You get into an elevator on the eighth floor and hit the button for the lobby on the ground floor. What is the direction of your acceleration right after the elevator starts moving and right before it stops at the ground floor?

22. You get into an elevator in the lobby on the ground floor and hit the button for the 10th floor. What is the direction of your acceleration right after the elevator starts moving and right before it stops at the 10th floor?

23. What force allows you to turn a bicycle while riding on a flat parking lot?

24. Most footraces take place on unbanked tracks. How do the racers turn the corners?

25. A monkey is swinging from tree to tree on vines. At the bottom of a swing, what force provides (or forces provide) the centripetal force required for the monkey to travel along a circular path?

26. A race car is traveling around a banked curve as described in the feature "Banking Corners." What force provides (or forces provide) the centripetal force required for the race car to travel along its circular path?

27. Imagine that you swing a bucket in a vertical circle at constant speed. Will you need to exert more force when the bucket is at the top of the circle or at the bottom? Explain.

28. A vine is just strong enough to support Tarzan when he is hanging straight down. However, when he tries to swing from tree to tree, the same vine breaks at the bottom of the swing. How could this happen?

29. Earth executes a nearly circular orbit around the Sun. What does this tell you about the speed of Earth along its orbit? Explain your reasoning carefully.

30. According to Newton's third law, Earth exerts a force on the Sun. Does the Sun move in a circular path as Earth goes around it each year? How can this idea be used to determine if nearby stars have planets?

31. You are driving your race car around a circular test track. Which would have a greater effect on the magnitude of your acceleration, doubling your speed or moving to a track with half the radius of curvature? Why?

32. A figure skater skates a figure-eight pattern with a small circle and a big circle, as shown in the following picture. The big circle has twice the radius of the small circle, and he skates it at twice the speed. Compare the magnitude of his centripetal acceleration on the two circles.

33. A playful astronaut decides to throw rocks on the Moon. What forces act on the rocks while they are in the "vacuum"? (We can't say "air"!)

34. A book slides along a frictionless table at a constant velocity and then sails off the edge. Draw a free-body diagram for the book while it is on the table and while it is in the air.

35. A left fielder throws a baseball toward home plate. At the instant the ball reaches its highest point, what are the directions of the ball's velocity, the net force on the ball, and the ball's acceleration?

36. The following figure shows the path of a thrown baseball. Draw arrows to indicate the directions of the ball's velocity and acceleration vectors at the three labeled points.

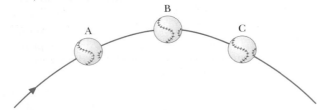

37. A hammer dropped on the surface of the Moon falls with an acceleration of 1.6 (meters per second) per second. Would its acceleration be smaller, larger, or the same if it was thrown horizontally at 6 meters per second? Why?

38. A rock dropped from 3.3 meters above the surface of the Moon requires 2.0 seconds to reach the ground. Would it require a shorter, a longer, or the same time if the rock was thrown horizontally from this height with a speed of 12 meters per second? Why?

39. Two identical balls roll off the edge of a table. One leaves the table traveling twice the speed of the other. Which ball hits the floor first? Why?

40. Two balls, one of mass 1 kilogram and one of mass 4 kilograms, roll off the edge of a table at the same time traveling at the same speed. Which ball hits the floor first? Why?

41. A physics student reports that upon arrival on planet X, she promptly sets up the "gorilla-shoot" demonstration. She does not realize that the gravity on planet X is stronger than it is on Earth. Will the demonstration work? Explain why or why not.

42. A fearless bicycle rider announces that he will jump the Beaver River Canyon. If he does not use a ramp, but simply launches himself horizontally, is there any way that he can succeed? Why?

43. In football and soccer, it is often desirable to give up some of the distance a kick travels to gain hang time, the time the ball remains in the air. How does the kicker do this?

44. The irons used in golf have faces that make different angles with the shaft of the club. How does this affect the distance traveled and maximum height of the golf ball?

© Cengage Learning/ George Semple

45. If Earth exerts a gravitational force of 5 newtons on the apple while it is in orbit, what force does the apple exert on Earth?

46. Is the size of the gravitational force that Earth exerts on the apple in orbit smaller than, larger than, or the same size as the force the apple exerts on Earth? Why?

47. We know that Earth travels around the Sun in a nearly circular orbit. Draw a free-body diagram for Earth.

48. Draw a free-body diagram for our fictitious apple in orbit near the Earth's surface.

49. A carpenter's square is tossed through the air. As it tumbles, only one point follows a simple parabolic path. In the following picture, which of the four labeled points most likely represents this point? Why?

50. A tennis ball and a softball are fastened together by a light rigid rod as shown. When this arrangement is thrown tumbling through the air, which of the labeled points is most likely to follow a parabolic path? Why?

Exercises

Important: Ignore the effects of air resistance in the following questions and exercises.

51. Find the size and direction of the change in velocity for each of the following initial and final velocities:
    a. 5 m/s west to 10 m/s west
    b. 10 m/s west to 5 m/s west
    c. 5 m/s west to 10 m/s east

52. What is the change in velocity for each of the following initial and final velocities?
    a. 75 km/h right to 100 km/h right
    b. 75 km/h right to 100 km/h left

53. What are the size and direction of the change in velocity if the initial velocity is 30 m/s south and the final velocity is 40 m/s west?

54. What is the change in velocity of a car that is initially traveling west at 50 km/h and then travels 120 km/h toward the north?

55. A migrating bird is initially flying south at 8 m/s. To avoid hitting a high-rise building, the bird veers and changes its velocity to 6 m/s east over a period of 2 s. What is the bird's average acceleration (magnitude and direction) during this 2-s interval?

56. A fox is chasing a bunny. The bunny is initially hopping east at 1 m/s when it first sees the fox. Over the next half second, the bunny changes its velocity to west at 4 m/s and escapes. What was the bunny's average acceleration (magnitude and direction) during this half-second interval?

57. A cyclist turns a corner with a radius of 50 m at a speed of 10 m/s.
    a. What is the cyclist's acceleration?
    b. If the cyclist and cycle have a combined mass of 120 kg, what is the force causing them to turn?

58. A 60-kg person on a merry-go-round is traveling in a circle with a radius of 3 m at a speed of 2 m/s.
    a. What acceleration does the person experience?
    b. What is the net force? How does it compare with the person's weight?

59. Earth orbits once around the Sun every 365¼ days at an average radius of 1.5×10^{11} m. Earth's mass is 6×10^{24} kg.
    a. How many seconds does it take Earth to orbit the Sun?
    b. What distance does Earth travel in 1 year?
    c. What is Earth's average centripetal acceleration?
    d. What is the average force that the Sun exerts on Earth?

60. Given that the average distance from Earth to the Moon is 3.8×10^{8} m, that the Moon takes 27 days to orbit Earth, and that the mass of the Moon is 7.4×10^{22} kg, what is the average centripetal acceleration of the Moon and the size of the attractive force between Earth and the Moon?

61. A baseball is hit with a horizontal speed of 45 m/s and a vertical speed of 18 m/s upward. What are these speeds 1 s later?

62. What are the horizontal and vertical speeds of the baseball in the previous exercise 2 s after it is hit?

63. An SUV accidentally drove off a cliff with a horizontal velocity of 40 m/s. Given that it took 5 s for the SUV to hit the ground, how far vertically and horizontally from the top of the cliff did the SUV land?

64. Angel Falls in southeastern Venezuela is the highest uninterrupted waterfall in the world, dropping 979 m (3212 ft). Ignoring air resistance, it would take 14 s for the water to fall from the lip of the falls to the river below. If the water lands 50 m from the base of the vertical cliff, what was its horizontal speed at the top?

© Kevin Schafer/Stone/Getty

65. A tennis ball is hit with a vertical speed of 10 m/s and a horizontal speed of 30 m/s. How long will the ball remain in the air? How far will the ball travel horizontally during this time?

66. If a baseball is hit with a vertical speed of 30 m/s and a horizontal speed of 6 m/s, how long will the ball remain in the air? How far from home plate will it land?

67. Given that the radius of Earth is 6400 km, calculate the acceleration of an apple in orbit just above Earth's surface.

68. The Moon orbits Earth at a distance of 3.8×10^{8} m and a speed of 1 km/s. What is the centripetal acceleration of the Moon?

5 Gravity

▶A popular legend has it that Newton had his most creative thought while watching an apple fall to the ground. He made a huge conceptual leap by equating the motion of the apple to the motion of the Moon and developing the concept of gravity.   How far does gravity reach? Do we see any evidence of gravity elsewhere in the universe?

(See page 93 for the answer to this question.)

NASA

This view of Earth greeted the *Apollo 11* astronauts as they orbited the Moon.

N the physicist's view of the world, there are four fundamental forces: the gravitational, the electromagnetic, the weak, and the strong. We begin our studies of forces with the most familiar force in our everyday lives. Every school-age child knows that objects fall because of gravity. But what is gravity? Saying that it is what makes things fall doesn't tell us much.

Is gravity a material like a fluid or a fog? Or is it something more ethereal? No one knows. Because we have given something a name doesn't mean we understand it completely. We do understand gravity in the sense that we can precisely describe how it affects the motion of objects. For instance, we have already seen how to use the concept of gravity to describe the motion of falling objects. We can do more. By looking carefully at the motions of certain objects, we can develop an equation that describes this attractive force between material objects and explore some of its consequences.

On the other hand, we cannot answer questions such as, "What is gravity?" or "Why does gravity exist?"

The Concept of Gravity

The concept of gravity hasn't always existed. It was conceived when changes in our world view required a new explanation of why things fell to Earth. When Earth was believed to be flat, gravity wasn't needed. Objects fell because they were seeking their natural places. A stone on the end of a string hung down because of its tendency to return to its natural place. "Up" and "down" were absolute directions.

The realization that Earth was spherical required a change in perspective. What happens to the unfortunate people on the other side of Earth who are upside down? But the change in thinking was made without gravity. The center of Earth was at the center of the universe, and things naturally moved toward this point. "Up" and "down" became relative, but the location of the center of the universe became absolute.

Gravity was also not needed to understand the motion of the heavenly bodies. The earliest successful scheme viewed Earth as the center of the universe, with the celestial bodies going around Earth in circular orbits. Perpetual, circular orbits were considered quite natural for celestial motions; little attention was given to the causes of these motions. Aristotle did not recognize any connections between what he saw as perfect, heavenly motion and imperfect, earthly motion. He stated that circular motion with constant speed was the most perfect of all motions, and thus the natural heavenly motion needed no further explanation.

This changed slightly with a new view of heavenly motion by Nicholas Copernicus, a 16th-century Polish scientist and clergyman. He proposed that the planets (including Earth) go around the Sun in circular orbits and that the Moon orbits Earth. This is essentially the scheme taught in schools today. A hint of a concept of gravity appears in Copernicus's work. He believed that the Sun and Moon would attract objects near their surfaces—each would have a local gravity—but he had no concept of that attractive influence spreading throughout space.

A hundred years later, Johannes Kepler, a German mathematician and astronomer, suggested that the planets move because of an interaction between them and the Sun. Kepler also moved us away from the assumption that the planets traveled in circular paths. After many years of trial and error, Kepler correctly deduced that the orbits of the planets were *ellipses*—but ellipses that are close to being circles. Furthermore, the planets do not have constant speeds in their journeys around their elliptical paths but speed up as they approach the Sun and slow down as they move farther away.

Kepler | *Music of the Spheres*

Johannes Kepler was born two days after Christmas in 1571. In his horoscope, which he later compiled, he noted that the family spelled their names in a variety of ways, that he was premature at birth, and that he was a sickly child. The village in which he was born is now part of greater Stuttgart in the German state of Baden-Würtemberg. His peasant origins could not mask his precocity and gift for mathematics. He was chosen by the local duke to receive a good education and later attended the new Lutheran seminary at Tübingen University. There he became acquainted with the techniques necessary to work on advanced astronomical problems.

Kepler's adult life centered on mathematical astronomy. A Lutheran, he spent most of his life in Prague in the employment of a Catholic Holy Roman emperor. He was a committed Copernican and sought to extend the accuracy of that new astronomy. His first work—*Cosmic Mysteries* (1597)—was a somewhat mystical and numerological theory relating distances and times of revolutions of the planets. His generation regarded mathematics as a language that revealed the inner harmony of creation and that celestial motions revealed physical harmony and unity in action. So his search for a "music" of the spheres truly was a search for a mathematical description of God's creation.

His mentor in astronomy, Tycho Brahe, was a Danish master of observational technique and precision. In fact, Brahe's work reached the limit of naked-eye observations. After Brahe's death, Kepler gained access to a wealth of astronomical data on the motions of Mars. This planet was the most perplexing in its apparently irregular movements. Kepler published his book on Mars in 1609 and ushered in an age of "new" astronomy. It is here that his first two laws of motion can be found.

His work was known throughout Europe, and he and Galileo (and even Galileo's father) engaged in a desultory correspondence concerning issues of astronomy. Again, music was frequently discussed in this context.

Johannes Kepler

Bettmann/CORBIS

Kepler found himself engulfed in a series of major European wars, in personal tragedy, and, when his mother was accused of witchcraft in Baden-Württemberg, in trouble with the law. His great ability continued to shine through in all this turmoil. He developed a technique—logarithms—to speed up calculations, and he continued his astronomical work. He desperately needed money and so wrote astrological predictions for the powerful. He developed the concept of satellites and even wrote a little fable on space travel.

Always seeking harmony in the cosmos, he published in 1618 a book similar to his earlier one on cosmic mysteries. In this later work, he laid out his third law, which related planetary orbits, times of revolution, and heliocentric theory. It was this third law upon which Isaac Newton so successfully built his physics. Again, a musical theme emerges: this book was titled *Harmonies of the World*.

Johannes Kepler provided impetus to the development of a new physics because his astronomy destroyed the idea of orbits as perfect circles and the customary ideas of natural and unnatural motions. When he died in 1630, his legacy was secure. He had created a new astronomy and demanded a new concept of physics to support it.

Sources: Max Caspar, *Kepler*, trans. and ed. C. Doris Hellman (New York: Abelard-Schuman, 1959); Arthur Koestler, *The Watershed: A Biography of Johannes Kepler* (Garden City, N.Y.: Anchor Books, 1960).

Influenced by early, important work on magnetism, Kepler postulated that the planets were "magnetically" driven along their paths by the Sun. Because this work occurred shortly before the acceptance of the idea of inertia, Kepler did not realize that a force was not needed to drive the planets along their orbits but to cause the orbits to be curved. Kepler postulated an interaction reaching from the Sun to the various planets and driving the planets, but he didn't consider the possibility of any interaction between the planets themselves; the Sun reigned supreme, a metaphor for his god, from which everything else gained strength.

Newton developed our present view of gravity. He started by saying there was nothing special about the rules of nature that he had developed for use on Earth. They should also apply to heavenly motions. As we learned in the previous chapter, anything traveling in a circle must be accelerating. The acceleration must be toward the center of the circle, and the object must therefore have a net force acting on it. Newton went searching for this force.

Newton's Gravity

Creativity often involves bringing together ideas or things from seemingly unrelated areas. After an artist or scholar has done it, the connection often seems obvious to others. Newton made such a synthesis between motions on Earth and motions in the heavens. Legend has it that he made his intellectual leap while contemplating such matters and seeing an apple fall. In the preceding chapter, we discussed the problem of launching an apple into orbit around Earth. The transition of the Earth-bound apple into heavenly orbits provides an analogy of Newton's intellectual leap.

Newton believed that the laws of motion that worked on Earth's surface should also apply to motion in the heavens. Because the Moon orbits Earth in a nearly circular orbit, it must be accelerating toward Earth. According to the second law, any acceleration requires a force. He believed that if this force could be shut off, the Moon would no longer continue to move along its circular path but would fly off in a straight line like a stone from a sling.

The genius of Newton was in relating the cause of this heavenly motion to earthly events. Newton felt that the Moon's acceleration was due to the force of gravity—the same gravity that caused the apple to fall from the tree. How could he demonstrate this? First, he calculated the acceleration of the Moon. Because the distance to the Moon and the time it took the Moon to make one revolution were already known, he was able to calculate that the Moon accelerated 0.002 72 (meter per second) per second. This is a very small acceleration. In 1 second the Moon moves about 1 kilometer along its orbit but falls only 1.4 millimeters (about $\frac{1}{20}$ inch in 0.6 mile).

In contrast to the Moon's acceleration, the apple has an acceleration of 9.80 (meters per second) per second and falls about 5 meters in its first second of flight. [In earlier discussions we rounded the acceleration off to 10 (meters per second) per second for ease of computation. The small difference is important here.] We can compare these two accelerations by dividing one by the other:

$$\frac{0.002\,72 \text{ m/s}^2}{9.80 \text{ m/s}^2} = \frac{1}{3600}$$

Why are these two accelerations so different? The mass of the Moon is certainly much larger than that of the apple. But that doesn't matter. As we saw in Chapter 2, free-falling objects all have the same acceleration independent of their masses.

However, the accelerations of the apple and the Moon weren't equal. Could Newton's idea that both motions were governed by the same gravity be wrong? Or could the rules of motion that he developed on Earth not apply to heavenly motion? Neither. Newton reasoned that the Moon's acceleration is smaller because Earth's gravitational attraction is smaller at larger distances; it is "diluted" by distance.

How did the force decrease with increasing distance? Retracing Newton's reasoning is impossible because he didn't write about how he arrived at his conclusions, but he may have used the following kind of reasoning.

Many things get less intense the farther you are from their source. Imagine a paint gun that can spray paint uniformly in all directions. Suppose the gun is in the center of a sphere of radius 1 meter, and at the end of 1 minute of spraying, the paint on the inside wall of the sphere is 1 millimeter thick. If we repeat the experiment with the same gun but with a sphere that is 2 meters in radius, the paint will be only $\frac{1}{4}$ millimeter thick because a sphere with twice the radius has a surface area that is four times the original (Figure 5-1). If the sphere has three times the radius, the surface is nine times bigger, and the paint is $(\frac{1}{3})^2 = \frac{1}{9}$ as thick. The thickness of the paint decreases as the square of the radius of the

Legend has it that Newton conceived his law of universal gravitation while observing an apple fall from a tree in his yard.

Jim Sugar/CORBIS

◄ $\dfrac{\text{acceleration of Moon}}{\text{acceleration of apple}}$

Figure 5-1 If the sphere's radius is doubled, the sphere's surface increases by a factor of 4.

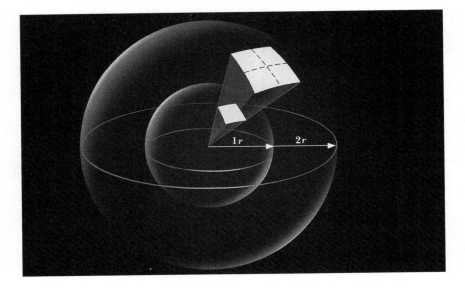

sphere increases. This is known as an **inverse-square** relationship. A force reaching into space could be diluted in a similar manner.

Are You On the Bus?

Q: If the sphere were 4 meters in radius, how thick would the paint be?

A: It would be $(\frac{1}{4})^2 \times 1$ millimeter $= \frac{1}{16}$ millimeter thick.

Newton may also have received encouragement for this explanation by working backward from observational results on the motion of the planets that had been developed by Kepler. Kepler found a relationship that connected the orbital periods of the planets with their average distances from the Sun. Using Kepler's results and an expression for the acceleration of an object in circular motion, we can show that the force decreases with the square of the distance. This means that if the distance between the objects is doubled, the force is only one-fourth as strong. If the distance is tripled, the force is one-ninth as strong, and so on. This relationship is shown in Figure 5-2.

What, exactly does it mean to say that the force is one-ninth as strong? We cannot be referring to the gravitational forces acting on different objects; the gravitational force on an automobile is obviously much larger than the gravitational force on a person. We must compare the gravitational forces acting on a single object, such as our apple, at different distances. When the apple is moved three times as far away from Earth's center, the gravitational force on the apple is one-ninth as strong.

The obvious test of the notion of gravity was to see whether the relationship between distance and force gave the correct relative accelerations for the apple on Earth and the orbiting Moon. Newton could use this rule and make the comparison. The distance from the center of Earth to the center of the Moon is about 60 times the radius of Earth. The Moon is therefore 60

Figure 5-2 The force on a 0.1-kilogram mass at various distances from Earth. Notice that the force decreases as the square of the distance.

times farther away from the center of Earth than the apple. Thus, the force at the Moon's location—and its acceleration—should be 60^2, or 3600, times smaller. This is in agreement with the previous calculation. The data available in Newton's time were not as good as those we have used here, but they were good enough to convince him of the validity of his reasoning. Modern measurements yield more precise values and agree that *the gravitational force is inversely proportional to the square of the distance.*

Q: What happens to the force of gravity if the distance between the two objects is cut in half?

A: The force becomes four times as strong.

Are *You* On the Bus?

Newton now knew how gravity changes with distance: the force of Earth's gravity exists beyond Earth and gets weaker the farther away you go. But there are other factors. He already knew that the force of gravity depends on the object's mass. His third law of motion says that the force exerted on the Moon by Earth is equal in strength to that exerted on Earth by the Moon; they attract each other. This symmetry indicates that both masses should be included in the same way. *The gravitational force is proportional to each mass.*

The Law of Universal Gravitation

◀ Extended presentation available in the *Problem Solving* supplement

Having made the connection between celestial motion and motion near Earth's surface with a force that reaches across empty space and pulls objects to Earth, Newton took another, even bolder, step. He stated that the force of gravity exists between *all* objects, that it is truly a *universal* law of gravitation.

The boldness of this assertion becomes apparent when one realizes that the force between two ordinary-sized objects is extremely small. Clearly, as you walk past a friend, you don't feel a gravitational attraction pulling you together. But that is exactly what Newton was claiming. Any two objects have a force of attraction between them; his rule for gravity is a **law of universal gravitation**.

Putting everything together, we arrive at an equation for the gravitational force:

$$F = G\frac{m_1 m_2}{r^2}$$

◀ law of universal gravitation

where m_1 and m_2 are the masses of the two objects, r is the distance between their centers, and G is a constant that contains information about the strength of the force.

Although Newton arrived at this conclusion when he was 24 years old, he didn't publish his results for more than 20 years. This was partly due to one unsettling aspect of his work. The distance that appears in the relationship is the distance from Earth's center. This means that Earth's mass is assumed to be concentrated at a point at its center. This may seem to be a reasonable assumption when considering the force of gravity on the Moon; Earth's size is irrelevant when dealing with these huge distances. But what about the apple on Earth's surface? In this case the apple is attracted by mass that is only a few meters away and mass that is 13,000 kilometers away, as well as all the mass between (Figure 5-3). It seems less intuitive that all of this mass would somehow act like a very compact mass located at Earth's center. But that is just what happens. Newton was eventually able to show mathematically that the sum of the forces due to each cubic meter of Earth is the same as if all of them were concentrated at its center.

Figure 5-3 The sum of all the forces on the apple exerted by all portions of Earth acts as if all the mass were located at Earth's center.

This result holds if Earth is spherically symmetric. It doesn't have to have a uniform composition; it need only be composed of a series of spherical shells, each of which has a uniform composition. In fact, 1 cubic meter of material near Earth's center has almost four times the mass of a typical cubic meter of surface material.

Newton applied the laws of motion and the law of universal gravitation extensively to explain the motions of the heavenly bodies. He was able to show that three observational rules developed by Kepler to describe planetary motion were a mathematical consequence of his work. Kepler's rules were the results of years of work reducing observational data to a set of simple patterns.

By the 18th century, scientists were so confident of Newton's work that when a newly discovered planet failed to behave "properly," they assumed that there must be other, yet to be discovered, masses causing the deviations. When Uranus was discovered in 1781, a great effort was made to collect additional data on its orbit. By going back to old records, scientists determined additional times and locations of its orbit. Although the main contribution to Uranus's orbit is the force of the Sun, the other planets also have their effects on Uranus. In this case, however, the calculations still differed from the actual path by a small amount. The deviations were explained in terms of the influence of an unknown planet. This led to the discovery of Neptune in 1846.

This still didn't completely account for the orbits of Uranus and Neptune; a search began for yet another planet. The discovery of Pluto in 1930 still left some discrepancies. Although the search for new planets continues, analysis of the paths of the known planets indicates that any additional planets must be very small and/or very far away.

The Value of G

Even though Newton had an equation for the gravitational force, he couldn't use it to actually calculate the force between two objects; he needed to know the value of the constant G. The way to get this is to measure the force between two known masses separated by a known distance. However, the force between two objects on Earth is so tiny that it couldn't be detected in Newton's time.

It was more than 100 years after the publication of Newton's results before Henry Cavendish, a British scientist, developed a technique that was sensitive enough to measure the force between two masses. Modern measurements yield the value

gravitational constant ▶
$$G = 0.000\,000\,000\,066\,7\,\frac{\text{N} \cdot \text{m}^2}{\text{kg}^2} = 6.67 \times 10^{-11}\,\frac{\text{N} \cdot \text{m}^2}{\text{kg}^2}$$

(See the Working It Out box in Chapter 1 for an explanation of this notation.) Putting this value into the equation for the gravitational force tells us that the force between two 1-kilogram masses separated by 1 meter is only 0.000 000 000 066 7 newton. This is minuscule compared with a weight of 9.8 newtons for each mass. The small value of G explains why two friends don't feel their mutual gravitational attraction when standing near each other.

Cavendish referred to his experiment as one that "weighed" Earth, although it would have been more accurate to claim that it "massed" Earth. His point, though, was important. By measuring the value of G, Cavendish made it possible to accurately determine Earth's mass for the first time. The acceleration of a mass near Earth's surface depends on the value of G and Earth's mass and radius. Because he now knew the values of all but Earth's mass, he could calculate it. Earth's mass is 5.98×10^{24} kilograms; that's about a million million million million times as large as your mass.

WORKING IT OUT   *Gravity*

Let's calculate the gravitational force between two friends. To make the calculation of this force easier, we make one unrealistic assumption: we assume that the friends are spheres! This allows us to use the distance between their centers as their separation and still get a reasonable answer. Assuming that the friends have masses of 70 and 86 kg (about 154 and 189 lb, respectively) and are standing 2 m apart, we have

$$F = G\frac{m_1 m_2}{r^2} = \left(6.67 \times 10^{-11}\frac{\text{N} \cdot \text{m}^2}{\text{kg}^2}\right)\frac{(70 \text{ kg})(86 \text{ kg})}{(2 \text{ m})^2} = 1.00 \times 10^{-7}\text{N}$$

This very tiny force is about one ten-billionth (10^{-10}) of either friend's weight.

The attraction between these spheroidal friends depends on the distance between their centers.

FLAWED REASONING

In the blockbuster movie *Armageddon,* the heroes land their space shuttle on a Texas-sized comet that is careening toward Earth. They then walk around the comet like construction workers here on Earth. **What is wrong with this picture?**

ANSWER A comet is not massive enough to provide the gravity required to walk around normally. The astronauts would have to tether themselves to keep from flying away because of the smallest exertion. The physics was much more accurate in the other comet-coming-to-destroy-Earth movie *Deep Impact.* In this movie, the spaceship is tethered to the comet and one of the astronauts is thrown into space by an explosion on the surface of the comet.

   Once Earth's mass is known, we can use the law of universal gravitation to calculate the acceleration due to gravity near Earth's surface:

$$g = \frac{F}{m} = \frac{GM_E}{R_E^2}$$

where M_E is Earth's mass and $R_E = 6370$ kilometers is Earth's radius. Plugging in the numerical values yields $g = 9.8$ (meters per second) per second, as expected.

   Because Earth orbits the Sun, the Sun's mass can also be calculated with the Cavendish results. In making these computations, the value of G measured on Earth is assumed to be valid throughout the Solar System. This cannot be proved. On the other hand, there is no evidence to the contrary, and this assumption gives consistent results. Newton made this same claim more than 100 years earlier.

Everyday Physics | *How Much Do You Weigh?*

In the 1600s people didn't imagine that anybody would one day travel to distant planets. Nevertheless, Newton's law of universal gravitation allowed them to predict what they would weigh if they ever found themselves on another planet.

According to Newton's law of universal gravitation, a person's weight on a planet depends on the mass and radius of the planet, as well as the mass of the person. Your weight on Jupiter compared with that on Earth depends on these factors. Assuming that Jupiter is the same size as Earth (it isn't) but that it has 318 times as much mass as Earth (it does) means that you would weigh 318 times as much on this fictitious Jupiter as on Earth. Actually, Jupiter's diameter is 11.2 times that of Earth. Because the law of universal gravitation contains the radius squared in the denominator, your weight is actually reduced by a factor of 11.2 squared, or 125. Combining these two factors means that you would tip a Jovian bathroom scale at $\frac{318}{125}$, or $2\frac{1}{2}$ times your weight on Earth. You would be lightest on Mercury or Mars, weighing only 38% of your Earth weight. Your weight on each of the planets is given in the following table.

On the Moon the force of gravity is only $\frac{1}{6}$ of that on Earth. Thus, an astronaut's weight on the Moon is only $\frac{1}{6}$ of what it would be on Earth. This means that astronauts on the Moon can jump higher and fall more slowly, as we have seen from the television images transmitted to Earth during the lunar explorations. Vehicles designed for lunar travel would collapse under their own weight on Earth.

1. Explain the concept of "relative weight." Why do we not calculate a "relative mass" for each of the planets?

2. How is the value for the relative weight on Mars calculated?

Weights on Each of the Planets		
Planet	**Weight on the Planet Relative to Earth**	**Weight of a 150-lb Person**
Mercury	0.38	57 lb
Venus	0.91	136
Earth	1.00	150
Mars	0.38	57
Jupiter	2.53	380
Saturn	1.07	160
Uranus	0.92	138
Neptune	1.18	177

NASA

The lunar rover would collapse under its own weight if used on Earth.

FLAWED REASONING

You read in a comic book that the gravity on the Moon is not as strong as on Earth because there is no atmosphere on the Moon. This doesn't seem right, so you do a little research. **What do you find?**

ANSWER The source of the gravitational attraction is mass, not air. The gravitational attraction on the Moon is less than on Earth because the Moon's mass is so much less than Earth's. Actually, the argument in the comic book is completely backward. The Moon does not have an atmosphere because its gravity is too weak to hold one.

Gravity near Earth's Surface

In earlier chapters we assumed that the gravitational force on an object is constant near Earth's surface. We were able to do this because the force changes so little over the distances in question. In fact, to assume otherwise would have unnecessarily complicated matters.

Near Earth's surface the gravitational force decreases by one part in a million for every 3 meters (about 10 feet) of gain in elevation. Therefore, an object that weighs 1 newton at Earth's surface would weigh 0.999 999 newton at an elevation of 3 meters. An individual with a mass of 50 kilograms weighs 500 newtons (110 pounds) in New York City; this person would weigh about 0.25 newton (1 ounce) less in mile-high Denver.

The variations in the gravitational force result in changes in the acceleration due to gravity. The value of the acceleration—normally symbolized as g—is nearly constant near Earth's surface. As long as one stays near the surface, the distance between the object and Earth's center changes very little. If an object is raised 1 kilometer (about $\frac{5}{8}$ mile), the distance changes from 6378 kilometers to 6379 kilometers, and g only changes from 9.800 (meters per second) per second to 9.797 (meters per second) per second.

However, even without a change in elevation, g is not strictly constant from place to place. Earth would need to be composed of spherical shells, with each shell being uniform; this is not the case. Underground salt deposits have less mass per cubic meter and give smaller values of g than average, whereas metal deposits produce larger g values. Therefore, measurements of g can be used to locate large-scale underground ore deposits. By noting variations, geologists can map regions for further exploration.

The passengers in this airplane weigh less because of their altitude.

Q: Given the fact that water has less mass per cubic meter than soil and rock, would you expect the value of g to be smaller or larger than average over a lake?

A: A cubic meter of water would provide less attraction than a cubic meter of soil and rock. Therefore, the value of g would be smaller.

Satellites

Newton's theory also predicts the orbits of artificial satellites orbiting Earth. By knowing how the force changes with distance from Earth, we know what accelerations—and, consequently, other orbital characteristics—to expect at different altitudes. For instance, a satellite at a height of 200 kilometers should orbit Earth in 88.5 minutes. This is close to the orbit of the satellite *Vostok 6*, which carried the first woman to enter outer space, Valentina Tereshkova, into Earth orbit in June 1963. Its orbit varied in height from 170 to 210 kilometers and had a period of a little more than 88 minutes.

The higher a satellite's orbit, the longer it takes to complete one orbit. The Moon takes 27.3 days; *Vostok 6* took 88 minutes. It is possible to calculate the height that a satellite would need to have a period of 1 day. With this orbit, if the satellite were positioned above the equator, it would appear to remain fixed directly above one spot on Earth—an orbit called *geosynchronous*. Such geosynchronous satellites have an altitude of 36,000 kilometers, or about $5\frac{1}{2}$ Earth radii, and are useful in establishing worldwide communications networks. Backyard satellite dishes that pick up television signals point to geosynchronous satellites. The first successful geosynchronous satellite was *Syncom II*, launched in July 1963. Some geosynchronous satellites are used to monitor the weather on Earth.

Any space probe requires the same computations as those done for satellites; the computers at the National Aeronautics and Space Administration (NASA) calculate the trajectories for space flights, using Newton's laws of motion and the law of gravitation. The forces on the spacecraft at any time depend on the positions of the other bodies in the Solar System. These can be calculated with the gravitation equation by inserting the distance to and the

Satellite dishes are aimed at communications satellites in geosynchronous orbits above Earth's equator.

GEOS weather satellites orbit Earth once each day, maintaining fixed locations above the equator.

NASA

mass of each body. The net force produces an acceleration of the spacecraft, which changes its velocity. From this the computer calculates a new position for the spacecraft. It also calculates new positions for the other celestial bodies, and the process starts over. In this manner the computer plots the path of the spacecraft through the Solar System.

Are You On the Bus?

Q: If you could spot a geosynchronous satellite in the sky, how could you distinguish it from a star?

A: The satellite remains in the same location in the sky, whereas the stars drift westward as Earth rotates under them.

Tides

Before Newton's work with gravity, no one was able to explain why we have tides. Some things were known. The tides are caused by bulges in the surface of Earth's oceans. There are two bulges, one on each side of Earth, as shown in Figure 5-4. The occurrence of tides at a given location is due to Earth's rotation. Imagine for simplicity that the bulges are stationary—pointing in some direction in space—and that Earth is rotating. Each point on Earth passes through both bulges in 24 hours, and we have high tides at these times. Low tides occur halfway between the bulges. So we have two low and two high tides each day.

What wasn't known was why Earth had these bulges. Newton claimed that they were due to the Moon's gravity. Earth exerts a gravitational force on the Moon that causes the Moon to orbit it. But the Moon exerts an equal and opposite force on Earth that causes Earth to orbit the Moon. Actually, both Earth and the Moon orbit a common point located between them. This point is the center of mass of the Earth–Moon system. Because Earth is so much more massive than the Moon, the center of mass is much closer to Earth. In fact, its location is inside Earth, as shown in Figure 5-5. Earth's orbital motion would look more like a wobble to somebody viewing its motion from high above the North Pole. But it is an orbit.

Because Earth has an orbital motion, we can use the conclusions developed for the Moon's motion to help us understand Earth's tides. Namely, because we concluded that the Moon is continually falling toward Earth, Earth then is continually falling toward the Moon. This centripetal acceleration toward the Moon is the key to understanding tidal bulges.

Forget momentarily that Earth is moving along its orbit and just consider Earth falling toward the Moon, as in Figure 5-6. This acceleration is the major

Figure 5-4 Exaggerated ocean bulges. As Earth rotates, the bulges appear to move around Earth's surface.

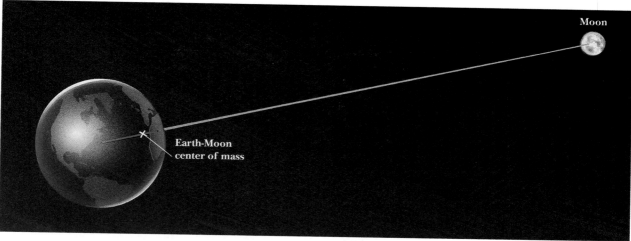

Figure 5-5 The center of mass of the Earth–Moon system is located inside Earth.

Q: Are the forces between Earth and the Moon a Newtonian third-law pair?

A: Yes, one of the forces is the force of *Earth on the Moon,* and the other is the force of the Moon on Earth.

contributor to the tides. Because the strength of the Moon's gravity gets weaker with increasing distance, the force on different parts of Earth is different. For example, on the side nearest the Moon, 1 kilogram of ocean water feels a stronger force than an equal mass of rock at Earth's center. Similarly, 1 kilogram of material on the far side of Earth feels a smaller force than both the kilogram on the near side and the one at the center.

If there are different-sized forces at different spots on Earth, there are different accelerations for different parts. Parts of Earth outrace other parts in their fall toward the Moon. Material on the side of Earth facing the Moon tries to get ahead, while the material on the other side lags behind. Of course, Earth has internal forces keeping it together that eventually balance these inequalities. But we do end up with a stretched-out Earth.

Although this reasoning accounts for the occurrence of the two high tides each day, it is too simple to get the details right. We observe that high tides do not occur at the same time each day. This happens because the Moon orbits the rotating Earth once a month. The normal time interval between successive high tides is 12 hours and 25 minutes. High tides do not occur when the Moon is overhead but later—as much as 6 hours later. This is due to such factors as the frictional and inertial effects of the water and the variable depth of the ocean.

Although the height difference between low and high tides in the middle of the ocean is only about 1 meter, the shape of the shoreline can greatly amplify the tides. The greatest tides occur in the Bay of Fundy, on the eastern seaboard between Canada and the United States; there the maximum range from low to high tide is 16 meters (54 feet)!

We would also expect to observe solar tides because the Sun also exerts a gravitational pull on Earth and Earth is "falling" toward the Sun. These do occur, but their heights are a little less than one-half those due to the Moon. This value may seem too small, taking into consideration that the Sun's gravitational force on Earth is about 180 times as large as the Moon's. The solar effect is so small because it is the difference in the force from one side of Earth to the other that matters and not the absolute size. The tides due to the planets are even smaller, that of Jupiter being less than one ten-millionth that due to the Sun.

Figure 5-6 Equal masses of Earth experience different gravitational forces due to their different distances from the Moon. The effect is exaggerated in the diagram.

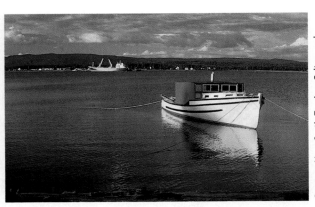

Courtesy Nova Scotia Tourism, Culture and Heritage (both)

Photographs of the same view in the Minas Basin, Nova Scotia, at low and high tides

The continents are much more rigid than the oceans. Even so, the land experiences measurable tidal effects. Land areas may rise and fall as much as 23 centimeters (9 inches). Because the entire area moves up and down together, we don't notice this effect.

Are You On the Bus?

Q: Is the height of the high tide related to the phase of the Moon? That is, is it higher when the Sun and Moon are on the same side of Earth (new moon), when they are on opposite sides (full moon), or when they are at right angles to each other (first- or third-quarter moon)?

A: The highest high tides and the lowest low tides occur near new and full moons, when Earth, the Moon, and the Sun are in a line.

How Far Does Gravity Reach?

The law of gravitation has been thoroughly tested within the solar system. It accounts for the planets' motions, including their irregularities due to the mutual attraction of all the other planets.

What about tests outside the solar system? We haven't sent probes out there. We are fortunate, however, because nature has provided us with ready-made probes. Astronomers observe that many stars in our galaxy revolve around a companion star. These binary star systems are the rule rather than the exception. These pairs revolve around each other in exactly the way predicted by Newton's laws.

Occasionally, a star is spotted that appears to be alone yet is moving in an elliptical path. Our faith in Newton's laws is so great that we assume a companion star is there; it is just not visible. Some of these invisible stars have later been detected because of signals they emit other than visible light.

Photographs of star clusters show that the gravitational interaction occurs between stars. In fact, measurements show that all the stars in the Milky Way Galaxy are rotating about a common point under the influence of gravity. This has been used to estimate the total mass of the galaxy and the number of stars in it. The Milky Way Galaxy is very similar in size and shape to our neighboring galaxy, the Andromeda Galaxy.

Such successes are a remarkable witness to Newton's genius. For more than two centuries, scientists applied his laws of motion and the law of gravitation without discovering any discrepancies. However, some exceptions to the Newtonian world view were eventually discovered. It should not take away from his fame to admit these exceptions. They occur only when we venture very far from the realm of our ordinary senses. In the world of very high velocities and extremely large masses, we must replace Newton's ideas with the theories

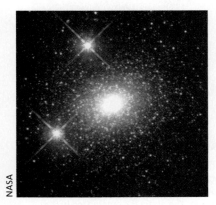

NASA

Star clusters provide evidence of gravity at work between stars.

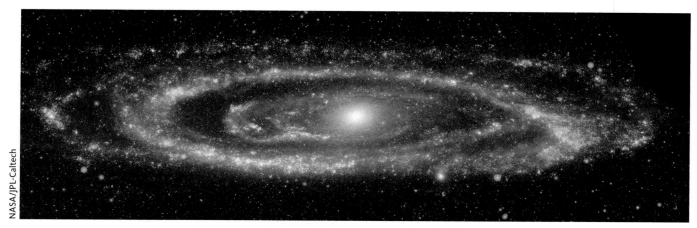

The Andromeda Galaxy is similar to our own Milky Way Galaxy.

of special and general relativity (Chapter 10). In the world of the extremely small, we must use the theories of quantum mechanics (Chapter 24). It should be noted, however, when these newer theories are applied in the realm where Newton's laws work, the new theories give the same results.

We also do not know whether the value of G varies with time. No such variation has been detected, but a small variation with time could exist. Because the measurements of G are still limited in accuracy, it has been suggested that NASA orbit two satellites about each other. Accurate knowledge of the satellites' masses as well as their orbital data would give a more accurate value for G.

The Field Concept

Implicitly, we have assumed the force between two masses to be the result of some kind of direct interaction—sort of an action-at-a-distance interaction. This type of interaction is a little unsettling because there is no direct pushing or pulling mechanism in the intervening space. Gravitational effects are evident even in situations in which there is a vacuum between the masses.

It is both conceptually and computationally useful to separate the gravitational interaction into two distinct steps using the **field** concept. First, one of the objects modifies, by virtue of its mass, the surrounding space; it produces a **gravitational field** at every point in space. Second, the other object interacts, by virtue of its mass, with this gravitational field to experience the force. The field concept divides the task of determining the force on a mass into two distinct parts: determining the field from the first mass and then calculating the force that this field exerts on the second mass.

◀ gravitational field

If this were the only purpose of the field idea, it would play a minor role in our physics world view. In fact, it probably seems as if we are trading one unsettling idea for another. However, as we continue our studies, we will find that the field takes on an identity of its own and is a valuable aid in understanding these and many other phenomena.

By convention the value of the gravitational field at any point in space is equal to the force experienced by a 1-kilogram mass if it were placed at that point. Then the gravitational force on any other object is the product of its mass and the gravitational field at that point. If you hold a 1-kilogram block near Earth's surface, it feels a gravitational force of 10 newtons. Therefore, the gravitational field has a magnitude of 10 newtons per kilogram at this point. If you replace this

"THAT WRAPS IT UP — THE MASS OF THE UNIVERSE."

© Sidney Harris. Used by permission. ScienceCartoonsPlus.com.

block with a 5-kilogram block, the gravitational force changes to 50 newtons; this is just the product of (5 kilograms) and (10 newtons per kilogram). We can also see that we do not need to use a 1-kilogram block to find the gravitational field. We could just as easily use the 5-kilogram block and divide the resulting force by the mass of the block: (50 newtons)/(5 kilograms) = 10 newtons per kilogram.

Because force is a vector quantity, the gravitational field is a vector field; it has a magnitude and a direction at each point in space. It is often convenient to talk about the gravitational field rather than the gravitational force. The strength of the gravitational force depends on the object being considered, whereas the strength of the gravitational field is independent of the object.

We saw in Chapter 3 that the gravitational force **W** could be expressed through Newton's second law as

$$\mathbf{W} = m\mathbf{g}$$

where **g** is the acceleration due to gravity, 9.8 (meters per second) per second. Because the gravitational force is equal to the mass times either the gravitational *field* or the gravitational *acceleration,* these two must be numerically the same. Indeed, we use the symbol **g** for both the gravitational field and the acceleration due to gravity. We can also show that newtons per kilogram may be rewritten as (meters per second) per second.

Summary

Although no one knows what gravity is or why it exists, we can accurately describe how gravity affects the motions of objects. The same laws of motion work on Earth and in the heavens. Newton's universal law of gravitation states that a gravitational attraction exists between every pair of objects and is given by

$$F = G\frac{m_1 m_2}{r^2}$$

where m_1 and m_2 are the masses of the two objects, r is the distance between their centers, and G is the gravitational constant. The value of G was first determined by Cavendish and is believed to be constant with time and space.

The higher a satellite's orbit, the longer it takes to complete one orbit. A satellite with a period of 1 day and positioned above the equator would appear to remain fixed in the sky. The Moon, a natural satellite, takes 27.3 days to complete one orbit around Earth.

The force of gravity can be considered constant when the motion occurs over short distances near Earth's surface. However, small variations occur in the acceleration due to gravity with latitude, elevation, and the types of surface material. At larger distances the force decreases as the square of the distance. Stars in binary systems revolving around each other and the motion of stars within galaxies support this idea.

The value of the gravitational field at any point in space is equal to the force experienced by a 1-kilogram mass placed at that point.

CHAPTER 5 *Revisited*

Studying the motions of distant celestial objects, as well as the distributions of mass within multiparticle systems such as globular clusters and galaxies, provides direct evidence that gravity extends throughout the entire visible universe.

Key Terms

field   A region of space that has a number or a vector assigned to every point.

gravitational field   The gravitational force experienced by a 1-kilogram mass placed at a point in space.

inverse-square   Describes a relationship in which a quantity is related to the reciprocal of the square of a second quantity. An example is the law of universal gravitation; the force is inversely proportional to the square of the distance. If the distance is doubled, the force decreases by a factor of 4.

law of universal gravitation   All masses exert forces on all other masses. The force F between any two objects is given by $F = Gm_1 m_2/r^2$, where G is a universal constant, m_1 and m_2 are the masses of the two objects, and r is the distance between their centers.

Questions and exercises are paired so that most odd-numbered are followed by a similar even-numbered.

Blue-numbered questions and exercises are answered in Appendix B.

🌶 indicates more-challenging questions and exercises.

Web**Assign** Many Conceptual Questions and Exercises for this chapter may be assigned online at WebAssign.

Conceptual Questions

1. What force (if any) drives the planets along their orbits?

2. What force (if any) causes the planets to execute (nearly) circular orbits?

3. Is the size of the gravitational force that Earth exerts on the Moon smaller than, larger than, or the same size as the force the Moon exerts on Earth? Why?

NASA

4. Earth exerts a gravitational force of more than 1 million newtons on the International Space Station. What force does the ISS exert on Earth?

5. How does the average acceleration of the Moon about the Sun compare with that of Earth about the Sun?

6. If an apple were placed in orbit at the same distance from Earth as the Moon, what acceleration would the apple have?

7. What happens to the surface area of a cube when the length of each side is doubled? How does this compare with what happens to the surface area of a sphere when you double its radius?

8. What happens to the volume of a cube if the length of each side is doubled? How does this compare with what happens to the volume of a sphere when you double its radius?

9. A future space traveler, Skip Parsec, lands on the planet MSU3, which has the same mass as Earth but twice the radius. If Skip weighs 800 newtons on Earth's surface, how much does he weigh on MSU3's surface?

10. Astronaut Skip visits planet MSU8, which is composed of the same materials as Earth but has twice the radius. If Skip weighs 800 newtons on Earth's surface, how much does he weigh on MSU8's surface?

11. Why didn't Newton have to know the mass of the Moon to obtain the law of universal gravitation?

12. Comment on the following statement made by a TV newscaster during an *Apollo* flight to the Moon: "The spacecraft has now left the gravitational force of Earth."

13. In a parallel universe, there is a planet with the same mass and radius as Earth. However, when an apple is dropped on this planet, it falls with an acceleration of 20 (meters per second) per second. What is the value of the gravitational constant *G* in this parallel universe?

14. For simplicity we use 10 (meters per second) per second for the acceleration due to gravity instead of the more accurate 9.8 (meters per second) per second. If Cavendish had made the same approximation, would his estimate for Earth's mass have been too high or too low?

15. If a satellite in a circular orbit above Earth is continually "falling," why doesn't it quickly return to Earth?

16. As a satellite orbits Earth, the gravitational force is constantly pulling the satellite inward. What counters this force?

17. Astronaut Story Musgrave spent a total of 1281 hours, 59 minutes, and 22 seconds in space on his six space shuttle missions. If Story's mass was 80 kilograms, then the gravitational force acting on him in orbit was approximately 730 newtons. Why did he feel weightless?

18. You are standing on a bathroom scale in an elevator when suddenly the cable breaks and the elevator falls freely down the shaft. How does the reading on the scale change from just before to just after the cable breaks? How does the force of gravity that Earth exerts on you change over the same time interval?

19. NASA uses the famous "Vomit Comet," a KC-135 cargo plane, to provide astronauts and scientists a simulated zero-gravity environment. The plane flies a series of parabolic arcs, as shown in the figure. Explain why the passengers feel "weightless" when the plane is near the top of its arc.

20. You have no doubt seen pictures of the astronauts floating around inside a space shuttle as it orbits some 300 kilometers above Earth's surface. Would the force of gravity Earth exerts on an astronaut be the same as, a little less than, or much less than the force that the astronaut would experience on Earth's surface? Why?

21. If Earth were hollow but still had the same mass and radius, would your weight be different? Why?

22. Astronomers believe that when Earth first formed, its composition was uniform. Over time, the heavier materials sank to the middle to create a dense iron core, with the less-dense materials toward the outside. How did the value of the acceleration due to gravity at Earth's surface change while this process was occurring?

23. The gravitational force between two books sitting on a table does not cause them to accelerate toward each other because of frictional forces. If these same two books were floating near each other in deep space, they would still not appear to accelerate toward each other. Why not?

24. Why do we use the form $\mathbf{W} = m\mathbf{g}$ for the gravitational force on an object near Earth, but the form $F = Gm_1m_2/r^2$ when the object is far from Earth?

25. Skylab caused quite a commotion when it returned to Earth in July 1979. Why would it suddenly return to Earth after it had been in orbit for many years?

26. When the Hubble Space Telescope (HST) was originally launched by the space shuttle *Discovery,* its approximately circular orbit was at an altitude of about 600 kilometers. However, over the next several years, the altitude decreased so that subsequent servicing missions were required to lift the HST back into the higher orbit. What is responsible for the orbital decay?

27. How could we determine the mass of a planet such as Venus, which has no moon?

28. Why can we not determine the mass of the Moon by noting that it orbits Earth in a nearly circular orbit? What can we do to determine the Moon's mass?

29. Would you expect the value of the acceleration due to gravity to be larger or smaller than normal over a large deposit of uranium ore? Why?

30. You are steaming across the Atlantic Ocean on a large cruise ship. What happens to your weight as the ship leaves the deep waters of the North Atlantic and enters the shallow coastal waters of the United States?

31. What changes would occur in the solar system if the gravitational constant *G* were slowly getting larger?

32. What do you think would happen to the Moon's orbit if the gravitational attraction between the Moon and Earth were slowly growing stronger?

33. Is it possible for an Earth satellite to remain "stationary" over Paris? Why or why not?

34. During the Gulf War with Iraq in 1991, a newspaper story reported that American spy satellites were in stationary orbits over Iraq, providing continuous intelligence information. Explain why this is impossible.

35. Assume that NASA fails in its attempt to put a communications satellite into geosynchronous orbit. If the orbit is too big, what apparent motion will the satellite have as seen from the rotating Earth?

36. Some National Oceanic and Atmospheric Administration (NOAA) Earth satellites remain above a single location on Earth. Why don't these geosynchronous satellites fall to Earth under the influence of gravity?

37. Newton's third law says that the gravitational force exerted on Earth by the Moon is equal to that exerted on the Moon by Earth. Why, then, doesn't Earth appear to orbit the Moon?

38. The Sun has a profound influence on Earth's motion. Does Earth influence the Sun's motion? Explain.

39. When the tide is high along the American western seaboard, is the tide in Japan nearer high tide or low tide? Japan is approximately 90 degrees west of San Francisco.

40. When it is high tide off the coast of Ecuador, is the tide off the coast of Indonesia, which is 180 degrees around the globe from Ecuador, nearer high tide or low tide?

41. Jupiter rotates once in 9 hours 50 minutes. How much time passes between high tides in its atmosphere?

42. The Moon is observed to keep the same side facing Earth at all times. If the Moon had oceans, how much time would elapse between its high tides?

43. Which position in the following figure corresponds to the new moon? This is when the Moon is above the horizon but cannot be seen because the lit side faces away from Earth. Why are high tides higher than normal during this phase?

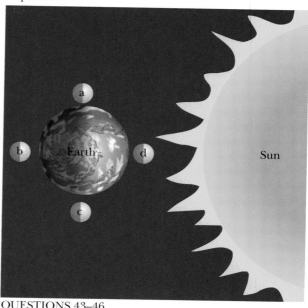

QUESTIONS 43–46.

44. Which position in the preceding figure corresponds to the full moon, which is when the Moon appears as a fully lit disk? Why are high tides higher than normal during this phase?

45. A classmate asserts that when the Moon is in position (b) in the preceding figure, the gravitational effects of the Sun and Moon tend to cancel, producing lower-than-normal high tides. What is wrong with this reasoning?

46. Which positions of the Moon in the preceding figure correspond to the smallest difference between high tide and low tide?

47. In *The Jupiter Effect,* authors John Gribbin and Stephen Plagemann claim that the additional tidal force produced when all the planets lie along one line might be enough to trigger an earthquake along the San Andreas Fault in California. What do you think about the possibility?

48. Why would the inertia and friction of water cause the tides to occur after the Moon passes overhead?

49. How does the magnitude of Earth's gravitational field change with increasing distance?

50. Show that units of newtons per kilogram are equivalent to units of (meters per second) per second.

51. Is it possible for the gravitational field to point in two different directions at the same location in space? Why or why not?

52. Which of the following are independent of the mass of an object falling freely near Earth's surface: the acceleration of the object, the gravitational force acting on the object, the gravitational force acting on Earth, and the magnitude of the gravitational field?

Exercises

53. Earth's speed in its orbit about the Sun is about 30 km/s. What is Earth's acceleration?

54. The Moon's speed in its orbit is approximately 1 km/s, and the Earth–Moon distance is 380,000 km. Show that these numbers yield an acceleration for the Moon that is very close to that given in the text.

55. What is the acceleration due to gravity at a distance of 2 Earth radii above Earth's surface?

56. If you were located halfway between Earth and the Moon, what acceleration would you have toward Earth? The Earth–Moon separation is 60 Earth radii. (Ignore the gravitational force of the Moon because it is much less than Earth's.)

57. A solid lead sphere of radius 10 m (about 66 ft across!) has a mass of about 57 million kg. If two of these spheres are floating in deep space with their centers 20 m apart, the gravitational attraction between the spheres is only 540 N (about 120 lb). How large would this gravitational force be if the distance between the centers of the two spheres were tripled?

58. Two spacecraft in outer space attract each other with a force of 20 N. What would the attractive force be if they were one-half as far apart?

59. How would the Sun's gravitational force on Earth change if Earth had one-half its present mass? Would Earth's acceleration change?

60. The gravitational force between two very large metal spheres in outer space is 50 N. How large would this force be if the mass of each sphere were cut in half?

61. By what factor is the ratio of the gravitational force on you when you are 6400 km above Earth's surface versus when you are standing on the surface? (Earth's radius is 6400 km.)

62. How does Earth's gravitational force on you differ when you are standing on Earth and when you are riding in a space shuttle 400 km above Earth's surface? (Earth's radius is 6400 km.)

63. A 320-kg satellite experiences a gravitational force of 800 N. What is the radius of the satellite's orbit? What is its altitude?

64. A 600-kg geosynchronous satellite has an orbital radius of 6.6 Earth radii. What gravitational force does Earth exert on the satellite?

65. What is the gravitational force between two 20-kg iron balls separated by a distance of 0.5 m? How does this compare with the weight of either ball?

66. The masses of the Moon and Earth are 7.4×10^{22} kg and 6×10^{24} kg, respectively. The Earth–Moon distance is 3.8×10^8 m. What is the size of the gravitational force between Earth and the Moon? Does the acceleration of the Moon produced by this force agree with the value given in the text?

67. If an astronaut in full gear has a weight of 1200 N on Earth, how much will the astronaut weigh on the Moon?

68. The acceleration due to gravity on Titan, Saturn's largest moon, is about 1.4 m/s². What would a 50-kg scientific instrument weigh on Titan?

69. Mercury has a radius of about 0.38 Earth radius and a mass of only 0.055 Earth mass. Estimate the acceleration due to gravity on Mercury.

70. Mars has a radius of about 0.53 Earth radius and a mass of only 0.11 Earth mass. Estimate the acceleration due to gravity on Mars.

71. A geosynchronous satellite orbits at a distance from Earth's center of about 6.6 Earth radii and takes 24 h to go around once. What distance (in meters) does the satellite travel in one day? What is its orbital velocity (in m/s)?

72. An 80-kg satellite orbits a distant planet with a radius of 4000 km and a period of 280 min. From the radius and period, you calculate the satellite's acceleration to be 0.56 m/s². What is the gravitational force on the satellite?

73. The radius of Venus's orbit is 0.72 times that of Earth's orbit. How much stronger is the Sun's gravitational field at Venus than at Earth?

NASA

74. By what factor is Earth's gravitational field reduced at a distance of 5 Earth radii from the center of Earth?

75. If Earth shrank until its diameter were only one-half its present size without changing its mass, what would a 1-kg mass weigh at its surface?

76. If Earth expanded to twice its diameter without changing its mass, find the resulting magnitude of the gravitational field.

The Big Picture

The Discovery of Invariants

One theme of the physics world view has been that change is an essential part of the universe. In fact, change has been so pervasive in our journey that you may be surprised to learn that some things don't change. A number of quantities are in fact constant, or invariant. For something to be *invariant*, the numerical value associated with it must be *constant*, meaning that we obtain the same value at all times. We say that the quantity is *conserved*. In our everyday language, we often use the word *conserve* to mean that something is saved, or at least used sparingly. In physics when something is conserved, it remains constant; its value does not change.

According to Swiss child psychologist Jean Piaget, an essential part of an individual's development is the establishment of invariants, or conservation rules. That is, as we process sensory data, we build constructs that are relatively permanent components of our personal world views. Most of this building occurs instinctively at an early age. We certainly don't go around consciously searching for invariants.

A striking example of such a learned invariant that occurs at an early age is the permanence of the physical size of objects. Although the size of the image on your retina gets smaller as an object moves away from you, common sense tells you that the object remains the same size. How do you know this? Piaget would claim that because you have seen objects return to their original sizes many times, you therefore have formed this invariant.

The concept that objects have a permanent size is so much a part of our common sense that our brain over-

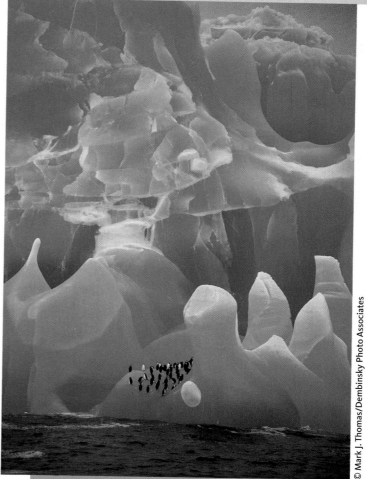

Chinstrap penguins on a rare blue iceberg.

© Mark J. Thomas/Dembinsky Photo Associates

rides the sensory information it receives. When you perceive objects as being far away through depth perception clues, you automatically compensate for the smaller retinal images. Artists can create nonsensical situations by providing our brain with conflicting information, as in the optical illusion shown in the photographs.

Which figure is taller?

mass does not change when chemical reactions take place in closed flasks. These investigations led to the generalization that the mass of any closed system is invariant. Lavoisier's generalization is known as a law of nature: the law of *conservation of mass*. (When we study relativity and nuclear physics, we will discover that this law has to be modified. However, for ordinary physical and chemical processes, this law is obeyed to a high degree of accuracy and is useful in analyzing many processes.)

Other invariants have been discovered. Imagine a moving billiard ball hitting a stationary billiard ball head on. After the collision, the first ball stops and the second has a velocity that is equal to the original velocity of the first ball. It appears that something is transferred from the first ball to the second. Christian Huygens, a contemporary of Galileo and Newton, suggested that a "quantity of motion" is invariant in this collision. This quantity of motion is transferred from one ball to the other.

Discovering invariants is difficult. A lot of effort has been spent and will continue to be spent searching for them because the rewards are worth it. Their discovery yields powerful generalizations in the physics world view.

In principle, we can use Newton's second law to predict the motion of any object, whether it be a leaf, an airplane, a billiard ball, or even a planet. All we have to do is determine the net force acting on the object and calculate the resulting acceleration. Solving these equations, however, often turns out to be quite difficult. To use the second law, we need to know all the forces acting on the object at all times. This is complicated for most motions because the forces are often not constant in direction or size.

The use of invariants often allows us to bypass the details of the individual interactions. In some cases, such as those involving nuclear forces, it is not a case of trying to avoid these troubling complications, but rather of taking the only possible path; we don't know the forces well enough to apply Newton's laws.

In the next three chapters, we will look at three important invariants—linear momentum, energy, and angular momentum. These three "quantities of motion" allow us to look at the behaviors of objects and systems of objects from an entirely new point of view.

One of the first invariants discovered was the quantity of matter. When a piece of wood burns, it loses most of its mass. This loss is quite obvious if it is burned on an equal-arm balance. As the wood burns, the balance rises continually, indicating a decrease in mass. Where does the lost mass go? Does it disappear, or does it simply escape into the air?

Burning wood in a closed container gives different results. Provided there is enough air to allow the wood to burn and the products of the burning—the ashes and smoke—don't escape, the equal-arm balance remains level. Thus, the mass of the closed system does not change.

Near the end of the 18th century, the chemist Antoine Lavoisier conducted many experiments showing that

As the wood burns, the left side rises.

Momentum

▶Collisions send objects scattering in seemingly random directions. Yet nature has a scheme—a rule—that governs the path of each individual particle in collisions and explosions, whether it is a group of billiard balls or colliding galaxies. If the rule predicts the future, can it also tell us about the past?

(See page 110 for the answer to this question.)

© Cengage Learning/David Rogers

The collisions of billiard balls illustrate the law of conservation of linear momentum.

The kayak and the ship have different momenta even when they are moving at the same speed, because of their different masses.

W E know that it is harder to get a more massive object moving from rest than a less massive object. This is the concept of inertia that we have already included in our world view. We also know intuitively that for two objects moving with the same speed, the more massive object would be the harder one to stop. For example, if a kayak and a cargo ship were both moving through the harbor at 5 kilometers per hour, you would clearly have an easier time stopping the kayak.

We need to add a new concept to our world view to address the question, "How hard would it be to stop an object?" We call this new concept *momentum,* and it depends on both the mass of the object and how fast it is moving.

Linear Momentum

The **linear momentum** of an object is defined as the product of its mass and its velocity. Momentum is a vector quantity that has the same direction as the velocity. Using the symbol **p** for momentum, we write the relationship as

$$\mathbf{p} = m\mathbf{v}$$

◀ linear momentum = mass × velocity

The adjective *linear* distinguishes this from another kind of momentum that we will discuss in Chapter 8. Unless there is a possibility of confusion, this adjective is usually omitted.

There is no special unit for momentum as there is for force; the momentum unit is simply that of mass times velocity (or speed)—that is, kilogrammeter per second (kg·m/s).

An object may have a large momentum due to a large mass, a large velocity, or both. A slow battleship and a rocket-propelled race car have large momenta.

Are You On the Bus?

Q: Which has the greater momentum, an 18-wheeler parked at the curb or a Volkswagen rolling down a hill?

A: Because the 18-wheeler has zero velocity, its momentum is also zero. Therefore, the VW has the larger momentum as long as it is moving.

The word *momentum* is often used in our everyday language in a much looser sense, but it is still roughly consistent with its meaning in the physics world view; that is, something with a lot of momentum is hard to stop. You have probably heard someone say, "We don't want to lose our momentum!" Coaches are particularly fond of this word.

▶ Extended presentation available in the *Problem Solving* supplement

Changing an Object's Momentum

The momentum of an object changes if its velocity and/or its mass changes. We can obtain an expression for the amount of change by rewriting Newton's second law ($\mathbf{F}_{net} = m\mathbf{a}$) in a more general form. Actually, Newton's original formulation is closer to the new form. Newton realized that mass as well as velocity could change. His form of the second law says that the net force is equal to the change in the momentum divided by the time required to make this change:

◀ net force = $\dfrac{\text{change in momentum}}{\text{time taken}}$

$$\mathbf{F}_{net} = \frac{\Delta(m\mathbf{v})}{\Delta t}$$

If we now multiply both sides of this equation by the time interval Δt, we get an equation that tells us how to produce a change in momentum:

$$\mathbf{F}_{net}\Delta t = \Delta(m\mathbf{v})$$

◀ impulse = net force × time = change in momentum

This relationship tells us that this change is produced by applying a net force to the object for a certain time interval. The interaction that changes an object's momentum—a force acting for a time interval—is called **impulse**. Impulse is a vector quantity that has the same direction as the net force.

Because impulse is a product of two things, there are many ways to produce a particular change in momentum. For example, two ways of changing an object's momentum by 10 kilogram-meters per second are to exert a net force of 5 newtons on the object for 2 seconds or to exert 100 newtons for 0.1 second. They each produce an impulse of 10 newton-seconds (N·s) and therefore a momentum change of 10 kilogram-meters per second. The units of impulse (newton-seconds) are equivalent to those of momentum (kilogram-meters per second).

Q: Which of the following will cause the larger change in the momentum of an object—a force of 2 newtons acting for 10 seconds or a force of 3 newtons acting for 6 seconds?

A: The larger impulse causes the larger change in the momentum. The first force yields an impulse of (2 newtons)(10 seconds) = 20 newton-seconds; the second yields (3 newtons)(6 seconds) = 18 newton-seconds. Therefore, the first impulse produces the larger change in momentum.

Are You On the Bus?

Although the momentum change may be the same, certain effects depend on the particular combination of force and time. Suppose you had to jump from a second-story window. Would you prefer to jump onto a wooden or a concrete surface? Intuitively, you would choose the wooden one. Our commonsense world view tells us that jumping onto a surface that "gives" is better. But why is this so?

You undergo the same change in momentum with either surface; your momentum changes from a high value just before you hit to zero afterward. The difference is in the time needed for the collision to occur. When a surface gives, the collision time is longer. Therefore, the average net force must be correspondingly smaller to produce the same impulse.

Because our bones break when forces are large, the particular combination of force and time interval is important. For a given momentum change, a short collision time could cause large enough forces to break bones. You may break a leg landing on the concrete. On the other hand, the collision time with wood may be large enough to keep the forces in a huge momentum change from doing any damage.

This idea has many applications. Dashboards in cars are covered with foam rubber to increase the collision time during an accident. New cars are built with shock-absorbing bumpers to minimize damage to cars and with air bags to minimize injuries to passengers. The barrels of water or sand in front of highway median strips serve the same purpose. Stunt people are able to leap from amazing heights by falling onto large air bags that increase their collision times on landing. Volleyball players wear knee pads. Small pieces of polystyrene foam are used as packing material in shipping boxes to smooth out the bumpy rides.

Even without a soft surface, we have learned how to increase the collision time when jumping. Instead of landing stiff-kneed, we bend our knees immediately on colliding with the ground. We are then brought to rest gradually rather than abruptly.

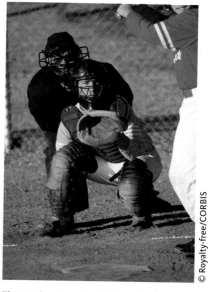
© Royalty-free/CORBIS

The catcher is protected from the baseball's momentum.

© Romilly Lockyer/The Image Bank/ Getty Images

Modern cars employ air bags to protect passengers by increasing their stopping time.

Conservation of Linear Momentum

A pole-vaulter lands on thick pads to increase the collision time and thus reduce the force.

Imagine standing on a giant skateboard that is at rest [Figure 6-1(a)]. What is the total momentum of you and the skateboard? It must be zero because everything is at rest. Now suppose that you walk on the skateboard. What happens to the skateboard? When you walk in one direction, the skateboard moves in the other direction, as shown in Figure 6-1(b). An analogous thing happens when you fire a rifle: the bullet goes in one direction, and the rifle recoils in the opposite direction.

These situations can be understood even though we don't know the values of the forces—and thus the impulses—involved. We start by assuming that there is no net external force to the objects. In particular, we assume that the frictional forces are negligible and that any other external force—such as gravity—is balanced by other forces.

When you walk on the skateboard, there is an interaction. The force you exert on the skateboard is, by Newton's third law, equal and opposite to the force the skateboard exerts on you. The time intervals during which these forces act on you and the skateboard must be the same because there is no way that one can touch the other without also being touched. Because you and the skateboard each experience the same force for the same time interval, you must each experience the same-size impulse and therefore the same-size change in momentum.

But impulse and momentum are vectors, so their directions are important. Because the impulses are in opposite directions, the changes in the momenta are also in opposite directions. Thus, your momentum and that of the skateboard still add to zero. In other words, even though you and the skateboard are

Q: Suppose the skateboard has half your mass and you walk at a velocity of 1 meter per second to the left. Describe the motion of the skateboard.

A: The skateboard must have the same momentum but in the opposite direction. Because it has half the mass, its speed must be twice as much. Therefore, its velocity must be 2 meters per second to the right.

Everyday Physics | *Landing the Hard Way: No Parachute*

An extreme example of minimizing the effects of momentum change occurred during World War II. A Royal Air Force rear gunner jumped (without a parachute!) from a flaming Lancaster bomber flying at 5500 meters (18,000 feet). He attained a terminal speed (no pun intended) of more than 54 meters per second (120 miles per hour) but survived because his momentum change occurred in a series of small impulses with some branches of a pine tree and a final impulse from 46 centimeters (18 inches) of snow. Because this took a longer time than hitting the ground directly, the forces were reduced. Miraculously, he suffered only scratches and bruises.

The record for surviving a fall without a parachute is held by Vesna Vulovic. She was serving as a flight attendant on a Yugosla-

vian DC-9 that blew up at 10,160 meters (33,330 feet) in 1972. She suffered many broken bones and was hospitalized for 18 months after being in a coma for 27 days.

Source: Guinness Book of Records (New York: Bantam Books, 1999).

1. If the airman had a mass of 80 kg, find the magnitude of the air drag acting on him when he reached terminal velocity of 54 m/s.

2. Would the total impulse imparted by the snow have been larger, the same, or smaller if he had not first hit the branches on the way down? Explain.

moving and, individually, have nonzero momenta, the total momentum of the system consisting of you and the skateboard remains zero. Notice that we arrived at this conclusion without considering the details of the forces involved. It is true for all forces between you and the skateboard.

Because the changes in the momenta of the two objects are equal in size and opposite in direction, the value of the total momentum does not change. We say that the total momentum of the system is **conserved**.

We can generalize these findings. Whenever any object is acted on by a force, there must be at least one other object involved. This other object may be in actual contact with the first, or it may be interacting at a distance of 150 million kilometers, but it is there. If we widen our consideration to include all of the interacting objects, we gain a new insight.

Consider the objects as a system. Whenever there is no net force acting on the system from the outside (that is, the system is isolated, or closed), the forces that are involved act only between the objects within the system. As a consequence of Newton's third law, the total momentum of the system remains constant. This generalization is known as the law of **conservation of linear momentum**.

◄ conservation of linear momentum

The total linear momentum of a system does not change if there is no net external force.

This means that if you add up all of the momenta now and leave for a while, when you return and add the momenta again, you will get the same vector even if the objects were bumping and crashing into each other while you were gone. In practice we apply the conservation of momentum to systems in which the net external force is zero or the effects of the forces can be neglected.

You experience conservation of momentum firsthand when you try to step from a small boat onto a dock. As you step toward the dock, the boat moves away from the dock, and you may fall into the water. Although the same effect occurs when we disembark from an ocean liner, the large mass of the ocean liner reduces the speed given it by our stepping off. A large mass requires a small change in velocity to undergo the same change in momentum.

(a)

(b)

Figure 6-1 (a) Person and skateboard at rest have zero momentum. (b) When the person walks to the right, the board moves to the left, keeping the total momentum zero.

WORKING IT OUT *Momentum*

Let's calculate the recoil of a rifle. A 150-grain bullet for a .30-06 rifle has a mass m of 0.01 kg and a muzzle velocity v of 900 m/s (2000 mph). Therefore, the magnitude of the momentum p of the bullet is

$$p = mv = (0.01 \text{ kg})(900 \text{ m/s}) = 9 \text{ kg} \cdot \text{m/s}$$

Because the total momentum of the bullet and rifle was initially zero, conservation of momentum requires that the rifle recoil with an equal momentum in the opposite direction. If the mass M of the rifle is 4.5 kg, the speed V of its recoil is given by

$$V = \frac{p}{M} = \frac{9 \text{ kg} \cdot \text{m/s}}{4.5 \text{ kg}} = 2 \text{ m/s}$$

If you do not hold the rifle snugly against your shoulder, the rifle will hit your shoulder at this speed (4.5 mph!) and hurt you.

Q: Why does holding the rifle snugly reduce the recoil effects?

Are *You* On the Bus?

A: Holding the rifle snugly increases the recoiling mass (your mass is now added to that of the rifle) and therefore reduces the recoil speed.

An isolated system

The total linear momentum of an isolated system is conserved. A system is isolated if no net external force is acting on it.

Although we don't notice it, the same effect occurs whenever we start to walk. Our momentum changes; the momentum of something else must therefore change in the opposite direction. The something else is Earth. Because of its enormous mass, Earth's speed need only change by an infinitesimal amount to acquire the necessary momentum change.

Collisions

Interacting objects don't need to be initially at rest for conservation of momentum to be valid. Suppose a ball moving to the left with a certain momentum crashes head-on with an identical ball moving to the right with the same-size momentum. Before the collision, the two momenta are equal in size but opposite in direction, and because they are vectors, they add to zero.

After the collision the balls move apart with equal momenta in opposite directions. Because the masses of the balls are the same, the speeds after the collision are also the same. These speeds depend on the type of ball. The speeds may be almost as large as the original speeds in the case of billiard balls, quite a bit smaller in the case of lead balls, or even zero if the balls are made of soft putty and stick together. In all cases the two momenta are the same size and in opposite directions. The total momentum remains zero.

Consider the following example. A boxcar traveling at 10 meters per second approaches a string of four identical boxcars sitting stationary on the track. The moving boxcar collides and links with the stationary cars, and the five boxcars move off together along the track. What is the final speed of the five cars immediately after the collision?

Conservation of momentum tells us that the total momentum must be the same before and after the collision. Before the collision, one car is moving at 10 meters per second. After the collision, five identical cars are moving with a common final speed. Because the amount of mass that is moving has increased by a factor of 5, the speed must decrease by a factor of 5. The cars will have a final speed of 2 meters per second. Notice that we did not have to know the mass of each boxcar, only that they all had the same mass.

We can use the conservation of momentum to measure the speed of fast-moving objects. For example, consider determining the speed of an arrow shot from a bow. We first choose a movable, massive target—a wooden block

---

FLAWED REASONING

A question on the final exam asks, "What do we mean when we claim that total momentum is conserved during a collision?" The following two answers are given.

> Answer 1: Total momentum of the system stays the same before and after the collision.
> Answer 2: Total momentum of the system is zero before and after the collision.
> **Which answer (if either) do you agree with?**

ANSWER Although we have considered several examples in which the total momentum of the system is zero, this is not the most general case. The momentum of a system can have any magnitude and any direction before the collision. If momentum is conserved, the momentum of the system always has the same magnitude and direction after the collision. Therefore, answer 1 is correct. This is a very powerful principle because of the word *always*.

Figure 6-2 Determining the speed of an arrow using conservation of momentum. The momentum of the block and arrow after the collision is equal to the momentum of the arrow before the collision.

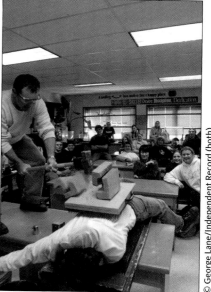

Figure 6-3 Your author demonstrates some physics by letting an assistant smash a concrete block on his chest while lying on a bed of nails. Please do not try this at home.

suspended by strings (Figure 6-2). Before the arrow hits the block, the total momentum of the system is equal to that of the arrow (the block is at rest). After the arrow is embedded in the block, the two move with a smaller, more measurable speed. The final momentum of the block and arrow just after the collision is equal to the initial momentum of the arrow. Knowing the masses, we can determine the arrow's initial speed.

Another example of a small, fast-moving object colliding with a much more massive object is graphically illustrated by one of your brave(?) authors, who lies down on a bed of nails, as shown in Figure 6-3. (This in itself may seem like a remarkable feat. However, your author does not have to be a fakir with mystic powers, because he knows that the weight of his upper body is supported by 500 nails so that each nail has to support only 0.4 pound. It takes approximately 1 pound of force for the nail to break the skin.) Once on the bed of nails, he places a plate of nails on his chest and tops that off with a concrete block. He then invites an assistant to smash the concrete block with a sledgehammer.

This dramatic demonstration illustrates several ideas. The board on the chest spreads out the blow so that the force on any one part of the chest is small. Because it takes time for the hammer to break through the concrete block, the collision time is increased, and the force is therefore decreased even further. The concrete block is serving the same function as an air bag in a car (strangely enough). Finally, momentum is conserved in the collision, but the much larger mass of your author ensures that the velocity imparted to his body is much less than the velocity of the hammer. This means that his body is only slowly pushed down onto the nails, and the additional force that each nail must exert to stop his body is small. Therefore, your author's back is not perforated, and he lives to teach another day.

The initial speeds of these cars can be determined by analyzing the collision.

Investigating Accidents

Accident investigators use conservation of momentum to reconstruct automobile accidents. Newton's laws can't be used to analyze the collision itself because we do not know the detailed forces involved. Conservation of linear momentum, however, tells us that regardless of the details of the crash, the total momentum of the two cars remains the same. The total momentum immediately before the crash must be equal to that immediately after the crash. Because the impact takes place over a very short time, we normally ignore frictional effects with the pavement and treat the collision as if there were no net external forces.

As an example, consider a rear-end collision. Assume that the front car was initially at rest and the two cars locked bumpers on impact. From an analysis of the length of the skid marks made *after* the collision and the type of surface, the total momentum of the two cars just after the collision can be calculated. (We will see how to do this in Chapter 7; for now, assume we know their total momentum.) Because one car was stationary, the total momentum before the crash must have been due to the moving car. Knowing that the momentum is the product of mass and velocity (*mv*), we can compute the speed of the car just before the collision. We can thus determine whether the driver was speeding.

WORKING IT OUT *Rear-Ended*

Let's use conservation of momentum to analyze this collision. For simplicity assume that each car has a mass of 1000 kg (a metric ton) and that the cars traveled along a straight line. Further assume that we have determined that the speed of the two cars locked together was 10 m/s (about 22 mph) just after the crash. The total momentum after the crash was equal to the total mass of the two cars multiplied by their combined speed:

$$p = (m_1 + m_2)\, v = (1000 \text{ kg} + 1000 \text{ kg})(10 \text{ m/s}) = 20,000 \text{ kg} \cdot \text{m/s}$$

But because momentum is conserved, this was also the value before the crash. Before the crash, however, only one car was moving. So if we divide this total momentum by the mass of the moving car, we obtain its speed:

$$v = \frac{p}{m} = \frac{20,000 \text{ kg} \cdot \text{m/s}}{1000 \text{ kg}} = 20 \text{ m/s}$$

The car was therefore traveling at 20 m/s (about 45 mph) at the time of the accident.

Q: If the stationary car were not stationary but slowly rolling in the direction of the total momentum, how would the calculated speed of the other car change if the final momentum remained the same?

Are *You* On the Bus?

A: Because the rolling car accounts for part of the total momentum before the collision, the other car had less initial momentum and therefore a lower speed.

Assuming that the cars stick together after the collision simplifies the analysis but is not required. Conservation of momentum applies to all types of collisions. Even if the two cars do not stick together, the original velocity can be

WORKING IT OUT *A General Collision*

Let's use the principle of conservation of momentum to solve a general collision problem. A 10-kg block is sliding to the right across a frictionless floor at 4 m/s. A 5-kg block is traveling left at 2 m/s such that it hits the other block head-on. After the collision, the 10-kg block is observed moving to the right at 1 m/s. Find the final speed of the 5-kg block.

To find the 5-kg block's final velocity, we use the fact that momentum is conserved during the collision. We have enough information to find the initial momentum of the system:

$$p_{initial} = (10 \text{ kg})(+4 \text{ m/s}) + (5 \text{ kg})(-2 \text{ m/s}) = +30 \text{ kg} \cdot \text{m/s}$$

(Positive means to the right.)

After the collision, we know the momentum of the 10-kg block:

$$p^f_{10 \, kg} = (10 \text{ kg})(+1 \text{ m/s}) = +10 \text{ kg} \cdot \text{m/s}$$

Because the total must still equal $+30 \text{ kg} \cdot \text{m/s}$, the final momentum of the 5-kg block must be $+20 \text{ kg} \cdot \text{m/s}$. This means that the 5-kg block is moving to the right after the collision with a speed of 4 m/s.

determined if the velocity of each car just after the collision can be determined. The cars do not even have to be going in the same initial direction. If the cars suffer a head-on collision, we must be careful to include the directions of the momenta, but the procedure of equating the total momenta before and after the accident remains the same.

Because momentum is a vector, this procedure can also be used in understanding two-dimensional collisions, such as when cars collide while traveling at right angles to each other. The total vector momentum must be conserved.

Airplanes, Balloons, and Rockets

Conservation of momentum also applies to flight. If we look only at the airplane, momentum is certainly not conserved. It has zero momentum before takeoff, and its momentum changes many times during a flight.

But if we consider the system of the airplane plus the atmosphere, momentum is conserved. In the case of a propeller-driven airplane, the interaction occurs when the propeller pushes against the surrounding air molecules,

Studying physics in a billiard parlor improves both your physics and your game.

In flight a Cessna 172's propeller blades push the air backward to go forward.

FLAWED REASONING

Two students are arguing about a collision between two gliders on an air track. Glider A hits glider B, which is twice as large and initially stationary.

Jose: "I think that glider B will have the largest final speed when glider A's final speed is zero. In this case, glider A gives *all* of its momentum to glider B."

Shaq: "You are forgetting that momentum is a vector. If glider A bounces backward during the collision, it experiences a greater change in momentum than if it stops. Glider B must always experience the same change in momentum (but in the opposite direction) as glider A, so it would have a faster final speed in this case."

With which student (if either) do you agree?

ANSWER Shaq was paying attention in class. Anyone who has credit cards knows that it is possible to lose more than everything you have. If glider A is initially moving at 3 meters per second in the positive direction and stops, its change in velocity is –3 meters per second. If, on the other hand, the same glider bounces back with a final velocity of –2 meters per second, the change in its velocity is –5 meters per second. Because the change in momentum is just the mass times the change in velocity, bouncing results in the greater change of momentum.

increasing their momenta in the backward direction. This is accompanied by an equal change in the airplane's momentum in the forward direction. If we could ignore the air resistance, the airplane would continually gain momentum in the forward direction.

Q: Why doesn't the airplane continually gain momentum?

A: As the airplane pushes its way through the air, it hits air molecules, giving them impulses in the forward direction. This produces impulses on the airplane in the backward direction. In straight, level flight at a constant velocity, the two effects cancel.

Release an inflated balloon, and it takes off across the room. Is this similar to the propeller-driven airplane? No, because the molecules in the atmosphere are not necessary. The air molecules in the balloon rush out, acquiring a change in momentum toward the rear. This is accompanied by an equal change in momentum of the balloon in the forward direction. The air molecules do not need to push on anything; the balloon can fly through a vacuum.

This is also true of rockets and explains why they can be used in spaceflight. Rockets acquire changes in momentum in the forward direction by expelling gases at very high velocities in the backward direction. By choosing the direction of the expelled gases, the resulting momentum changes can also be used to change the direction of the rocket. An interesting classroom demonstration of this is often done using a modified fire extinguisher as the source of the high-velocity gas, as shown in Figure 6-4.

Figure 6-4 A fire extinguisher provides the impulse for a hallway rocket. Why does the person move backward?

Jet airplanes lie somewhere between propeller-driven airplanes and rockets. Jet engines take in air from the atmosphere, heat it to high temperatures, and then expel it at high speed out the back of the engine. The fast-moving gases impart a change in momentum to the airplane as they leave the engine.

Noether | *The Grammar of Physics*

Upon entering Plato's Academy, the novice of science was enjoined: "Let None Enter Here Who Know Not Geometry." Mathematics has always been the language of the physical sciences. In recent centuries science students have sometimes been frustrated by the bewildering array of symbols and manipulations unique to mathematics; to provide the most general statements for a range of phenomena, it has become necessary to state the rules and theories in mathematical terms. So students of science have undertaken increasingly complex studies of mathematics simply to master the basic grammar of science.

For centuries it was nearly impossible for half the human race—women—even to aspire to a life in science or mathematics. Fortunately, that changed dramatically in the 20th century. An example of hard struggle, sheer brilliance, and determination can be found in the life of Amalie Emmy Noether (1882–1935). She came from a family of distinguished parents and siblings. She attended Erlangen University, where her father was a research mathematician, and then Göttingen, but was not allowed to matriculate formally. (It was just not done by women in the Germany of that day.) The quality of her work drew high praise, however, and she finally obtained a Ph.D. cum laude from Erlangen in 1907.

She was already attracting attention from the finest minds in Germany. Her work on algebraic invariants sparked an invitation by David Hilbert to teach at Göttingen in 1915. There she supplied some elegant formulations for Albert Einstein's general theory of relativity. But she was never one of the "boys." Her official title was unofficial associate professor. Yet she was a widely lauded, paid instructor in algebra.

Amalie Emmy Noether

Courtesy of the Special Collections Dept., Bryn Mawr College Library

Her work on the relationships between the symmetries of space and time and the conservation laws is widely used in modern theoretical physics: If the equations of the theory do not explicitly contain time, energy is conserved. If the theory does not depend on translations in space, linear momentum is conserved. And, likewise, if the theory does not depend on rotations of space, angular momentum is conserved.

When Adolf Hitler came to power in 1933, Noether, like many Jews, was expelled. She migrated to the United States, where she acquired a regular professorship at Bryn Mawr College in Pennsylvania. She was also an associate of Albert Einstein at the Institute for Advanced Study in Princeton.

Her unexpected and untimely death occurred in the course of a routine surgical procedure. Today she is regarded as the finest female mathematician in the history of that discipline.

—*Pierce C. Mullen, historian and author*

Sources: Auguste Dick, *Emmy Noether, 1882–1935*, trans. H. I. Blocher (Boston: Birkhaeuser, 1981). For a general study of women in science, see Merelene F. Rayner-Canham, *A Devotion to Their Science: Pioneer Women of Radioactivity* (Montreal: McGill-Queen's University Press, 1997); and Evelyn Fox Keller, *Reflections on Gender and Science* (New Haven, Conn.: Yale University Press, 1985).

Although the gases do not push on the atmosphere, jet engines require the atmosphere as a source of oxygen for combustion.

Summary

The momentum of an object changes if its velocity or its mass changes. This change is produced by an impulse, a net force acting on the object for a certain time $\mathbf{F}\Delta t$. Impulse is a vector quantity with the same direction as the force; this is also the direction of the change in momentum. There are many ways of producing a particular change in momentum by changing the strength of the force and the time interval during which it acts.

The momentum of a system is the vector sum of all the momenta of the system's particles. Assuming that no net external force is acting on the system, the total momentum does not change. This generalization is known as the law of conservation of linear momentum. Conservation of momentum applies to many systems, such as balloons and billiard balls.

The rule—called the conservation of linear momentum—*is* valid in both directions of time. If we know the velocities and masses of all objects at any time, we can back up the equations to see where the objects came from, and we can go forward to see where they are headed.

Key Terms

conservation of linear momentum   If the net external force on a system is zero, the total linear momentum of the system does not change.

conserved   This term is used in physics to mean that a number associated with a physical property does not change over time.

impulse   The product of the force and the time during which it acts, $F\Delta t$. This vector quantity is equal to the change in momentum.

linear momentum   A vector quantity equal to the product of an object's mass and its velocity, $\mathbf{p} = mv$.

Questions and exercises are paired so that most odd-numbered are followed by a similar even-numbered.

Blue-numbered questions and exercises are answered in Appendix B.

🌶 indicates more challenging questions and exercises.

Web**Assign** Many Conceptual Questions and Exercises for this chapter may be assigned online at WebAssign.

Conceptual Questions

1. What does it mean in physics to say that something is conserved?

2. Under what conditions is mass conserved?

3. Two identical carts with identical speeds collide head-on and stick together. Sydney argues, "Momentum for this system is conserved because the momentum of the first cart cancels the momentum of the second cart to give zero." Toby responds, "No, momentum is conserved because it's zero both before and after the collision." Which student do you agree with, and why?

4. A 2-kilogram cart moving at 6 meters per second hits a stationary 2-kilogram cart. The two move off together at 3 meters per second. Lee contends, "Momentum is conserved in this collision because the momentum of the system has the same value before and after the collision." Jackie counters, "The momentum of the system before the collision is 12 kilogram-meters per second, not zero, so momentum is not conserved." With which student do you agree, and why?

5. Why are supertankers so hard to stop? To turn?

6. Which has the greater momentum, a parked cement truck or a child on a skateboard moving slowly down the street? Why?

7. State Newton's second law in terms of momentum.

8. State Newton's first law in terms of momentum.

9. How does the padding (or air pockets) in the soles of running shoes reduce the forces on your legs? Explain your answer in terms of impulse and momentum.

10. How does padding dashboards in automobiles make them safer? Explain your answer in terms of impulse and momentum.

11. An astronaut training at the Craters of the Moon in Idaho jumps off a platform in full space gear and hits the surface at 5 meters per second. If later, on the Moon, the astronaut jumps from the landing vehicle and hits the surface at the same speed, will the impulse be larger than, smaller than, or the same as that on Earth? Why?

12. Why is skiing into a wall of deep powder less hazardous to your health than skiing into a wall of bricks? Assume in both cases that you have the same initial speed and come to a complete stop. Explain your answer in terms of impulse and momentum.

13. Assume that a friend jumps from the roof of a garage and lands on the ground. How will the impulses the ground exerts on your friend compare if the landing is on grass or on concrete?

14. Why does an egg break when it is dropped onto a kitchen tile floor but not when it lands on a living room carpet?

15. A 2-kilogram sack of flour falls off the counter and lands on the floor. Just before hitting the floor, the sack has a speed of 4 meters per second. What impulse (magnitude and direction) does the floor exert on the sack?

16. A 2-kilogram rubber ball falls off a counter and lands on the floor. Just before hitting the floor, the ball has a speed of 4 meters per second. If the ball bounces, is the magnitude of the impulse the floor exerts on the ball less than, equal to, or greater than 8 kilogram-meters per second? Why?

17. Greg and Jeff are walking down the sidewalk when identical flowerpots fall out of a window above. One flowerpot lands on Greg's head and does not bounce; the other lands on Jeff's head and bounces. Which flowerpot experiences the greater impulse? Assuming that the collision time is the same for both cases, who ends up with the worse headache? Explain.

18. Two balls are dropped on the floor from the same height. The balls are made of different types of rubber so that one bounces back to nearly the same height while the other does not bounce at all. Assuming both balls have the same mass, which ball experiences the greater impulse in colliding with the floor? Why?

19. Explain why the 12-ounce boxing gloves used in amateur fights hurt less than the 6-ounce gloves used in professional fights.

20. You kick a soccer ball 15 meters without hurting your foot much. You then pump the ball up until it is really hard (the extra air does not significantly change the ball's mass) and again kick it 15 meters. This time it hurts a lot. Using the concept of impulse, explain why it hurts more in the second case.

21. Two people are playing catch with a ball. Describe the momentum changes that occur for the ball, the people, and Earth. Is momentum conserved at all times?

22. Describe the momentum changes that occur when you dribble a basketball.

23. Which produces the larger change in momentum: a force of 3 newtons acting for 5 seconds or a force of 4 newtons acting for 4 seconds? Explain.

24. Which produces the larger impulse: a force of 3 newtons acting for 3 seconds or a force of 4 newtons acting for 2 seconds? Explain.

25. How can you explain the recoil that occurs when a rifle is fired?

26. How might you design a rifle that does not recoil?

27. Young Bill loves to fly model rockets. In his current project, however, he worries that once the rocket leaves the launch pad it will have nothing left to push on. To fix this, Bill fastens to the rocket, directly below its engine, an aluminum pie plate that will travel with the rocket. Explain why Bill will be sorely disappointed with the results.

28. A student who recently studied the law of conservation of linear momentum decides to propel a go-cart by having a fan blow on a board as shown in the following figure. This idea won't work very well. Why not?

29. While a ball is falling toward the floor, it is continually speeding up and therefore increasing its momentum. Why is this not a violation of the law of conservation of linear momentum?

30. A cue ball hits a stationary eight ball on a pool table. For which of the following systems is there a change in momentum during the collision? Explain why.
   a. the cue ball
   b. the eight ball
   c. both balls

31. Two identical objects moving at the same speed collide with each other as shown in the following figure. If the two objects stick together after the collision, will they be moving to the left, to the right, or not at all? Justify your answer using the concept of linear momentum.

32. An object of mass m and an object of mass $3m$, both moving at the same speed, collide with each other as shown in the following figure. If the two objects stick together after the collision, will they be moving to the left, to the right, or not at all? Justify your answer using the concept of linear momentum.

33. An object of mass m and an object of mass $3m$ collide with each other as shown in the following figure. The lighter object is initially moving twice as fast as the heavier one. If the two objects stick together after the collision, will they be moving to the left, to the right, or not at all? Justify your answer using the concept of linear momentum.

34. Two identical objects, one moving twice as fast as the other, collide with each other as shown in the following figure. If the two objects stick together after the collision, will they be moving to the left, to the right, or not at all? Justify your answer using the concept of linear momentum.

35. Your teacher runs across the front of the classroom with a momentum of 250 kilogram-meters per second and foolishly jumps onto a giant skateboard. The skateboard is initially at rest and has a mass equal to your teacher's. If you ignore friction with the floor, what is the total momentum of your teacher and the skateboard before and after the landing?

36. A friend is standing on a giant skateboard that is initially at rest. If you ignore frictional effects with the floor, what is the momentum of the skateboard if your friend walks to the right with a momentum of 150 kilogram-meters per second? What is the momentum of the skateboard–person system?

37. The following figure shows two air-track gliders held together with a string. A spring is tightly compressed between the gliders and is released by burning the string. The mass of the glider on the left is twice that of the glider on the right, and they are initially at rest. What is the total momentum of both gliders after the release?

© Cengage Learning/David Rogers

38. If the glider on the right in Question 37 has a speed of 2 meters per second after the release, how fast will the glider on the left be moving?

39. Use conservation of momentum to explain why people who try to jump from rowboats onto docks often end up getting wet.

40. An astronaut in the space shuttle pushes off a wall to float across the room. What effect (if any) does this have on the motion of the shuttle?

41. A firecracker is initially at rest on a horizontal, frictionless surface. It explodes into two pieces of unequal mass that move in opposite directions. Is the momentum of the firecracker conserved during the explosion? Explain why or why not.

42. After the firecracker in Question 41 explodes, is the speed of the small piece larger than, equal to, or smaller than the speed of the large piece? Explain your answer.

43. An astronaut is floating in the center of a space station with no translational motion relative to the station. Is it possible for the astronaut to move to the floor? Explain why or why not.

44. During his last trip, Al the Astronaut happened on an enormous bag of gold coins floating in space. He quickly brought his spaceship to a halt, put on his space suit, tied a rope around his waist, and pushed off in the direction of the gold. But problems developed; as he reached the bag of gold, the rope broke. Devise a way of getting Al back to his spaceship before his oxygen runs out. Although Al cares most about his life, the creative problem solver can get Al back alive with money.

45. Two identical objects, one moving north and the other moving east, collide and stick together. If the northbound object is initially moving twice as fast as the eastbound object, which of the indicated paths in the following figure represents the most likely final motion of the pair? Justify your answer using the concept of linear momentum.

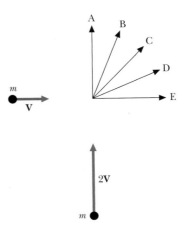

46. Two objects with the same speed, one moving north and the other moving east, collide and stick together. If the northbound object has twice the mass of the eastbound object, which of the indicated paths in the following figure represents the most likely final motion of the pair? Justify your answer using the concept of linear momentum.

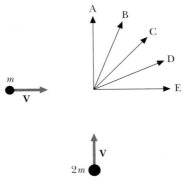

47. Two objects, one moving north and the other moving east, collide and stick together. If the eastbound object has three times the mass and is initially moving half as fast as the northbound object, which of the indicated paths in the following figure represents the most likely

final motion of the pair? Justify your answer using the concept of linear momentum.

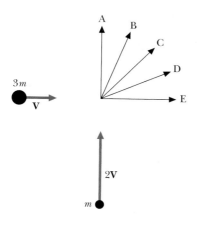

48. Two objects, one moving north and the other moving east, collide and stick together. If the eastbound object has twice the mass and is initially moving half as fast as the northbound object, which of the indicated paths in the following figure represents the most likely final motion of the pair? Justify your answer using the concept of linear momentum.

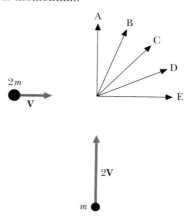

Exercises

49. What is the momentum of a 1200-kg sports car traveling down the road at a speed of 30 m/s?

50. Does a defensive end with a mass of 120 kg running at 6 m/s have a larger or smaller momentum than a running back with a mass of 100 kg running at 8 m/s?

51. How fast would you have to throw a baseball ($m = 145$ g) to give it the same momentum as a 10-g bullet traveling at 900 m/s?

52. How fast (in miles per hour) would a person with a mass of 80 kg have to run to have the same momentum as an 18-wheeler ($m = 24,000$ kg) rolling along at 1 mph?

53. What average net force is needed to accelerate a 1500-kg car to a speed of 30 m/s in a time of 8 s?

54. A jet plane takes about 30 s to go from rest to the takeoff speed of 100 mph (44.7 m/s). What is the average horizontal force that the seat exerts on the back of a 60-kg passenger during takeoff? How does this force compare to the weight of the passenger?

55. What impulse is needed to stop a 1400-kg car traveling at 25 m/s?

56. A soft rubber ball ($m = 0.5$ kg) was falling vertically at 6 m/s just before it hit the ground and stopped. What was the impulse experienced by the ball? If the ball had bounced, would the impulse have been less than, equal to, or greater than what you calculated?

57. A 1500-kg car has a speed of 30 m/s. If it takes 8 s to stop the car, what are the impulse and the average force acting on the car?

58. A coach is hitting pop flies to the outfielders. If the baseball ($m = 145$ g) stays in contact with the bat for 0.04 s

and leaves the bat with a speed of 50 m/s, what is the average force acting on the ball?

59. A very hard rubber ball ($m = 0.6$ kg) is falling vertically at 6 m/s just before it bounces on the floor. The ball rebounds back at essentially the same speed. If the collision with the floor lasts 0.04 s, what is the average force exerted by the floor on the ball?

60. A tennis ball ($m = 0.2$ kg) is thrown at a brick wall. It is traveling horizontally at 16 m/s just before hitting the wall and rebounds from the wall at 8 m/s, still traveling horizontally. The ball is in contact with the wall for 0.04 s. What is the magnitude of the average force of the wall on the ball?

61. A 150-grain .30-06 bullet has a mass of 0.01 kg and a muzzle velocity of 900 m/s. If it takes 1 ms (millisecond) to travel down the barrel, what is the average force acting on the bullet?

62. A .30-06 rifle fires a bullet with a mass of 10 g at a velocity of 800 m/s. If the rifle has a mass of 4 kg, what is its recoil speed?

63. A father ($m = 80$ kg) and son ($m = 40$ kg) are standing facing each other on a frozen pond. The son pushes on the father and finds himself moving backward at 3 m/s after they have separated. How fast will the father be moving?

64. A woman with a mass of 50 kg runs at a speed of 6 m/s and jumps onto a giant skateboard with a mass of 30 kg. What is the combined speed of the woman and the skateboard?

65. A 3-kg ball traveling to the right with a speed of 4 m/s collides with a 4-kg ball traveling to the left with a speed

of 3 m/s. What is the total momentum of the two balls before and after the collision?

66. A 4-kg ball traveling to the right with a speed of 4 m/s collides with a 5-kg ball traveling to the left with a speed of 2 m/s. What is the total momentum of the two balls before they collide? After they collide?

67. A 1200-kg car traveling north at 14 m/s is rear-ended by a 2000-kg truck traveling at 25 m/s. What is the total momentum before and after the collision?

68. If the truck and car in Exercise 67 lock bumpers and stick together, what is their speed immediately after the collision?

69. A boxcar traveling at 10 m/s approaches a string of three identical boxcars sitting stationary on the track. The moving boxcar collides and links with the stationary cars, and the four move off together along the track. What is the final speed of the four cars immediately after the collision?

70. If the boxcar in Exercise 69 instead bounces back with a speed of 1 m/s after the collision, find the speed of the three boxcars that were initially stationary.

7

Energy

▶Energy is an important commodity. People and countries that know how to obtain and use energy are generally the wealthiest and most powerful. But what is energy? And what does it mean to say that energy is conserved?

(See page 135 for the answer to this question.)

© Thomas Del Brase/Stone/Getty

Windmills are being developed to transform the energy in the wind to electrical energy.

N the preceding chapter, we came to understand momentum by considering the effect of a force acting for a certain time. If, instead, we look at an object's motion after the force has acted for a certain distance, we find another quantity that is sometimes invariant: the energy of motion. This is, however, only one form of a more general and much more profound invariant known as energy.

conservation of energy ▶   **The total energy of an isolated system does not change.**

Energy is one of the most fundamental and far-reaching concepts in the physics world view. It took nearly 300 years to fully develop the ideas of energy and energy conservation. In view of its importance and popularity, you may think that it would be easy to give a precise definition of energy. Not so.

What Is Energy?

Nobel laureate Richard Feynman, in his *Lectures on Physics,* captures the essential character of energy and its many forms when he discusses the law of conservation of energy:

> There is a certain quantity, which we call energy, that does not change in the manifold changes which nature undergoes. That is a most abstract idea, because it is a mathematical principle; it says that there is a numerical quantity which does not change when something happens. It is not a description of a mechanism, or anything concrete; it is just a strange fact that we can calculate some number and when we finish watching nature go through her tricks and calculate the number again, it is the same. (Something like the bishop on a red square, and after a number of moves—details unknown—it is still on some red square. It is a law of this nature.) Since it is an abstract idea, we shall illustrate the meaning of it by an analogy.
>
> Imagine a child, perhaps "Dennis the Menace," who has blocks which are absolutely indestructible, and cannot be divided into pieces. Each is the same as the other. Let us suppose that he has 28 blocks. His mother puts him with his 28 blocks into a room at the beginning of the day. At the end of the day, being curious, she counts the blocks very carefully, and discovers a phenomenal law—no matter what he does with the blocks, there are always 28 remaining! This continues for a number of days, until one day there are only 27 blocks, but a little investigating shows that there is one under the rug—she must look everywhere to be sure that the number of blocks has not changed. One day, however, the number appears to change—there are only 26 blocks. Careful investigation indicates that the window was open, and upon looking outside, the other two blocks are found. Another day, careful count indicates that there are 30 blocks! This causes considerable consternation, until it is realized that Bruce came to visit, bringing his blocks with him, and he left a few at Dennis' house. After she has disposed of the extra blocks, she closes the window, does not let Bruce in, and then everything is going along all right, until one time she counts and finds only 25 blocks. However, there is a box in the room, a toy box, and the mother goes to open the toy box, but the boy says "No, do not open my toy box," and screams. Mother is not allowed to open the toy box. Being extremely curious, and somewhat ingenious, she invents a scheme! She knows that a block weighs three ounces, so she weighs the box at a time when she sees 28 blocks, and it weighs 16 ounces. The next time she wishes to check, she weighs the box again, subtracts sixteen ounces and divides by three. She discovers the following:

$$\binom{\text{number of}}{\text{blocks seen}} + \frac{(\text{weight of box}) - 16 \text{ ounces}}{3 \text{ ounces}} = \text{constant}$$

There then appear to be some new deviations, but careful study indicates that the dirty water in the bathtub is changing its level. The child is throwing blocks into the water, and she cannot see them because it is so dirty, but she can find out how many blocks are in the water by adding another term to her formula. Since the original height of the water was 6 inches and each block raises the water a quarter of an inch, this new formula would be:

$$\begin{pmatrix}\text{number of} \\ \text{blocks seen}\end{pmatrix} + \frac{(\text{weight of box}) - 16\ \text{ounces}}{3\ \text{ounces}} + \frac{(\text{height of water}) - 6\ \text{inches}}{\frac{1}{4}\ \text{inch}} = \text{constant}$$

In the gradual increase in the complexity of her world, she finds a whole series of terms representing ways of calculating how many blocks are in places where she is not allowed to look. As a result, she finds a complex formula, a quantity which *has to be computed,* which always stays the same in her situation. . . .

The analogy has the following points. First, when we are calculating the energy, sometimes some of it leaves the system and goes away, or sometimes some comes in. In order to verify the conservation of energy, we must be careful that we have not put any in or taken any out. Second, the energy has a large number of *different forms,* and there is a formula for each one. . . . If we total up the formulas for each of these contributions, it will not change except for energy going in or out.

It is important to realize that in physics today, we have no knowledge of what energy is. We do not have a picture that energy comes in little blobs of a definite amount. It is not that way. However, there are formulas for calculating some numerical quantity, and when we add it all together it gives "28"—always the same number. It is an abstract thing in that it does not tell us the mechanisms or the *reasons* for the various formulas.

Energy of Motion

Extended presentation available in the *Problem Solving* supplement

The most obvious form of energy is the one an object has because of its motion. We call this quantity of motion the **kinetic energy** of the object. Like momentum, kinetic energy depends on the mass and the motion of the object. But the expression for the kinetic energy is different from that for momentum. The kinetic energy *KE* of an object is

◀ kinetic energy
= $\frac{1}{2}$ mass × speed squared

$$KE = \tfrac{1}{2}mv^2$$

where the factor of $\frac{1}{2}$ makes the kinetic energy compatible with other forms of energy, which we will study later.

Notice that the kinetic energy of an object increases with the square of its speed. This means that if an object has twice the speed, it has four times the kinetic energy; if it has three times the speed, it has nine times the kinetic energy; and so on.

Q: What happens to the kinetic energy of an object if its mass is doubled while its speed remains the same?

A: Because the kinetic energy is directly proportional to the mass, the kinetic energy doubles.

Are You On the Bus?

---

* R. P. Feynman, R. B. Leighton, and M. Sands, *The Feynman Lectures on Physics* (Reading, Mass.: Addison-Wesley, 1963), 1: 4-1–4-2.

The units for kinetic energy, and therefore for all types of energy, are kilograms multiplied by (meters per second) squared $(\text{kg} \cdot \text{m}^2/\text{s}^2)$. This energy unit is called a **joule** (rhymes with *tool*). Kinetic energy differs from momentum in that it is not a vector quantity. An object has the same kinetic energy regardless of its direction as long as its speed does not change.

A typical textbook dropped from a height of 10 centimeters (about 4 inches) hits the floor with a kinetic energy of about 1 joule (J). The kinetic energy of a 70-kilogram (154-pound) person running at a speed of 8 meters per second is

$$KE = \tfrac{1}{2}mv^2 = \tfrac{1}{2}(70 \text{ kg})(8 \text{ m/s})^2 = (35 \text{ kg})(64 \text{ m}^2/\text{s}^2) = 2240 \text{ J}$$

Conservation of Kinetic Energy

The search for invariants of motion often involved collisions. In fact, early in their development, the concepts of momentum and kinetic energy were often confused. Things became much clearer when these two were recognized as distinct quantities. We have already seen that momentum is conserved during collisions. Under certain, more restrictive conditions, kinetic energy is also conserved.

Consider the collision of a billiard ball with a hard wall. Obviously, the kinetic energy of the ball is not constant. At the instant the ball reverses its direction, its speed is zero and therefore its kinetic energy is zero. As we will see, even if we include the kinetic energy of the wall and Earth, the kinetic energy of the system is not conserved.

However, if we don't concern ourselves with the details of what happens during the collision and look only at the kinetic energy before the collision and after the collision, we find that the kinetic energy is nearly conserved. During the collision the ball and the wall distort, resulting in internal frictional forces that reduce the kinetic energy slightly. But once again we ignore some things to get at the heart of the matter. Let's assume we have "perfect" materials and can ignore these frictional effects. In this case the kinetic energy of the ball after it leaves the wall equals its kinetic energy before it hit the wall. Collisions in which kinetic energy is conserved are known as **elastic** collisions.

Actually, many atomic and subatomic collisions are perfectly elastic. A larger-scale approximation of an elastic collision would be one between air-hockey pucks that have magnets stuck on top so that they repel each other.

Courtesy of Arbor Scientific, Inc.

Figure 7-1 The collision-ball apparatus demonstrates the conservation of momentum and kinetic energy.

Are You On the Bus?

Q: The apparatus shown in Figure 7-1 may look familiar. It has five balls of equal mass suspended in a line so that they can swing along the direction of the line. The balls collide with nearly elastic collisions. If you pull on a ball at one end and release it, the last ball at the other end leaves with the velocity of the original. What will happen if you pull back two balls and release them together?

A: The only way that momentum and kinetic energy of the system can be conserved is for the original balls to stop and for two balls on the opposite side to leave with the same velocity the original balls had before the collision. They will rise to nearly the starting height before stopping to turn around.

Collisions in which kinetic energy is lost are known as **inelastic** collisions. The loss in kinetic energy shows up as other forms of energy, primarily in the form of heat, which we will discuss in Chapter 13. Collisions in which the objects move away with a common velocity are never elastic. We will see this in the billiard ball example presented at the end of this section.

The outcomes of collisions are determined by the conservation of momentum and the extent to which kinetic energy is conserved. We know that the collisions of billiard balls are not perfectly elastic because we hear them collide. (Sound is a form of energy and therefore carries off some of the energy.)

We can determine the relative elasticities of balls made of various materials by dropping them from a uniform height onto a very hard surface, such as a concrete floor or a thick steel plate. The more elastic the material, the closer the ball will return to its original height. It is somewhat surprising to find that a glass marble is much more elastic than a rubber ball. This suggests that the physics definition of *elastic* is very different than the everyday usage of the word.

WORKING IT OUT *Conservation of Kinetic Energy*

Collisions between very hard objects such as billiard balls are nearly elastic. Consider the head-on collision of a moving billiard ball with a stationary billiard ball [Figure 7-2(a)]. The moving ball stops, and the stationary one acquires the initial velocity of the moving ball [Figure 7-2(b)]. Suppose the mass of each ball is 0.2 kg, and the initial velocity is 4 m/s. The total momenta before and after the collision are

$$p \text{ (before)} = m_1 v_1 + m_2 v_2 = (0.2 \text{ kg})(4 \text{ m/s}) + 0 = 0.8 \text{ kg} \cdot \text{m/s}$$

and

$$p \text{ (after)} = m_1 v_1 + m_2 v_2 = 0 + (0.2 \text{ kg})(4 \text{ m/s}) = 0.8 \text{ kg} \cdot \text{m/s}$$

Figure 7-2 (a) A moving billiard ball collides head-on with a stationary one. Which possibility occurs? (b) The moving ball stops, and the stationary ball takes on the initial velocity. (c) The balls move off together with one-half the initial velocity.

Therefore, momentum is conserved. Likewise, the kinetic energies before and after the collision are

$$KE \text{ (before)} = \tfrac{1}{2}m_2 v_1^2 = \tfrac{1}{2}m_2 v_1^2 = \tfrac{1}{2}(0.2 \text{ kg})(4 \text{ m/s})^2 + 0 = 1.6 \text{ J}$$

and

$$KE \text{ (after)} = \tfrac{1}{2}m_1 v_1^2 + \tfrac{1}{2}m_2 v_2^2 = 0 + \tfrac{1}{2}(0.2 \text{ kg})(4 \text{ m/s})^2 = 1.6 \text{ J}$$

Therefore, kinetic energy is also conserved.

But this is not the only possibility that conserves momentum. Another is for the two balls to both move in the forward direction, each with one-half of the initial velocity [Figure 7-2(c)]. Both possibilities conserve the total momentum of the two-ball system. Yet when we do the experiment, the first possibility is the one that we always observe.

Why does one occur and not the other? The first possibility conserves kinetic energy, but the second does not. To see that the second doesn't conserve kinetic energy, we again calculate the kinetic energy after the collision:

$$KE \text{ (after)} = \tfrac{1}{2}m_1 v_1^2 + \tfrac{1}{2}m_2 v_2^2 = \tfrac{1}{2}(0.2 \text{ kg})(2 \text{ m/s})^2 + \tfrac{1}{2}(0.2 \text{ kg})(2 \text{ m/s})^2 = 0.8\text{J}$$

One-half of the kinetic energy is lost in this case.

Changing Kinetic Energy

A cart rolling along a frictionless, horizontal surface has a certain kinetic energy because it is moving. A net force on the cart can change its speed and thus its kinetic energy. If you push in the direction the cart is moving, you increase the cart's kinetic energy. Pushing in the opposite direction slows the cart and decreases its kinetic energy (Figure 7-3).

When a force acts for a certain time, the force produces an impulse that changes the momentum of the object. In contrast, the *distance* through which the force acts determines how much the kinetic energy changes. The product of the force *F in the direction of motion* and the distance moved *d* is known as the **work** *W:*

work = force × distance moved ▶

$$W = Fd$$

From the definition of work, we conclude that the units of work are new-ton-meters. If we substitute our previous expression for a newton (kg · m/s²), we find that a newton-meter (N · m) equals a kg · m²/s², which is the same as the units for energy—that is, a joule. Therefore, the units of work are the same as those of energy. In the U.S. customary system, the units of work are foot-pounds.

Newton's second law and our expressions for acceleration and distance traveled can be combined with the definition of work to show that the work done on an object is equal to the change in its kinetic energy:

work done = change in kinetic energy ▶

$$W = \Delta KE$$

If the net force is in the same direction as the velocity, the work is positive, and the kinetic energy increases. If the net force and velocity are oppositely directed, the work is negative, and the kinetic energy decreases.

(a)

(b)

Figure 7-3 (a) A force in the direction of motion increases the kinetic energy. (b) A force opposite the direction of motion decreases the kinetic energy.

Everyday Physics | *Stopping Distances for Cars*

The data for stopping distances given in driver's manuals do not seem to have a simple pattern, other than the faster you drive, the longer it takes to stop. The driver's manual for the state of Montana, for example, states that it takes at least 186 feet to stop a car traveling at 50 miles per hour and only 65 feet for a car that is traveling at 25 miles per hour. Why is the stopping distance one-third as much when the speed is one-half as much? Furthermore, these tables make no mention of the size of the car.

A car's kinetic energy depends on its speed and its mass. When the brakes are applied, the brake pads slow the wheels, which in turn apply a force on the highway. The reaction force of the highway on the car does work on the car, reducing its kinetic energy to zero. To stop the car, the total work done on the car must be equal to its initial kinetic energy.

The frictional force between the tires and the road depends on the car's mass and whether it is rolling or skidding. (Skidding greatly reduces the frictional force.) The force of friction does not change very much for various tire designs or road surfaces, providing the roads are not wet or icy. Because both the frictional force and kinetic energy are proportional to the car's mass, the stopping distance is independent of the car's mass. So how far the car travels after the brakes are applied depends almost entirely on its initial speed.

The distances in driver's manual tables include the distance traveled during approximately 1 second of reaction time in addition to the distance required for the brakes to stop the car. At 50 miles per hour the car is traveling at 74 feet per second, so the distance required to stop the car once the brakes are applied is 186 feet – 74 feet = 112 feet.

Because the car's kinetic energy changes as the square of its speed, a car that is traveling at 25 miles per hour has only one-fourth the kinetic energy of one traveling at 50 miles per hour. Because only one-fourth as much work is required to stop the car,

the force must act through only one-fourth the distance. A car traveling at 25 miles per hour can stop in $\frac{1}{4} \times 112$ feet = 28 feet. During the 1-second reaction time, the car travels an additional $\frac{1}{2} \times 74$ feet = 37 feet, so the total stopping distance is 28 feet + 37 feet = 65 feet. The stopping distances for other speeds are given in the following table.

Stopping Distances for Automobiles Traveling at Selected Speeds				
Speed		**Stopping Distances**		
(mph)	**(ft/s)**	**Reaction (ft)**	**Braking (ft)**	**Total (ft)**
10	15	15	5	20
20	29	29	18	47
30	44	44	40	84
40	59	59	72	131
50	74	74	112	186
60	88	88	161	249
70	103	103	220	323
80	117	117	287	404
90	132	132	363	495
100	147	147	448	595

1. The car in front of you suddenly brakes to avoid hitting an elk standing in the middle of the road. If that car was traveling 40 miles per hour before hitting the brakes, and you were 120 feet behind moving at a speed of 60 miles per hour when you saw the brake lights, would you have time to safely stop?

2. If you had been following the car in Question 1, also with a speed of 40 miles per hour, how many feet should you keep between the cars to safely stop when you see the brake lights go on?

Forces That Do No Work

The meaning of *work* in physics is different from the common usage of the word. Commonly, people talk about "playing" when they throw a ball and "working" when they study physics. The physics definition of work is quite precise—work occurs when the product of the force and the distance is nonzero. When you throw a ball, you are actually doing work on the ball; its kinetic energy is increased because you apply a force through a distance. Although you may move pages and pencils as you study physics, the amount of work is quite small.

Similarly, if you hold a suitcase above your head for 30 minutes, you would probably claim it was hard work. According to the physics definition, however, you did not do any work on the suitcase; the 30 minutes of straining and groaning did not change the suitcase's kinetic energy. A table could hold

It takes no work to hold a cheerleader in the air.

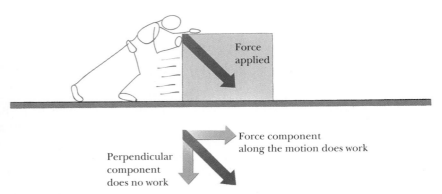

Figure 7-4 Any force can be replaced by two perpendicular component forces. Only the component along the direction of motion does work on the box.

Figure 7-5 When the force is perpendicular to the velocity, the force does no work.

Figure 7-6 The gravitational force of the Sun does work on Earth whenever the force is not perpendicular to Earth's velocity. The elliptical nature of the orbit has been exaggerated to show the parallel component.

up the suitcase just as well. Your body, however, is doing physiological work because the muscles in your arms do not lock in place but rather twitch in response to nerve impulses. This work shows up as heat (as evidenced by your sweating) rather than as a change in the kinetic energy of the suitcase.

There are other situations in which a net force does not change an object's kinetic energy. If the force is applied in a direction perpendicular to its motion, the velocity of the object changes, but its speed doesn't. Therefore, the kinetic energy does not change. The definition of work takes this into account by stating that it is only the force that acts along the direction of motion that can do work.

Often, a force is neither parallel nor perpendicular to the displacement of an object. Because force is a vector, we can think of it as having two components, one that is parallel and one that is perpendicular to the motion as illustrated in Figure 7-4. The parallel component does work, but the perpendicular one does not do any work.

Consider an air-hockey puck moving in a circle on the end of a string attached to the center of the table shown in Figure 7-5. Because the speed is constant, the kinetic energy is also constant. The force of gravity is balanced by the upward force of the table. These vertical forces cancel and do no work. The tension that the string exerts on the puck is not canceled but does no work because it always acts perpendicular to the direction of motion.

If Earth's orbit were a circle with the Sun at the center, the gravitational force the Sun exerts on Earth would do no work, and Earth would have a constant kinetic energy and therefore a constant speed. However, because the orbit is an ellipse, the force is not always perpendicular to the direction of motion (Figure 7-6) and therefore does work on Earth. During one-half of each orbit, a small component of the force acts in the direction of motion, increasing Earth's kinetic energy and speed. During the other half of each orbit, the component is opposite the direction of motion, and Earth's kinetic energy and speed decrease.

Gravitational Potential Energy

When a ball is thrown vertically upward, it has a certain amount of kinetic energy that disappears as it rises. At the top of its flight, it has no kinetic energy, but as it falls, the kinetic energy reappears. If we believe that energy is an invariant, we must be missing one or more forms of energy.

The loss and subsequent reappearance of the ball's kinetic energy can be understood by examining the work done on the ball. As the ball rises, the force of gravity performs negative work on the ball, reducing its kinetic energy until it reaches zero. On the way back down, the force of gravity increases the ball's kinetic energy by the same amount it lost on the way up. Rather than simply saying that the kinetic energy temporarily disappears, we can retain the idea of the conservation of energy by defining a new form of energy. Kinetic energy is then transformed into this new form and later transformed back. This new energy is called **gravitational potential energy.**

We have some clues about the expression for this new energy. Its change must also be given by the work done by the force of gravity, and it must increase when the kinetic energy decreases, and vice versa. Therefore, we define the gravitational potential energy of an object at a height h above some zero level as equal to the work done by the force of gravity on the object as it falls to height zero. The gravitational potential energy GPE of an object near Earth's surface is then given by

$$GPE = mgh$$

◀ gravitational potential energy = force of gravity × height

As an example, we can calculate the gravitational potential energy of a 6-kilogram ball located 0.5 meter above the level that we choose to call zero:

$$GPE = mgh = (6 \text{ kg})(10 \text{ m/s}^2)(0.5 \text{ m}) = 30 \text{ J}$$

Notice that only the vertical height is important. Moving an object 100 meters sideways does not change the gravitational potential energy because the force of gravity is perpendicular to the motion and therefore does no work on the object.

The amount of gravitational potential energy an object has is a relative quantity. Its value depends on how we define the height—that is, what height we take as the zero value. We can choose to measure the height from any place that is convenient. The only thing that has any physical significance is the *change* in gravitational potential energy. If a ball gains 20 joules of kinetic energy as it falls, it must lose 20 joules of gravitational potential energy. It does not matter how much gravitational potential energy the ball had at the beginning; only the amount lost has any meaning in physics.

Q: What is the change in gravitational potential energy of a 50-kilogram person who climbs a flight of stairs with a height of 3 meters and a horizontal extent of 5 meters?

A: The change in gravitational potential energy is

$$\Delta GPE = mgh_{\text{final}} - mgh_{\text{initial}} = (50 \text{ kg})(10 \text{ m/s}^2)(3 \text{ m}) = 1500 \text{ J}$$

The horizontal extent has no effect on the answer.

Are You On the Bus?

FLAWED REASONING

Bill and Will are calculating the gravitational potential energy of a 5-newton ball held 2 meters above the floor of their classroom.

Bill says, "This is easy. Gravitational potential energy is mgh, where mg is the weight of the ball. We just multiply the 5 newtons by the 2 meters to get the gravitational potential energy of 10 joules."

Will says, "You are forgetting that our classroom is on the second floor. We are going to have to find out how high the ball is relative to the ground."

Do you agree with either of these students?

ANSWER It is only the difference in gravitational potential energy that matters. Either we can say that the ball fell from a height of 2 meters to a height of zero, or we can say that it fell from a height of 5 meters (relative to the ground) to a height of 3 meters. Either way, we get the same decrease in gravitational potential energy and the same increase in kinetic energy. Both students would get correct answers. In general, it is often easiest to take the lowest point in each problem to be the zero for height.

Conservation of Mechanical Energy

mechanical energy = kinetic energy + ▶
gravitational potential energy

The sum of the gravitational potential and kinetic energies is conserved in some situations. This sum is called the **mechanical energy** of the system:

$$ME = KE + GPE = \tfrac{1}{2}mv^2 + mgh$$

When frictional forces can be ignored and the other nongravitational forces do not perform any work, the mechanical energy of the system does not change. The simplest example of this circumstance is free fall (Figure 7-7). Any decrease in the gravitational potential energy shows up as an increase in the kinetic energy, and vice versa.

Let's use the conservation of mechanical energy to analyze an idealized situation originally discussed by Galileo. Galileo released a ball that rolled down a ramp, across a horizontal track, and up another ramp, as shown in Figure 7-8. He remarked that the ball always returned to its original height. From an energy point of view, he gave the ball some initial gravitational potential energy by placing it at a certain height above the horizontal ramp. As the ball rolled down the first ramp, its gravitational potential energy was transformed into kinetic energy. Going up the opposite ramp merely reversed this process; the ball's kinetic energy was transformed back into gravitational potential energy. The ball continued to move until its kinetic energy was entirely transformed back to gravitational potential energy, which occurred when it once again reached its original height. This result is independent of the slopes of the ramps.

During the horizontal portion of the ball's trip, the gravitational potential energy remained constant. Hence, the kinetic energy did not change and the speed remained constant, in agreement with Newton's first law.

The pendulum bob shown in Figure 7-9 cyclically gains and loses kinetic and gravitational potential energy. Note that the tension exerted by the string does no work. Therefore, if we ignore frictional forces, the total mechanical energy is conserved. Suppose that at the beginning the bob has zero speed at point A and a gravitational potential energy of 10 joules. (We have chosen the zero for gravitational potential energy to be at the lowest point of the bob's path.) Because the kinetic energy is zero, the total mechanical energy is the same as the gravitational potential energy—that is, 10 joules.

max PE
no KE

$\tfrac{1}{2}PE + \tfrac{1}{2}KE$

h

max KE
no PE

Figure 7-7 As a ball falls, its mechanical energy is conserved; any loss in gravitational potential energy shows up as a gain in kinetic energy.

Starting height

The bob is released. As it swings down, it loses gravitational potential energy and gains kinetic energy. The gravitational potential energy is zero at the lowest point of the swing, and so the mechanical energy is all kinetic and equal to 10 joules. The bob continues to rise up the other side until all the kinetic energy is transformed back to gravitational potential energy. Because the gravitational potential energy depends on the height, the bob must return to its original height. And the motion repeats.

$PE = 10\,\text{J}$
$KE = 0\,\text{J}$

$PE = 0$
$KE = 10\,\text{J}$

$PE = 6\,\text{J}$
$KE = 4\,\text{J}$
Total $10\,\text{J}$

Figure 7-9 The total mechanical energy (kinetic plus gravitational potential) remains equal to 10 joules.

Q: Suppose the bob is released at twice the height. What is the maximum kinetic energy?

A: The initial gravitational potential energy is now twice as big, so the maximum kinetic energy will also be twice as big—that is, 20 joules.

Are You On the Bus?

Even when the bob is someplace between the highest and lowest points of the swing, the total mechanical energy is still 10 joules. If at this point we determine from the height of the bob that it has 6 joules of gravitational potential energy, we can immediately declare that it has 4 joules of kinetic energy. Because we know the expression for the kinetic energy, we can calculate the speed of the bob at this point.

WORKING IT OUT *Conservation of Mechanical Energy*

A simple pendulum consists of a small ball of mass $m = 0.2$ kg swinging on a string of length 7 m. The pendulum is released from the horizontal position, as shown in Figure 7-10. A short time later, the pendulum reaches the bottom of its swing and encounters a nail sticking out of the page a distance 6 m below the pivot point, and the pendulum begins to wrap around the nail. How fast is the ball moving when it is at the top of its first rotation around the nail?

If we define the zero of height to be the lowest point in the ball's path, then the ball starts with gravitational potential energy at a height of 7 m. When it reaches the top of its first rotation, it is 2 m above the lowest point, and has kinetic energy as well as gravitational potential energy. Because the string always pulls on the ball in a direction perpendicular to its motion, no work is done and the mechanical energy is conserved.

$$PE_i = KE_f + PE_f$$

$$mgh_i = \tfrac{1}{2}mv_f^2 + mgh_f$$

We can divide both sides of this equation by the mass of the ball and see that the final speed is independent of the mass. Using the given values for the initial and final height of the ball, our equation becomes

$$(10 \text{ m/s/s})(7 \text{ m}) = \tfrac{1}{2}v_f^2 + (10 \text{ m/s/s})(2 \text{ m})$$

Solving for v_f, we have

$$v_f^2 = 2(10 \text{ m/s/s})(7\text{m} - 2\text{m}) = 100 \text{ m}^2/\text{s}^2$$

$$v_f = 10 \text{ m/s}$$

Figure 7-10 How fast is the ball moving at the top of the first rotation?

Roller Coasters

Imagine trying to determine the speed of a roller coaster inside a thrilling loop-the-loop as it traverses the track's hills and valleys. Determining the speed at any spot using Newton's second law is difficult because the forces are continually changing magnitude and direction. We can, however, use conservation of mechanical energy to determine the speed of an object without knowing the details of the net forces acting on it, providing we can ignore the frictional forces. If we know the mass, speed, and height of the roller coaster at some spot, we can calculate its mechanical energy. Now determining its speed at any other spot is greatly simplified. The height gives us the gravitational potential energy. Subtracting this from the total mechanical energy yields the kinetic energy from which we can obtain the roller coaster's speed. Notice that we don't need to worry about all the energy transformations that occurred earlier in the ride.

Suppose the roller-coaster ride was designed like the one in Figure 7-11, and on reaching the top of the lower hill, your car almost comes to rest. Assuming that there are no frictional forces to worry about (not true in real situations), is there any possibility that you can get over the higher hill? The answer is no. Your gravitational potential energy at the top of the hill is nearly equal to the mechanical energy. This energy is not enough to get you over the higher hill. You will gain speed and thus kinetic energy as you coast down the hill, but as you start up the other hill, you will find that you cannot exceed the height of the original hill.

The mechanical energy of a roller coaster on a loop-the-loop is conserved if the frictional forces can be neglected.

© Bob Torrez/Stone/Getty

Figure 7-11 The car does not have enough gravitational potential energy to coast over the higher hill.

Q: Is there any way that the roller-coaster car can make it over the second hill when starting on top of the first hill?

A: Yes. If your car has some kinetic energy at the top of the first hill, it might be possible to make it over the second hill. You would need enough kinetic energy to equal or exceed the extra gravitational potential energy required to climb the second hill.

Are *You* On the Bus?

Our use of the conservation of mechanical energy is limited because we ignore losses due to frictional forces, and in many cases this is not realistic. These frictional forces do work on the car and thus drain away some of the car's mechanical energy. If we know the magnitude of these frictional forces, however, and the distances through which they act, we can calculate the energy transformed to other forms and allow for the loss of mechanical energy.

Where does the energy go that is thus drained away? The answer to this question is still ahead of us. To maintain the notion that energy is an invariant, we will have to search for other forms of energy.

FLAWED REASONING

The following question appears on the final exam: "Three bears are throwing identical rocks from a bridge to the river below. Papa Bear throws his rock upward at an angle of 30 degrees above the horizontal. Mama Bear throws hers horizontally. Baby Bear throws the rock at an angle of 30 degrees below the horizontal. Assuming that all three bears throw with the same speed, which rock will be traveling fastest when it hits the water?" Three students meet after the exam and discuss their answers.

Emma says, "Baby Bear's rock will be going the fastest because it starts with a downward component of velocity."

Hector says, "But Papa Bear's rock will stay in the air the longest, so it will have more time to speed up. I think his rock will be traveling the fastest."

M'Lynn says, "Papa Bear's rock does stay in the air longer, but part of that time it is moving upward and slowing down. I think Mama Bear's rock will be traveling fastest when it hits the water because it is in the air longer than Baby Bear's and it is speeding up all of the time."

With which student (if any) do you agree?

ANSWER All three students are wrong. They are making an easy problem much too difficult by ignoring the power of the energy approach to problem solving. Because each of the three rocks started with the same kinetic energy (same speed) and the same gravitational potential energy (same height), they must all end up with the same final kinetic energy before hitting the water. All three rocks must therefore hit the water with the same speed. Note that the three rocks will not hit the water at the same time, with the same direction of velocity, or at the same distance from the bridge. However, the energy method will not give us this information.

Other Forms of Energy

We can identify other places where kinetic energy is temporarily stored. For example, a moving ball can compress a spring and lose its kinetic energy (Figure 7-12). While the spring is compressed, it stores energy much like the ball in the gravitational situation. If we latch the spring while it is in the compressed state, we can store its energy indefinitely as elastic potential energy. Releasing it at some future date will transform the spring's elastic potential energy back into the ball's kinetic energy.

Q: When a ball is hung from a vertical spring, it stretches the spring. As it drops, it loses gravitational potential energy, but this does not all show up as kinetic energy. What happens to the gravitational potential energy?

A: The gravitational potential energy is converted to kinetic energy and to elastic potential energy of the spring. At the bottom, it is all elastic potential energy.

Figure 7-12 The gravitational potential energy (a) is converted to kinetic energy (b), which is then converted to elastic potential energy of the spring (c).

A thrilling example of this is bungee jumping. A nylon rope is securely fastened to the ankles of the jumper, who then dives headfirst from a very high platform. As the jumper falls, gravitational potential energy is converted to kinetic energy. As the rope tightens and stretches, both the kinetic energy and some additional gravitational potential energy are converted to elastic potential energy. When the rope reaches its maximum stretch, the jumper bounces back up into the air as much of the elastic potential energy is converted back into kinetic and gravitational potential energy. After several bounces the bungee jumper is lowered to the ground. Notice that if there were no loss of mechanical energy, the bungee jumper would bounce forever!

Many objects or materials when distorted by a force hold some elastic potential energy as a result of the distortions. A floor "gives" when we jump on it. Our kinetic energy on impact is transformed, in part, to elastic potential energy of the floor. As the floor springs back to its original shape, we regain some of the kinetic energy. In these cases some of the mechanical energy (maybe most) is lost to the dissipating effects of the distortion.

We have described a gravitational potential energy that is associated with the gravitational force. Other potential energies are associated with other forces in nature. The elastic potential energy in a spring is due to electromagnetic (electric and magnetic) forces. There are also other forms of electromagnetic potential energy. Chemical energy is really just a potential energy associated with the electromagnetic force. We will see in Chapter 26 that nuclear potential energy is associated with the nuclear force.

The transformation of these various forms of potential energy to kinetic energy is what powers our civilization. The gravitational potential energy of the water behind dams powers hydroelectric plants; that of a weight runs grandfather clocks. Most of the energy we use every day is the result of releasing the chemical potential energy of fossil fuels. Nuclear power plants are

As the bungee jumper falls, gravitational potential energy is converted to kinetic energy.

The gravitational potential energy of the water behind Hoover Dam is converted to electric energy.

The Sun is a great storehouse of nuclear potential energy.

Everyday Physics | *Exponential Growth*

America's energy usage has grown at a rate of about 5% per year for the past few decades. At first glance this sounds relatively harmless, but don't be fooled. Anything that grows in proportion to its current size gets out of control. This kind of growth is called *exponential growth,* and whether it's a bank account, energy usage, or population, it follows the same pattern. This pattern can be illustrated with a simple story of a creative mathematician.

Legend has it that a mathematician in ancient India invented the game of chess. The ruler of India was so pleased that the mathematician was allowed to choose his own reward. This clever fellow shunned the obvious bounties of gold and jewels and instead requested grains of wheat. His plan was to put one grain on the first square of the chessboard and then double the number of grains on each successive square, so there would be two grains of wheat for the second square, four grains for the third square, and so on. You can determine the number on each square with a simple calculator. Just start with the first grain and keep multiplying by two, 64 times.

At the end of the first row, there are 128 grains on the 8th square (Figure A). But that's not very many. By the end of the second row, the mathematician has only enough wheat to make a few loaves of bread. But what happens if we continue this pattern of doubling? The photo in Figure B is only symbolic, as the number of grains on the 64th square is about 9 billion billion! And the grand total is twice this! This is roughly 500 times the amount of wheat grown in the entire world in a year's time and probably more wheat than has been grown during the entire history of humankind.

The pattern is very important. The doubling process starts out very slowly—a gentle, rather innocent, growth that eventually gets totally out of hand. The legend ends with the mathematician being beheaded around the 35th square when the granaries of India were exhausted.

Figure A The 8th square has 128 grains of wheat.

There are many other examples of exponential growth. Consider the compound interest that you earn on a savings account. Suppose the bank offers a guaranteed interest rate of 5% per year. The growth pattern is the same as for the chess example. Your money will double after a certain number of years and double again each time this number of years elapses. The *doubling time* is obtained by dividing 70 by the percentage rate. For your savings account this would be 70/(5%/year), or 14 years. If you put $1 in the bank today, it will double every 14 years. In 100 years your descendants will collect $128. If they choose to leave the money in the bank for their descendants—say, for another 400 years—your initial $1 becomes $64 billion! (Bankers are not worried about 500-year-old accounts, but they are well aware of this pattern and have rules about inactive accounts.)

Historically, the growth of the world's population has been roughly exponential. According to a United Nations estimate, the world's population reached 6 billion in October 1999 and was increasing at a rate of 1.3% per year. Most people are not worried because

designed to release nuclear potential energy. Nuclear potential energy is the ultimate source of the energy we receive from the Sun.

We have defined a variety of potential energies to explain the temporary losses of kinetic energy. However, this explanation doesn't apply to situations that include friction. To understand the physics of friction, consider the following. We start a box sliding across the floor with a given amount of kinetic energy. As it slows and comes to a halt, its kinetic energy decreases and finally reaches zero. We may imagine that this energy was stored in some form of "frictional potential energy." If this were the case, we could somehow release this frictional potential energy, and the box would move across the floor, continually gaining kinetic energy. This does not happen. As we will discuss in Chapter 13, when frictional forces act, some of the energy changes form and appears as thermal energy.

Figure B The 64th square has 9 billion billion grains of wheat. The photograph is of course only symbolic.

© First Light/CORBIS

they multiply by dividing. Assume that you have a colony of bacteria in which each bacterium divides exactly in two at the end of each minute. Therefore, the doubling time is 1 minute. Let's further assume that the bottle the bacteria are in will be exactly full at the end of 1 hour; that is, it will take 60 doublings to fill the bottle. When will the bottle be half-filled? In 59 minutes. If you were a bacterium, when would you realize that you were running out of space? Because you're reading this essay, assume that at 55 minutes you've recognized that the growth is exponential and that there will be a major problem. At this time the bottle is only 3% full. How successful would you be in convincing your fellow bacteria that there are only 5 minutes left?

Imagine that during this debate an adversary argues that with proper government funding the colony can find more space, and in fact they do. After a great deal of expense and effort, they locate three new bottles. The resources have quadrupled. How much longer can the growth continue? The answer is only 2 more minutes!

The point of this discussion is to understand our current usage of energy. Clearly, we cannot let our usage grow exponentially. Although there are sizable variations in the estimates of the world's nonrenewable energy resources, all estimates are rather inconsequential if exponential growth continues. Does it really make a difference if we have underestimated these resources by a factor of 2, or even 4? No. If we continue to increase our use of these resources by even a small percentage each year, they will quickly be exhausted. During the next doubling time, we will use as much energy as has been used by humankind up to the present.

1. Many real estate agents claim that the average home price in the United States doubles every 10 years. What average yearly increase in home value is assumed in this claim?

2. If you were lucky enough to receive a guaranteed salary raise of 10% each year, how many years would it take to double your salary?

Source: Bartlett, A.A. *American Journal of Physics,* 46, page 876, 1978. Reprinted with permission from *American Journal of Physics,* 46. Copyright © 1978, American Association of Physics Teachers.

this seems to be a low rate of increase. However, if this rate were to continue, the world's population would double in 55 years to 12 billion by 2054. Few people believe that the world's population will actually reach this number by the middle of the century because the rate of increase has been getting smaller and may be as low as 0.5% (a doubling time of 140 years) by 2050. If this occurs, the world's population will be somewhere between 9 and 10 billion in 2050.

Exponential growth can have very serious consequences because the big changes come on very fast. We can look at another example to illustrate the nature of the problem. Bacteria are rather curious in that

Is Conservation of Energy a Hoax?

It may seem strange that whenever we discover a situation in which energy does not appear to be conserved, we invent a new form of energy. How can the conservation law have any validity if we keep modifying it whenever it appears to be violated? In fact, the whole procedure would be worthless if it were not internally consistent, meaning that the total amount of energy stays the same no matter what sequence of changes takes place. This idea of internal consistency places a rather strong constraint on the law of conservation of energy.

Imagine a rotating wheel that has 10 joules of kinetic energy. If we extract this energy by stopping the wheel with a brake, we end up with 10 joules of thermal energy, as shown in Figure 7-13(a). Instead imagine that we convert the 10 joules of kinetic energy to electric energy by turning a generator to

Figure 7-13 The two different paths for the transformation of the kinetic energy of the wheel yield the same amount of thermal energy.

(a) Wheel with brake

(b) Electric charging of battery

Battery warms heating element

charge a battery, then use the battery's chemical energy to produce the electricity to run a heater (Figure 7-13[b]). We still get 10 joules of thermal energy. This result is reassuring. If the result were different, we would not have a valid scientific principle. Conservation of energy is a useful concept because the mathematical expressions for the various kinds of energy are independent of the specific situations in which they occur.

Conservation of energy is one of the most powerful principles in the physics world view, partly because of its generality; its only restriction is that the system be isolated, or closed. Conservation of energy can be applied to a wide range of problems in physics and in our everyday world.

Power

$$\text{power} = \frac{\text{energy converted}}{\text{time taken}}$$

In previous chapters we discussed how various quantities change with time. For example, speed is the change of position with time, and acceleration is the change of velocity with time. The change of energy with time is called **power.** Power P is equal to the amount of energy converted from one form to another ΔE divided by the time Δt during which this conversion takes place:

$$P = \frac{\Delta E}{\Delta t}$$

Power is measured in units of joules per second, a metric unit known as a **watt** (W). One watt of power would raise a 1-kilogram mass (with a weight of 10 newtons) a height of 0.1 meter each second.

The English unit for electric power is the watt, but a different English unit is used for mechanical power. A *horsepower* is defined as 550 foot-pounds per second. This definition was proposed by the Scottish inventor James Watt because he found that an average strong horse could perform 550 foot-pounds of work each second during an entire working day. One horsepower is equal to 746 watts.

We get electric energy from our local power company. But power is not an amount of energy. The energy transformed during a period of time is given by the power multiplied by the time during which this power is expended. Power companies usually bill us for the amount of energy we use, not the rate of consumption.

Everyday Physics | *Human Power*

A human can generate 1500 watts (2 horsepower) for short periods of time, such as in weight lifting. The maximum average human power for an 8-hour day is more like 75 watts (0.1 horsepower). Each person in a room generates thermal energy equivalent to that of a 75-watt lightbulb. That's one of the reasons why crowded rooms warm up!

Achieving human-powered flight has been a dream for centuries. People failed because it has been difficult to create enough aerodynamic lift with the power output that's humanly possible. In 1979 an American team led by Paul MacCready designed, built, and flew a "winged bicycle" that convincingly demonstrated that human-powered flight is possible. Professional cyclist Bryan Allen crossed the English Channel in the *Gossamer Albatross*. During this flight he generated an average power output of 190 watts (0.25 horsepower).

1. Two hundred people are crowded into a concert hall for a 3-hour performance. How much thermal energy does the audience generate in this time?

2. A laborer produces 65 watts of power moving furniture, and at the same time generates 75 watts of thermal energy. Find the total energy generated by the laborer in an 8-hour workday.

The human-powered aircraft *Gossamer Albatross*, which successfully flew across the English Channel in 1979.

© CORBIS

WORKING IT OUT | *Power*

A compact car traveling at 27 m/s (60 mph) on a level highway experiences a resistive force of about 300 N due to the air resistance and the friction of the tires with the road. Therefore, the car must obtain enough energy by burning gasoline to compensate for the work done by the frictional forces each second:

$$\Delta E = W = Fd = (300 \text{ N})(27 \text{ m}) = 8100 \text{ J}$$

and

$$P = \frac{\Delta E}{\Delta t} = \frac{8100 \text{ J}}{1 \text{ s}} = 8100 \text{ W}$$

This means that the power needed is 8100 W, or 8.1 kW. This is equivalent to a little less than 11 horsepower.

How much electrical energy does a motor running at 1000 W for 8 h require?

$$\Delta E = P\Delta t = (1000\text{ W})(8\text{ h}) = 8000\text{ Wh}$$

This is usually written as 8 kilowatt-hours (kWh). Although this doesn't look like an energy unit, it is—(energy/time) × time = energy.

The energy used by the motor in 1 h is

$$\Delta E = P\Delta t = (1000\text{ W})(1\text{ h}) = (1000\text{ J/s})(3600\text{ s}) = 3{,}600{,}000\text{ J}$$

In other words, 1 kWh = 3.6 million J.

Are *You* On the Bus?

Q: How much energy is required to leave a 75-W yard light on for 8 h?

A: $\Delta E = P\Delta t = (75\text{ W})(8\text{ h}) = 600\text{ Wh} = 0.6\text{ kWh}$

Summary

Energy is an abstract quantity that is conserved whenever a system is closed. Independent of the kinds of transformations that take place within a closed system, the total amount of energy remains the same.

Kinetic energy is the energy of motion and is defined to be one-half the mass times the speed squared, $KE = \frac{1}{2}mv^2$. If the kinetic energy before a collision is the same as that after the collision, the collision conserves kinetic energy and is said to be elastic. Kinetic energy is transformed into other forms of energy in inelastic collisions.

Work is equal to the product of the force in the direction of motion and the distance traveled—that is, $W = Fd$. If the force is perpendicular to the displacement of the object, or if the object does not move, no work is done by the force. The change in kinetic energy of an object is equal to the work done on the object.

The gravitational potential energy is equal to the work done by the force of gravity when an object falls through a height h—that is, $GPE = mgh$. The location for the zero value of gravitational potential energy is arbitrary because only the change in gravitational potential energy has any physical meaning. If gravity is the only force that does work on an object, the total mechanical energy (kinetic plus gravitational potential) is conserved. Therefore, any loss in gravitational potential energy shows up as a gain in the kinetic energy, and vice versa.

Other forms of potential energy can be associated with the electromagnetic and nuclear forces. However, a potential energy cannot be associated with the frictional force. This force transforms mechanical energy into thermal energy.

Power is the rate at which energy is transformed from one form into another, $P = \Delta E/\Delta t$. A kilowatt-hour is an energy unit because it is power multiplied by time.

CHAPTER 7 *Revisited*

We really don't know what energy *is*, but we know the many forms it takes and we have an accounting system for determining the amount of energy in a system. When we say that energy is conserved, we are acknowledging that although energy changes from one form into other forms, the total amount of energy in the system stays the same. Because of this, we can keep track of the energy.

Key Terms

elastic   A collision or interaction in which kinetic energy is conserved.

gravitational potential energy   The work that would be done by the force of gravity if an object fell from a particular point in space to the location assigned the value of zero, $GPE = mgh$.

inelastic   A collision or interaction in which kinetic energy is not conserved.

joule   The SI unit of energy and work, equal to 1 newton acting through a distance of 1 meter.

kinetic energy   The energy of motion, $KE = \frac{1}{2}mv^2$, where m is the object's mass and v is its speed.

mechanical energy   The sum of the kinetic energy and various potential energies, which may include the gravitational and the elastic potential energies.

power   The rate at which energy is converted from one form to another, $P = \Delta E/\Delta t$. Power is measured in joules per second, or watts.

watt   The SI unit of power, 1 joule per second.

work   The product of the force along the direction of motion and the distance moved, $W = Fd$. Work is measured in energy units, joules.

Questions and exercises are paired so that most odd-numbered are followed by a similar even-numbered.

Blue-numbered questions and exercises are answered in Appendix B.

🌶 indicates more challenging questions and exercises.

WebAssign Many Conceptual Questions and Exercises for this chapter may be assigned online at WebAssign.

Conceptual Questions

1.   Two identical cars traveling at the same speed collide head-on and come to rest in a mangled heap. At first glance it appears that energy is not conserved in this collision. However, like Dennis's mother in Richard Feynman's story at the beginning of the chapter, we find the energy "hidden" in many different forms. The initial kinetic energy is transformed into sound energy, thermal energy, and deformation energy. Where does the initial momentum of the system hide?

2.   Energy is always conserved for an isolated system. However, this chapter begins by stating that the energy of motion is only sometimes invariant, even for an isolated system. How can total energy be conserved while energy of motion is not?

3.   You have been asked to analyze a collision at a traffic inter-section. Will you be better off to begin your analysis using conservation of momentum or conservation of kinetic energy? Why?

4.   A sports car with a mass of 1200 kilograms travels down the road with a speed of 20 meters per second. Why can't we say that its momentum is smaller than its kinetic energy?

5.   If a system has zero kinetic energy, does it necessarily have zero momentum? Give an example to illustrate your answer.

6.   If a system has zero momentum, does it necessarily have zero kinetic energy? Give an example to illustrate your answer.

7.   Which has the greater kinetic energy, a supertanker berthed at a pier or a motorboat pulling a water skier? Why?

8.   Two pickup trucks with the same mass are driving on the freeway. If the Chevy has twice the speed of the Ford, does the Chevy have twice as much kinetic energy as the Ford? Explain your answer.

9.   Assume that a minivan has a mass of 2000 kilograms and a sports car has a mass of 1000 kilograms. If both vehicles are

traveling at the same speed, which vehicle has the higher kinetic energy? Why?

10. If the sports car in Question 9 has twice the speed of the minivan, which vehicle has the higher kinetic energy? Why?

11. A silver Camry is driving on the freeway at a constant 70 mph. Another Camry, identical but white, is on the on-ramp and is speeding up at a rate of 5 mph per second. Compare their kinetic energies at the instant the white Camry reaches 70 mph.

12. A jet is circling above the Salt Lake City airport at constant speed and elevation. How does the jet's kinetic energy change, if at all, as it circles? How does the jet's momentum change, if at all, as it circles?

13. What will happen if you pull two balls from the same side of the collision-ball apparatus in Figure 7-1 and let them go?

14. What will happen if the end balls of the collision-ball apparatus in Figure 7-1 are pulled out the same distances and let go?

15. A bowler lifts a bowling ball from the floor and places it on a rack. If you know the weight of the ball, what else must you know to calculate the work he does on the ball?

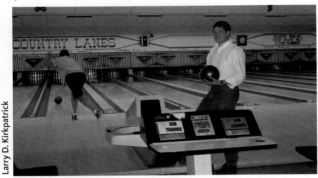

Larry D. Kirkpatrick

16. Bill's job is to lift bags of flour and place them in the back of a truck, which is parked right next to him. Sally is loading the same bags of flour into a similar truck that is located 10 meters away. Sally wants a raise because she says that she is doing more work than Bill. Does the physics definition of work support her claim?

17. An airplane is flying due south when it experiences a wind gust that exerts a force on the airplane acting due north. Will the kinetic energy of the airplane initially increase, decrease, or stay the same? Explain.

18. A bowling ball is rolling directly north along a smooth floor. Using a hammer, you tap the ball such that the force is directed east. How does the tap affect the ball's kinetic energy and its momentum?

19. In tryouts for the national bobsled team, each competing team pushes a sled along a level, smooth surface for 5 meters. One team brings a sled that is much lighter than all the others. Assuming that each team pushes with the same net force, compare the kinetic energy of the light sled to that of the others after 5 meters. Compare the momentum of the light sled to that of the others after 5 meters. (*Hint:* Think about the times involved.)

20. Suppose the rules were changed in Question 19 so that the teams pushed for a fixed time of 5 seconds rather than a fixed distance of 5 meters. Compare the momentum of the light sled to that of the others after 5 seconds. Compare the kinetic energy of the light sled to that of the others after 5 seconds. (*Hint:* Think about the distances involved.)

21. The tractor of an 18-wheeler performs work on its trailer when the truck is traveling along a level highway with a constant velocity. Why doesn't the trailer continually gain kinetic energy—that is, continually speed up?

22. The Chandra X-ray satellite orbits Earth in a highly elliptical orbit, as shown in the following figure. The force that Earth exerts on the satellite is always directed toward Earth. Is the satellite's kinetic energy increasing, decreasing, or staying the same at each of the points indicated? Explain your reasoning. (*Note:* The velocity vectors on the figure are *not* drawn to scale.)

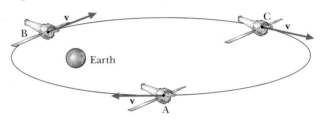

23. Two forces are used to move a block 2 meters across a level surface as shown in the following figure. Is the work done by force A greater than, equal to, or less than the work done by force B? (*Note:* The force vectors are drawn to scale.)

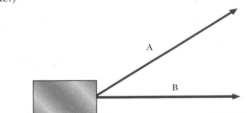

24. Two forces are used to move a block 2 meters across a level surface as shown in the following figure. Is the work done by force A greater than, equal to, or less than the work done by force B? (*Note:* The force vectors are drawn to scale.)

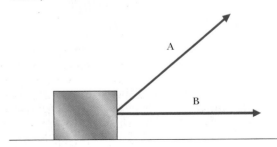

25. Two cars have different masses but the same kinetic energies. If the same frictional force is used to stop each car, which car, if either, will stop in the shorter distance?

26. Two cars have different masses but the same linear momenta. If the same frictional force is used to stop each car, which car, if either, will stop in the shorter distance? (*Hint:* Which car has the larger kinetic energy?)

27. We can use Newton's third law to demonstrate that the momentum lost by one object is gained by another. Can you also do this for kinetic energy? Explain why or why not.

28. Is it possible to change an object's momentum without changing its kinetic energy? What about the reverse situation?

29. Which of the following, if either, does more work: a force of 5 newtons acting through a distance of 5 meters or a force of 4 newtons acting through a distance of 7 meters?

30. Which of the following, if either, produces the larger change in kinetic energy: a force of 6 newtons acting through a distance of 3 meters or a force of 3 newtons acting through a distance of 6 meters?

31. On a test, the physics teacher asks, "What is the gravitational potential energy of a 10-newton ball resting on a shelf 2 meters above the floor?" Jamie got no points for responding that the answer was zero. What argument could Jamie use to convince the teacher that zero could be the right answer?

32. As the firefighter in the following picture slides down the pole, he initially speeds up to some terminal velocity, which he maintains until reaching the bottom. Gravitational potential energy is constantly decreasing during this process. What happens to the energy?

33. The kinetic energy of a free-falling ball is not conserved. Why is this not a violation of the law of conservation of mechanical energy?

34. Which of the following is conserved as a ball falls freely in a vacuum: the ball's kinetic energy, gravitational potential energy, momentum, or mechanical energy?

35. At which point in the swing of an ideal pendulum (ignoring friction) is the gravitational potential energy at its maximum? At which point is the kinetic energy at its maximum?

36. As an ideal pendulum (ignoring friction) swings from the bottom to the top of its arc, the string is always exerting a force on the ball. Why then is the gravitational potential energy at the top not greater than the kinetic energy at the bottom?

37. If we do not ignore frictional forces, what can you say about the height to which a pendulum bob swings on consecutive swings?

38. A block of wood, released from rest, loses 100 joules of gravitational potential energy as it slides down a ramp. If it has 90 joules of kinetic energy at the bottom of the ramp, what can you conclude?

39. Describe the energy transformations that occur as a satellite orbits Earth in a highly elliptical orbit.

40. Imagine a giant catapult that could hurl a spaceship to the Moon. Describe the energy transformations that would take place on such a journey.

41. Describe the energy changes that take place when you dribble a basketball.

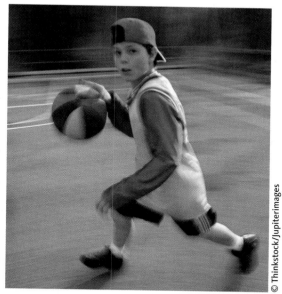

© Thinkstock/Jupiterimages

42. An elephant, an ant, and a professor jump from a lecture table. Assuming no frictional losses, which of the following could be said about their motion just before they hit the floor?
   a. They all have the same kinetic energy.
   b. They all started with the same gravitational potential energy.
   c. They will all experience the same force on stopping.
   d. They all have the same speed.

© Creasource/Corbis

43. Magnets mounted on top of air-hockey pucks allow the pucks to "collide" without touching each other. Describe the energy transformations that take place when one puck collides head-on with another.

44. An air-hockey puck is fastened to the table with a spring so that it oscillates back and forth on top of the table. Describe the energy transformations that take place. How would your description change if the puck were suspended from the ceiling by the spring?

45. Mountain highways often have emergency ramps for truckers whose brakes fail. Why are these covered with soft dirt or sand rather than pavement?

David Frazier Photolibrary

46. Athletes sometimes run along the beach to increase the effect of their workouts. Why is running on soft sand so tiring?

47. What happens to the chemical potential energy in the batteries used to power electric socks?

48. Describe the energy changes that take place when you stop your car using the brakes.

49. Why can we not associate a potential energy with the frictional force as we did with the gravitational force?

50. A physics textbook is launched up a rough incline with a kinetic energy of 200 joules. When the book comes momentarily to rest near the top of the incline, it has gained 180 joules of gravitational potential energy. How much kinetic energy will it have when it returns to the launch point?

51. The winch on your pickup truck is rated at 600 watts. Is it possible to do more than 600 joules of work with this winch? Explain.

52. A company advertises a new battery, which it claims is twice as powerful as anything else on the market. If you were to put this new battery in your flashlight, would you expect the light to be brighter? Would you expect it to last longer?

53. When you get your power bill, you are charged for the number of kilowatt-hours that you have used. Is the kilowatt-hour a unit of power or a unit of energy?

54. Valerie can do 1200 joules of work in 10 seconds. Brett can do 5000 joules of work in 50 seconds. Who is more powerful?

55. Which of the following is an energy unit: newton, kilowatt, kilogram-meter per second, or kilowatt-hour?

56. Which of the following is not a unit of energy: joule, newton-meter, kilowatt-hour, or watt?

Exercises

57. What is the kinetic energy of a 1400-kg sports car traveling down the road with a speed of 30 m/s?

58. What is the kinetic energy of an 87-kg sprinter running at 9 m/s?

59. In reviewing his lab book, a physics student finds the following description of a collision: "A 4-kg air-hockey puck with an initial speed of 6 m/s to the right collided head-on with a 1-kg puck moving to the left at the same speed. After the collision, both pucks traveled to the right, the 4-kg puck at 3 m/s and the 1-kg puck at 12 m/s." Is momentum conserved in this description? Is kinetic energy conserved in this description? Could this collision actually have taken place as described?

60. In reviewing her lab book, a physics student finds the following description of a collision: "A 4-kg air-hockey puck with an initial speed of 6 m/s to the right collided head-on with a 1-kg puck moving to the left at the same speed. After the collision, both pucks traveled to the right, the 4-kg puck at 2 m/s and the 1-kg puck at 10 m/s." Is momentum conserved in this description? Is kinetic energy conserved in this description? Could this collision actually have taken place as described?

61. A 4-kg toy car with a speed of 5 m/s collides head-on with a stationary 1-kg car. After the collision, the cars are locked together with a speed of 4 m/s. How much kinetic energy is lost in the collision?

62. A 3-kg toy car with a speed of 6 m/s collides head-on with a 2-kg car traveling in the opposite direction with a speed of 4 m/s. If the cars are locked together after the collision with a speed of 2 m/s, how much kinetic energy is lost?

63. A 0.5-kg air-hockey puck is initially at rest. What will its kinetic energy be after a net force of 0.8 N acts on it for a distance of 2 m?

64. A 20-N block lifted straight upward by a hand applying a force of 20 N has an initial kinetic energy of 16 J. If the block is lifted 1 m, how much work does the hand do? What is the block's final kinetic energy?

65. A radio-controlled car increases its kinetic energy from 4 J to 12 J over a distance of 2 m. What was the average net force on the car during this interval?

66. A toy car has a kinetic energy of 12 J. What is its kinetic energy after a frictional force of 0.6 N has acted on it for 5 m?

67. Earth, which orbits the Sun in an elliptical path, reaches its closest point to the Sun on about January 4 each year. Will the work done by the gravitational force of the Sun on Earth be positive, negative, or zero over the next 6 months? Over the next year?

68. How much work is performed by the gravitational force on a satellite in near-Earth orbit during one revolution?

69. How much work does a 55-kg person do against gravity in walking up a trail that gains 720 m in elevation?

70. A woman with a mass of 65 kg climbs a set of stairs that are 3 m high. How much gravitational potential energy does she gain?

71. A 145-g baseball is thrown straight upward with kinetic energy 8.7 J. When the ball has risen 6 m, find (a) the work done by gravity, (b) the ball's kinetic energy, and (c) the ball's speed.

72. What is the gravitational potential energy of a ball with a weight of 50 N when it is sitting on a shelf 1.5 m above the floor? What assumption do you need to make to get your answer?

73. If a 0.5-kg ball is dropped from a height of 6 m, what is its kinetic energy when it hits the ground?

74. A 2-kg block is released from rest at the top of a 20-m-long frictionless ramp that is 4 m high. At the same time, an identical block is released next to the ramp so that it drops straight down the same 4 m. What are the values for each of the following for the blocks just before they reach ground level? Which quantities are the same for the two blocks?
    a. gravitational potential energy
    b. kinetic energy
    c. speed
    d. momentum

75. A 1200-kg frictionless roller coaster starts from rest at a height of 24 m. What is its kinetic energy when it goes over a hill that is 12 m high?

76. You reach out a second-story window that is 5 m above the sidewalk and throw a 0.1-kg ball straight upward with 6 J of kinetic energy.
    a. What is the ball's gravitational potential energy when it is released?
    b. What is the ball's gravitational potential energy just before hitting the sidewalk?
    c. What is the ball's kinetic energy just before hitting the sidewalk?
    d. How would the answer to part (c) change if the ball had initially been thrown straight down with 6 J of kinetic energy?

77. What average power does a weight lifter need to lift 300 lb a distance of 4 ft in 0.8 s?

78. If an 80-kg sprinter can accelerate from a standing start to a speed of 10 m/s in 3 s, what average power is generated?

79. If a CD player uses electricity at a rate of 15 W, how much energy does it use during an 8-h day?

80. If a hair dryer is rated at 1500 W, how much energy does it require in 5 min?

11

Structure of Matter

▶ Hot air allows balloonists to enjoy the scenery from a high vantage point. But what do the macroscopic properties of gases—volume, pressure, and temperature—tell us about the existence of atoms and molecules and their microscopic properties of size, mass, and speed?

(See page 235 for the answer to this question.)

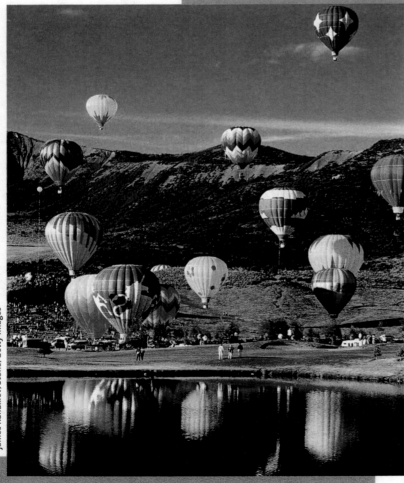

James Randklev/Stone/Getty Images

Hot-air balloons are a beautiful illustration of the ideal gas law.

WHEN we talk about the properties of objects, we usually think about their bulk, or **macroscopic**, properties such as size, shape, mass, color, surface texture, and temperature. For instance, a gas has mass, occupies a volume, exerts a pressure on its surroundings, and has a temperature. But a gas is composed of particles that have their own characteristics, such as velocity, momentum, and kinetic energy. These are the **microscopic** properties of the gas. It seems reasonable that connections should exist between these macroscopic and microscopic properties. At first glance, you may assume that the macroscopic properties are just the sum, or maybe the average, of the microscopic ones; however, the connections provided by nature are much more interesting.

We could begin our search for the connections between the macroscopic and the microscopic by examining a surface with a conventional microscope and discovering that rather than being smooth, as it appears to the naked eye, the surface has some texture. The most powerful electron microscope shows even more structure than an optical microscope reveals. But until recently, instruments could not show us the basic underlying structure of things. To gain an understanding of matter at levels beyond which they could observe directly, scientists constructed models of various possible microscopic structures to explain their macroscopic properties.

Building Models

By the middle of the 19th century, the body of chemical knowledge had pretty well established the existence of **atoms** as the basic building blocks of matter. All the evidence was indirect but was sufficient to create some idea of how atoms combine to form various substances. This description of matter as being composed of atoms is a model. It is not a model in the sense of a scale model, such as a model railroad or an architectural model of a building, but rather a theory or mental picture.

To illustrate this concept of a model, suppose someone gives you a tin can without a label and asks you to form a mental picture of what may be inside. Suppose you hear and feel something sloshing when you shake the can. You guess that the can contains a liquid. Your model for the contents is that of a liquid, but you do not know whether this model actually matches the contents. For example, it is possible (but unlikely) that the can contains an electronic device that imitates sloshing sounds. However, you can use your knowledge of liquids to test your model. You know, for example, that liquids freeze. Therefore, your model predicts that cooling the can would stop the sloshing sounds. In this way you can use your model to help you learn more about the contents. The things you can say, however, are limited. For example, there is no way to determine the color of the liquid or its taste or smell.

Sometimes a model takes the form of an analogy. For example, the flow of electricity is often described in terms of the flow of water through pipes. Similarly, little, solid spheres might be analogous to atoms. We could use the analogy to develop rules for combining these spheres and learn something about how atoms combine to form molecules. However, it is important to distinguish between a collection of atoms that we cannot see and the spheres that we can see and manipulate. Sooner or later the analogy breaks down because electricity is not water and atoms are not tiny spheres.

Sometimes a model takes the form of a mathematical equation. Physicists have developed equations that describe the structure and behavior of atoms. Although mathematical models can be abstract, they can also be accurate descriptions of the way nature behaves. Suppose that in examining a tin can you hear a sliding sound followed by a clunk whenever you tilt the can. You

(a)

Gerald F. Wheeler

(b)

Gerald F. Wheeler

(c)

Gerald F. Wheeler

An electron micrograph of (a) a fly's head (27×) and its eye at (b) 122× and (c) 1240×.

predict that the can contains a metal rod. You may devise an equation for the length of time it takes a rod of a certain length to slide along the wall before hitting an end. This mathematical model would allow you to predict the time delays for different-length rods.

Regardless of its form, a model should summarize and account for the known data. It must agree with the way nature behaves. And to be useful, it must also be able to make predictions about new situations. The model for the contents of the tin can allows you to predict that liquid will flow out of the can if it is punctured. Our model for the structure of matter must allow us to make predictions that can be tested by experiment. If the predictions are borne out, they strengthen our belief in the model. If they are not borne out, we must modify our model or abandon it and invent a new one.

Early Chemistry

Experimental techniques, as opposed to careful observation, became of increasing interest to scientists around the end of the first millennium. They flourished especially in the Arabic-speaking world, where so many protochemical ideas found a home. We now unconsciously use many chemical terms—*alcohol, alchemy, alembic,* and so on—that derive from Arabic.

Alchemy was the search for the magic formula for perpetual life—the elixir (note again the Arabic)—and for the secret of how to make gold. Alchemists set the stage for our modern atomic world view, but their contributions are often overlooked. We hear only that these strange pseudoscientists spent all their time foolishly trying to turn lead into gold. But they weren't so strange. In the Aristotelian world view that all matter was composed of various proportions of four elements—earth, water, fire, and air—and that matter was continuous, it made sense that different materials could be made by changing the proportions of the four elementary substances. Making solutions and heating materials were natural beginnings because water and fire were two of the elements. A cooked egg was not the same as a raw one. Adding fire changed its characteristics. Adding fire to other materials released some air even though the original material gave no hint of containing air. Adding water to sugar changed it from a white solid to a clear liquid. This kind of experimentation—heating, dissolving in water, stirring, blending, grinding, and so on—was the work of the alchemists.

But the alchemists' role in the building of our world view was more than just finding ways of manipulating materials. By showing that the common materials were not all that existed, alchemists set the stage for the experimenters who followed. These new experimenters began with a different background of knowledge, a different common sense about the world, and a different expectation of what was possible. They went beyond cataloging ways of manipulating substances and began considering the structures of matter that could be causing these results—they built models. Although it is difficult to pinpoint the transition from alchemy to chemistry, the early chemists were part of the Newtonian age and had a different way of looking at the world. Isaac Newton himself spent many arduous hours in a little wooden laboratory outside his rooms at Trinity College, Cambridge, performing alchemical experiments.

By the latter half of the 17th century, the existence of the four Aristotelian elements was in doubt. They were of little or no help in making sense out of the chemical data that had been accumulated. However, the idea that all matter was composed of some basic building blocks was so appealing that it persisted. This belief fueled the development of our modern atomic model of matter.

The simplest, or most elementary, substances were known as **elements.** These elements could be combined to form more complex substances, the

This early 19th-century Japanese woodcut shows an artisan extracting copper from its ore.

Table 11–1	*Lavoisier's List of the Elements (1789)*		
Lavoisier's Name	**Modern English Name**	**Lavoisier's Name**	**Modern English Name**
Lumiere	Light*	Etain	Tin
Calorique	Heat*	Fer	Iron
Oxygene	Oxygen	Manganese	Manganese
Azote	Nitrogen	Mercure	Mercury
Hydrogene	Hydrogen	Molybdene	Molybdenum
Soufre	Sulfur	Nickel	Nickel
Phosphore	Phosphorus	Or	Gold
Carbone	Carbon	Platine	Platinum
Radical muriatique†	—	Plomb	Lead
Radical fluorique†	—	Tungstene	Tungsten
Radical boracique†	—	Zinc	Zinc
Antimoine	Antimony	Chaux	Calcium oxide (lime)‡
Argent	Silver	Magnésie	Magnesium oxide‡
Arsenic	Arsenic	Baryte	Barium oxide‡
Bismuth	Bismuth	Alumina	Aluminum oxide‡
Cobalt	Cobalt	Silice	Silicon dioxide (sand)‡
Cuivre	Copper		

*Not elements.
†No elements with these properties have ever been found.
‡These "elements" are really compounds.

compounds. By the 1780s, French chemist and physicist Antoine Lavoisier and his contemporaries had enough data to draw up a tentative list of elements (Table 11-1). Something was called an element if it could not be broken down into simpler substances. (The modern periodic table of the elements is printed on the inside front cover of this book.)

A good example of an incorrectly identified element is water. It was not known until the end of the 18th century that water is a compound of the elements hydrogen and oxygen. Hydrogen had been crudely separated during the early 16th century, but oxygen was not discovered until 1774. When a flame is put into a test tube of hydrogen, it "pops." One day while popping hydrogen, an experimenter noticed some clear liquid in the tube. This liquid was water. This was the first hint that water was not an element. The actual decomposition of water was accomplished at the end of the 18th century by a technique known as *electrolysis,* in which an electric current passing through a liquid or molten compound breaks it down into its respective elements.

The radically different properties of elements and their compounds are still striking in the modern world view. Hydrogen, an explosive gas, and oxygen, the element required for all burning, combine to form a compound that is great for putting out fires! Similarly, sodium, a very reactive metal that must be kept in oil to keep it from reacting violently with moisture in the air, combines with chlorine, a poisonous gas, to form a compound that tastes great on mashed potatoes—common table salt!

© Cengage Learning/Charles D. Winters

The electrolysis of water breaks up the water molecules into hydrogen and oxygen. Notice that the volume of hydrogen collected in the right-hand tube is twice that of the oxygen collected in the left-hand tube.

Q: Could these early chemists know for sure that a substance was an element?

A: They never really knew whether the substance was an element or whether they had not yet figured out how to break it down.

Are You On the Bus?

Chemical Evidence of Atoms

Another important aspect of elements and compounds was discovered around 1800. Suppose a particular compound is made from two elements and when you combine 10 grams of the first element with 5 grams of the second, you get 12 grams of the compound and have 3 grams of the second element remaining. If you now repeat the experiment, only this time adding 10 grams of each element, you still get 12 grams of the compound, but now have 8 grams of the second element remaining. This result was exciting. It meant that, rather than containing some random mixture of the two elements, the compound had a definite ratio of their masses. This principle is known as the **law of definite proportions.**

Are You On the Bus?

Q: How much of the compound would you get if you added only 1 gram of the second element?

A: Because 10 grams of the first element require 2 grams of the second, 1 gram of the second will combine with 5 grams of the first. The total mass of the compound is just the sum of the masses of the two elements, so 6 grams of the compound will be formed.

This law is difficult to explain using the Aristotelian model, which viewed matter as a continuous, smooth substance. In the continuum model, one would expect a range of masses in which the elements could combine to form the compound. An atomic model, on the other hand, provides a simple explanation: The atoms of one element can combine with atoms of another element to form **molecules** of the compound. It may be that one atom of the first element combines with one atom of the second to form one molecule of the compound, which contains two atoms. Or it may be that one atom of the first element combines with two atoms of the second element. In any case the ratio of the masses of the combining elements has a definite value. (Note that the value of the ratio doesn't change with differing amounts. The mass ratio of 10 baseballs and 10 basketballs is the same as that of 1 baseball and 1 basketball. It doesn't matter how many balls we have as long as there is one baseball for *each* basketball.)

The actual way in which the elements combined and what caused them to always combine in the same way was unknown. English mathematician and physicist John Dalton hypothesized that the elements might have hooks (Figure 11-1) that control how many of one atom combine with another. Dalton's hooks can be literal or metaphorical; the actual mechanism is not important.

Figure 11-1 Dalton's atomic model used hooks to explain the law of definite proportions.

The essential point of his model was that different atoms have different capacities for attaching to other atoms. Regardless of the visual model we use, atoms combine in a definite ratio to form molecules. One atom of chlorine combines with one atom of sodium to form salt. The ratio in salt is always one atom to one atom.

In retrospect it may seem that the law of definite proportions was a minor step and that it should have been obvious once mass measurements were made. However, seeing this relationship was difficult because some processes did not obey this law. For instance, *any* amount of sugar (up to some maximum) dissolves completely in water. One breakthrough came when it was recognized that this process was distinctly different. The sugar–water solution was not a compound with its own set of properties but simply a mixture of the two substances. Mixtures had to be recognized as different from compounds and eliminated from the discussion.

Another complication occurred because some elements can form more than one compound. Carbon atoms, for example, could combine with one or two oxygen atoms to form two compounds with different characteristics. When this happened in the same experiment, the final product was not a pure compound but a mixture of compounds. This result yielded a range of mass ratios and was quite confusing until chemists were able to analyze the compounds separately.

Fortunately, the atomic model makes predictions about situations in which two elements form more than one compound. Imagine for a moment that atoms can be represented by nuts and bolts. Suppose a hypothetical molecule of one compound consists of one nut and one bolt and another consists of two nuts and one bolt. Because there are twice as many nuts for each bolt in the second compound (Figure 11-2), it has a mass ratio of nuts to bolts that is twice that of the first compound. This prediction was confirmed for actual compounds and provided further evidence for the existence of atoms.

(a)   (b)

Figure 11-2 A simple model of two compounds: (a) "bolt-mononut" and (b) "bolt-dinut." The ratio of the mass of nuts to the mass of bolts is twice as large for "bolt-dinut."

Masses and Sizes of Atoms

Extended presentation available in the *Problem Solving* supplement

Even with their new information, the 18th-century chemists did not know how many atoms of each type it took to make a specific molecule. Was water composed of one atom of oxygen and one atom of hydrogen, or was it one atom of oxygen and two atoms of hydrogen, or two of oxygen and one of hydrogen? All that was known was that 8 grams of oxygen combined with 1 gram of hydrogen. These early chemists needed to find a way to establish the relative masses of atoms.

The next piece of evidence was an observation made when gaseous elements were combined; the gases combined in definite *volume* ratios when their temperatures and pressures were the same. This statement was not surprising except that the volume ratios were always simple fractions. For example, 1 liter of hydrogen combines with 1 liter of chlorine (a ratio of 1 to 1), 1 liter of oxygen combines with 2 liters of hydrogen (a ratio of 1 to 2), 1 liter of nitrogen combines with 3 liters of hydrogen (a ratio of 1 to 3), and so on (Figure 11-3).

It was tempting to propose an equally simple underlying rule to explain these observations. Italian physicist Amedeo Avogadro suggested that under identical conditions, each liter of any gas contains the same number of molecules. Although it took more than 50 years for this hypothesis to be accepted, it was the key to unraveling the question of the number of atoms in molecules.

Once the number of atoms in each molecule was known, the data on the mass ratios could be used to calculate the relative masses of different atoms. For example, an oxygen atom has about 16 times the mass of a hydrogen atom.

Figure 11-3 Gases combine completely to form compounds when the ratios of their volumes are equal to the ratios of small whole numbers. One liter of nitrogen combines completely with 3 liters of hydrogen to form 2 liters of ammonia.

Are You On the Bus?

Q: Given that oxygen and hydrogen gases are both composed of molecules with two atoms each, how many atoms of oxygen and hydrogen combine to form water?

A: The observation that 2 liters of hydrogen gas combine with 1 liter of oxygen means that there are two hydrogen molecules for each oxygen molecule. Therefore, there are four hydrogen atoms for every two oxygen atoms. The simplest case would be for these to form two water molecules with two hydrogen atoms and one oxygen atom in each water molecule. This is confirmed by the observation that 2 liters of water vapor are produced.

To avoid the use of ratios, a mass scale was invented by choosing a value for one of the elements. An obvious choice was to assign the value of 1 to hydrogen because it is the lightest element. However, setting the value of carbon equal to 12 **atomic mass units** (amu) makes the relative masses of most elements close to whole-number values. These values are known as **atomic masses** and keep the value for hydrogen very close to 1. Even though the values for the actual atomic masses are now known, it is still convenient to use the relative atomic masses.

Are You On the Bus?

Q: What is the atomic mass of carbon dioxide, a gas formed by combining two oxygen atoms with each carbon atom?

A: We have 12 atomic mass units for the carbon atom and 16 atomic mass units for each oxygen atom. Therefore, 12 atomic mass units + 32 atomic mass units = 44 atomic mass units.

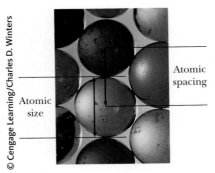
© Cengage Learning/Charles D. Winters

Atomic spacing

Atomic size

Figure 11-4 Assuming that the atoms are "touching" like marbles, the spacing between their centers is the same as their diameters.

The problem of determining the masses and diameters of individual atoms required the determination of the number of atoms in a given amount of material. Diffraction experiments, like those described in Chapter 19 but using X rays, determined the distance between individual atoms in solids to be about 10^{-10} meter, one 10-billionth of a meter. If we assume that atoms in a solid can be represented by marbles like those in Figure 11-4, the diameter of an atom is about equal to their spacing. This means that it would take 10 billion atoms to make a line 1 meter long. Stated another way, if we imagine expanding a baseball to the size of Earth, the individual atoms of the ball would only be the size of grapes!

A useful quantity of matter for our purposes is the **mole.** If the mass of the molecule is some number of atomic mass units, 1 mole of the substance is this same number of grams. For example, 1 mole of carbon is 12 grams. Further experiments showed that 1 mole of any substance contains the same number of molecules—namely, 6.02×10^{23} molecules, a number known as **Avogadro's number.** With this number we can calculate the size of the atomic mass unit in terms of kilograms. Because 12 grams of carbon contain Avogadro's number of carbon atoms, the mass of one atom is

$$m_{carbon} = \frac{12 \text{ g}}{6.02 \times 10^{23} \text{ molecules}} = 2 \times 10^{-23} \text{ g/molecule}$$

Because one carbon atom also has a mass of 12 atomic mass units, we obtain

$$\frac{2 \times 10^{-23} \text{ g}}{12 \text{ amu}} = 1.66 \times 10^{-24} \text{ g/amu}$$

Therefore, 1 atomic mass unit equals 1.66×10^{-27} kilogram, a mass so small that it is hard to imagine. This is the approximate mass of one hydrogen atom. The most massive atoms are about 260 times this value.

FLAWED REASONING

Two students are arguing after class about ideal gases.

Hyrum: "If two ideal gases are both at the same temperature and pressure, then equal volumes will contain equal numbers of atoms. This means that 1 mole of ammonia (NH_3) would take up twice as much volume as 1 mole of nitrogen (N_2) because each ammonia molecule has four atoms and each nitrogen molecule has only two."

Brielle: "No, equal volumes will contain equal numbers of *molecules*, not *atoms*. One mole of ammonia would contain the same number of molecules as 1 mole of nitrogen—namely, Avogadro's number—so they would take up the same volume."

Do you agree with either of these students?

ANSWER Brielle was paying attention in class. Avogadro found that the number of molecules determined the volume of a gas for a given temperature and pressure. The number of atoms in each molecule does not matter.

The Ideal Gas Model

Many macroscopic properties of materials can be understood from the atomic model of matter. In many situations the behavior of real gases is closely approximated by an **ideal gas.** The gas is assumed to be composed of an enormous number of tiny particles separated by relatively large distances. These particles are assumed to have no internal structure and to be indestructible. They also do not interact with each other except when they collide, and then they undergo elastic collisions much like air-hockey pucks. Although this model may not seem realistic, it follows in the spirit of Galileo in trying to get at essential features. Later we can add the complications of real gases.

For this model to have any validity, it must describe the macroscopic behavior of gases. For instance, we know that gases are easily compressed. This makes sense; the model says that the distance between particles is much greater than the particle size, and they don't interact at a distance. There is, then, a lot of space in the gas, and it should be easily compressed. This aspect of the model also accounts for the low mass-to-volume ratio of gases.

Because a gas completely fills any container and the particles are far from one another, the particles must be in continual motion. Is there any other evidence that the particles are continually moving? We may ask, "Is the air in the

Pollen grains suspended in a liquid exhibit continual, erratic motion known as Brownian motion.

room moving even with all the doors and windows closed to eliminate drafts?" The fact that you can detect an open perfume bottle across the room indicates that some of the perfume particles have moved through the air to your nose.

More direct evidence for the motion of particles in matter was observed in 1827 by Scottish botanist Robert Brown. To view pollen under a microscope without it blowing away, Brown mixed the pollen with water. He discovered that the pollen grains were constantly jiggling. Brown initially thought that the pollen might be alive and moving erratically on its own. However, he observed the same kind of motion with inanimate objects as well.

Brownian motion is not restricted to liquids. Observation of smoke under a microscope shows that the smoke particles have the same erratic motion. This motion never ceases. If the pollen and water are kept in a sealed container and put on a shelf, you would still observe the motion years later. The particles are in continual motion.

It was 78 years before Brownian motion was rigorously explained. Albert Einstein demonstrated mathematically that the erratic motion was due to collisions between water molecules and pollen grains. The number and direction of the collisions occurring at any time is a statistical process. When the collisions on opposite sides have equal impulses, the grains are not accelerated. But when more collisions occur on one side, the pollen experiences an abrupt acceleration that is observed as Brownian motion.

Pressure

Let's take a look at one of the macroscopic properties of an ideal gas that is a result of the atomic motions. **Pressure** is the force exerted on a surface divided by the area of the surface—that is, the force per unit area:

$$\text{pressure} = \frac{\text{force}}{\text{area}} \blacktriangleright$$

$$P = \frac{F}{A}$$

This definition is not restricted to gases and liquids. For instance, if a crate weighs 6000 newtons and its bottom surface has an area of 2 square meters, what pressure does it exert on the floor under the crate?

$$P = \frac{F}{A} = \frac{6000 \text{ N}}{2 \text{ m}^2} = 3000 \text{ N/m}^2$$

Therefore, the pressure is 3000 newtons per square meter. The SI unit of pressure [newton per square meter (N/m^2)] is called a pascal (Pa). Pressure in the U.S. customary system is often measured in pounds per square inch (psi) or atmospheres (atm), where 1 atmosphere is equal to 101 kilopascals or 14.7 pounds per square inch.

Are You On the Bus?

Q: Susan asks politely whether it would be all right with you if she pushes on your arm with a force of 5 newtons (about 1 pound). Should you let her?

A: That depends. If she pushes on your arm with the palm of her hand, you will hardly notice a force of 5 newtons. If, on the other hand, she pushes on your arm with a sharp hatpin, you will definitely notice! The damage to your arm does not depend on the force but on the pressure.

Imagine a cubic container of gas in which particles are continually moving around and colliding with each other and with the walls (Figure 11-5). In each collision with a wall, the particle reverses its direction. Assume a head-on collision as in the case of the yellow particle in Figure 11-5. (If the collision is a glancing blow, only the component of the velocity perpendicular to the wall would be reversed.) This means that its momentum is also reversed. The change in momentum means that there must be an impulse on the particle and an equal and opposite impulse on the wall (Chapter 6). Our model assumes that an enormous number of particles strike the wall. The average of an enormous number of impulses produces a steady force on the wall that we experience as the pressure of the gas.

We can use this application of the ideal gas model to make predictions that can be tested. Suppose, for example, that we shrink the volume of the container. This means that the particles have less distance to travel between collisions with the walls and should strike the walls more frequently, increasing the pressure. Therefore, decreasing the volume increases the pressure, provided that the average speeds of the molecules do not change. Similarly, if we increase the number of particles in the container, we expect the pressure to increase because there will be more frequent collisions with the walls.

Figure 11-5 Gas particles are continually colliding with each other and with the walls of the container.

Atomic Speeds and Temperature

Presumably, the atomic particles making up a gas have a range of speeds due to their collisions with the walls and with each other. The distribution of these speeds can be calculated from the ideal gas model and a connection made with temperature. Therefore, a direct measurement of these speeds would provide additional support for the model. This is not an easy task. Imagine trying to measure the speeds of a large group of invisible particles. One needs to devise a way of starting a race and recording the order of the finishers.

One creative approach led to a successful experiment in 1920. The gas leaves a heated vessel and passes through a series of small openings that select only those particles going in a particular direction [Figure 11-6(a)]. Some of these particles enter a small opening in a rapidly rotating drum, as shown in Figure 11-6(b). This arrangement guarantees that a group of particles start across the drum at the same time. Particles with different speeds take different times to cross the drum and arrive on the opposite wall after the drum has rotated by different amounts. The locations of the particles are recorded by a film of sensitive material attached to the inside of the drum [Figure 11-6(d)]. A drawing of the film's record is shown in Figure 11-7(a). A graph of the number of particles versus their position along the film is shown in Figure 11-7(b).

Most of the particles have speeds near the average speed, but some move very slowly and some very rapidly. The average speed is typically about 500 meters per second, which means that an average gas particle could travel the length of five football fields in a single second. This high value may seem counter to your experience. If the particles travel with this speed, why does it take several minutes to detect the opening of a perfume bottle on the other side of the room? This delay is due to collisions between the gas particles. On average a gas particle travels a distance of only 0.0002 millimeter before it collides with another gas particle. (This distance is about 1000 particle diameters.) Each particle makes approximately 2 billion collisions per second. During these collisions the particles can radically change directions, resulting in zigzag paths. So although their average speed is quite fast, it takes them a long time to cross the room because they travel enormous distances.

When the speeds of the gas particles are measured at different temperatures, something interesting is found. As the temperature of the gas increases, the speeds of the particles also increase. The distributions of speeds for three temperatures are given in Figure 11-8. The calculations based on the model

Atomic particles in air travel in zigzag paths because of numerous collisions with air molecules.

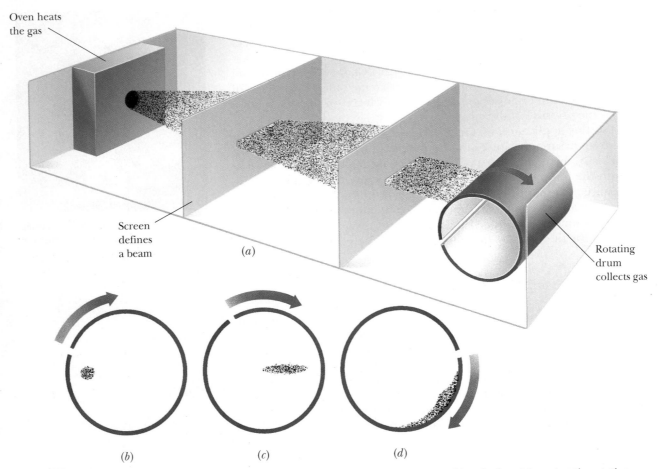

Figure 11-6 (a) An apparatus for measuring the speeds of atomic particles in a gas. (b) A small bunch of particles enters the rotating drum through the narrow slit. (c) The particles spread out as they move across the drum because of their different speeds. (d) The particles are deposited on a sensitive film at locations determined by how much the drum rotates before they arrive at the opposite wall.

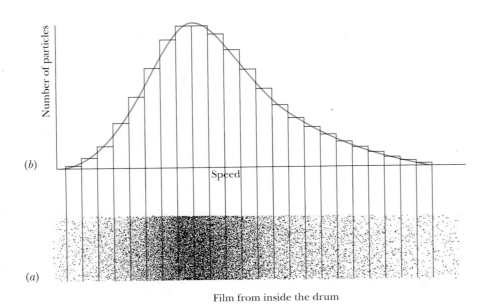

Film from inside the drum

Figure 11-7 (a) The distribution of atomic particles as recorded by the film along the circumference of the drum. (b) The distribution of atomic speeds calculated from this experiment.

Figure 11-8 The distribution of atomic speeds changes as the temperature of the gas changes.

agree with these results. A relationship can be derived that connects temperature, a macroscopic property, with the average kinetic energy of the gas particles, a microscopic property. However, the simplicity of this relationship is apparent only with a particular temperature scale.

Temperature

We generally associate temperature with our feelings of hot and cold; however, our subjective feelings of hot and cold are not very accurate. Although we can usually say which of two objects is hotter, we can't state just how hot something is. To do this we must be able to assign numbers to various temperatures.

Assigning numbers to various temperatures turns out to be a difficult task that has occupied some of the greatest scientific minds. Just as it is not possible to define time in a simple way, it is not possible to define temperature in a simple way. In Chapters 12 through 14, we will return to the subject of temperature a number of times with the aim of helping you incorporate the concept of temperature into your world view. We start with a familiar concept: measuring temperature with a thermometer.

Galileo was the first person to develop a thermometer. He observed that some of an object's properties change when its temperature changes. For example, with only a few exceptions, when an object's temperature goes up, it expands. Galileo's thermometer (Figure 11-9) was an inverted flask with a little water in its long neck. As the enclosed air got hotter, it expanded and forced the water down the flask's neck. Conversely, the air contracted on cooling, and the water rose. Galileo completed his thermometer by marking a scale on the neck of the flask.

Unfortunately, the water level also changed when atmospheric pressure changed. The alcohol-in-glass thermometer, which is still popular today, replaced Galileo's thermometer. The column is sealed so that the rise and fall of the alcohol is due to its change in volume and not the atmospheric pressure. The change in height is amplified by adding a bulb to the bottom of the column, as shown in Figure 11-10. When the temperature rises, the larger volume in the bulb expands into the narrow tube, making the expansion much more obvious.

In 1701 Newton proposed a method for standardizing the scales on thermometers. He put the thermometer in a mixture of ice and water, waited for the level of the alcohol to stop changing, and marked this level as zero. He used the temperature of the human body as a second fixed temperature, which he called 12. The scale was then marked off into 12 equal divisions, or degrees.

Shortly after this, German physicist Gabriel Fahrenheit suggested that the zero point correspond to the temperature of a mixture of ice and salt. Because this was the lowest temperature producible in the laboratory at that time, it avoided the use of negative numbers for temperatures. The original 12 degrees were later divided into eighths and renumbered so that body temperature became 96 degrees.

It is important that the fixed temperatures be reliably reproducible in different laboratories. Unfortunately, neither of Fahrenheit's reference temperatures could be reproduced with sufficient accuracy. Therefore, the reference temperatures were changed to those of the freezing and boiling points of pure water at standard atmospheric pressure. To get the best overall agreement with the previous scale, these temperatures were defined to be 32°F and 212°F, respectively. This is how we ended up with such strange numbers on the **Fahrenheit temperature scale.** On this scale, normal body temperature is 98.6°F.

At the time the metric system was adopted, a new temperature scale was defined with the freezing and boiling points as 0°C and 100°C. The name of this centigrade (or 100-point) scale was changed to the **Celsius temperature**

Figure 11-9 The height of the liquid in Galileo's first thermometer indicated changes in temperature.

Figure 11-10 A comparison of the Fahrenheit and Celsius temperature scales.

Courtesy of Patrick Harman

The Fahrenheit and Celsius scales give the same reading at −40°.

scale in 1948 in honor of Swedish astronomer Anders Celsius, who devised the scale. A comparison of the Fahrenheit and Celsius scales is given in Figure 11-10. This figure can be used to convert temperatures from one scale to the other. If the temperature outside is −40° (which sometimes happens in Montana) it does not matter which temperature scale you are using. The Celsius and Fahrenheit scales yield the same value at this temperature.

Q: What are room temperature (68°F) and body temperature (98.6°F) on the Celsius scale?

A: Using Figure 11-10, we see that room temperature is about 20°C and body temperature is about 37°C.

Assume that we have a quantity of ideal gas in a special container designed to always maintain the pressure of the gas at some constant low value. When the volume of the gas is measured at a variety of temperatures, we obtain the graph shown in Figure 11-11. If the line on the graph is extended down to the left, we find that the volume goes to zero at a temperature of −273°C

Figure 11-11 When the graph of volume versus temperature of an ideal gas is extrapolated to zero volume, the temperature scale reads −273°C.

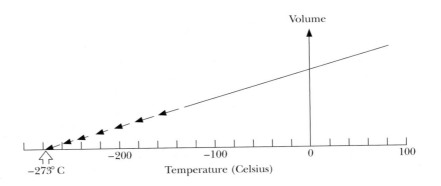

(−459°F). Although we could not actually do this experiment with a real gas, this very low temperature arises in several theoretical considerations and is the basis for a new, more fundamental temperature scale.

The **Kelvin temperature scale,** also known as the **absolute temperature scale,** has its zero at −273°C and the same-size degree marks as the Celsius scale. The difference between the Celsius and Kelvin scales is that temperatures are 273 degrees higher on the Kelvin scale. Water freezes at 273 K and boils at 373 K. (Notice that the degree symbol is dropped from this scale. The freezing point of water is read "273 kelvin," or "273 kay.") The Kelvin scale is named for British physicist William Thomson, who is more commonly known as Lord Kelvin.

◀ K = C + 273

Q: What is normal body temperature (37°C) on the Kelvin scale?

A: Body temperature is 37 + 273 = 310 K.

It would seem that all temperature scales are equivalent and which one we use would be a matter of history and custom. It is true that these scales are equivalent because conversions can be made between them. However, the absolute temperature scale has a greater simplicity for expressing physical relationships. In particular, the relationship between the volume and temperature of an ideal gas is greatly simplified using absolute temperatures. *The volume of an ideal gas at constant pressure is proportional to the absolute temperature.* This means that if the absolute temperature is doubled while keeping the pressure fixed, the volume of the gas doubles.

The volume of an ideal gas at constant pressure can be used as a thermometer. All we need to do to establish the temperature scale is measure the volume at one fixed temperature. Of course, thermometers must be made of real gases. But real gases behave like the ideal gas if the pressure is kept low and the temperature is well above the temperature at which the gas liquefies.

This new scale also connects the microscopic property of atomic speeds and the macroscopic property of temperature. *The absolute temperature is directly proportional to the average kinetic energy of the gas particles.* This means that if we double the average kinetic energy of the particles, the absolute temperature of a gas doubles. Remember, however, that the average speed of the gas particles does not double, because the kinetic energy depends on the square of the speed (Chapter 7).

◀ T is proportional to KE_{ave}

FLAWED REASONING

When you wake, the temperature outside is 40°F, but by noon it is 80°F. **Why is it not reasonable to say that the temperature doubled?**

ANSWER The zero point for the Fahrenheit scale was arbitrarily chosen as the temperature of a mixture of ice and salt. If this zero point had been chosen differently—say, 30 degrees higher—the temperature during the morning would have changed from 10°F to 50°F, an increase of a factor of 5! Clearly, we can attach no physical significance to the doubling of the temperature reading on the Fahrenheit scale (or the Celsius scale). If, on the other hand, the temperature of a gas doubles on the Kelvin scale, we can say that the average kinetic energy of the gas particles has also doubled.

WORKING IT OUT *Constant Pressure*

An ideal gas at room temperature is contained in a cylinder by a piston that is free to move. Sand is placed on top of the piston (as shown in Figure 11-12) until the pressure of the gas is 2 atmospheres. The gas is then heated and, as the temperature is raised, the volume of the gas increases while the pressure of the gas remains constant. At what temperature will the volume of the gas be three times its initial volume?

At constant pressure, the volume of an ideal gas is directly proportional to the absolute temperature of the gas. Room temperature is $20 + 273 = 293$ K. The volume of the gas will triple when the absolute temperature is tripled, at $3(293$ K$) = 879$ K. This is a temperature of $879 - 273 = 606°C$, which is $1120°F$.

Figure 11-12 The piston is free to move, allowing the volume of the gas to change as it is heated. The pressure of the gas remains constant.

The Ideal Gas Law

The three macroscopic properties of a gas—volume, temperature, and pressure—are related by a relationship known as the **ideal gas law.** This law states that

ideal gas law ▶

$$PV = nRT$$

where P is the pressure, V is the volume, n is the number of moles, T is the *absolute* temperature, and R is a number known as the gas constant.

This relationship is a combination of three experimental relationships that had been discovered to hold for the various pairs of these three macroscopic properties. For example, if we hold the temperature of a quantity of gas constant, we can experimentally determine what happens to the pressure as we compress the gas. Or we can vary the pressure and measure the change in volume. This experimentation leads to a relationship known as *Boyle's law,* which states that the product of the pressure and the volume is a constant. This is equivalent to saying that they are inversely proportional to each other; as one increases, the other must decrease by the same factor.

In a similar manner, we can investigate the relationship between temperature and volume while holding the pressure constant. The results for a gas at one pressure are shown in Figure 11-11. As stated in the previous section, the volume in this case is directly proportional to the absolute temperature.

The third relationship is between temperature and pressure at a constant volume. The pressure in this case is directly proportional to the absolute temperature.

Each of these relationships can be obtained from our model for an ideal gas. For example, let's take a qualitative look at Boyle's law. As we decrease the volume while keeping the temperature the same, the molecules will be moving at the same average speed as before but will now hit the sides more frequently; therefore, the pressure increases in agreement with the statement of Boyle's law.

Q: How does the ideal gas model explain the rise in pressure of a gas as its temperature is raised without changing its volume?

A: Raising the temperature of the gas increases the kinetic energies of the particles. The increased speeds of the particles mean not only that they have larger momenta but also that they hit the walls more frequently.

Everyday Physics | *Evaporative Cooling*

Many of the concepts of the ideal gas model apply to liquids as well as gases. One big difference between liquids and gases is the strength of the attractive forces between the molecules. In our ideal gas model, we assumed that these forces could be neglected until the molecules collided. As a consequence a gas expands to fill its container. However, liquids have a definite volume. It is intermolecular forces that hold the molecules in the liquid together.

If we assume that a model for liquids is similar to that for gases, we can begin to understand the evaporation of liquids. Assume that the kinetic energies of the molecules in liquids have a distribution similar to that in gases and that the average kinetic energy of the particles increases with increasing temperature. The intermolecular forces perform work on molecules that try to escape the liquid, allowing only those with large enough kinetic energies to succeed. Therefore, the molecules that leave the liquid have higher-than-average kinetic energies. Conservation of energy requires that the average kinetic energy of the molecules left behind is lower, and the liquid is cooled. Water is often carried in canvas bags on desert trips to keep the water cool. The canvas bag is kept wet by water seeping through the canvas. The evaporation of the water from the wet canvas keeps the rest of the water cool.

Evaporative cooling can be demonstrated with a simple experiment. Wrap the end of a thermometer in cotton soaked in room-temperature water and observe the temperature as the water evaporates.

Your body uses evaporative cooling to maintain your body temperature on hot days or during strenuous exercise. The evaporating sweat cools our bodies. If you didn't sweat, you might die of heat prostration! A more effective way of cooling your body is the alcohol rub used to reduce fevers.

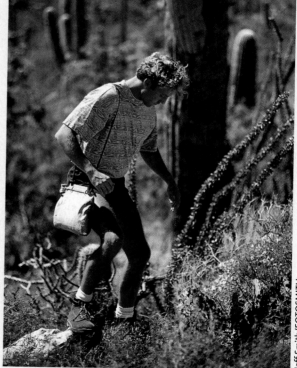

Jeff Smith/FOTOSMITH

Canvas bags like the one shown here keep water cool, even in a hot desert.

1. Why might hikers get hypothermia during wet weather even when the temperature is above freezing?

2. The temperature of boiling water does not increase even if the heat is turned on high. Use the microscopic model to explain this.

WORKING IT OUT *Ideal Gas Law*

A volume of 200 cm³ of an ideal gas has an initial temperature of 20°C and an initial pressure of 1 atm. What is the final pressure if the volume is reduced to 100 cm³ and the temperature is raised to 100°C?

An ideal gas in a closed container will have a fixed number of molecules. The number of moles of gas, n, will remain unchanged as temperature, volume, and pressure are varied. In this case, the ideal gas law takes the form

$$\frac{PV}{T} = nR = \text{constant}$$

This combination of pressure, volume, and absolute temperature always stays the same for the gas, so we can write

$$\frac{P_f V_f}{T_f} = \frac{P_i V_i}{T_i}$$

The initial temperature is 20 + 273 = 293 K. The final temperature is 100 + 273 = 373 K. Substituting the values given, we have

$$\frac{P_f(100 \text{ cm}^3)}{373 \text{ K}} = \frac{(1 \text{ atm})(200 \text{ cm}^3)}{293 \text{ K}}$$

The final pressure is therefore

$$P_f = (1 \text{ atm})\frac{(200 \text{ cm}^3)}{(100 \text{ cm}^3)}\frac{373 \text{ K}}{293 \text{ K}} = 2.55 \text{ atm}$$

Summary

The most elementary substances are elements, which are composed of a large number of very tiny, identical atoms. These atoms combine in definite mass ratios to form molecules according to the law of definite proportions. Observations that gaseous elements combine in definite volume ratios allowed the determination of the relative masses of different atoms. The mass of a carbon atom is set equal to 12 atomic mass units.

Modern experiments show that atoms have diameters of approximately 10^{-10} meter and masses ranging from 1 to more than 260 atomic mass units, where 1 atomic mass unit is equal to 1.66×10^{-27} kilogram.

The Celsius temperature scale is defined with the freezing and boiling points of water as 0°C and 100°C, respectively. The absolute, or Kelvin, temperature scale has its zero point at −273°C and the same-size degree as the Celsius scale.

The ideal gas model assumes that the gas is composed of an enormous number of tiny, indestructible spheres with no internal structure, separated by relatively large distances, and interacting only via elastic collisions. The pressure exerted by a gas is due to the average of the impulses exerted by the gas particles on the walls of the container. Most of the particles have speeds near the average speed, but some move very slowly and some very rapidly. The average speed is typically about 500 meters per second. The absolute temperature of a gas is proportional to the average kinetic energy of the gas particles.

The ideal gas law states the relationship between the pressure, the volume, the number of moles, and the absolute temperature of a gas as $PV = nRT$, where R is the gas constant.

CHAPTER 11 *Revisited*

Observation of the motion of smoke particles provided evidence of the existence of atoms and molecules. Knowledge of the volumes of gases that combined chemically to form other gases helped establish the masses of individual atoms and molecules. The ideal gas law tells us that the average kinetic energy of the gas particles is determined by the absolute temperature.

Key Terms

absolute temperature scale   The temperature scale with its zero point at absolute zero and degrees equal to those on the Celsius scale. Also called the Kelvin temperature scale.

atom   The smallest unit of an element that has the chemical and physical properties of that element.

atomic mass   The mass of an atom in atomic mass units.

atomic mass unit   One-twelfth of the mass of a carbon atom. One amu is equal to 1.66×10^{-27} kilogram.

Avogadro's number 6.02×10^{23} molecules, the number of molecules in 1 mole of any substance.

Celsius temperature scale   The temperature scale with values of 0°C and 100°C for the temperatures of freezing and boiling water, respectively.

compound   A combination of chemical elements that forms a substance with its own properties.

element   Any chemical species that cannot be broken up into other chemical species.

Fahrenheit temperature scale   The temperature scale with values of 32°F and 212°F for the temperatures of freezing and boiling water, respectively. Its degree is five-ninths of that on the Celsius or Kelvin scales.

ideal gas   An enormous number of tiny particles separated by relatively large distances. The particles have no internal structure, are indestructible, and do not interact with each other except when they collide; all collisions are elastic.

ideal gas law $PV = nRT$, where P is the pressure, V is the volume, T is the absolute temperature, n is the number of moles, and R is the gas constant.

Kelvin temperature scale   The temperature scale with its zero point at absolute zero and a degree equal to that on the Celsius scale. Also called the *absolute temperature scale*.

law of definite proportions   When two or more elements combine to form a compound, the ratios of the masses of the combining elements have fixed values.

macroscopic   Describes the bulk properties of a substance, such as mass, size, pressure, and temperature.

microscopic   Describes properties not visible to the naked eye, such as atomic speeds or the masses and sizes of atoms.

mole   The amount of a substance that has a mass in grams numerically equal to the mass of its molecules in atomic mass units.

molecule   A combination of two or more atoms.

pressure   The force per unit area of surface. Pressure is measured in newtons per square meter, or pascals.

Conceptual Questions

1. The two essential elements of a good model are insight and predictive power. Many ancient cultures explained natural phenomena in terms of the actions of their gods. Did these models fail primarily because of lack of insight or lack of predictive power?

2. The two essential elements of a good model are insight and predictive power. Choose a model with which you are familiar and point out how it meets these two criteria.

3. A friend has created a model of how a candy vending machine works. His theory says that a little blue person (LBP) lives inside each vending machine. This person takes your coins and gives you candy in return. Although this LBP theory may not seem reasonable to you, can you suggest ways of disproving it without opening the machine?

4. For most of human history, we believed that Earth was stationary and the Sun and planets orbited Earth (the geocentric model). Beginning about 500 years ago, a second model emerged in which Earth orbits the Sun (the heliocentric model). When we read in the paper that the Sun rose at 6:40 this morning, which of these two models is being used?

5. Your friend notices that a brown can of diet cola floats, whereas a green can of lemon–lime soda and a can of orange soda both sink. He postulates a model in which only nonbrown cans of soda sink. To prove his model, he tries a brown can of diet root beer and finds that it floats as expected. Has he proven that his model is correct? In general, can a model ever be proven true?

6. Following the experiments described in Question 5, your friend tries a brown can of nondiet root beer and finds that it sinks. He rightfully discards his original model and proposes an alternative. What would this new model be? Has it been proven correct?

7. Alchemists held a model in which matter was continuous. Atomists showed the fallacy of this model and replaced it with a model in which all objects are made of small, discrete particles. In your day-to-day living, which model do you appeal to more often? Is the most complete model always the most useful?

8. What role did the alchemists play in the development of an atomistic world view?

9. Which of the following are not elements: hydrogen, salt, nitrogen, granite, sodium, chlorine, water?

10. Would you expect carbon monoxide to be an element or a compound? Why?

11. When the element mercury is heated in air, a red powder is formed. Careful measurement shows that the mass of the resulting powder is greater than the mass of the original mercury. Is this powder an element or a compound? How do you account for this additional mass?

12. Give an example that clearly illustrates the meaning of the law of definite proportions.

13. What are the basic differences between mixtures and compounds?

14. Do water and salt form a compound or a mixture?

15. The atomic mass of ammonia is 17 atomic mass units. What is the atomic mass of nitrogen if a molecule of ammonia consists of one atom of nitrogen and three atoms of hydrogen?

16. The atomic mass of iron oxide (rust) is 160 atomic mass units. What is the atomic mass of iron if a molecule of iron oxide consists of two atoms of iron and three atoms of oxygen?

17. Silver has an atomic mass of 108. Which, if either, contains more atoms: 1 gram of silver or 1 gram of hydrogen?

18. Silver has an atomic mass of 108. Which, if either, contains more atoms: 1 mole of silver or 1 mole of hydrogen?

19. How does the number of molecules in 1 liter of oxygen compare with the number of molecules in 1 liter of carbon dioxide if they are both at the same temperature and pressure?

20. Oxygen molecules contain two oxygen atoms, and carbon dioxide molecules contain one atom of carbon and two atoms of oxygen. How does the total number of atoms in 1 liter of oxygen compare with the total number of atoms in 1 liter of carbon dioxide if they are both at the same temperature and pressure?

21. The atomic mass of sulfur is 32 atomic mass units. How many grams of sulfur are needed to have an Avogadro's number of sulfur atoms?

22. How many grams of water are needed to have an Avogadro's number of water molecules?

23. If a gas condenses to a liquid, the liquid occupies a much smaller volume than the gas. How does the ideal gas model account for this?

24. The ideal gas model accounts very well for the behavior of gases at standard temperature and pressure. Would the ideal gas model begin to fail for very large pressures or for very small pressures? Explain your answer.

25. A cube and a spherical ball are made of the same material and have the same mass. Which exerts the larger pressure while resting on a floor?

26. You can apply enough force to the head of a pushpin to push it into a plaster wall with your thumb. However, it is not a good idea to try to do this with a needle. Use the concept of pressure to explain the difference between these two situations.

27. If you screw the cap of an empty plastic drinking bottle on tightly while walking in the mountains, why are the sides of the bottle caved in when you return to the valley?

28. Your right rear tire has to support a weight of 3000 newtons. Normally, the contact area of your tire with the road is 200 square centimeters. If the pressure in your tire is suddenly reduced from 32 pounds per square inch to 16 pounds per square inch, what must be the new contact area to support the car?

29. Use the microscopic model of a gas to explain why the pressure in a tire increases as you add more air.

30. As you drive your car down the road, the friction of the rubber with the road causes the air inside the tire to increase in temperature, resulting in an increase in pressure. Use the microscopic model of a gas to explain why the pressure increases.

31. If the average speed of a perfume molecule is 500 meters per second, why does it take several minutes before you smell the perfume from a bottle opened across the room?

32. What happens to the average speed of the molecules of a gas as it is heated?

33. Why does an alcohol-in-glass thermometer have a bulb at the bottom?

34. It is possible to cut the very top off an alcohol-in-glass thermometer without any of the alcohol spilling out. However, it will no longer function as a good thermometer. Why not?

35. It takes longer to boil an egg in the mountains than it does at sea level. What conditions must be imposed before the boiling point of water can be used as a fixed temperature?

36. Why is body temperature not a good fixed temperature for establishing a temperature scale?

37. Two students are sick in bed with 2-degree fevers. One has a temperature of 39.0°C; the other, 100.6°F. Which student has the higher fever?

38. Is a sauna at a temperature of 190°F hotter or colder than one at 90°C?

39. At what temperature should you set your new Celsius thermostat so that your hot tub stays at a comfortable 102°F?

40. You move to Canada and find that the thermostat in your home is in Celsius degrees. You normally like your house at about 72°F. To what temperature should you set your new thermostat?

41. What is the freezing point of water on the Kelvin scale?

42. Nitrogen boils at 77 K. At what Celsius temperature does it boil?

43. What microscopic property of an ideal gas doubles when the absolute temperature is doubled?

44. What temperature change would be needed to double the average speed of the molecules in an ideal gas?

45. Air is a mixture of several gases, primarily nitrogen and oxygen. Is the average kinetic energy of the nitrogen molecules greater than, equal to, or less than the average kinetic energy of the oxygen molecules?

46. Consider a mixture of helium and neon gases. The atomic masses of helium and neon are 4 atomic mass units and 20 atomic mass units, respectively. Is the average speed of a helium atom greater than, equal to, or less than the average speed of a neon atom?

47. If you heat a gas in a container with a fixed volume, the pressure increases. Use the ideal gas model to explain this.

48. If the volume of an ideal gas is held constant, what happens to the pressure if the absolute temperature is cut in half?

49. What macroscopic property of an ideal gas doubles when the absolute temperature is doubled while the pressure remains constant?

50. If you put a sealed plastic bottle partially filled with hot tea in the refrigerator, the sides of the bottle will cave in as the tea cools. Why?

51. What happens to the temperature of an ideal gas if you reduce its volume to one-fourth while holding the pressure constant?

52. Why does the pressure inside the tires increase after a car has been driven?

53. Use the microscopic gas model to explain why the pressure of a gas rises as the volume is reduced while the temperature remains constant.

54. If you hold the temperature of an ideal gas constant, what happens to its volume when you triple its pressure?

55. The water in a canvas water bag placed in front of your car when driving across the desert stays cooler than the surrounding air. Explain this in terms of the average

kinetic energies of the water molecules that leave and those that stay behind.

Gerald F. Wheeler

56. How does an alcohol rub cool your body?

Exercises

57. If 1 g of hydrogen combines completely with 8 g of oxygen to form water, how many grams of hydrogen are needed to combine completely with 24 g of oxygen?

58. In ammonia, 14 g of nitrogen combines completely with 3 g of hydrogen. How many grams of hydrogen are needed to combine completely with 56 g of nitrogen?

59. Given that 1 g of hydrogen combines completely with 8 g of oxygen to form water, how many grams of water can you make with 8 g of hydrogen and 16 g of oxygen?

60. Given that 12 g of carbon combines completely with 16 g of oxygen to form carbon monoxide, how many grams of carbon monoxide can be made from 48 g of carbon and 48 g of oxygen?

61. A ham sandwich consists of one slice of ham (10 g) and two slices of bread (25 g each). You have 1 kg of ham and 1 kg of bread. You make as many sandwiches as you can. How many sandwiches did you make? What is the mass of the sandwiches? Which ingredient is left over? What is the mass of the ingredient that is left over?

62. One mole of water molecules consists of 1 mole of oxygen (16 g) and 2 moles of hydrogen (1 g each). You combine 1 kg of oxygen with 1 kg of hydrogen to make water. How many moles of water did you make? What is the mass of the water? What is the mass of the element that is left over?

63. Given that the carbon atom has a mass of 12 amu, how many carbon atoms are there in a diamond with a mass of 1 g?

64. Given that the sulfur molecule has a mass of 32 amu, how many sulfur molecules are in 1 g of sulfur?

65. One liter of water has a mass of 1 kg, and the mass of a water molecule is 18 amu. How many molecules of water are in 1 L of water?

66. One liter of oxygen has a mass of 1.4 g, and the oxygen molecule has a mass of 32 amu. How many oxygen molecules are in 1 L of oxygen?

67. One liter of nitrogen combines with 3 L of hydrogen to form 2 L of ammonia. If the molecules of nitrogen and hydrogen have two atoms each, how many atoms of hydrogen and nitrogen are in one molecule of ammonia?

68. One liter of oxygen combines with 1 L of hydrogen to form 1 L of hydrogen peroxide. Given that the molecules of hydrogen and oxygen contain two atoms each, how many atoms are in one molecule of hydrogen peroxide?

69. About how many atoms would it take to deposit a single layer on a surface with an area of 1 cm²?

70. About how many atoms would you expect there to be in a cube of material 1 cm on each side?

71. You exert a force of 30 N on the head of a thumbtack. The head of the thumbtack has a radius of 5 mm. What is the pressure on your thumb?

72. The pressure in each of your car tires is 2.5×10^5 Pa. The mass of your car is 1600 kg. Assuming that each of your tires bears one-quarter of the total load, what is the contact area of each tire with the road?

Blue-numbered answered in Appendix B   🌶 = more challenging questions

73. What happens to the volume of 1 L of an ideal gas when the pressure is tripled while the temperature is held fixed?

74. An ideal gas at 27°C is contained in a piston that ensures that its pressure will always be constant. Raising the temperature of the gas causes it to expand. At what temperature will the gas take up twice its original volume?

75. A helium bottle with a pressure of 100 atm has a volume of 3 L. How many balloons can the bottle fill if each balloon has a volume of 1 L and a pressure of 1.25 atm?

76. When the temperature of an automobile tire is 20°C, the pressure in the tire reads 32 psi on a tire gauge. (The gauge measures the difference between the pressures inside and outside the tire.) What is the pressure when the tire heats up to 40°C while driving? You may assume that the volume of the tire remains the same and that atmospheric pressure is a steady 14 psi.

77. A volume of 150 cm^3 of an ideal gas has an initial temperature of 20°C and an initial pressure of 1 atm. What is the final pressure if the volume is reduced to 100 cm^3 and the temperature is raised to 40°C?

78. An ideal gas has the following initial conditions: $V_i = 500$ cm^3, $P_i = 3$ atm, and $T_i = 100$°C. What is its final temperature if the pressure is reduced to 1 atm and the volume expands to 1000 cm^3?

States
of Matter

▶Many materials exist as solids, liquids, or gases, each with its own characteristics. However, different materials share many characteristics; for instance, many materials form crystalline shapes in the solid form.   What does the shape of a crystal tell us about the underlying structure of the material?

(See page 255 for the answer to this question.)

© Ted Kinsman/Photo Researchers, Inc.

Ice crystals

ALL matter is composed of approximately 100 different elements. Yet the material world we experience—say, in a walk through the woods—holds a seemingly endless variety of forms. This variety arises from the particular combinations of elements and the structures they form, which can be divided into four basic forms, or states: solid, liquid, gas, and plasma.

Many materials can exist in the solid, liquid, and gaseous states if the forces holding the chemical elements together are strong enough that their melting and vaporization temperatures are lower than their decomposition temperatures. Hydrogen and oxygen in water, for example, are so tightly bonded that water exists in all three states. Sugar, on the other hand, decomposes into its constituent parts before it can turn into a gas.

If we continuously heat a solid, the average kinetic energy of its molecules rises and the temperature of the solid increases. Eventually, the intermolecular bonds break, and the molecules slide over one another (the process called melting) to form a liquid. The next change of state occurs when the substance turns into a gas. In the gaseous state, the molecules have enough kinetic energy to be essentially independent of each other. In a plasma, individual atoms are ripped apart into charged ions and electrons, and the subsequent electrical interactions drastically change the resulting substance's behavior.

Atoms

At the end of the previous chapter, we established the evidence for the existence of atoms. It would be natural to ask, "Why stop there?" Maybe atoms are not the end of our search for the fundamental building blocks of matter. In fact, they are not. Atoms have structure, and we will devote two chapters near the end of the book to further developing our understanding of this structure. For now it is useful to know a little about this structure so that we can understand the properties of the states of matter.

A useful model for the structure of an atom for our current purposes is the solar-system model developed early in the 20th century. In this model the atom is seen as consisting of a tiny central nucleus that contains almost all of the atom's mass. This nucleus has a positive electric charge that binds very light, negatively charged electrons to the atom in a way analogous to the Sun's gravitational attraction for the planets. The electrons' orbits define the size of the atom and give it its chemical properties.

The basic force that binds materials together is the electrical attraction between atomic and subatomic particles. As we will see in later chapters, the gravitational force is too weak and the nuclear forces are too short-ranged to have much effect in chemical reactions. We live in an electrical universe when it comes to the states of matter. How these materials form depends on these electric forces. And the form they take determines the properties of the materials.

Density

◀ Extended presentation available in the *Problem Solving* supplement

One characteristic property of matter is its **density.** Unlike mass and volume, which vary from one object to another, density is an inherent property of the material. A ton of copper and a copper coin have drastically different masses and volumes but identical densities. If you were to find an unknown material and could be assured that it was pure, you could go a long way toward identifying it by measuring its density.

Density is defined as the amount of mass in a standard unit of volume and is expressed in units of kilograms per cubic meter (kg/m^3):

$$\text{density} = \frac{\text{mass}}{\text{volume}} \blacktriangleright \qquad D = \frac{M}{V}$$

For example, an aluminum ingot is 3 meters long, 1 meter wide, and 0.3 meter thick. If it has a mass of 2430 kilograms, what is the density of aluminum? We calculate the volume first and then the density:

$$V = lwh = (3 \text{ m})(1 \text{ m})(0.3 \text{ m}) = 0.9 \text{ m}^3$$

$$D = \frac{M}{V} = \frac{2430 \text{ kg}}{0.9 \text{ m}^3} = 2700 \text{ kg/m}^3$$

Therefore, the density of aluminum is 2700 kilograms per cubic meter. Densities are also often expressed in grams per cubic centimeter. Thus, the density of aluminum is also 2.7 grams per cubic centimeter (g/cm^3). Table 12-1 gives the densities of a number of common materials.

Are You On the Bus?

Q: Which has the greater density, 1 kilogram of iron or 2 kilograms of iron?

A: They have the same density; the density of a material does not depend on the amount of material.

Are You On the Bus?

Q: If a hollow sphere and a solid sphere are both made of the same amount of iron, which sphere has the greater average density?

A: The solid sphere has the greater average density because it occupies the smaller volume for a given mass of iron.

Table 12-1	Densities of Some Common Materials

Material	Density (g/cm³)
Air*	0.0013
Ice	0.92
Water	1.00
Magnesium	1.75
Aluminum	2.70
Iron	7.86
Copper	8.93
Silver	10.5
Lead	11.3
Mercury	13.6
Uranium	18.7
Gold	19.3
Osmium	22.5

*At 0°C and 1 atm.

The densities of materials range from the small for a gas under normal conditions to the large for the element osmium. One cubic meter of osmium has a mass of 22,480 kilograms (a weight of nearly 50,000 pounds), about 22 times as large as the same volume of water. It is interesting to note that the osmium atom is less massive than a gold atom. Therefore, the higher density of osmium indicates that the osmium atoms must be packed closer together.

The materials that we commonly encounter have densities around the density of water, 1 gram per cubic centimeter. A cubic centimeter is about the volume of a sugar cube. The densities of surface materials on Earth average approximately 2.5 grams per cubic centimeter. The density at Earth's core is about 9 grams per cubic centimeter, making Earth's average density about 5.5 grams per cubic centimeter.

Solids

Solids have the greatest variety of properties of the four states of matter. The character of a solid substance is determined by its elemental constituents and its particular structure. This underlying structure depends on the way it was formed. For example, slow cooling often leads to solidification with the atoms in an ordered state known as a **crystal.**

Crystals grow in a variety of shapes. Their common property is the orderliness of their atomic arrangements. The orderliness consists of a basic arrangement of atoms that repeats throughout the crystal, analogous to the repeating geometric patterns in some wallpapers.

The microscopic order of the atoms is not always obvious in macroscopic samples. For one thing, few perfect crystals exist; most samples are aggregates

Everyday Physics | *Density Extremes*

Which weighs more: a pound of feathers or a pound of iron? The answer to this junior high school puzzle, of course, is that they weigh the same. The key difference between the two materials is *density*; although the two weigh the same, they have very different volumes. Density is a comparison of the masses of two substances with the same volumes. Obviously, 1 cubic meter of iron has much more mass than 1 cubic meter of feathers.

The densities of objects vary over a large range. The densities of common Earth materials pale in comparison to those of some astronomical objects. After a star runs out of fuel, its own gravitational attraction causes it to collapse. The collapse stops when the outward forces due to the pressure in the star balance the gravitational forces. The resultant stellar cores can have astronomically large densities. White dwarf stars are the death stage of most stars. They can have masses up to 1.4 times that of our Sun compressed to a size about that of Earth, with resulting densities a million times the density of water. Neutron stars are cores left after a star explodes and can have densities a billion times those of white dwarfs. A teaspoon of material from a neutron star would weigh a billion tons on Earth!

A very low-density solid, called *silica aerogel*, was created at the Lawrence Livermore National Laboratory in California. This solid is made from silicon dioxide and has a density that is only

U of CA, Lawrence Livermore National Laboratory & the U.S. Dept. of Energy

Silica aerogel has an extremely low density but still can support 1600 times its own weight.

three times that of air. Because of this very low density, it is sometimes known as "solid smoke." Because silica aerogel is a solid, it holds its shape. In fact, it can support 1600 times its own weight!

The density of interstellar space is much smaller than that of air; there is about one atom per cubic centimeter, resulting in a density about one-billion-trillionth that of air at 1 atmosphere of pressure, or about one-trillion-trillionth (10^{-24}) that of water.

WORKING IT OUT | *Density*

Suppose you find a chunk of material that you cannot identify. You find that the chunk has a mass of 87.5 g and a volume of 50 cm³. What is the material, and what is the mass of a 6-cm³ piece of this material?

We could easily find the mass of 6 cm³, if only we knew the mass of 1 cm³. This is just the density. We can find the density from the measurements made on the original chunk:

$$D = \frac{M}{V} = \frac{87.5 \text{ g}}{50 \text{ cm}^3} = 1.75 \text{ g/cm}^3$$

This density is the same as that of magnesium. Therefore, the material could be magnesium, but we would need to look at other characteristics to be sure.

The 6-cm³ piece has a mass six times as large as the mass of 1 cm³:

$$M = DV = (1.75 \text{ g/cm}^3)(6 \text{ cm}^3) = 10.5 \text{ g}$$

---

of small crystals. However, macroscopic evidence of this underlying structure does exist. A common example in northern climates is a snowflake (Figure 12-1). Its sixfold symmetry is evidence of the structure of ice. Another example is mica (Figure 12-2), a mineral you may find on a hike in the woods. Shining flakes of mica can be seen in many rocks. Larger pieces can be easily separated into thin sheets. The thinness of the sheets seems (at least on the macroscopic

Figure 12-1 The sixfold symmetry exhibited by snowflakes is evidence that ice crystals have hexagonal shapes.

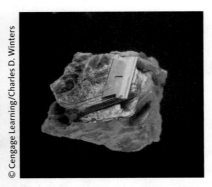

Figure 12-2 Samples of mica exhibit a two-dimensional crystalline structure as evidenced by the fact that one can peel thin sheets from the larger crystal.

Figure 12-3 Synthetic diamonds and finely divided graphite are two different crystalline forms of carbon.

Figure 12-4 A glass filled with milk beyond the brim is evidence of surface tension.

scale) to be limitless. It is easy to convince yourself that the atoms in mica are arranged in two-dimensional sheets with relatively strong bonds between atoms within the sheet and much weaker bonds between the sheets.

In contrast to mica, ordinary table salt exhibits a three-dimensional structure of sodium and chlorine atoms. If you dissolve salt in water and let the water slowly evaporate, the salt crystals that form have obvious cubic structures. If you try to cut a small piece of salt with a razor blade, you find that it doesn't separate into sheets like mica but fractures along planes parallel to its faces. (Salt from a saltshaker displays this same structure, but the grains are usually much smaller. A simple magnifying glass allows you to see the cubic structure.) Precious stones also have planes in their crystalline structure. A gem cutter studies the raw gemstones carefully before making the cleavages that produce a fine piece of jewelry.

Some substances have more than one crystalline structure. A common example is pure carbon. Carbon can form diamond or graphite crystals (Figure 12-3). Diamond is a very hard substance that is treasured for its optical brilliance. Diamond has a three-dimensional structure. Graphite, on the other hand, has a two-dimensional structure like that of mica, creating sheets of material that are relatively free to move over each other. Because of its slippery nature, graphite is used as a lubricant and as the "lead" in pencils.

Liquids

When a solid melts, interatomic bonds break, allowing the atoms or molecules to slide over each other, producing a **liquid.** Liquids fill the shape of the container that holds them, much like the random stacking of a bunch of marbles.

The temperature at which a solid melts varies from material to material simply because the bonding forces are different. Hydrogen is so loosely bound that it becomes a liquid at 14 K. Oxygen and nitrogen—the constituents of the air we breathe—melt at 55 K and 63 K, respectively. The fact that ice doesn't melt until 273 K (0°C) tells us that the bonds between the molecules are relatively strong.

Water is an unusual liquid. Although water is abundant, it is one of only a few liquids that occur at ordinary temperatures on Earth. The bonding between the water molecules is relatively strong, and a high temperature is required to separate them into the gaseous state.

The intermolecular forces in a liquid create a special "skin" on the surface of the liquid. This can be seen in Figure 12-4, in which a glass has been filled with milk beyond its brim. What is keeping the extra liquid from flowing over the edge?

Everyday Physics | *Solid Liquids and Liquid Solids*

Many substances exist between the ordinary boundaries of solids and liquids. When materials such as glass or wax cool, the molecules are frozen in space without arranging themselves into an orderly crystalline structure. These solids are *amorphous*, meaning that they retain some of the properties of liquids. A common example of an amorphous solid is a clear lollipop made by rapidly cooling liquid sugar. The average intermolecular forces in an amorphous material are weaker than those in a crystalline structure.

Despite the solid rigidity of an amorphous material, this form is more like a liquid than a solid because of its lack of order. In addition, the melting points of amorphous materials are not clearly defined. An amorphous material simply gets softer and softer, passing into the fluid state. Another characteristic of these solid liquids is that they actually do flow like a liquid, although many flow on time scales that make the flow difficult or impossible to detect. Although it is widely believed that old church windows in Europe are thicker at the bottom than at the top because of centuries of flow, they were most likely assembled with the thicker edges at the bottom.

Other substances are liquids that retain some degree of orderliness, characteristic of solids. Liquid crystals can be poured like regular liquids. They lack positional order, but they possess an orientational order. Small electric fields can align the rodlike molecules along a particular direction.

Liquid crystals have some interesting applications because polarized light behaves differently depending on whether it is traveling parallel or perpendicular to the alignment direction. For example, the orientation can be manipulated electrically to produce the numbers in a digital watch or an electronic calculator. You can verify that the light emerging from a liquid crystal display is polarized by looking at the display through polarized sunglasses. Changing the orientation of the sunglasses will vary the intensity of the image.

1. What properties of liquids are shared by amorphous materials?

2. What characteristic of solids is exhibited by liquid crystals?

The display on this LCD TV is an application of liquid crystal technology.

Figure 12-5 Surface tension minimizes the surface area of the soap film forming these bubbles.

Imagine two molecules, one on the surface of a liquid and one deeper into the liquid. The molecule beneath the surface experiences attractive forces in all directions because of its neighbors. The molecule on the surface only feels forces from below and to the sides. This imbalance tends to pull the "surface" molecules back into the liquid.

Surface tension also tries to pull liquids into shapes with the smallest possible surface areas. The shapes of soap bubbles are determined by the surface tension trying to minimize the surface area of the film (Figure 12-5). If there are no external forces, the liquid forms into spherical drops. In fact, letting liquids cool in space has been proposed as a way of making nearly perfect spheres. In the free-fall environment of an orbiting space shuttle, liquid drops are nearly spherical.

Surface tensions vary among liquids. Water, as you may expect, has a relatively high surface tension. If we add soap or oil to the water, its surface tension is reduced, meaning that the water molecules are not as attracted to each other. It is probably reasonable to infer that the new molecules in the solution are somehow shielding the water molecules from each other.

Gases

When the molecules separate totally, a liquid turns into a **gas.** (See Chapter 11 for a discussion of an ideal gas.) The gas occupies a volume about 1000 times as large as that of the liquid. In the gaseous state, the molecules have enough kinetic energy to be essentially independent of each other. A gas fills the container holding it, taking its shape and volume. Because gases are mostly empty space, they are compressible and can be readily mixed with each other.

Gases and liquids have some common properties because they are both "fluids." All fluids are able to flow, some more easily than others. The **viscosity** of a fluid is a measure of the internal friction within the fluid. You can get a qualitative feeling for the viscosity of a fluid by pouring it. Fluids that pour easily, such as water and gasoline, have low viscosities. Those that pour slowly, such as molasses, honey, and egg whites, have high viscosities. Glass is a fluid with an extremely high viscosity. In the winter, drivers put lower-viscosity oils in their cars so that the oils will flow better on cold mornings.

The viscosity of a fluid determines its resistance to objects moving through it. A parachutist's safe descent is due to the viscosity of air. Air and water have drastically different viscosities. Imagine running a 100-meter dash in water 1 meter deep!

Honey is a very viscous fluid.

Q: How might you explain the observation that the viscosities of fluids decrease as they are heated?

A: The increased kinetic energy of the molecules means that the molecules are more independent of each other.

Plasmas

At around 4500°C, all solids have melted. At 6000°C, all liquids have been turned into gases. And at somewhere above 100,000°C, most matter is ionized into the **plasma** state. In the transition between a gas and a plasma, the atoms break apart into electrically charged particles.

Although the fourth state of matter, plasma, is more rare on Earth than the solid, liquid, and gaseous states, it is actually the most common state of matter in the universe (more than 99%). Examples of naturally occurring plasmas on Earth include fluorescent lights and neon-type signs. Fluorescent lights consist of a plasma created by a high voltage that strips mercury vapor of some of its electrons. "Neon" signs employ the same mechanism but use a variety of gases to create the different colors.

Perhaps the most beautiful naturally occurring plasma effect is the aurora borealis, or northern lights. Charged particles emitted by the Sun and other stars are trapped in Earth's upper atmosphere to form a plasma known as the Van Allen radiation belts. These plasma particles can interact with atoms of nitrogen and oxygen over both magnetic poles, causing them to emit light as discussed in Chapter 23.

Plasmas are important in nuclear power as well as in the interiors of stars. An important potential energy source for the future is the "burning" of a plasma of hydrogen ions at extremely high temperatures to create nuclear energy. We will discuss nuclear energy more completely in Chapter 26.

The aurora borealis results from the interaction of charged particles with air molecules.

Pressure

A characteristic feature of a fluid—either a gas or a liquid—is its change in pressure with depth. As we saw in Chapter 11, pressure is the force per unit area exerted on a surface, measured in units of newtons per square meter (N/m^2), a unit known as a pascal (Pa).

When a gas or liquid is under the influence of gravity, the weight of the material above a certain point exerts a force downward, creating the pressure at that point. Therefore, the pressure in a fluid varies with depth. You have probably felt this while swimming. As you go deeper, the pressure on your eardrums increases. If you swim horizontally at this depth, you notice that the pressure doesn't change. In fact, there is no change if you rotate your head; the pressure at a given depth in a fluid is the same in all directions.

Consider the box of fluid shown in Figure 12-6. Because the fluid in the box does not move, the net force on the fluid must be zero. Therefore, the fluid below the box must be exerting an upward force on the bottom of the box that is equal to the weight of the fluid in the box plus the force of the atmosphere on the top of the box. The pressure at the bottom of the box is just this force per unit area.

Our atmosphere is held in a rather strange container: Earth's two-dimensional surface. Gravity holds the atmosphere down so that it doesn't escape. There is no definite top to our atmosphere; it just gets thinner and thinner the higher you go above Earth's surface.

The air pressure at Earth's surface is due to the weight of the column of air above the surface. At sea level the average atmospheric pressure is about 101 kilopascals. This means that a column of air that is 1 square meter in cross section and reaches to the top of the atmosphere weighs 101,000 newtons and has a mass of 10 metric tons. A column of air 1 square inch in cross section weighs 14.7 pounds; therefore, atmospheric pressure is also 14.7 pounds per square inch.

We can use these ideas to describe what happens to atmospheric pressure as we go higher and higher. You may think that the pressure drops to one-half the surface value halfway to the "top" of the atmosphere. However, this is not true, because the air near Earth's surface is much denser than that near the top of the atmosphere. This means that there is much less air in the top half compared to the bottom half. Because the pressure at a given altitude depends on the weight of the air above that altitude, the pressure changes more quickly near the surface. In fact, the pressure drops to half at about 5500 meters (18,000 feet) and then drops by half again in the next 5500 meters.

The pressure on a scuba diver increases with depth.

Q: Why doesn't the large force on the surface of your body crush you?

A: You aren't crushed because the pressure inside your body is the same as the pressure outside. Therefore, the inward force is balanced by the outward force.

Figure 12-6 The force on the bottom of the box of fluid is equal to the weight of the fluid in the box plus the force of the atmosphere on the top of the box.

This means that commercial airplanes flying at a typical altitude of 36,000 feet experience pressures that are only one-fourth those at the surface.

Like fish living on the ocean floor, we land-lovers are generally unaware of the pressure due to the ocean of air above us. Although the atmospheric pressure at sea level may not seem like much, consider the total force on the surface of your body. A typical human body has approximately 2 square meters (3000 square inches) of surface area. This means that the total force on the body is about 200,000 newtons (20 tons!).

An ingenious experiment conducted by a contemporary of Isaac Newton demonstrated the large forces that can be produced by atmospheric pressure. German scientist Otto von Guericke joined two half spheres (Figure 12-7) with just a simple gasket (no clamps or bolts). He then pumped the air from the sphere, creating a partial vacuum. Two teams of eight horses could not pull the hemispheres apart!

In weather reports, atmospheric pressure is often given in units of millimeters or inches of mercury. A typical pressure is 760 millimeters (30 inches) of mercury. Because pressure is a force per unit area, reporting pressure in units of length must seem strange. This scale comes from the historical method of measuring pressure. Early pressure gauges were similar to the simple mercury barometer shown in Figure 12-8. A sealed glass tube is filled with mercury and inverted into a bowl of mercury. After inversion the column of mercury does not pour out into the bowl but maintains a definite height above the pool of mercury. Because the mercury is not flowing, we know that the force due to atmospheric pressure at the bottom of the column is equal to the weight of the mercury column. This means that the atmospheric pressure is the same as the pressure at the bottom of a column of mercury 760 millimeters tall if there is a vacuum above the mercury. Therefore, atmospheric pressure can be characterized by the height of the column of mercury it will support.

Atmospheric pressure also allows you to drink through a straw. As you suck on the straw, you reduce the pressure above the liquid in the straw, allowing the pressure below to push the liquid up. In fact, if you could suck hard enough to produce a perfect vacuum above water, you could use a straw 10 meters (almost 34 feet) long! So although we often talk of sucking on a soda straw and pulling the soda up, in reality we are removing the air pressure on the top of the soda column in the straw, and the atmospheric pressure is pushing the soda up.

Q: How high a straw could you use to suck soda?

A: Because soda is mostly water, we assume that it has the same density as water. Therefore, the straw could be 10 meters high—but only if you have very strong lungs. A typical height is more like 5 meters.

Figure 12-7 Two teams of eight horses could not separate von Guericke's evacuated half spheres.

Underwater explorers must use vessels such as this bathysphere at the great depths of the ocean floor.

Rod Catanach, Woods Hole Oceanographic Institution

Figure 12-8 In a mercury barometer, the atmospheric pressure is balanced by the pressure due to the weight of the mercury column.

As you dive deeper in water, the pressure increases for the same reasons as in air. Because atmospheric pressure can support a column of water 10 meters high, we have a way of equating the two pressures. The pressure in water must increase by the equivalent of 1 atmosphere (atm) for each 10 meters of depth. Therefore, at a depth of 10 meters, you would experience a pressure of 2 atmospheres, 1 from the air and 1 from the water. The pressures are so large at great depths that very strong vessels must be used to prevent the occupants from being crushed.

Q: What is the pressure on a scuba diver at a depth of 30 meters (100 feet)?

A: The pressure would be (30 meters)/(10 meters per atmosphere) = 3 atmospheres because of the water plus 1 atmosphere because of the air above the water, for a total of 4 atmospheres.

Are You On the Bus?

FLAWED REASONING

Jeff designs a new scuba setup that is so profoundly simple he is surprised that no one has thought of this before. He has attached one end of a long garden hose to a large block of Styrofoam to keep the hose above the water level. He will breathe through the other end of the hose as he explores the depths. **What is wrong with Jeff's simple design?**

ANSWER If Jeff dives 10 meters below the surface, the water will push inward on him with 2 atmospheres of pressure. Therefore, the air in his lungs will be at a pressure of 2 atmospheres. Because the air in the hose will be at a pressure of 1 atmosphere, air will be expelled from his lungs and he will not be able to breathe!

Sink and Float

Floating is so commonplace to anyone who has gone swimming that it may not have occurred to you to ask, "Why do things sink or float?" "Why does a golf ball sink and an ocean liner float?" "How is a hot-air balloon similar to an ocean liner?"

Anything that floats must have an upward force counteracting the force of gravity, because we know from Newton's first law of motion (Chapter 3) that an object at rest has no unbalanced forces acting on it. To understand why things float therefore requires that we find the upward buoyant force opposing the gravitational force.

The buoyant force exists because the pressure in the fluid varies with depth. To understand this, consider the cubic meter of fluid in Figure 12-9. The pressure on the bottom surface is greater than on the top surface, resulting in a net upward force. The downward force on the top surface is due to the weight of the fluid above the cube. The upward force on the bottom surface is equal to the weight of the column of fluid above the bottom of the cube. The difference between these two forces is just the weight of the fluid in the cube. Therefore, the net upward force must be equal to the weight of the fluid in the cube.

These pressures do not change if a cube of some other material replaces the cube of fluid. Therefore, the net upward force is still equal to the weight of the fluid that was replaced. This result is known as **Archimedes' principle,** named for the Greek scientist who discovered it.

Archimedes' principle ▶

The buoyant force is equal to the weight of the displaced fluid.

When you lower an object into a fluid, it displaces more and more fluid as it sinks lower into the liquid, and the buoyant force therefore increases. If the buoyant force equals the object's weight before it is fully submerged, the object floats. This occurs whenever the density of the object is less than that of the fluid.

We can change a "sinker" into a "floater" by increasing the amount of fluid it displaces. A solid chunk of steel equal in weight to an ocean liner clearly sinks in water. We can make the steel float by reshaping it into a hollow box. We don't throw away any material; we only change its volume. If we make the volume big enough, it will displace enough water to float.

Figure 12-9 The buoyancy force on the cube is due to the larger pressure on the lower surface.

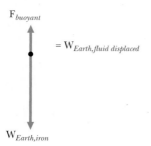

$F_{buoyant}$

$= W_{Earth, fluid\ displaced}$

$W_{Earth, iron}$

Figure 12-10 Free-body diagram for the piece of iron right after it is released under the surface of the lake.

WORKING IT OUT *Buoyant Force*

A piece of iron with a mass of 790 grams displaces 100 grams of water when it is submerged. If we lower the piece of iron under the surface of a lake and then release it from rest, what will its initial acceleration be as it sinks to the bottom of the lake?

The free-body diagram for the piece of iron will initially have two forces, the gravitational force (true weight) and the buoyant force, as shown in Figure 12-10. (After the iron begins moving, there will also be a drag force, but we are calculating the initial acceleration, right after release.) The gravitational force is given by $mg = (0.79$ kilogram)[10 (meters per second) per second] = 7.9 newtons. The buoyant force will be equal to the weight of the water that is displaced by the piece of iron. The volume of the water displaced will be equal to the volume of the iron, 100 cm³, and this much water will have a mass of 100 g = 0.1 kg. The buoyant force is equal to the weight of 0.1 kg of water, or 1 newton. The acceleration is caused by the net force, which is 7.9 newtons − 1.0 newton = 6.9 newtons. Newton's second law yields an acceleration of

$$a = \frac{F_{net}}{m} = \frac{6.9\ \text{N}}{0.79\ \text{kg}} = 8.7\ \text{m/s}^2$$

Changing the shape of the steel makes it a "floater." A solid piece of steel with the mass of the ocean liner sinks.

T. Nakamura/Superstock, Inc.

Bernoulli's Effect **251**

Everyday Physics | *How Fatty Are You?*

Exercising does not automatically reduce your weight. One outcome of exercising is the conversion of fatty tissue into muscle without changing your weight. Because healthy people have more muscle, it is important to be able to determine the percentage of body fat. That's a question that just stepping on the bathroom scale won't tell you. However, a 2000-year-old technique developed by Archimedes does work.

Around 250 BC, Archimedes was chief scientist for King Hiero of Syracuse (now modern Sicily). As the story goes, the king was concerned that his crown was not made of pure gold but had some silver hidden under its surface. Not wanting to destroy his crown to find out whether he had been cheated, he challenged his scientist to find an alternative procedure. Everybody knows the legend of Archimedes leaping from his bathtub and shouting, "Eureka, I have found it!"

The key to Archimedes' solution is determining the average density of the crown or, in our case, your body. There is no problem getting your weight. A simple bathroom scale will do. The tricky part is determining your volume. You, like the king's crown, are oddly shaped (sorry!), not matching any of the geometric volumes you studied in school.

Archimedes discovered that an object immersed in water feels an upward buoyant force. If you were to stand on a scale while totally submerged, the reading on the scale (your apparent weight) would be less than your true weight because the buoyant force supports part of your weight. This buoyant force is equal to the weight of the water your body displaces. From your measured weights in air and under water, your volume can be calculated.

Human performance scientists consider the body to be made of fat and "muscle." (Everything but fat—skin, bone, and

© Taxi/Getty Images

organs—is grouped as muscle.) From the study of cadavers, the density of human fat is found to be about 90% the density of water, whereas the density of "muscle" is about 110% the density of water. The more fat you have, the lower your average density will be. The percentage of fat for healthy adults should be between 15% and 20% for men and between 22% and 28% for women. Champion distance runners and bicyclists have about 5–8% fat.

1. If your percentage of body fat is 20%, do you float or sink in a pond of fresh water?

2. Your friend claims that she weighs less when she is submerged in water. You understand what she means, but her physics is not correct. Use the concept of apparent weight as you explain the error in your friend's comment.

Ice floats because of a **buoyant force.** When water freezes, the atoms arrange themselves in a way that actually takes up more volume. As a result, ice has a lower density than liquid water and floats on the surface. This is fortunate; otherwise, ice would sink to the bottom of lakes and rivers, freezing the fish and plants.

The buoyant force is present even when the object sinks! For example, any object appears to weigh less in water than in air. You can verify this by hanging a small object by a rubber band. As you lower it into a glass of water, the rubber band is stretched less because the buoyant force helps support the object.

Bernoulli's Effect

The pressure in a stationary fluid changes with depth but is the same if you move horizontally. If the fluid is moving, however, the pressure can also change in the horizontal direction. Suppose we have a pipe that has a narrow section

FLAWED REASONING

Two wooden blocks with the same size and shape are floating in a bucket of water. Block A floats low in the water, and block B floats high, as shown in the following figure.

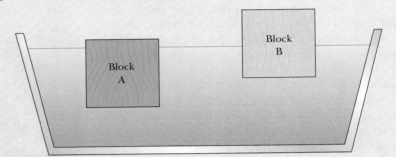

Three students have just come from an interesting lecture on Archimedes' principle and are discussing the buoyant forces on the blocks.

Aubrey: "Block B is floating higher in the water. It must have the greater buoyant force acting on it."

Mary: "You are forgetting Newton's first law. Neither block is moving, so the buoyant force must balance the gravitational force in both cases. The buoyant forces must be equal to each other."

Cassandra: "Archimedes taught us that the buoyant force is *always* equal to the weight of the fluid displaced. Block A is displacing a lot more water than block B, so block A has the larger buoyant force."

Do you agree with any of these students?

ANSWER Cassandra is correct. Archimedes' principle *always* applies, regardless of whether an object sinks or floats. Block A displaces the most water, so it experiences the larger buoyant force. Mary starts out with correct ideas but then draws a faulty conclusion. The buoyant force on either block must equal the gravitational force on that block (by Newton's first law), so block A must be heavier. Because both blocks have the same volume, block A must be made of a denser wood. Perhaps block A is made of oak, and block B is made of pine.

like the one shown in Figure 12-11. If we put pressure gauges along the pipe, the surprising finding is that the pressure is lower in the narrow region of the pipe. If the fluid is not compressible, the fluid must be moving faster in the narrow region; that is, the same amount of fluid must pass by every point in the pipe, or it would pile up. Therefore, the fluid must flow faster in the narrow regions. This may lead one to conclude incorrectly that the pressure would be higher in this region. Swiss mathematician and physicist Daniel Bernoulli stated the correct result as a principle.

Figure 12-11 The pressure is smaller in the narrow region of the pipe where the velocity of the fluid is greater.

Figure 12-12 The "cube" of fluid entering the narrow region of the pipe must experience a net force to the right.

The pressure in a fluid decreases as its velocity increases.

◄ Bernoulli's principle

We can understand **Bernoulli's principle** by "watching" a small cube of fluid flow through the pipe (Figure 12-12). The cube must gain kinetic energy as it speeds up entering the narrow region. Because there is no change in its gravitational potential energy, there must be a net force on the cube that does work on it. Therefore, the force on the front of the cube must be less than on the back. That is, the pressure must decrease as the cube moves into the narrow region. As the cube of fluid exits from the narrow region, it slows down. Therefore, the pressure must increase again.

There are many examples of Bernoulli's effect in our everyday activities. Smoke goes up a chimney partly because hot air rises but also because of the Bernoulli effect. The wind blowing across the top of the chimney reduces the pressure and allows the smoke to be pushed up. This effect is also responsible for houses losing roofs during tornadoes (or attacks by big bad wolves). When a tornado reduces the pressure on the top of the roof, the air *inside* the house lifts the roof off.

A fluid moving past an object is equivalent to the object moving in the fluid, so the Bernoulli effect should occur in these situations. A tarpaulin over the back of a truck lifts up as the truck travels down the road because of the reduced pressure on the outside surface of the tarpaulin produced by the truck moving through the air. This same effect causes your car to be sucked toward a truck as it passes you going in the opposite direction. The upper surfaces of airplane wings are curved so that the air has to travel farther to get to the back edge of the wing. Therefore, the air on top of the wing must travel faster than that on the underside and the pressure on the top of the wing is less, providing lift to keep the airplane in the air.

A tornado caused the difference in the air pressures inside and outside this house that tore off the roof.

Summary

Density is an inherent property of a substance and is defined as the amount of mass in one unit of volume.

Elements combine into substances that can exist in four states of matter: solids, liquids, gases, and plasmas. The transitions between states occur when energy is supplied to or taken from substances. When a solid is heated above its melting point, interatomic bonds break to form a liquid in which atoms and molecules are free to move about. Upon further heating, the molecules totally separate to form a gas. In the plasma state, the atoms have been torn apart, producing charged ions and electrons. Although plasma is rare on Earth, it is the most common state in the universe.

The electric forces between atoms bind all materials together. If the atoms are ordered, a crystalline structure results. Liquids take the shape of their container, and most lack an ordered arrangement of their molecules. The intermolecular forces in a liquid create a surface tension that holds the molecules

Everyday Physics | *The Curve Ball*

When a ball moves through air, strange things can happen. Perhaps the most common examples are the curve in baseball, the slice in golf, and the topspin serve in tennis. Isaac Newton wrote about the unusual behavior of spinning tennis balls, and baseball players and scientists have debated the behavior of the curve ball since the first baseball was thrown more than a hundred years ago. The balls naturally follow projectile paths because of gravity (Chapter 5), but it's the extra motion that is the bane of all batters.

Early on, the debate centered on whether the curve ball even existed. Scientists, believing that the only forces on the ball were gravity and air resistance (drag), argued that the curve ball must be just an optical illusion. "Not true!" retorted the baseball players. "It's like the ball rolled off a table just in front of the plate."

When there's a debate about the material world, the best procedure is to devise an experiment—that is, to ask the question of the material world itself. In the early 1940s, *Life* magazine commissioned strobe photos of a curve ball and concluded that the scientists were right: The curve ball is an optical illusion. Not to be outdone, *Look* magazine commissioned its own photos and concluded that the scientists (and *Life* magazine!) were wrong.

More recently, three scientists reexamined the question. They found a dark warehouse, a bank of strobe lights, and—most important—a professional pitcher. After careful analysis the verdict was clear: the ball does indeed curve away from the projectile path, and the deviation is created by the ball's spin. If the ball travels at 75 mph, it takes about $\frac{1}{2}$ second to travel the 60 feet to the batter. During this time, the ball rotates about 18 times and deviates from the projectile path by about 1 foot.

The direction of the deviation depends on the orientation of the spin. The deviation is always perpendicular to the axis of the spin; therefore, spin around a vertical axis moves the ball left or right. Because this would only change the point of contact with a horizontal bat, it is not very effective. Spin around a horizontal axis causes the ball to move up or down. Backspin (the bottom of the ball moving toward the batter) causes the ball to stay above the projectile path, which helps the batter. The best situation (for the pitcher!) is topspin. This increases the drop as the ball approaches the batter.

The spinning ball causes the airflow to deflect upward, imparting a downward force on the ball.

The question of when the ball deviates was also answered by this experiment. Batters have claimed for years that the ball travels along its normal path and then breaks at the last moment. Scientists claim that the forces—both gravity and the one caused by the spin—are constant; thus, the path is a continuous curve. Our study of projectile motion has shown that a ball falls farther during each succeeding second. This is compounded by the extra drop due to the spin. Thus, the vertical speed is much faster near the plate. But the drop is a continuous one; it does not abruptly change.

We are now left with the question of what causes the downward force. The baseball's cotton stitches—216 on a regulation ball—grab air, creating a layer of air that is carried around the spinning ball. (An insect sitting on the spinning ball would feel no wind—just as dust on a fan's blades is undisturbed by the fan's rotation.) To account for the force, we must look at the turbulence, or wake, behind the ball. The airflow over the top of the ball has a larger speed relative to the surrounding air and breaks up sooner than the airflow under the bottom, as shown in the figure. This causes the wake behind the ball to be shifted upward. According to Newton's third law, the momentum imparted to the wake in the upward direction causes an equal momentum to be imparted downward on the ball, and the curve ball drops.

Let the games begin!

The blue curve represents the ball's path due to its spin without any gravity. The green curve is the ball's path due to gravity without spin. The red curve shows the combined effect of spin and gravity.

to the liquid. A gas fills the container holding it, assuming its shape and volume. All gases are compressible and can be readily mixed with each other. The viscosity of a fluid determines how easily it pours and what resistance it offers to objects moving through it.

The pressure in a liquid or gas varies with depth because of the weight of the fluid above that point. At sea level the average atmospheric pressure is about 101 kilopascals (14.7 pounds per square inch).

An object in a fluid experiences a buoyant force equal to the weight of the fluid displaced; therefore, all objects appear to weigh less in water than in air. The buoyant force exists because the pressure in a fluid varies with depth. The pressure on the bottom surface of an object is greater than on its top surface. Objects less dense than the fluid float. The pressure in a moving fluid decreases with increasing speed.

CHAPTER 12 : *Revisited*

The crystal's macroscopic shape results from a growth process that adds to the overall structure of the crystal, atom by atom. Study of these crystalline shapes gives scientists clues about the way atoms combine.

Key Terms

Archimedes' principle   The buoyant force is equal to the weight of the displaced fluid.

Bernoulli's principle   The pressure in a fluid increases as its velocity decreases.

buoyant force   The upward force exerted by a fluid on a submerged or floating object. (See *Archimedes' principle*.)

crystal   A material in which the atoms are arranged in a definite geometric pattern.

density   A property of a material equal to the mass of the material divided by its volume. Density is measured in kilograms per cubic meter.

gas   Matter with no definite shape or volume.

liquid   Matter with a definite volume that takes the shape of its container.

plasma   A highly ionized gas with equal numbers of positive and negative charges.

solid   Matter with a definite size and shape.

viscosity   A measure of the internal friction within a fluid.

Questions and exercises are paired so that most odd-numbered are followed by a similar even-numbered.

Blue-numbered questions and exercises are answered in Appendix B.

ʃ indicates more challenging questions and exercises.

WebAssign Many Conceptual Questions and Exercises for this chapter may be assigned online at WebAssign.

Conceptual Questions

1. What are the four states of matter?

2. Is the average kinetic energy of the molecules in a liquid greater or smaller than in a solid of the same material? Why?

3. Does the aluminum in a soda can or in an automobile engine have the larger density? Why?

4. Which has a greater density, a tiny industrial diamond used in grinding powders or a 3-carat diamond in a wedding ring? Explain.

5. Aluminum and magnesium have densities of 2.70 and 1.75 grams per cubic centimeter, respectively. If you have equal masses of each, which one will occupy the larger volume? Explain.

6. Gold and silver have densities of 19.3 and 10.5 grams per cubic centimeter, respectively. If you have equal volumes of each, which one will have the larger mass? Explain.

7. Although the uranium atom is more massive than the gold atom, gold has the larger density. What does this tell you about the two solids?

8. People who live in cold climates know better than to turn off the heat in their homes in the winter without first draining the water out of the copper pipes. Explain why the pipes will burst if this is not done.

9. Are the crystal structures of ice and table salt the same? How do you know?

10. How does the crystal structure of mica differ from that of table salt?

11. How does the structure of diamond differ from that of graphite?

12. What does the observation that mica can be separated into thin sheets tell you about the crystal structure of mica?

13. When solids melt to form liquids, they retain their chemical identity. Which is stronger, the forces holding atoms together to form molecules or the forces holding the molecules together in a solid? Why?

14. Which is stronger in a typical liquid, the intermolecular forces (those between molecules) or the interatomic forces (those holding the molecule together)? What evidence do you have for your answer?

15. What shape would you expect a drop of water to take if it were suspended in the air in the space shuttle?

16. Why does water bead up when it is spilled on a waxed floor?

17. If you fill a glass with water level with the top of the glass, you can carefully drop several pennies into the glass without spilling any water. How do you explain this?

18. Why does soapy water bead up less than plain water on a countertop?

19. When you half-fill a glass with water, the water creeps up where it meets the glass. What can you conclude about the relative strengths of the intermolecular forces between the water molecules and the adhesive forces between the water molecules and the glass?

© Cengage Learning/Charles D. Winters

Questions 19 and 20

20. When you half-fill a glass with mercury, the mercury curls down where it meets the glass. What can you conclude about the relative strengths of the interatomic forces between the mercury atoms and the adhesive forces between the mercury atoms and the glass? (*Caution:* Mercury is toxic and should *not* be handled.)

21. How does a gas differ from a plasma?

22. What state of matter forms the Van Allen belts?

23. Use the concept of pressure to explain why it is more comfortable to walk in bare feet across a paved driveway than across a gravel driveway.

24. A closed book that is initially lying flat on a desk is turned to balance on its spine. Compare the forces exerted by the book on the table in the two orientations. Compare the pressures exerted by the book on the table in the two orientations.

25. You place a small amount of water in a 1-gallon can and bring it to a rapid boil. You take the can off the stove and screw the cap on tightly (the order is important here!). As the steam inside cools, it condenses back into water, causing the can to collapse. Why?

26. At sea level each square inch of surface experiences a force of 14.7 pounds due to air pressure. You are carrying a cookie sheet loaded with chocolate chip cookies. The surface area of the cookie sheet is 250 square inches, which means that the downward force exerted by the air column above the cookie sheet is 3675 pounds! Why doesn't the cookie sheet feel this heavy?

27. You repeat von Guericke's experiment (see Figure 12-7) using somewhat smaller half spheres and find that two teams of eight horses are just strong enough to pull the half spheres apart. You then transport the apparatus to Denver, which is at an elevation of 1 mile. Will you need more or fewer horses to pull the half spheres apart? Why?

28. A classmate explains that if your bathroom scale reads 150 pounds when you stand on it at sea level, it will read only 75 pounds on the top of an 18,000-foot mountain, where atmospheric pressure is reduced by one-half. What is wrong with your classmate's reasoning? Would you expect the scale reading to be reduced at all as a result of the decrease in atmospheric pressure?

29. On a weather map, you see areas of low pressure marked with an L and areas of high pressure marked with an H. By convention, the pressures reported are always corrected to the value they would have at sea level. If this were not the case, what letter would you see permanently above the mile-high city of Denver? Explain.

⌇ 30. Mountaineers often carry altimeters that measure altitude by measuring atmospheric pressure. If a low-pressure weather system moves in, will the altimeter report an altitude that is higher or lower than the true altitude? Explain.

31. Are your ears going to hurt more because of water pressure if you are swimming 12 feet down in your swimming pool or 12 feet down in the middle of Lake Superior? Explain.

32. How was it possible for the little Dutch boy to hold back the entire North Sea by putting his finger in the dike?

33. Compare the pressures at the bottom of the two glasses shown in the following figure. Assume that both are filled to the same depth with the same fluid.

34. Two identical wooden barrels are fitted with long pipes extending out their tops. The pipe on the first barrel is 1 foot in diameter, and the pipe on the second barrel is only $\frac{1}{2}$ inch in diameter. When the larger pipe is filled with water to a height of 20 feet, the barrel bursts. To burst the second barrel, will water have to be added to a height less than, equal to, or greater than 20 feet? Explain.

35. Fresh water has a density of 1000 kilograms per cubic meter at 4°C and 998 kilograms per cubic meter at 20°C. In which temperature water would you feel the greater pressure at a depth of 10 meters? Why?

36. Salt water is more dense than fresh water. This means that the mass of 1 cubic centimeter of salt water is larger than that of 1 cubic centimeter of fresh water. Would a scuba diver have to go deeper in salt water or in fresh water to reach the same pressure? Why?

37. Why can't water be "sucked" to a height greater than 10 meters even with a very good suction pump?

38. At sea level even a perfect vacuum can raise water only 10 meters up a straw. At an elevation of 5000 feet in Bozeman, Montana, can water be raised to a height greater than, equal to, or less than 10 meters? Explain your reasoning.

39. If you have a water well that is much deeper than about 5 meters, you put the pump at the bottom of the well and have it push the water up. Why is this better than placing the pump at the top?

40. You place a long straw in a glass of water and find that, no matter how hard you suck, you cannot drink the water. You place the same straw in an unknown liquid X and find that you can drink. If you combine liquid X and water together in a glass, which one will float on the surface?

41. Some toys contain two different-colored liquids that do not mix. If the purple liquid always sinks in the clear liquid as shown in the toy in the following figure, what can you say about the densities of the liquids?

© Cengage Learning/George Semple

42. Spilled gasoline can sometimes be seen as a colorful film on the top of rain puddles. What does this tell you about the density of gasoline?

43. Salt water is slightly more dense than fresh water. Will a boat float higher in salt water or fresh water?

44. Use Archimedes' principle to explain why an empty freighter sits higher in the water than a loaded one.

45. Salt water is slightly more dense than fresh water. Will a 50-ton ship feel a greater buoyant force floating in a freshwater lake or in the ocean?

46. Salt water is slightly more dense than fresh water. Will a 12-pound bowling ball feel a greater buoyant force sitting on the bottom of a freshwater lake or on the bottom of the ocean?

47. When you blow air from your lungs, you are changing both your mass and your volume. Which of these effects explains why this causes you to sink to the bottom of a swimming pool?

48. What happens to the depth of a scuba diver who takes a particularly deep breath?

49. Use the data in Table 12-1 to determine whether a block of lead would float in liquid mercury. What about a block of gold? (*Caution:* Do *not* try this experiment; mercury is very toxic!)

50. Use the data in Table 12-1 to determine which would float higher in liquid mercury, a block of copper or a block of silver.

51. A submarine could be made to surface by either increasing the buoyant force or decreasing the weight. When a submarine's ballast tanks are blown out, which is happening?

52. A scuba diver achieves neutral buoyancy by adjusting the volume of air in her air vest so that the buoyant force equals her weight. If she then kicks her feet and swims down an additional 20 feet, will the net force now be upward, zero, or downward? Explain.

53. You have two cubes of the same size, one made of aluminum and the other of lead. Both cubes are allowed to sink to the bottom of a water-filled aquarium. Which cube, if either, experiences the greater buoyant force? Why?

54. You have two cubes of the same size, one made of wood and the other of aluminum. Both cubes are placed in a water-filled aquarium. The wooden block floats, and the aluminum block sinks. Which cube, if either, experiences the greater buoyant force?

55. An ice cube is floating in a glass of water. Will the water level in the glass rise, go down, or stay the same as the ice cube melts? Why?

56. You are sitting in a boat in your swimming pool. There are six gold bricks in your boat. (You are rich!) If you throw the gold into the swimming pool, does the water level in the pool rise, fall, or stay the same? Explain.

57. You place a dime flat on a tabletop a couple of inches from the edge. With your mouth near the edge of the table, you blow sharply across the top of the dime. Why does the dime pop up in the air? Try this.

58. Why does your car get pulled sideways when a truck passes you going in the opposite direction on a two-lane highway?

59. Why do table-tennis players put a lot of topspin on their shots?

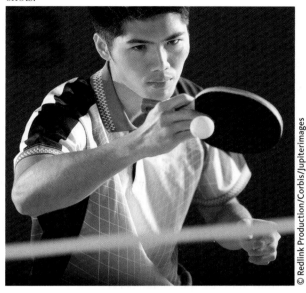

60. A partial vacuum can be created by installing a pipe at a right angle to a water faucet and turning on the water, as shown in the following figure. What is the physics behind this?

Vacuum created

61. The Green Building at the Massachusetts Institute of Technology (MIT) is a tall tower built on an inverted U-shaped base that is open to the Charles River Basin. Why might the doors in the opening have opened "by themselves" on windy days before revolving doors were installed to correct the design flaw?

62. Why would an aneurysm (a widening of an artery) be especially subject to rupturing?

Exercises

63. What is the density of a substance that has a mass of 27 g and a volume of 10 cm³? Use Table 12-1 to identify this substance.

64. A small ball has a mass of 6.75 g and a volume of 0.3 cm³. Can you identify the material using Table 12-1?

65. A bowling trophy has a mass of 180 g. When placed in water, the trophy displaces 600 cm³. What is the average density of the trophy?

66. If Archimedes' crown had a mass of 1 kg and a volume of 120 cm³, was the crown made of pure gold? Explain.

67. A solid ball with a volume of 0.4 m³ is made of a material with a density of 3000 kg/m³. What is the mass of the ball?

68. What is the mass of a lead sinker with a volume of 3 cm³?

69. Given that most people are just about neutrally buoyant, it is reasonable to estimate the density of the human body to be about that of water. Use this assumption to find the volume of a 70-kg person.

70. A cube with a mass of 48 g is made from a metal with a density of 6 g/cm^3. What is the volume of the cube and the length of each edge?

71. If 1000 cm^3 of a gas with a density of 0.0009 g/cm^3 condenses to a liquid with a density of 0.9 g/cm^3, what is the volume of the liquid?

72. A cube of ice, 10 cm on each side, is melted into a measuring cup. What is the volume of the liquid water?

73. Calculate the weight of a column of fresh water with cross-sectional area 1 m^2 and height 10 m. What pressure does this create at the bottom of the column of water? How does this compare to atmospheric pressure?

74. Calculate the height of a column of mercury with cross-sectional area 1 m^2 such that it has the same weight as the column of water in Exercise 73.

75. Given that atmospheric pressure drops by a factor of 2 for every gain in elevation of 18,000 ft, what is the height of a mercury column in a barometer located in an unpressurized compartment of an airliner flying at 36,000 ft?

76. Two barometers are made with water and mercury. If the mercury column is 30 in. tall, how tall is the water column?

77. Each cubic inch of mercury has a weight of 0.5 lb. What is the pressure at the bottom of a column of mercury 30 in. tall if there is a vacuum above the mercury?

78. If 1 m^3 of water has a mass of 1000 kg, what is the pressure at a depth of 150 m? Is the atmospheric pressure important?

79. An object has a mass of 150 kg and a volume of 0.2 m^3. What is its average density? Will this object sink or float in water?

80. Will an object with a mass of 1000 kg and a volume of 1.6 m^3 float?

81. A 500-g wooden block is lowered carefully into a completely full beaker of water and floats. What is the weight of the water, in newtons, that spills out of the beaker?

82. A 400-cm^3 block of aluminum ($D = 2.7$ g/cm^3) is lowered carefully into a completely full beaker of water. What is the weight of the water, in newtons, that spills out of the beaker?

83. A ball of wax is lowered carefully into a completely full beaker of water, where it floats. This causes 18 cm^3 of water to spill out. The same ball of wax is then lowered carefully into a completely full beaker of ethyl alcohol ($D = 0.79$ g/cm^3), where it sinks, causing 20 cm^3 of alcohol to spill out. Which of these two experiments allows you to find the wax's mass, and which allows you to find its volume? Find the density of the wax.

84. A yellow object is lowered carefully into a completely full beaker of water, where it floats. This causes 28 cm^3 of water to spill out. The same object is then lowered carefully into a completely full beaker of gasoline ($D = 0.68$ g/cm^3), where it sinks, causing 40 cm^3 of gasoline to spill out. In which liquid does the yellow object experience the greater buoyant force?

85. A cubic meter of copper has a mass of 8930 kg. The block of copper is lowered into a lake by a strong cable until the block is completely submerged. Draw a free-body diagram for the block. Find the buoyant force on the block and the tension in the cable.

86. A ball fully submerged in a bathtub has a volume of 5 cm^3 and a mass of 30 g. Draw a free-body diagram for the ball. What is the normal force of the tub on the ball?

13

Thermal Energy

▶ This false-color thermograph of a human head shows the temperature variations of the surface. What factors control the rate at which radiation is emitted or absorbed and the resulting temperature changes?

(See page 278 for the answer to this question.)

Mark Harmel/FPG/Getty

False-color thermograph of a human head.

F we examine any system of moving objects carefully or if we look at it long enough, we find that mechanical energy is not conserved. A pendulum bob swinging back and forth does in fact come to rest. Its original mechanical energy disappears.

Other examples show the same thing. Rub your hands together. You are doing work—applying a force through a distance—but clearly your hands do not fly off with some newfound kinetic energy. Similarly, take a hammer and repeatedly strike a metal surface. The moving hammer has kinetic energy, but on hitting the surface, its kinetic energy disappears. What happens to the energy? It is not converted to potential energy as happened in Chapter 7 because the energy doesn't reappear. So either the kinetic energy truly disappears and total energy is not conserved, or it is transferred into some form of energy that is not a potential energy.

There are similarities in the preceding examples. When you rub your hands together, they feel hot. The metal surface and the hammer also get hotter when they are banged together. The pendulum bob is not as obvious; the interactions are between the bob and the surrounding air molecules and between the string and the support. But closer examination shows that, once again, the system gets hotter.

At first glance it is tempting to suggest that temperature, or maybe the change in temperature, could be equated with the lost energy. However, neither of these ideas works. If the same amount of energy is expended on a collection of different objects, the resulting temperature increases are not equal. Suppose, for example, that we rub two copper blocks together. The temperature of the copper blocks increases. If we repeat the experiment by expending the same amount of mechanical energy with two aluminum blocks, the change in temperature will not be the same. The temperature change is an indication that something has happened, but it is not equal to the lost energy.

The Nature of Heat

Early ideas about the nature of heat centered on the existence of a fluid that was supposedly transferred between objects at different temperatures. Over centuries people had noted that a kettle of water could become hot, boil, and turn to steam or that a snowbank could absorb the Sun's heat for an extended period and slowly melt. Fire transferred something to the hot water to make it boil; sunshine imparted something to the snow to liquefy it over time. This "fluid" was studied intensively in the era of early steam-engine technology. It became known as *caloric,* from the Latin *calor,* meaning "heat." It was invisible and presumably massless because experimenters could not detect any changes in the mass of an object that was heated.

Count Rumford, an 18th-century British scientist, pioneered a study of work and heat. At that time he was in charge of boring cannons at a military arsenal in Munich and was struck by the enormous amount of heat produced during the boring process. Rumford decided to investigate this. He placed a dull boring tool and a brass cylinder in a barrel filled with cold water. The boring tool was forced against the bottom of the cylinder and rotated by two horses. These are the results described by Rumford:

> At the end of 2 hours and 30 minutes it [the water in the barrel] actually boiled! . . . It would be difficult to describe the surprise and astonishment expressed by the countenances of the by-standers, on seeing so large a quantity of cold water heated, and actually made to boil without any fire.

Rumford showed that large quantities of heat could be produced by mechanical means without fire, light, or chemical reaction. (This is a large-scale version of the simple hand-rubbing experiment.) The importance of his experiment was the demonstration that the production of heat seemed inexhaustible. As long as the horses turned the boring tool, heat was generated without any limitation. He concluded that anything that could be produced without limit could not possibly be a material substance. Heat was not a fluid but something generated by motion.

In our modern physics world view, **heat** is energy *flowing* between two objects because of a difference in temperature. We measure the amount of energy gained or lost by an object by the resulting temperature change in the object. By convention, 1 **calorie** is defined as the amount of heat that raises the temperature of 1 gram of water by 1°C. In the U.S. customary system, the unit of heat, called a **British thermal unit** (Btu), is the amount of energy needed to change the temperature of 1 pound of water by 1°F. One British thermal unit is approximately equal to 252 calories.

Q: How many calories are required to raise the temperature of 8 grams of water by 5°C?

A: To raise the temperature of 1 gram by 5°C requires 5 calories. Therefore, 8 grams requires 5 calories/gram × 8 grams = 40 calories.

Are You On the Bus?

Mechanical Work and Heat

Extended presentation available in the *Problem Solving* supplement

The Rumford experiment used the *work* supplied by the horses to raise the temperature of the water, clearly demonstrating an equivalent way of "heating" the water. The water got hotter as if it were heated by a fire, but there was no fire.

There is a close connection between work and heat. Both are measured in energy units, but neither resides in an object. In Chapter 7 we saw that work was a measure of the energy "flowing" from one form to another. For example, the gravitational force does work on a free-falling ball, causing its kinetic energy to increase—thus, potential energy changes to kinetic energy. Similarly, heat does not reside in an object but flows into or out of an object,

Q: Because joules and calories are both energy units, do we need to retain both of them?

A: No. However, both are currently used for historical reasons. Europeans are much further along than Americans in converting from Calories to kilojoules in the labeling of food.

1 calorie = 4.2 joules ▶

Figure 13-1 Joule's apparatus for determining the equivalence of work and heat. The decrease in the gravitational potential energy of the falling mass produces an increase in the energy of the water.

changing the internal energy of the object. This internal energy is sometimes known as **thermal energy**, and the area of physics that deals with the connections between heat and other forms of energy is called **thermodynamics**.

Although Rumford's experiment hinted at the equivalence between mechanical work and heat, James Joule uncovered the quantitative equivalence 50 years later. Joule's experiment used a container of water with a paddle-wheel arrangement like that shown in Figure 13-1. The paddles are connected via pulleys to a weight. As the weight falls, the paddle wheel turns, and the water's temperature goes up. The potential energy lost by the falling weight results in a rise in the temperature of the water. Because Joule could raise the water temperature by heating it or by using the falling weights, he was able to establish the equivalence between the work done and the heat transferred. Joule's experiment showed that 4.2 joules of work are equivalent to 1 calorie of heat.

There are other units of energy. The Calorie used when referring to the energy content of food is not the same as the calorie defined here. The food Calorie (properly designated by the capital C to distinguish it from the one used in physics) is equal to 1000 of the physics calories. A piece of pie rated at 400 Calories is equivalent to 400,000 calories of thermal energy, or nearly 1.7 million joules of mechanical energy.

Temperature Revisited

If we bring two objects at different temperatures into contact with each other, there is an energy flow between them, with energy flowing from the hotter object to the colder. We know from the structure of matter (Chapter 11) that the molecules of the hotter object have a higher average kinetic energy. Therefore, on the average, the more-energetic particles of the hotter object lose some of their kinetic energy when they collide with the less-energetic particles of the colder object. The average kinetic energy of the hotter object's particles decreases and that of the colder object's particles increases until they become equal. On a macroscopic scale, the temperature changes for each object: the hotter object's temperature drops, and the colder object's temperature rises. The flow of energy stops when the two objects reach the same temperature, a condition known as **thermal equilibrium**. Atomic collisions still take place, but on average the particles do not gain or lose kinetic energy.

Suppose that we have two objects, labeled A and B, that cannot be placed in thermal contact with each other. How can we determine whether they would be in thermal equilibrium if we could bring them together? Let's assume that we have a third object, labeled C, that can be placed in thermal contact with A and that A and C are in thermal equilibrium. If C is now placed in thermal contact with B and if B and C are also in thermal equilibrium, then we can conclude that A and B are in thermal equilibrium. This is summarized by the statement of the **zeroth law of thermodynamics**:

zeroth law of thermodynamics ▶

If objects A and B are in thermal equilibrium with object C, then A and B are in thermal equilibrium with each other.

Joule : *A New View of Energy*

James Prescott Joule (1818–1889) was the second son of a family of wealthy brewers in the village of Salford near Manchester, England, an industrial region. Joule was tutored as a youngster by John Dalton, a noted, elderly chemist, and determined early in life to pursue physical science as a serious hobby.

Today Joule is remembered for his experiments and theories on work, energy, and heat. Michael Faraday was an early, major source of inspiration, and Joule conducted experiments on heating in conducting wires—mostly made of copper or platinum. His energy sources were voltaic batteries and the dynamo. This early work earned him the reputation of a pioneer in battery design and efficiency. His views on the equivalence of heat and work derived from these early electrochemical experiments.

Manchester was a city of steam engines, so it was natural that Joule and his friends in the local scientific and engineering group would discuss theories of heat and means of improving engine performance. His own work led in a general way to the concept of thermal efficiency, but he contributed little to steam technology because he thought engineers were better suited to technical improvements. The physicists' job was to extend the power of theory.

Joule excelled in precision measurement. He was fortunate that a local Manchester firm could manufacture calorimeters and thermometers for him. By 1840 he had calibrated temperature differences to an accuracy of 0.01°F, and in later experiments he used

The Royal Society, London/The Bridgeman Art Library

James Prescott Joule

calibrations one-half this size. Many of his peers at that time underestimated the importance of precision measurements. Joule's appointment to the first major British commission on scientific standards validated his efforts.

As his conceptual grasp of the issues evolved, Joule recognized and measured the equivalence between mechanical work and heat in several ways. One of the most striking was the accurate measurement of the increase in temperature of water as it fell over a waterfall into a pool. His first observations were in France, but he speculated about Niagara Falls, which he had not visited, and accurately predicted the increase.

At a meeting during which Joule delivered a paper that most in the audience did not understand completely, he met an important collaborator, William Thomson, known later as Lord Kelvin. Kelvin's mathematical skills combined with Joule's careful measurements solidified understanding of the first and second laws of thermodynamics. Among other contributions, their work presaged the rise of mechanical refrigeration later in the 19th century. Current concepts of energy and the use of precision measurements owe much to the quiet and unassuming work performed by James Prescott Joule.

—*Pierce C. Mullen, historian and author*

Sources: Mary B. Hesse, *Forces and Fields: The Concept of Action at a Distance in the History of Physics* (New York: Philosophical Library, 1962); Henry John Steffens, *James Prescott Joule and the Concept of Energy* (New York: Science History Publications, 1979).

Although this statement may seem to be so obvious that it is not worth elevating to the stature of a law, it plays a fundamental role in thermodynamics because it is the basis for the definition of temperature. Two objects in thermal equilibrium have the same temperature. On the other hand, if two objects are not in thermal equilibrium, they must have different temperatures. The zeroth law was developed later in the history of thermodynamics but labeled with a zero because it is more basic than the other laws of thermodynamics.

Heat, Temperature, and Internal Energy

Heat and temperature are not the same thing. Heat is an energy, whereas temperature is a macroscopic property of an object. Two objects can be at the same temperature (the same average atomic kinetic energy) and yet transfer vastly different amounts of energy to a third object. For example, a swimming pool of water and a coffee cup of water at the same temperature can melt very different amounts of ice.

When we consider the total microscopic energy of an object—such as translational and rotational kinetic energies, vibrational energies, and the energy stored in molecular bonds—we are talking about the **internal energy** of the object. There are two ways of increasing the internal energy of a system. One way is to heat the system; the other is to do work on the system. The law of conservation of energy tells us that the total change in the internal energy of the system is equal to the change due to the heat added to the system plus that due to the work done on the system. This is called the **first law of thermodynamics** and is really just a restatement of the law of conservation of energy.

first law of thermodynamics ▶ | The increase in the internal energy of a system is equal to the heat added plus the work done on the system.

This law sheds more light on the nature of internal energy. Let's assume that if 10 calories of heat are added to a sample of gas, its temperature rises by 2°C. If we add the same 10 calories to a sample of the same gas that has twice the mass, we discover that the temperature rises by only 1°C (Figure 13-2). Adding the same amount of heat does not produce the same rise in temperature. This makes sense because the larger sample of gas has twice as many particles, and therefore each particle receives only half as much energy on the average. The average kinetic energy, and thus the temperature, should increase by half as much. An increase in the temperature is an indication that the internal energy of the gas has increased, but the mass must be known to predict how much it increases.

Figure 13-2 Adding equal amounts of heat to different amounts of a material produces different temperature changes.

Absolute Zero

The temperature of a system can be lowered by removing some of its internal energy. Because there is a limit to how much internal energy can be removed, it is reasonable to assume that there is a lowest possible temperature. This temperature is known as **absolute zero** and has a value of −273°C, the same temperature used to define the zero of the Kelvin scale.

The existence of an absolute zero raised the challenge of experimentally reaching it. The feasibility of doing so was argued extensively during the first three decades of the 20th century, and scientists eventually concluded that it was impossible. This belief is formalized in the statement of the **third law of thermodynamics**:

third law of thermodynamics ▶ | Absolute zero may be approached experimentally but can never be reached.

There appears to be no restriction on how close experimentalists can get, only that it cannot be reached. Small systems in low-temperature laboratories have reached temperatures closer than a few billionths of a degree to absolute zero.

A substance at absolute zero has the lowest possible internal energy. Originally, it was thought that all atomic motions would cease at absolute zero. The development of quantum mechanics (Chapter 24) showed that all motion does not cease; the atoms sort of quiver with the minimum possible motion. In this state the atoms are packed closely together. Their mutual binding forces arrange them into a solid block.

Specific Heat

Suppose we have the same number of molecules of two different gases, and each gas is initially at the same temperature. If we add the same amount of heat

Figure 13-3 Three forms of internal energy for a diatomic molecule.

to each gas, we find that the temperatures do not rise by the same amount. Even though the gases undergo the same change in their internal energies, their molecules do not experience the same changes in their average translational kinetic energies. Some of the heat appears lost. Actually, the heat is transformed into other forms of energy. If the gas molecules have more than one atom, part of the internal energy is transformed into rotational kinetic energy of the molecules and part of it can go into the vibrational motion of the atoms (Figure 13-3). Only a small fraction of the increase in internal energy for most real gases goes into increasing the average kinetic energy that shows up as an increase in temperature.

The amount of heat it takes to increase the temperature of an object by 1°C is known as the **heat capacity** of the object. The heat capacity depends on the amount and type of material used to construct the object. An object with twice the mass will have twice the heat capacity, provided both objects are made of the same material.

We can obtain an intrinsic property of the material that does not depend on the size or shape of an object by dividing the heat capacity by the mass of the object. This property is known as the **specific heat** and is the amount of heat required to increase the temperature of 1 gram of the material by 1°C.

By definition, the specific heat of water is numerically 1; that is, 1 calorie raises the temperature of 1 gram of water by 1°C. The specific heat for a given material in a particular state depends slightly on the temperature but is usually assumed to be constant. The specific heats of some common materials are given in Table 13-1. Notice that the SI units for specific heat are joules per kilogram-kelvin. These are obtained by multiplying the values in calories per gram-degree Celsius by 4186. Note also that the value for water is quite high compared with most other materials.

Table 13-1 *Specific Heats for Various Materials*

Material	Specific Heat (cal/g · °C)	Specific Heat (J/kg · K)
Solids		
Aluminum	0.215	900
Copper	0.092	385
Diamond	0.124	519
Gold	0.031	130
Ice	0.50	2090
Silver	0.057	239
Liquids		
Ethanol	0.75	3140
Mercury	0.033	138
Water	1.00	4186
Gases		
Air	0.24	1000
Helium	1.24	5190
Nitrogen	0.25	1040
Oxygen	0.22	910

Q: What is the rise in temperature when 20 calories is added to 10 grams of ice at −10°C?

A: This is the same as adding 2 calories to each gram. Because $\frac{1}{2}$ calorie is required to raise the temperature of 1 gram of ice by 1°C, the 2 calories will raise its temperature by 4°C.

Are You On the Bus?

When we bring two different objects into thermal contact with each other, they reach thermal equilibrium but don't normally experience the same changes in temperature because they typically have different heat capacities. For example, when unequal amounts of hot and cold water are mixed together, the equilibrium temperature will not be midway between the hot

WORKING IT OUT *Specific Heat*

The specific heat c is obtained by dividing the heat Q added to the material by the product of the mass m and the resulting change in temperature ΔT:

$$c = \frac{Q}{m\Delta T}$$

For example, if 11 cal is required to raise the temperature of an 8-g copper coin 15°C, we can calculate the specific heat of copper:

$$c = \frac{Q}{m\Delta T} = \frac{11\ \text{cal}}{(8\ \text{g})(15°\text{C})} = 0.092\ \text{cal/g} \cdot °\text{C}$$

Note that this agrees with the entry in Table 13-1.

We can rearrange our definition of specific heat to obtain an expression for the heat required to change the temperature of an object by a specific amount. For instance, suppose that you have a cup of water at room temperature that you want to boil. How much heat will this require? Let's assume that the cup contains $\frac{1}{4}$ L of water at 20°C and that we can ignore the heating of the cup itself. The mass of the water is 250 g, and the boiling point of water is 100°C at 1 atm. Therefore, the temperature change is 80°C, and we have

$$Q = cm\Delta T = \left(1\frac{\text{cal}}{\text{g} \cdot °\text{C}}\right)(250\ \text{g})(80°\text{C}) = 20,000\ \text{cal} = 20\ \text{kcal}$$

The 20 kcal of energy must be supplied by the stove or microwave oven.

FLAWED REASONING

A lab manual asks students to mix 400 grams of warm ethanol at 60°C with 300 grams of room-temperature water at 20°C. Before performing this experiment, two students are making predictions for the final temperature of the mixture.

Christian: "The final temperature will be *higher* than 40°C. If the masses were equal, the final temperature would be halfway between 20°C and 60°C, but there is more ethanol than water."

Shannon: "No, the final temperature of the mixture will be *lower* than 40°C. Water has a higher specific heat, so the water will have a smaller change in temperature."

Both of these students are wrong. **Find the flaw in their reasoning.**

ANSWER Christian is focusing on the relative masses, whereas Shannon is focusing on the relative specific heats. Both factors play a role in determining the heat capacity of an object. The equilibrium temperature will be closer to the initial temperature of the fluid with the larger heat capacity. The heat capacities C of the ethanol and water are as follows:

$$C_{ethanol} = c_{ethanol}m_{ethanol} = \left(0.75\frac{\text{cal}}{\text{g} \cdot °\text{C}}\right)(400\ \text{g}) = 300\frac{\text{cal}}{°\text{C}}$$

$$C_{water} = c_{water}m_{water} = \left(1\frac{\text{cal}}{\text{g} \cdot °\text{C}}\right)(300\ \text{g}) = 300\frac{\text{cal}}{°\text{C}}$$

In this example both fluids have the same heat capacity, so the final equilibrium temperature will be 40°C, midway between the two initial temperatures.

and cold temperatures, but closer to the initial temperature of the larger sample. However, conservation of energy tells us that the heat lost by the hotter object is equal to the heat gained by the colder object. (We're assuming that no energy is lost to the environment.)

The specific heats of the materials on Earth's surface account for the temperature extremes lagging behind the season changes. The first day of summer in the Northern Hemisphere usually occurs on June 21. On this day the soil receives the largest amount of solar radiation because it is the longest day of the year and the sunlight arrives closest to the vertical. And yet the hottest days of summer typically occur several weeks later. It takes time for the ground to warm up because it requires a lot of energy to raise its temperature each degree.

Change of State

We continue our investigation of internal energy by continually removing energy from a gas and watching its temperature. If we keep the pressure constant, the volume and temperature of the gas decrease rather smoothly until the gas reaches a certain temperature. At this temperature there is a rapid drop in volume and *no* change in temperature. Drops of liquid begin to form in the container. As we continue to remove energy from the gas, more and more liquid forms, but the temperature remains the same. When all the gas has condensed into liquid, the temperature drops again (Figure 13-4). The change from the gaseous state to the liquid state (or from the liquid to the solid), or vice versa, is known as a **change of state.**

While the gas was condensing into a liquid, energy was continually leaving the system, but the temperature remained the same. Most of this energy came from the decrease in the electric potential energy between the molecules as they got closer together to form the liquid. This situation is analogous to the release of gravitational potential energy as a ball falls toward Earth's surface. The energy that must be released or gained per unit mass of material is known as the **latent heat.** The values of the latent heat for melting and vaporization are given in Table 13-2.

The same processes occur when you heat a liquid. If you place a pan of water on the stove, the temperature rises until the water begins to boil. The temperature then remains constant as long as the water boils. It doesn't matter whether the water boils slowly or rapidly. (Because the rate at which foods cook depends only on the temperature of the water, you can conserve energy

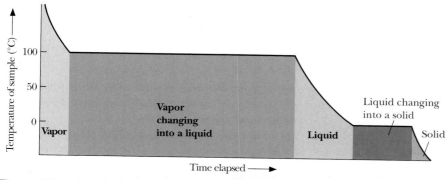

Figure 13-4 A graph of the temperature of water versus time as thermal energy is removed from the water. Notice that the temperature remains constant while the steam condenses to liquid water and while the liquid water freezes to form ice.

Table 13-2 *Melting Points, Boiling Points, and Latent Heats for Various Materials*

Material	Melting Point (°C)	Latent Heat (Melting)		Boiling Point (°C)	Latent Heat (Vaporization)	
		(kJ/kg)	(cal/g)		(kJ/kg)	(cal/g)
Nitrogen	−210	25.7	6.14	−196	199	47.5
Oxygen	−218	13.8	3.3	−183	213	50.9
Water	0	334	79.8	100	2257	539
Aluminum	660	396	94.6	2467	10,900	2600
Gold	1064	63	15	2807	1710	409

by turning the heat down as low as possible while still maintaining a boil.) During the change of state, the additional energy goes into breaking the bonds between the water molecules and not into increasing the average kinetic energy of the molecules. Each gram of water requires a certain amount of energy to change it from liquid to steam without changing its temperature. In fact, this is the same amount of energy that must be released to convert the steam back into liquid water. Furthermore, the temperature at which steam condenses to water is the same as the boiling point. The melting and boiling points for some common substances are given in Table 13-2.

Are You On the Bus?

Q: At the boiling temperature, what determines whether the liquid turns into gas or the gas turns into liquid?

A: If heat is being supplied, the liquid will boil to produce additional gas. However, if heat is being removed, some of the gas will condense to form additional liquid.

The melting of snow and ice in Glacier National Park is a slow process because of the latent heat required to change the ice to liquid water.

A similar change of state occurs when snow melts. The snow does not suddenly become water when the temperature rises to 0°C (32°F). Rather, at that temperature the snow continues to take in energy from the surroundings, slowly changing into water as it does. Incidentally, we are fortunate that it behaves this way; otherwise, we would have gigantic floods the moment the temperature rose above freezing! The latent heat required to melt ice explains why ice can keep a drink near freezing until the last of the ice melts.

On nights when the temperature is predicted to drop below freezing, owners of fruit orchards in California and Florida turn on sprinklers to keep the fruit from freezing. As the water freezes, heat is given off that maintains the temperature of the fruit at 0°C, a temperature above the value at which the fruit freezes. Once the water is completely frozen, the fruit is still protected because ice does not conduct heat very well. The ice serves as a "sweater" for the fruit. However, if the air temperature drops too low, the fruit will be ruined.

Conduction

Thermal energy is transported from one place to another via three mechanisms: conduction, convection, and radiation. Each of these is important in some circumstances and can be ignored in others.

If temperature differences exist within a single isolated object such as a branding iron held in a campfire, thermal energy will flow until thermal equilibrium is achieved. We say that the thermal energy is conducted through the material. **Conduction** takes place via collisions between the particles of the material. The molecules and electrons at the hot end of the branding iron collide with their neighbors, transferring some of their kinetic energy, on the

Figure 13-5 The left end of this ceramic dish is ice-cold while the right end is very hot. This occurs because ceramic is a poor conductor of thermal energy.

Thermal energy is transported along the branding iron by conduction because the brand is hotter than the handle.

average. This increased kinetic energy is passed along the rod via collisions until the end in your hand gets hot.

The rate at which energy is conducted varies from substance to substance. Solids, with their more tightly packed particles, tend to conduct thermal energy better than liquids and gases. The mobility of the electrons within materials also affects the thermal conductivity. Metals such as copper and silver are good thermal **conductors** as well as good electrical conductors. Conversely, electrical insulators such as glass and ceramic are also good thermal **insulators.** A glassblower can hold a glass rod in a flame for a long time without getting burned. The ceramic bowl in Figure 13-5 has regions at drastically different temperatures.

The differences in the conductivity of materials explain why aluminum and wooden benches in a football stadium do not feel the same on a cold day. Before you sit on either bench, they are at the same temperature. When you sit down, some of the thermal energy in your bottom flows into the bench. Because the wooden bench does not conduct the heat well, the spot you are sitting on warms up and feels more comfortable. On the other hand, the aluminum bench continually conducts heat away from your bottom, making your seat feel cold.

The rate at which thermal energy is conducted through a slab of material depends on many physical parameters besides the type of material. You may correctly guess that it depends on the area and thickness of the material. A larger area allows more thermal energy to pass through, and a greater thickness allows less. Experimentation tells us that the difference in the temperatures on the two sides of the slab also matters—the greater the difference, the greater the flow. Table 13-3 gives the thermal conductivities of a variety of common materials.

In our everyday lives, we are more concerned with reducing the transfer of thermal energy than increasing it. We wear clothing to retain our body heat, and we insulate our houses to reduce our heating and air-conditioning bills. Figure 13-6 shows regions of a roof after a snowstorm. The unmelted patches exist where there is better insulation or, in the case of an unheated porch or garage, where there is little or no temperature difference. Table 13-3 allows us

Table 13-3 *Thermal Conductivities for Various Materials*

Material	(W/m · °C)
Solids	
Silver	428
Copper	401
Aluminum	235
Stainless steel	14
Building Materials	
Polyurethane foam	0.024
Fiberglass	0.048
Wood	0.08
Window glass	0.8
Concrete	1.1
Gases	
Air (stationary)	0.026
Helium	0.15

Figure 13-6 Melting snow patterns reveal differences in thermal conduction. For example, the old garage has been converted to living space, but it appears not to have been insulated.

Two climbers prepare their gear while sitting in a snow cave on Passo Superior.

to compare the heat loss through slabs of different materials of the same size and thickness for the same difference in temperatures.

Examination of Table 13-3 reveals that static air is a pretty good insulator. This insulating property of air means that porous substances with many small air spaces are good insulators, and it explains why the goose down used in sleeping bags keeps you so warm. It also explains how thermal-knit long underwear keeps you warm. The air trapped in the holes keeps your body heat from being conducted away.

Snow contains a lot of air space between the snowflakes, which makes snow a good thermal insulator. Mountaineers often dig snow caves to escape from severe weather. Likewise, snow-covered ground does not freeze as deep as bare ground.

Q: Why do people who spend time outdoors in cold weather wear many layers of clothing?

A: In addition to the flexibility of adding and removing layers to get the required insulation, the air spaces between the layers contribute to the overall insulation.

Convection

Thermal energy can also be transferred in fluids by **convection.** In convection the energy is transported by the movement of the fluid. This movement could be forced, as in heating systems or the cooling system in an automobile, or it could happen because of the changes that occur in the density of the fluid when it is heated or cooled. As the gas near the flame of a candle is heated, it becomes less dense and rises because of the buoyant force (Chapter 12).

Convection in Earth's atmosphere plays a fundamental role in our global climate as well as our daily weather. Convection currents arise from the uneven heating of Earth's surface. Glider pilots, hang-glider fliers, and birds of prey (such as hawks and eagles) use convection currents called *thermals* to provide them with the lift they need to keep aloft.

Local winds near a large body of water can be caused by temperature differences between the water and the land. The specific heat of water is much greater than that of rock and soil. (Convection currents in the water also moderate the changes in the water temperature.) During the morning, the land warms up faster than the water. The hotter land heats the air over it, causing the air to rise. The result is a pleasant "sea breeze" of cooler air coming from

Glider pilots search for convective thermals to gain altitude.

70°F 60°F

50°F 60°F

Figure 13-7 The difference in temperature between the land and the water causes breezes to blow (a) onshore during the morning and (b) offshore during the evening.

the water [Figure 13-7(a)]. During the evening, the land cools faster, reversing the convection cycle [Figure 13-7(b)].

Q: What role does convection play in bringing a pot of water to a boil?

A: As the flame or heating element warms the water near the bottom of the pan, it becomes less dense and rises. This circulation causes all of the water to warm up at the same time.

Radiation

The third mechanism for the transfer of thermal energy involves electromagnetic waves. As we will see in Chapter 22, these waves can travel through a vacuum, and thus **radiation** is still effective in situations in which the conduction and convective processes fail. The electromagnetic radiation travels through space and is converted back to thermal energy when it hits other objects. Most of the heat that you feel from a cozy fire is transferred by radiation, especially if the fire is behind glass or in a stove.

All objects emit radiation. Although radiation from objects at room temperature is not visible to the human eye, it can be viewed with special "night glasses" or recorded on infrared-sensitive film. When colors are artificially added, we can distinguish the different temperatures of objects as shown in Figure 13-8.

As the temperature of the object rises, more and more of the radiation becomes visible. Objects such as the heating coils on kitchen stoves glow with a red-orange color. Betelgeuse, the red star marking the right shoulder of the constellation Orion, has a surface temperature of approximately 3000 K. As an object gets hotter, the color shifts to yellow and then white. Our Sun appears white (above Earth's atmosphere) with a temperature of 5800 K. The hottest stars appear blue and have temperatures exceeding 30,000 K.

The radiation that Earth receives from the Sun is typical of an object at 5800 K. If Earth absorbed all of this energy, it would continue to get hotter and hotter until life as we know it could not exist. However, Earth also radiates. Its temperature rises until it radiates as much energy into space as it receives; it reaches equilibrium.

Figure 13-8 This infrared photograph of a toaster shows the different temperatures of its parts.

A phenomenon known as the *greenhouse effect* can have a major effect on the equilibrium condition. Visible light easily passes through the windows of a greenhouse or a car, heating the interior. However, the infrared radiation given off by the interior does not readily pass through the glass and is trapped inside. Only when the temperature reaches a high value is equilibrium established. This is one reason why we are warned to never leave children or pets in a car with the windows rolled up on a hot day.

A similar thing happens with Earth. The atmosphere is transparent to visible light, but the water vapor and carbon dioxide in the atmosphere tend to block the infrared radiation from escaping, causing Earth's temperature to increase. The high surface temperatures on Venus are due to the greenhouse effect of its thick atmosphere. It is feared that increases in the carbon dioxide concentration in Earth's atmosphere (due to such things as the burning of fossil fuels) will cause global warming that in turn will cause unwanted changes in Earth's climate. Such alterations in the climate could change the types of crops that will grow and melt the polar ice caps, flooding coastal cities!

Wind Chill

The meteorologist on television announces the temperatures for the day and then adds that it's going to feel even colder because of the wind. If the air is a certain temperature, why does it matter whether the wind is blowing?

Your body is constantly producing heat that must be released to the environment to keep your body from overheating. The primary way that your body gets rid of excess heat is through evaporation. For each liter (about 1 quart) of water that evaporates, roughly 600 kilocalories of heat are absorbed from your body. Although most of us correctly associate this mechanism with sweating, a surprising 25% of the heat lost by evaporation in a resting individual is due to the evaporation of water from the linings of our lungs into the air we exhale. Vigorous activity can produce sweating at a rate in excess of 2 liters per hour, removing 1200 kilocalories per hour, a rate tens of times larger than for a resting individual.

Another form of heat loss is due to convection of air away from the body. Even with no wind, the air leaves your skin on a cold day because its density changes as your body warms the air. A third form of heat loss is radiation loss. If your body is warmer than the surrounding objects such as the walls in a room, your body radiates energy to the walls. This is why you feel cold some mornings even though the air in the room has been heated to normal room temperature. The walls take some time to warm up, and you will continue to radiate to them until they warm up.

The wind near your body greatly alters the effectiveness of these heat transfers. In stationary air, the layer of air next to your skin becomes warm and moist, reducing the further loss of heat to this layer of air. However, if there is a wind, the wind brings new air to your skin that is colder and drier. Warming and adding moisture to this new air requires additional heat from your body.

In the mid-1940s a single index was created—the *windchill factor*—to express the cooling effects for various ambient temperatures and wind speeds in terms of an "equivalent" temperature with no wind. We can use the data shown in Table 13-4 to find the windchill temperature for a thermometer reading of 25°F on a day when the wind is blowing at 40 mph. Look along the top of the table until you locate the 25°F and then move down this column to the row labeled 40 mph along the left-hand side of the table. The entry at the intersection of this row and this column is the equivalent temperature. Therefore, the cooling effects are equivalent to a temperature of 6°F on a calm day.

Table 13-4 · *Windchill*

Temperature (°F)

Wind (mph) \ Calm	40	35	30	25	20	15	10	5	0	−5	−10	−15	−20	−25	−30	−35	−40	−45
5	36	31	25	19	13	7	1	−5	−11	−16	−22	−28	−34	−40	−46	−52	−57	−63
10	34	27	21	15	9	3	−4	−10	−16	−22	−28	−35	−41	−47	−53	−59	−66	−72
15	32	25	19	13	6	0	−7	−13	−19	−26	−32	−39	−45	−51	−58	−64	−71	−77
20	30	24	17	11	4	−2	−9	−15	−22	−29	−35	−42	−48	−55	−61	−68	−74	−81
25	29	23	16	9	3	−4	−11	−17	−24	−31	−37	−44	−51	−58	−64	−71	−78	−84
30	28	22	15	8	1	−5	−12	−19	−26	−33	−39	−46	−53	−60	−67	−73	−80	−87
35	28	21	14	7	0	−7	−14	−21	−27	−34	−41	−48	−55	−62	−69	−76	−82	−89
40	27	20	13	6	−1	−8	−15	−22	−29	−36	−43	−50	−57	−64	−71	−78	−84	−91
45	26	19	12	5	−2	−9	−16	−23	−30	−37	−44	−51	−58	−65	−72	−79	−86	−93
50	26	19	12	4	−3	−10	−17	−24	−31	−38	−45	−52	−60	−67	−74	−81	−88	−95
55	25	18	11	4	−3	−11	−18	−25	−32	−39	−46	−54	−61	−68	−75	−82	−89	−97
60	25	17	10	3	−4	−11	−19	−26	−33	−40	−48	−55	−62	−69	−76	−84	−91	−98

Frostbite times: 30 minutes · 10 minutes · 5 minutes

Table 13-5 · *Heat Index*

Relative Humidity (%)

Air Temperature (°F)	0	5	10	15	20	25	30	35	40	45	50	55	60	65	70	75	80	85	90	95	100
140	125																				
135	120	128																			
130	117	122	131																		
125	111	116	123	131	141																
120	107	111	116	123	130	139	148														
115	103	107	111	115	120	127	135	143	151												
110	99	102	105	108	112	117	123	130	137	143	150										
105	95	97	100	102	105	109	113	118	123	129	135	142	149								
100	91	93	95	97	99	101	104	107	110	115	120	126	132	138	144						
95	87	88	90	91	93	94	96	98	101	104	107	110	114	119	124	130	136				
90	83	84	85	86	87	88	90	91	93	95	96	98	100	102	106	109	113	117	122		
85	78	79	80	81	82	83	84	85	86	87	88	89	90	91	93	95	97	99	102	105	108
80	73	74	75	76	77	77	78	79	79	80	81	81	82	83	85	86	86	87	88	89	91
75	69	69	70	71	72	72	73	73	74	74	75	75	76	76	77	77	78	78	79	79	80
70	64	64	65	65	66	66	67	67	68	68	69	69	70	70	70	71	71	71	71	71	72

Apparent temperature
is how hot the heat–humidity combination makes it feel.

Extreme danger · Danger · Extreme caution · Caution

Just as wind can increase the rate at which heat leaves the human body, relative humidity slows the evaporation of perspiration, decreasing the rate. The *heat index* in Table 13-5 combines the effects of temperature and relative humidity to yield an apparent temperature, similar to the windchill factor. Heatstroke is highly likely with continued exposure to heat indexes of 130°F or greater. Sunstroke, heat cramps, and heat exhaustion are likely for heat indexes between 105°F and 130°F, and possible for values between 90°F and 105°F. It should be noted that these values are based on shady, light-wind conditions. Exposure to full sunshine can increase the heat-index values by up to 15°F.

Thermal Expansion

All objects change size as they change temperature. When the temperature increases, nearly all materials expand. But not all materials expand at the same rate. Solids, being most tightly bound, expand the least. All gases expand at

FLAWED REASONING

Susan has just completed sculpting a figurine out of candle wax when she notices that the melting point for the wax is only 125°F. The weather report predicts tomorrow's high temperature to be 100°F with 65% relative humidity, giving an apparent temperature on the heat index of 136°F. **Should Susan store her masterpiece in the refrigerator?**

ANSWER The heat index applies only to humans and other animals that use the sweating process to cool themselves. Susan's statue will not melt.

Edward M. Wheeler

Expansion slots allow bridges to change length with temperature changes without damage.

the same rate, following the ideal gas equation developed in Chapter 11. Each material's characteristic **thermal expansion** is reflected in a number called its *coefficient of expansion*. The coefficient of expansion gives the fractional change in the size of the object per degree change in temperature.

Thermal expansion has many consequences. Civil engineers avoid the possibility of a bridge buckling by including expansion slots and by mounting one end of the bridge on rollers. The gaps between sections of concrete in highways and sidewalks allow the concrete to expand and contract without breaking or buckling. The romantic "clickety-clack" of train rides is due to expansion joints between the rails.

WORKING IT OUT   *Thermal Expansion*

Because the coefficient of thermal expansion tells us how much a unit length of material expands as the temperature is raised 1°C, the expansion for a particular object is given by

$$\Delta L = \alpha l \Delta T$$

where ΔL is the change in length, α is the coefficient of thermal expansion, L is the original length, and ΔT is the change in temperature. There is a similar expression for the volume expansion of liquids.

As an example, the coefficient of expansion for steel is 0.000 011 m for each meter of length for each degree Celsius rise in temperature. This means that a bridge that is 50 m long expands by 0.000 55 m, or 0.55 mm, for each degree of temperature increase. If the temperature increases by 40°C from night to day, the bridge expands by 22 mm (almost 1 in.).

We can also obtain this answer using the relationship for thermal expansion:

$$\Delta L = \alpha L \Delta T = \left(\frac{0.000011}{°C} \right) (50 \text{ m})(40\,°C) = 0.22 \text{ m}$$

Q: Why are telephone wires higher in winter than in summer?

A: The wires expand with the hotter temperatures in summer and therefore hang lower.

We use the differences in the thermal expansions of various materials to our advantage. Some thermostats are constructed of two different metal strips bonded together face-to-face, as shown in Figure 13-9. Because the metals have different coefficients of expansion, they expand by different amounts, causing the bimetallic strip to bend. Placing electric contacts in appropriate places allows the thermostat to function as an electric switch to turn a furnace, heater, or air conditioner on and off at specified temperatures.

Q: How does running hot water on a jar lid loosen it?

A: The metal expands more than the glass, and the lid pulls away from the jar.

Summary

The law of conservation of mechanical energy does not apply whenever frictional effects are present. Often the transformation of mechanical energy to thermal energy is accompanied by temperature changes that produce observable changes in the object. Heat and temperature are not the same thing. Heat is an energy, and temperature is a macroscopic property of the object. The number of calories required to raise the temperature of 1 gram of a substance by 1°C is known as its specific heat.

The first law of thermodynamics tells us that the total change in the internal energy of a system is the sum of the heat added to the system and the work done on the system. This is just a restatement of the law of conservation of energy. Performing 4.2 joules of work on a system is equivalent to adding 1 calorie of thermal energy. Part of this energy increases the average kinetic energy of the atoms; the absolute temperature is directly proportional to this average kinetic energy. Other parts of this energy break the bonds between the molecules and cause substances to change from solids to liquids to gases. At higher temperatures, molecules, atoms, and even nuclei break apart.

There is a limit to how much internal energy can be removed from an object, and thus there is a lowest possible temperature—absolute zero, or −273°C—the same as the zero on the Kelvin scale. A substance at absolute zero has the lowest possible internal energy.

The temperature of a substance does not change while it undergoes a physical change of state. The energy that is released or gained per gram of material is known as the latent heat.

The natural flow of thermal energy is always from hotter objects to colder ones. In the process called conduction, thermal energy is transferred by collisions between particles; in convection the transfer occurs through the movement of the particles; and in radiation the energy is carried by electromagnetic waves.

© Cengage Learning/Charles D. Winters

Figure 13-9 A bimetallic strip is used in some thermostats to control furnaces.

Everyday Physics | *Freezing Lakes*

Life as we know it depends on the unique thermal expansion properties of water. All materials change size when their temperatures change. Because density is the ratio of mass to volume and because the mass of an object does not change when heated or cooled, a change in size means a change in the object's density.

As stated in the chapter, most objects expand when heated and contract when cooled. The behavior of water is not so simple. Over most of its liquid range, water behaves as expected, decreasing in volume as its temperature decreases. As the water is cooled below 4°C, however, it expands!

This unusual property affects the way lakes freeze. While cooling toward 4°C, the surface water becomes denser and therefore sinks (Chapter 12), cooling the entire lake. However, once the entire lake becomes 4°C, the surface water expands as it cools further and becomes less dense. Therefore, the cooler water floats on the top and continues to cool until it freezes. Lakes freeze from the top down. However, because ice is a good thermal insulator, most lakes do not freeze to the bottom. If water were like most other materials, the very cold water would sink, and lakes would freeze from the bottom up, creating a challenging evolutionary problem for all aquatic and marine life.

1. What unusual property of water causes ponds and lakes to freeze from the top down?

2. What property of ice keeps lakes from freezing clear to the bottom, even in the coldest of winters?

Because water is denser than ice, the ice floats and covers lakes and ponds in winter. The layer of ice insulates the water below from freezing air temperatures.

CHAPTER 13 : *Revisited*

The rate at which an object radiates energy is determined by the difference in temperature between the object and its surroundings, the surface area of the object, and characteristics of the surface. The change in temperature of the object depends primarily on the amount of energy radiated away, the mass of the object, and the specific heat of the material.

Key Terms

absolute zero   The lowest possible temperature; 0 K, −273°C, or −459°F.

British thermal unit   The amount of heat required to raise the temperature of 1 pound of water by 1°F.

calorie   The amount of heat required to raise the temperature of 1 gram of water by 1°C.

change of state   The change in a substance between solid and liquid or liquid and gas.

conduction   The transfer of thermal energy by the collisions of the atoms or molecules within a substance.

conductor   A material that allows the easy flow of thermal energy. Metals are good conductors.

convection   The transfer of thermal energy in fluids by means of currents such as the rising of hot air and the sinking of cold air.

heat   Energy flowing because of a difference in temperature.

heat capacity   The amount of heat required to raise the temperature of an object by 1°C.

insulator   A material that is a poor conductor of thermal energy. Wood and stationary air are good thermal insulators.

internal energy   The total microscopic energy of an object, which includes its atomic and molecular translational and rotational kinetic energies, its vibrational energy, and the energy stored in the molecular bonds.

latent heat   The amount of heat required to melt (or vaporize) 1 gram of a substance. The same amount of heat is released when 1 gram of the same substance freezes (or condenses).

radiation   The transport of energy via electromagnetic waves.

specific heat   The amount of heat required to raise the temperature of 1 gram of a substance by 1°C.

thermal energy   Internal energy.

thermal equilibrium   A condition in which there is no net flow of thermal energy between two objects. This occurs when the two objects obtain the same temperature.

thermal expansion   The increase in size of a material when heated.

thermodynamics   The area of physics that deals with the connections between heat and other forms of energy.

thermodynamics, first law of   The increase in the internal energy of a system is equal to the heat added plus the work done on the system.

thermodynamics, third law of   Absolute zero may be approached experimentally but can never be reached.

thermodynamics, zeroth law of   If objects A and B are each in thermodynamic equilibrium with object C, then A and B are in thermodynamic equilibrium with each other. All three objects are at the same temperature.

Questions and exercises are paired so that most odd-numbered are followed by a similar even-numbered.

Blue-numbered questions and exercises are answered in Appendix B.

ʃ indicates more challenging questions and exercises.

WebAssign Many Conceptual Questions and Exercises for this chapter may be assigned online at WebAssign.

Conceptual Questions

1. In an avalanche, the snow and ice begin at rest at the top of the mountain and end up at rest at the bottom. What happens to the gravitational potential energy that is lost in this process?

2. What happens to the sound energy from your stereo speakers?

3. What evidence did Rumford have that heat was not a fluid?

4. Suppose a student was careless in re-creating Joule's experiment and allowed the masses to speed up quickly as they dropped toward the floor. If he equated the change in gravitational potential energy with the change in thermal energy, would he have found 1 calorie to be greater than or less than 4.2 joules? Explain.

5. How are the concepts of work and heat the same? How are they different?

6. What would you expect to find if you measure the temperature of the water at the top and bottom of Niagara Falls? Explain your reasoning.

7. It could be argued that the only time you measure the undisturbed temperature of a system is when the reading on the thermometer does not change when it is placed in thermal contact with the system. Use the zeroth law to explain why this is so.

8. Imagine a universe where the zeroth law of thermodynamics was not valid. Would the concept of temperature still make sense in this universe? Why or why not?

9. Could two objects be touching but not be in thermal equilibrium? Explain.

10. Is it possible for a bucket of water in Los Angeles and a bucket of water in New York City to be in thermal equilibrium? Explain.

11. Why is it incorrect to talk about the flow of temperature from a hot object to a colder object?

12. On the inside back cover of this textbook are conversion factors between different units. Why is there no conversion factor between joules and kelvin?

13. What is the difference between heat and temperature?

14. The same amount of heat flows into two different buckets of water, which are initially at the same temperature. Will both buckets necessarily end up at the same temperature? Explain.

15. Patrick claims, "Two buckets of water must have the same heat if they are at the same temperature." Victoria counters, "That's true only if both buckets contain the same amount of water." With which, if either, of these students do you agree? Explain.

16. How do the internal energies of a cup of water and a gallon of water at the same temperature compare?

17. Under what conditions is the first law of thermodynamics valid?

18. Work is done on a system without changing the internal energy of the system. Does heat enter or leave the system during this process? Use the first law of thermodynamics to justify your answer.

19. Does it take more thermal energy to raise the temperature of 5 grams of water or 5 grams of ice by 6°C? Explain.

20. Which of the following does not affect the amount of internal energy of an object: its temperature, the amount of material, its state, the type of material, or its shape?

21. Liquid X and gas Y have identical specific heats. Would 100 calories of heat raise the temperature of 1 liter of liquid X by the same amount as 1 liter of gas Y? Explain your reasoning.

22. One kilogram of material A at 80°C is brought into thermal contact with 1 kilogram of material B at 40°C. When the materials reach thermal equilibrium, the temperature is 68°C. Which material, if either, has the greater specific heat? Explain.

23. A hot block of aluminum is dropped into 500 grams of water at room temperature in a thermally insulated container, where it reaches thermal equilibrium. If 1000 grams of water had been used instead, would the amount of heat transferred to the water be greater than, equal to, or less than it was before? Why?

24. A hot block of iron is dropped into room-temperature water in a thermally insulated container, where it reaches thermal equilibrium. If twice as much water had been used, would the water's temperature change be greater than, equal to, or less than it was before? Why?

25. Why do climates near the coasts tend to be more moderate than those in the middle of the continent?

26. Why does the coldest part of winter occur during late January and February, when the shortest day is near December 21?

27. Given that the melting and freezing temperatures of water are identical, what determines whether a mixture of ice and water will freeze or melt?

28. If you make the mistake of removing ice cubes from the freezer with wet hands, the ice cubes stick to your hands. Why does the water on your hands freeze rather than the ice cubes melt?

29. Why can an iceberg survive for several weeks floating in seawater that's above freezing?

30. The boiling point for liquid nitrogen at atmospheric pressure is 77 K. Is the temperature of an open container of liquid nitrogen higher than, lower than, or equal to 77 K? Explain.

31. One hundred grams of ice at 0°C is added to 100 grams of water at 80°C. The system is kept thermally insulated from its environment. Will the equilibrium temperature of the mixture be greater than, equal to, or less than 40°C? Explain your reasoning.

32. An ice cube at 0°C is placed in a foam cup containing 200 grams of water at 60°C. When the system reaches thermal equilibrium, its temperature is 30°C. Was the mass of the ice cube greater than, equal to, or less than 200 grams? Explain your reasoning.

33. Why is steam at 100°C more dangerous than water at 100°C?

34. A new liquid is discovered that has the same boiling point and specific heat as water but a latent heat of vaporization of 10 calories per gram. Assuming that this new liquid is safe to drink, would it be more or less convenient than water for boiling eggs? Why?

35. A system is thermally insulated from its surroundings. Is it possible to do work on the system without changing its internal energy? Is it possible to do work on the system without changing its temperature? Explain.

36. In Washington, D.C., the weather report sometimes states that the temperature is 95°F and the humidity is 95%. Why does the high humidity make it so uncomfortable?

37. Use a microscopic model to explain how a metal rod transports thermal energy from the hot end to the cold end.

38. Why would putting a rug on a tiled bathroom floor make it feel less chilly to bare feet?

39. Rank the following materials in terms of their insulating capabilities: static air, glass, polyurethane foam, and concrete.

40. Which of the following is the best thermal conductor: fiberglass, stainless steel, wood, or silver?

41. If the temperature is 35°F and the wind is blowing at 20 mph, the equivalent windchill temperature is 24°F. Will a glass of water freeze in this situation? Explain your reasoning.

42. You hear on the morning weather report that the outside temperature is −5°F with a windchill equivalent temperature of −40°F. You know that your old car, which is parked outside, will not start if the temperature of the battery drops below −15°F. Will your car start this morning? Why or why not?

43. The respective thermal conductivities of iron and stainless steel are 79 W/m · °C and 14 W/m · °C. Use these data to explain why you need to use potholders for pots with iron handles but not for pots with stainless steel handles.

44. Why might a cook put large aluminum nails in potatoes before baking them?

45. In northern climates drivers often encounter signs that read, "BRIDGE FREEZES BEFORE ROADWAY." Why does this occur?

46. You have just made yourself a hot cup of coffee and are about to add the cream, which is at room temperature. Suddenly the phone rings and you have to leave the room for a while. Is it better to add the cream to the coffee before you leave or after you get back if you want your coffee as hot as possible? Why?

47. When pilots fly under clouds, they often experience a downdraft. Why is this?

Richard Hamilton Smith/ Dembinsky Photo Assoc.

48. It is midafternoon and you are canoeing down a river that empties into a large lake. You are having a hard time making progress because of a stiff wind in your face. Is this situation likely to get better or worse at sunset? Explain.

49. A black car and a white car are parked next to each other on a sunny day. The surface of the black car gets much hotter than the surface of the white car. Which mode of energy transport is responsible for this difference?

50. Earth satellites orbit Earth in a very good vacuum. Would you expect these satellites to cool off when they enter Earth's shadow? Explain.

51. A Thermos bottle is usually constructed from two nested glass containers with a vacuum between them, as shown in the following figure. The walls are usually silvered as well. How does this construction minimize the loss of thermal energy?

Vacuum

Silvered

52. Will a Thermos bottle (shown in the preceding figure) keep something cold as well as it keeps it hot? Explain.

53. The metal roof on a wooden shed makes noises when a cloud passes in front of the Sun. Why?

---

54. Why might a glass dish taken from the oven and put into cold water shatter?

55. Suppose the column in an alcohol-in-glass thermometer is not uniform. How would the spacing between the degrees on a wide portion of the thermometer compare with those on a narrow portion?

56. When a mercury thermometer is first put into hot water, the level of the mercury drops slightly before it begins to climb. Why?

Exercises

57. How much heat is required to raise the temperature of 500 g of water from 20°C to 30°C?

58. If the temperature of 600 g of water drops by 8°C, how much heat is released?

59. How much work is required to push a crate with a force of 200 N across a floor a distance of 4 m? How many calories of thermal energy does the friction produce?

60. How many joules of gravitational potential energy are converted to kinetic energy when 100 g of lead shot falls from a height of 50 cm? How many calories are released to the surroundings if none of this kinetic energy is converted to other forms when the shot hits the floor?

61. A physics student foolishly wants to lose weight by drinking cold water. If he drinks 1 L (1000 cm³) of water at 10°C below body temperature, how many Calories will it take to warm the water?

62. A typical jogger burns up food energy at the rate of about 40 kJ per minute. How long would it take to run off a piece of cake if it contains 400 Calories?

63. During a process, 28 J of heat are transferred into a system, while the system itself does 12 J of work. What is the change in the internal energy of the system?

64. What is the change in the internal energy of a system if 15 J of work are done on the system and 6 J of heat are removed from the system?

65. When an ideal gas was compressed, its internal energy increased by 180 J and it gave off 150 J of heat. How much work was done on this gas?

66. If the internal energy of an ideal gas increases by 150 J when 240 J of work is done to compress it, how much heat is released?

67. It takes 250 cal to raise the temperature of a metallic ring from 20°C to 30°C. If the ring has a mass of 90 g, what is the specific heat of the metal?

68. If it takes 3400 cal to raise the temperature of a 500-g statue by 44°C, what is the specific heat of the material used to make the statue?

69. How many calories will it take to raise the temperature of a 50-g gold chain from 20°C to 100°C?

70. How many calories would it take to raise the temperature of a 300-g aluminum pan from 293 K to 373 K?

71. Six grams of liquid X at 35°C are added to 3 grams of liquid Y at 20°C. The specific heat of liquid X is 2 cal/g · °C, and the specific heat of liquid Y is 1 cal/g · °C. If each gram of liquid X gives up 2 cal to liquid Y, find the change in temperature of each liquid.

72. In Exercise 71, imagine that liquid X continues to transfer energy to the other liquid 12 calories at a time. How many transfers would be required to reach a common temperature? What is this equilibrium temperature?

73. Eighty grams of water at 70°C is mixed with an equal amount of water at 30°C in a completely insulated container. The final temperature of the water is 50°C.

    a.   How much heat is lost by the hot water?

    b.   How much heat is gained by the cold water?

    c.   What happens to the total amount of internal energy of the system?

74. If 200 g of water at 100°C is mixed with 300 g of water at 50°C in a completely insulated container, what is the final equilibrium temperature?

75. A kettle containing 3 kg of water has just reached its boiling point. How much energy, in joules, is required to boil the kettle dry?

76. How much heat would it take to melt a 1-kg block of ice?

77. You wish to melt a 3-kg block of aluminum, which is initially at 20°C. How much energy, in joules, is required to heat the block to its melting point of 660°C? How much energy, in joules, is required to melt the aluminum without changing its temperature?

78. How much heat is required to convert 400 g of ice at −5°C to water at 5°C? (*Hint:* review Exercise 77.)

79. What is the change in length of a metal rod with an original length of 2 m, a coefficient of thermal expansion of 0.00002/°C, and a temperature change of 20°C?

80. A steel railroad rail is 24.4 m long. How much does it expand during a day when the low temperature is 50°F (18°C) and the high temperature is 91°F (33°C)?

14

Available Energy

▶ The energy of the hot water and steam from a geyser can be used to run engines, yet the internal energy of a pail of room-temperature water, although quite high, isn't as useful as an energy source. Why is this true, and what would it take to extract this internal energy and make it available for performing useful work?

(See page 296 for the answer to this question.)

© Brand X Pictures/Jupiterimages

Old Faithful Geyser in Yellowstone National Park, Wyoming. Hot water from geysers can be used as an energy source.

MECHANICAL energy can be completely converted into the internal energy of an object. This is clearly demonstrated every time an object comes to rest because of frictional forces. In the building of a scientific world view, the belief in a symmetry often leads to interesting new insights. In this case it seems natural to ask whether the process can be reversed. Is it possible to recover this internal energy and get some mechanical energy back? The answer is yes, but it is not an unqualified yes.

Imagine that we try to run Joule's paddle-wheel experiment (Figure 14-1) backward. Suppose we start with hot water and wait for the weight to rise up from the floor. Clearly, we don't expect this to happen. The water can be heated to increase its internal energy, but the hot water will not rotate the paddle wheel. Water can be very hot and thus contain a lot of internal energy, but mechanical work does not spontaneously appear.

The first law of thermodynamics doesn't exclude this; it places no restrictions on which energy transformations are possible. As long as the internal energy equals or exceeds that needed to raise the weight, there would be no violation of the first law if some of the internal energy were used to do this. And yet it doesn't happen. The energy is there, but it is not available. Apparently the availability of energy depends on the form that it takes.

This aspect of nature, unaccounted for so far in the development of our world view, is addressed in the second law of thermodynamics. It's a subtle law that has a rich history, resulting in three different statements of the law. The first form deals with heat engines.

Figure 14-1 Why won't the weight rise when we heat the water?

Heat Engines

We can extract some of an object's internal energy under certain circumstances. Internal energy naturally flows from a higher-temperature region to one of lower temperature. A hot cup of coffee left on your desk cools off as some of its thermal energy flows to the surroundings. The coffee continues to cool until it reaches thermal equilibrium with the room. Energy flows out of the hot region because of the presence of the cold one. But no work—no mechanical energy—results from this flow.

Many schemes have been proposed for taking part of the heat and converting it to useful work. Any device that does this is called a **heat engine**. The simplest and earliest heat engine is traced back to Hero of Alexandria. About AD 50, Hero invented a device similar to that shown in Figure 14-2. Heating the water-filled container changes the water into steam. The steam, escaping through the two tubes, causes the container to rotate. From our modern perspective, we may judge this to be more of a toy than a machine. The importance of this device, however, was that mechanical energy was in fact obtained—the container rotated. Apparently, Hero did do something useful with his heat engine; there are stories that he used pulleys and ropes to open a temple door during a religious service. (Probably much to the shock of the worshippers!)

The Industrial Revolution began with the move away from animal power toward machine power. The first heat machines were used to pump water out of mines in England. Waste coal was cheap at the pithead, and the engines were inefficient. By the 1790s James Watt had developed steam engines that were more thermally efficient and were increasingly used in applications requiring powerful and reliable energy sources, such as water systems, mills, and forges.

Figure 14-3 shows the essential features of the most successful of the early steam engines. Invented by Watt in 1769, it had a movable piston in a cylinder. The flow of heat was not a result of direct thermal contact between objects but of the transfer of steam that was heated by the hot region and cooled by the

Figure 14-2 A modern version of Hero's heat engine rotates under the action of the escaping steam.

Figure 14-3 The essential features of Watt's steam engine. Opening valve 1 lets steam into the chamber, raising the piston. Closing valve 1 and opening valve 2 reduces the pressure, allowing the piston to fall.

The steam locomotive is a heat engine.

The internal combustion engine in a Honda automobile is an example of a modern heat engine.

cold region. Alternately heating and cooling the steam in the cylinder drove the piston up and down, producing mechanical work.

The opening of the American West was helped by another heat engine: "the iron horse," or steam locomotive. A more modern heat engine is the internal combustion engine used in automobiles. Replacing the wood and steam, gasoline explodes to produce the high-temperature gas. The explosions move pistons, allowing the engine to extract some of this energy to run the automobile. The remaining hot gases are exhausted to the cooler environment.

Although there are many types of heat engines, all can be represented by the same schematic diagram. To envision this, recall that all heat engines involve the flow of energy from a hotter region to a cooler one. Figure 14-4(a) shows how we represent this flow. A heat engine is a device placed in the path of this flow to extract mechanical energy. Figure 14-4(b) shows heat flowing from the hotter region; part of this heat is converted to mechanical work, and the remainder is exhausted to the colder region. You can verify that heat is thrown away by an automobile's engine by putting your hand near (but not directly on) the exhaust pipe. The exhausted gases are hotter than the surrounding air. Without the cool region, the flow would stop, and no energy would be extracted.

Q: How much work does a heat engine perform if it extracts 100 joules of energy from a hot region and exhausts 60 joules to a cold region?

A: Conservation of energy requires that the work be equal to 40 joules, the difference between the input and the output.

Ideal Heat Engines

The first law of thermodynamics requires that the sum of the mechanical work and the exhausted heat be equal to the heat extracted from the hot region. But exhausted heat means wasted energy. When engines were developed at the beginning of the Industrial Revolution, engineers asked, "What kind of engine will get the maximum amount of work from a given amount of heat?"

In 1824 French army engineer Sadi Carnot published the answer to this question. Carnot found the best possible engine by imagining the whole process as a thought experiment. He assumed that his engine would use idealized gases and there would be no frictional losses due to parts rubbing against each other. His results were surprising. Even under these ideal conditions, the heat engine must exhaust some heat.

Carnot's work led to one version of the **second law of thermodynamics** and an understanding of why internal energy cannot be completely converted to mechanical energy. That is, even the best theoretical heat engine cannot convert 100% of the incoming heat into work.

◄ heat-engine form of the second law of thermodynamics

> It is impossible to build a heat engine to perform mechanical work that does not exhaust heat to the surroundings.

Stated slightly differently, the fact that the engine must throw away heat means that no heat engine can run between regions at the same temperature. This is unfortunate, because it would be a boon to civilization if we could "reach in" and extract some internal energy from a single region. For example, think of all the energy that would be available in the oceans. If we could build a single-temperature heat engine, it could be used to run an ocean liner. Its engine could take in ocean water at the front of the ship, extract some of the water's internal energy, and drop ice cubes off the back, as shown in Figure 14-5. Notice that this hypothetical engine isn't intended to get something for nothing; it only tries to get out what is there. But the second law says that this is impossible.

Perpetual-Motion Machines

Since the beginning of the Industrial Revolution, inventors and tinkerers have tried to devise a machine that would run forever. This search was fueled by the desire to get something for nothing. A machine that did some task without requiring energy would have countless applications in society, not to mention the benefits that would accrue to its inventor. No such machine has ever been invented; in fact, the only "successful" perpetual-motion machines have been devices that were later shown to be hoaxes.

Figure 14-4 (a) A schematic showing the natural flow of thermal energy from a higher-temperature region to a lower-temperature region. (b) The white square represents the many types of heat engine. The heat engine converts part of the heat from the hot region to mechanical work and exhausts the remaining heat to the cold region.

Figure 14-5 An impossible engine in an ocean liner extracts energy from ocean water, makes ice cubes, and propels the ship.

Figure 14-6 Fludd's proposed perpetual-motion machine cannot work because it would violate the first law of thermodynamics.

Recycled water

Grindstone

In the 17th century, English physician Robert Fludd proposed the device shown in Figure 14-6. Fludd wanted to turn the waterwheel to move the millstone and at the same time return the water to the upper level. It didn't work. Machines like Fludd's violate the first law of thermodynamics by trying to get more energy out of a device than is put into it. Every machine is cyclic by definition. After doing something the machine returns to its original state to start the whole process over again. Fludd's scheme was to have the machine start with a certain amount of energy (the potential energy of the water on the upper level), do some work, and then return to its original state. But if it did some work, some energy was spent. Returning the machine to its original state would require getting something for nothing, a violation of the first law.

Other perpetual-motion machines have been less ambitious. They were simply supposed to run forever without extracting any useful output. They also didn't work. Even if we ask nothing of such a machine except to run—and thus stay within the confines of the first law of thermodynamics—it still will not run forever. These machines failed because they were attempting to violate the second law of thermodynamics.

To illustrate this, suppose we invent such a machine as a thought experiment. Figure 14-7 is a diagram of one possibility. A certain amount of energy is initially put into the system, and somehow, the system is started. The paddle wheel turns in the water and, as the Joule experiment showed us, produces an increase in the temperature of the water. Mechanical energy has been converted to internal energy, as indicated by the water's temperature rise. Thermal energy from the hot water flows to the pad below the water. The pad is a heat engine designed to capture this energy and convert it to mechanical energy. The mechanical energy, in turn, is used to drive gears that rotate the paddles.

Does the machine continue to run? Notice that this is a perfect machine, and we can assume that there are no frictional losses; so if it fails, it does so because of some fundamental reason. Such engines have been built—and they have all failed. The problem lies with the heat engine in the pad. Any heat engine, no matter how good, returns less mechanical energy than the thermal energy it receives. Therefore, our machine runs down, and we have failed. We do not fail because we try to create energy but because we cannot use the energy that we have. Once the energy is converted to internal energy, it is no longer fully available.

Heat engine

Figure 14-7 This proposed perpetual-motion machine would violate the second law of thermodynamics.

Even in a simpler machine without a heat engine, there would still be moving parts. Any rubbing of moving parts increases the machine's internal energy. All this internal energy would have to be converted back into mechanical energy if the machine were to run forever. The second law tells us that this cannot happen.

All machines and heat engines, no matter how complicated, cannot avoid the constraints imposed by the first and second laws of thermodynamics.

FLAWED REASONING

A classmate claims to suddenly understand the second law of thermodynamics. He explains, "All machines necessarily have moving parts that experience friction. This is why a heat engine can never transform its entire heat input into work." **Do you think your classmate understands the second law?**

ANSWER Your classmate seems to understand the first law of thermodynamics (energy conservation) but is missing the key point of the second law. The second law of thermodynamics states that even in the absence of any friction, it is still impossible to transform thermal energy entirely into mechanical energy. Some of the input energy must be exhausted.

Real Engines

Extended presentation available in the *Problem Solving* supplement

All engines can be rated according to their efficiencies. In general, efficiency can be defined as the ratio of the output to the input. In the case of heat engines, the **efficiency** η of an engine is equal to the ratio of the work W produced divided by the heat Q extracted from the hot region:

$$\eta = \frac{W}{Q}$$

◀ efficiency $= \dfrac{\text{work out}}{\text{heat in}}$

Carnot's ideal heat engine has the maximum theoretical efficiency. This efficiency can be expressed as a simple relationship using Kelvin temperatures:

$$\eta = 1 - \frac{T_c}{T_h}$$

◀ efficiency of an ideal heat engine

The Carnot efficiency depends only on the temperatures of the two regions. It is larger (that is, closer to 1) if the temperature of the exhaust region T_c is low or that of the hot region T_h is high. The largest efficiency occurs when the temperatures are as far apart as possible. Usually, this efficiency is multiplied by 100 so that it can be stated as a percentage. Real engines produce more waste heat than Carnot's ideal engine, but his relationship is still important because it sets an upper limit for the efficiencies of real engines.

Today, steam engines are used primarily to drive electric generators in power plants. Although these engines are certainly not Carnot engines, their efficiencies can be increased by making their input temperatures as high as possible and their exhaust temperatures as low as possible. Because of constraints imposed by the properties of the materials used in their construction, these engines cannot be run much hotter than 550°C. Exhaust temperatures cannot be much lower than about 50°C. Given these constraints, we can calculate the maximum theoretical efficiency of the steam engines used in electricity-

A coal-fired electricity-generating plant.

Cooling towers at the nuclear power generating facility on Three Mile Island near Harrisburg, Pennsylvania.

generating plants. Remembering to convert the temperatures from Celsius to Kelvin, we have

$$\eta = 1 - \frac{T_c}{T_h} = 1 - \frac{323 \text{ K}}{823 \text{ K}} = 0.61 = 61\%$$

Actual steam engines have efficiencies closer to 50%. Besides the efficiency of the engine, we must consider the efficiency of the boiler for converting the chemical energy of the fuel into heat. Typical oil- or coal-fired power plants have overall efficiencies of about 40% or less. Nuclear power plants use uranium as a fuel to make steam. Safety regulations require that they run at lower temperatures, so they are 5–8% less efficient than the oil- or coal-fired plants. In other words, nuclear plants exhaust more waste heat for each unit of electricity generated.

Are You On the Bus?

Q: What is the maximum theoretical efficiency for a heat engine running between 127°C and 27°C?

A: Being careful to convert these temperatures to the Kelvin scale, we have

$$\eta = 1 - \frac{T_c}{T_h} = 1 - \frac{300 \text{ K}}{400 \text{ K}} = 1 - 0.75 = 0.25 = 25\%$$

Not all engines are heat engines. Those that don't go through a thermodynamic process are not heat engines and can have efficiencies closer to 100%. Electric motors are close to being 100% efficient if you look only at the electric motor. However, the overall efficiency of the electric motor must include the energy costs of generating the electricity and the energy losses that occur in the electric transmission lines. Because most electricity is produced by steam engines, the overall efficiency is quite low.

WORKING IT OUT *Efficiency*

An engine operates with an efficiency of 20%. If the engine has a power output of 15 kW (i.e., does 15,000 J of work each second), how many joules does the engine exhaust each second to the cold region?

The efficiency is defined as the ratio of useful work to the heat extracted from the hot region:

$$\eta = \frac{W}{Q_h}$$

An efficiency of 20% means that only 20% of the extracted heat is converted to useful work and the rest is exhausted to the cold region. For every joule of work done, 4 J of heat are exhausted. If 15,000 J of work are done each second, then 60,000 J of heat are exhausted.

Refrigerators

Figure 14-8 A refrigerator uses mechanical work to transfer heat from a colder region to a hotter one.

There are devices that extract heat from a cooler region and deposit it into a hotter one. These are essentially heat engines running backward. **Refrigerators** and air conditioners are common examples of such devices. By reversing the directions of the arrows in Figure 14-4(b), we produce the schematic of the process shown in Figure 14-8.

Refrigerators move heat in a direction opposite to its natural flow. The inside of your refrigerator is the cold region, and your room is the hot region. The refrigerator removes heat from the inside and exhausts it to the warm region outside. If you put your hand by the base of the refrigerator, you can feel the heat being transferred to the kitchen. Similarly, if you walk by the external part of an air conditioner, you feel heat being expelled outside the house. In both cases the warm region gets hotter, and the cool gets colder.

The icebox of olden days was a simple device: place a block of ice on the top shelf and your produce on the bottom shelf. The modern refrigerator is almost as simple. It is constructed from the four basic parts illustrated in Figure 14-9. The *evaporator* (1) is a long metal tube that allows the low-pressure, cold liquid refrigerant to have thermal contact with the air inside the refrigerator. When a liquid evaporates, it absorbs heat from its surroundings. A refrigerant, such as Freon-12, is often used because it evaporates at a low temperature (–29°C). The *compressor* (2) pressurizes the gas, and the resulting high-pressure gas passes through the *condenser* (3), a second long metal tube outside the refrigerator. Thermal energy is released into the kitchen as the gas condenses back into a liquid. The liquid cools as it passes through the *expansion valve* (4), and the process repeats.

Refrigerators require work to move energy from a lower-temperature region to a higher-temperature region. Therefore, the heat delivered to the higher-temperature region is larger than that extracted from the lower-temperature region. The extra heat is the amount of work done on the system. In fact, this process leads to an equivalent statement of the second law of thermodynamics.

◀ refrigerator form of the second law of thermodynamics

> It is impossible to build a refrigerator that can transfer heat from a lower-temperature region to a higher-temperature region without using mechanical work.

A **heat pump** is used in some homes to both cool during the summer and heat in the winter. It is simply a reversible heat engine. In the summer the heat pump functions as an air conditioner by extracting heat from inside your house. In the winter it runs in reverse (like a refrigerator), extracting energy from the outside air and putting it into your house.

It may seem strange to be able to extract heat from the cold air. Remember that heat and temperature are not the same. Even on the coldest days, the outside air still has enormous amounts of internal energy. The heat pump transfers some of this internal energy from outside to inside. Actually, there is an economical limit. If the outside temperature is too low, it takes more energy to run the heat pump than it does to use electric baseboard heaters. This situation can be corrected if a warmer region such as underground water is available as a source of the internal energy.

Order and Disorder

Whether it is a heat engine or a refrigerator, the second law of thermodynamics essentially says that we can't break even. If we are trying to get mechanical

Figure 14-9 A schematic for a refrigerator.

Q: How much energy does an air conditioner exhaust if it requires 200 joules of mechanical work to extract 1000 joules of energy from a house?

A: Conservation of energy requires that the exhaust equal 1200 joules, the sum of the inputs.

Are You On the Bus?

Heat pumps can be used to cool houses in summer and heat them in winter. They can extract energy from the outside air even in cold weather.

A new deck of cards has a high degree of order.

energy from a thermal source, we have to throw some heat away. If we want to cool something, we have to do work to counteract the natural flow of energy. But these two forms of the second law offer little insight into why this is so. The reason lies in the microscopic behavior of the many-particle systems that make up matter. To understand the second law of thermodynamics, we need to look at the details of such systems.

We begin by looking at systems in general. Any system consists of a collection of parts. Assuming that these parts can be shifted around, the system has a number of possible arrangements. We will examine systems and determine their organization because, as we will see, it is their organizational properties that lie at the heart of the second law. An organizational property can be such a thing as the position of objects in a box or the height of people in a community.

Suppose your system is a new deck of playing cards. The cards are arranged according to suits and within suits according to value. The organizational property in this case is the position of the card within the deck. If somebody handed you a card, you would have no trouble deciding where it came from. If the deck were arranged by suit but the values within the suits were mixed, you would not be able to pinpoint the location of the card, but you could say from which fourth of the deck it came. In a completely shuffled deck, you would have no way of knowing the origin of the card. The first arrangement is very organized, the second less so, and the last one very disorganized. We call a system that is highly organized an **ordered system** and one that shows no organization a **disordered system.**

Another way of looking at the amount of organization is to ask how many equivalent arrangements are possible in each of these situations. When the deck is arranged by suit and value, there is only one arrangement. When the values are shuffled within suits, there are literally billions of ways the cards could be arranged within each suit. When we completely shuffle the deck, the number of possible arrangements becomes astronomically large (8×10^{67}). The more disordered system is the one with the larger number of equivalent arrangements.

A simple system of three coins on a tray can model microscopic order. An obvious organizational property of this system is the number of heads (or tails) facing up. How many different arrangements are there? If the coins are different, there are eight different arrangements for the three coins, as shown in Figure 14-10.

In real situations the macroscopic property—for example, the total energy—doesn't depend on the actual properties of a particular atom. To apply our coin analogy to this situation, we would examine the different arrangements without identifying individual coins. In other words, we should simply count the number of heads and tails. How many different arrangements are there now? Four. Of the eight original arrangements, two groups of three are now indistinguishable, as shown in Figure 14-11. Each of these groups has three *equivalent states*. The arrangements with all heads or all tails occur in only one way and thus have only one state each. Arrangements with only one state have the highest order. The arrangements that occur in three ways have the lowest order.

HHH

THH

HTH

HHT

HTT

THT

TTH

TTT

Figure 14-10 The eight possible arrangements of three different coins.

Q: How many different arrangements can you make with three colored blocks: one *red*, one *blue*, and one *green*?

A: The arrangements are *rbg, rgb, bgr, brg, grb*, and *gbr*. Therefore, there are six possible arrangements.

Understanding the microscopic form of the second law of thermodynamics depends on realizing that the order of real systems is constantly changing. This dynamic nature of systems occurs because macroscopic objects are composed of an immense number of atomic particles that are continually moving and therefore changing the order of the system. We can use our coins to understand this.

To model the passage of time, we shake the tray to flip the three coins. If we repeatedly start our system with a particular high-order arrangement, we observe that after one flip, the amount of order in the system usually decreases. Each of the eight states has an equal probability of occurring. However, the probability of a low-order arrangement occurring is not equal to that of a high-order arrangement, simply because there are more low-order possibilities. Of the eight possible states, only two yield high-order arrangements. The chance is only 2 in 8, or 25%, of obtaining a high-order arrangement. The other 75% of the time we expect to obtain two heads and a tail or two tails and a head.

If we increase the number of coins, the probabilities change. With four coins, for example, the number of possible states grows to 16. Adding one additional coin doubles the total number of combinations because we are adding a head or a tail to each of the previous combinations. But again, there are only two arrangements with the highest order, those having all heads or all tails. The probability now of obtaining a high-order arrangement is 2 in 16, or about 12%. As the number of elements in the system increases, the probability of obtaining a high-order arrangement gets smaller. Conversely, the probability of obtaining a low-order arrangement increases with the size of the system.

For example, if we have 8 coins, the probability of obtaining the highest-order distribution (all heads or all tails) is less than 1%. If we increase the number of coins to 21, we only have one chance in a million of getting all heads or all tails. If the number of coins increases to that of the number of air molecules in a typical bedroom (10^{27}), the probability decreases to 1 in 10^{82}.

Figure 14-11 The four possible arrangements of three indistinguishable coins.

Q: What is the probability of obtaining all heads or all tails with five coins?

A: Adding the additional coin doubles the total number of possible combinations to 32. Therefore, we have 2 chances out of 32, or 6%, of obtaining all tails or all heads.

As another example, let's consider the air molecules in your room. Presumably, any air molecule can be anywhere in the room. A high-order arrangement may have all the molecules in one small location; a low-order arrangement would have them spread uniformly throughout the available space in one of many equivalent states. The number of low-order arrangements is astronomically higher than the number of high-order ones.

Entropy

We now introduce a new concept, called **entropy**, as a measure of a system's organization. A system that has some recognizable order has low entropy—the more disorganized the system, the higher its entropy. As with gravitational potential energy, we are concerned only with changes in entropy; the actual numerical value of the entropy does not matter.

WORKING IT OUT *Probability*

Suppose you had three dice: one red, one blue, and one green. If you roll all three dice at the same time, what is the probability that the sum of the dice will be 12? We expect the probability to be low. You certainly would not want to bet your house that you could do it on the first roll. But how improbable is it?

There are six possible values for each die, so there are 6 × 6 × 6 = 216 states, or possible ways for the three dice to roll. We need to find all of the states that add up to 12. Suppose that the red die rolls a 1. The blue and the green dice would have to add up to 11. That yields two possible states: 1 5 6 and 1 6 5. There are no other ways to roll a sum of 12 if the red die is only 1. We next let the red die roll 2, and find all the possible ways to get a total of 12. We find through this process a total of 25 different states that yield a total of 12:

1 5 6	2 4 6	3 3 6
1 6 5	2 5 5	3 4 5
	2 6 4	3 5 4
		3 6 3

4 2 6	5 1 6	6 1 5
4 3 5	5 2 5	6 2 4
4 4 4	5 3 4	6 3 3
4 5 3	5 4 3	6 4 2
4 6 2	5 5 2	6 5 1
	5 6 1	

There are 25 possible ways, out of 216 total, to roll a sum of 12 with three dice. The probability of rolling a 12 is therefore $\frac{25}{216}$ = 0.116 = 11.6%. Don't bet your house on it!

FLAWED REASONING

Sandra is explaining a get-rich-quick scheme to her friend Marie.

Sandra: "I have been flipping this nickel and the last 19 flips have all been heads! We should take this nickel to Las Vegas and bet someone that the next flip will be tails. We can't lose! The chance of a nickel landing heads 20 times in a row is less than one in a million."

Marie: "Let's stay home. If that nickel is not rigged, then any toss is as likely to give tails as heads. You would only have a 50% chance of winning your bet, even if the last 19 tosses turned up heads."

Should they go to Vegas?

ANSWER Marie is right. The results of past tosses of the nickel do not affect the next toss. Unless the nickel is weighted, every toss has an equal probability of being heads or tails. It's amazing that Sandra was able to get heads 19 times in a row. The odds against that are nearly 500,000 to 1.

We argued in the last section that the order of a system of many particles tends to decrease with time. Therefore, the entropy of the system tends to increase. This tendency is expressed by another version of the second law of thermodynamics.

entropy form of the second law of ▶
thermodynamics

The entropy of an isolated system tends to increase.

The word *tends* needs to be stressed. There is nothing that says the entropy *must* increase. It happens because of the overwhelming odds in its favor. It is still *possible* for all the air molecules to be in one corner of your room. (That would leave you gasping for air if you weren't in that corner!) Nothing prohibits this from happening. Fortunately, the likelihood of this actually happening is vanishingly small.

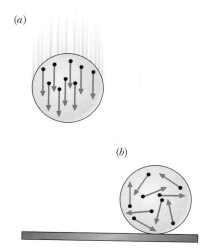

Figure 14-12 (a) The motions of atoms in a falling ball have a high degree of order. (b) The motions after the ball hits the ground are random with a low degree of order.

This entropy version looks so different from the forms of the second law developed for heat engines and refrigerators that you may think that they are different laws. The connection between them is that the motion associated with internal energy has a low-order arrangement, whereas macroscopic motion requires a high degree of organization. Atomic motions are random; at any instant, particles are moving in many directions with a wide range of speeds. If we could take a series of snapshots of these particles and scramble the snapshots, we could not distinguish them; one snapshot is just like another. There appears to be no order to the motions or positions of the particles.

The macroscopic motion of an object, on the other hand, gives organization to the motions of the object's atoms. Although the particles are randomly moving in all directions on the atomic level, they also have the macroscopic motion of the object. All the atoms in a ball falling with a certain kinetic energy are essentially moving in the same direction (Figure 14-12). When the ball strikes the ground, the energy is not lost; the collective motions are randomized. The atoms move in all directions with equal likelihood.

Although it is certainly physically possible for all the particles to once again move in a single direction at some future time (without somebody picking it up), it is so unlikely that we never expect to see it happen. If it did happen, however, some of the internal energy of the ball's particles would then be available to do some mechanical work. This would be bizarre; at some moment, a ball initially at rest would suddenly fly off with some kinetic energy.

Decreasing Entropy

Although the overall entropy of a system tends to increase, isolated pockets of activity can show a decrease in entropy—that is, an increase in order. But these increases in order are paid for with an overall decrease in the order of the universe. Life, for example, is a glorious example of increasing order. It begins as a simple fertilized egg and evolves into a complicated human being. Clearly, order is increasing. But this increase is accomplished through the use of energy. To bring about the increase in human order, a great deal of energy becomes less available for performing useful work. The order of the environment decreases accordingly.

As a simpler example, imagine that you have two buckets of water, one hot and the other cold. If you place a heat engine between the buckets, you can do some work. After the heat engine operates for a while, the buckets reach the same temperature. The system now has a higher entropy. You can decrease the entropy by heating one bucket and cooling the other. But you can only do this by increasing the entropy of the universe. We saw this with refrigerators. You can make the hot region hotter and the cold region colder but only after mechanical work is put in. That use of work lowers the availability of the energy in the universe to do mechanical work.

Entropy and Our Energy Crisis

Save energy! That is the battle cry these days. But why do we have to *save* energy? According to the first law of thermodynamics, energy is neither created nor destroyed. The total amount of energy is constant. There is no need

Everyday Physics | *Arrow of Time*

The concept of entropy gives insight into the character of time. Imagine a motion picture of a soccer game played backward. It looks silly because the order of the events doesn't match our experiences. There is a direction to time. But why do things happen in a particular sequence? According to most of the laws of nature we have established, other sequences of events *could* happen: The parabolic curve of the thrown football is the same regardless of which direction the film is played. Yet the game—played backward—is unreal. Even a repeating, cyclic event like a child on a swing has clues telling you whether the movie is running forward or backward. If the child ceases pumping, the swinging dies down because of frictional effects. Left without an input of energy, everyday motions stop when the macroscopic energy gets transformed into internal energy.

In principle, a stationary pendulum bob could transfer some of its internal energy into swinging the bob. This would (simply) require that the bob's atoms all move in the same direction at the same time. They are all moving but in random directions. It could happen that at some time they could all be moving in the same direction. It could happen, but it doesn't. The entropy form of the second law of thermodynamics tells us why: all systems tend toward disorder. This fact of nature gives a direction to time.

The direction of time is implicit in many events. Bright new paint has a radiant color because of the particular molecules and their particular arrangement. As time passes, there is a mixing of the molecules with the air and a rearranging of the molecules in the paint—and the paint fades. Structures crumble; their original shapes are very ordered, but as time passes, they deteriorate. Order is lost. The culprit is chance, and the consequence is increasing entropy. Time has a direction because the universe has a natural tendency toward disorder.

1. Which law of thermodynamics allows us to define a direction to time?

2. Why do you not violate the second law of thermodynamics when you clean your room?

© Cengage Learning/Charles D. Winters (both)

A desk work area is an example of a system that tends toward disorder.

to conserve energy; it happens naturally. The crux of the matter is really the second law. There are pockets of energy in our environment that are more valuable than others. Given the proper conditions, we can use this energy to do some useful work for society. But if we later add up all the energy, we still have the same amount. The energy used to drive our cars around town is transformed via frictional interactions and exhaust, the consumption of food results in our body temperatures being maintained as well as moving us around, and so on. All the energy is present and accounted for.

Water naturally flows downhill. The water is still there at the bottom of the hill, but it is less useful. If all the water is at the same level, there is no further flow. Similarly, if all the energy in the universe is spread out uniformly, we can get no more "flow" from one pocket to another.

The real energy issue is the preservation of the valuable pockets of energy. We can burn a barrel of oil only once. When we burn it, the energy becomes less useful for doing work, and the entropy of the universe goes up. The second law doesn't tell us how fast entropy should increase. It does not tell us how fast we must use up our pockets of energy; it only says that it will occur. The decision of finding an acceptable rate is left to us. The battle cry for the future should be, "Slow down the increase in entropy!"

Everyday Physics | *Quality of Energy*

With the exceptions of a geothermal source of hot water heating a home or a windmill pumping water out of a well, most energy that we use in our lives has been converted from its original form to the form we use. An obvious example is the electrical energy that lights our homes. This energy often comes from a fossil fuel that is burned to heat water to form steam, which turns a generator to produce the electricity that lights our homes.

It is not possible to convert, or even transfer, energy without "losing" some of the energy in the process. (By *losing energy* we mean that some of the energy is transformed to some form other than the one desired.) Even charging a battery—storing energy for later use—requires more energy than gets stored in the battery. A more obvious loss is that due to frictional forces. If we use a waterwheel to turn an electric generator, some of the input energy is transformed to thermal energy by the frictional forces in the rubbing of the mechanical parts. Losses also occur when electrical energy is transported from one place to another. A final loss comes from the thermal bottleneck described by the second law of thermodynamics. Whenever we convert thermal energy to *any* other form of energy, the second law of thermodynamics tells us that we always lose some energy to other forms: It is impossible to make the conversion with 100% efficiency.

All conversions of energy are not equivalent. Some energy conversions are less efficient and thus more costly to our world's energy budget. To obtain an accurate measure of the efficiency of the energy you use, you need to go back to the original source and calculate the overall efficiency of delivering it to your home and its final use. Suppose, for example, that the initial conversion is 50% efficient. For every unit of energy you start with, you have to give up one-half unit. If the process of transporting the energy to your home is also 50% efficient, the amount that you get to use is one-half of one-half, or one-fourth, of the original. To calculate the total efficiency, you multiply the fractional efficiencies of each stage together.

The table shows the relative efficiencies of two methods of heating water in your home. The efficiency of each step is shown as well as the overall efficiency through that step. Notice that the lowest efficiency occurs during the generation of electricity in the coal-fired plant. The extra steps of making and transporting the electricity reduce the efficiency of electric heat by a factor of $2\frac{1}{2}$ compared to heating with natural gas. Therefore, the quality of natural gas for heating is much higher than that of electricity produced from coal.

1. Heating household water with natural gas is only 64% efficient, while heating it with electricity is 92% efficient. Why is it better to use gas?

2. Give an example of energy lost as it is transformed from one form to another.

Gas water heaters are more efficient than electric ones.

© Cengage Learning/George Semple

Efficiency for Heating Water

	Step Efficiency (%)	Accumulative Efficiency (%)
Electric		
Mining of coal	96	96
Transportation of coal	97	96
Generation of electricity	33	31
Electrical transmission	85	26
Heating	92	24
Gas		
Production of natural gas	96	96
Transportation	97	93
Heating	64	60

Summary

Mechanical energy can be completely converted into internal energy, but it is not possible to recover all this internal energy and get the same amount of mechanical energy back. The energy is there, but it is not available.

We can extract some of an object's internal energy using a heat engine. Part of the heat is converted to mechanical work, and the remainder is emitted as exhaust. The first law of thermodynamics requires that the sum of the mechanical work and the exhaust heat be equal to the heat extracted. The second law of thermodynamics says that we cannot build a heat engine without an exhaust. As a consequence, it is impossible to completely convert heat to mechanical energy.

People have tried to construct two kinds of perpetual-motion machines. The first kind violates the first law of thermodynamics by trying to get more energy out of a device than is put into it. The perpetual-motion machines of the second kind do not try to do any useful work but just try to run forever. They fail because they violate the second law of thermodynamics.

The efficiency of an ideal heat engine—a Carnot engine—depends on the temperatures of the input heat source and the exhaust region. The maximum theoretical efficiency increases as the difference in these two temperatures increases and is given by $\eta = 1 - T_c/T_h$, where T_h is the input temperature and T_c is the exhaust temperature in kelvin. The efficiencies of real engines are always less than this.

Refrigerators and air conditioners are essentially heat engines running backward. These devices require work to move energy from a lower-temperature region to a higher-temperature region. An equivalent form of the second law says that we cannot build a refrigerator that moves energy from a colder region to a hotter region without work being done.

The third form of the second law says that the entropy of an isolated system tends to increase. Entropy is a measure of the order of a system. Entropy increases as the order within a system decreases. The entropy of a system can decrease, but the decrease is paid for with an overall increase in the entropy of the universe. Life is an example of decreasing entropy (increasing order).

Our energy crisis is really related to the using up of the pockets of energy in our environment that are more valuable than others. We still have the same amount of energy as we've always had.

CHAPTER 14 *Revisited*

Performing useful work with the internal energy in an object requires that the energy be able to flow from the object to a region at a lower temperature. Regardless of the amount of internal energy in an object, if there is no colder region nearby, the energy is useless for running an engine. On the other hand, the colder the exhaust region, the larger is the fraction of the internal energy that can be used.

Key Terms

disordered system   A system with an arrangement equivalent to many other possible arrangements.

efficiency   The ratio of the work produced to the energy input. For an ideal heat engine, the Carnot efficiency is given by $1 - T_c/T_h$.

entropy   A measure of the order of a system. The second law of thermodynamics states that the entropy of an isolated system tends to increase.

heat engine   A device for converting heat into mechanical work.

heat pump   A reversible heat engine that acts as a furnace in winter and an air conditioner in summer.

ordered system   A system with an arrangement belonging to a group with the smallest number (possibly one) of equivalent arrangements.

refrigerator   A heat engine running backward.

thermodynamics, second law of   There are three equivalent forms: (1) It is impossible to build a heat engine to perform mechanical work that does not exhaust heat to the surroundings. (2) It is impossible to build a refrigerator that can transfer heat from a lower-temperature region to a higher-temperature region without using mechanical work. (3) The entropy of an isolated system tends to increase.

Questions and exercises are paired so that most odd-numbered are followed by a similar even-numbered.

Blue-numbered questions and exercises are answered in Appendix B.

🌶 indicates more challenging questions and exercises.

WebAssign Many Conceptual Questions and Exercises for this chapter may be assigned online at WebAssign.

Conceptual Questions

1. What does a heat engine do?

2. In a heat engine, 140 joules of energy are extracted from a hot region. According to the first law of thermodynamics, what is the maximum amount of work that can be done by this engine? Is this result consistent with the second law of thermodynamics? Explain.

3. Why is it not possible to run an ocean liner by taking in seawater at the bow of the ship, extracting internal energy from the water, and dropping ice cubes off the stern?

4. One possible end to the universe is for it to reach thermal equilibrium; that is, it would have a uniform temperature. Would this temperature be absolute zero? Explain.

5. In an ideal heat engine, 1000 joules of energy are extracted from the hot region at 800 K. One of the laws of thermodynamics requires that if the cold region is at 320 K, the engine must exhaust 400 joules of energy. Which law of thermodynamics requires this?

6. If the ideal heat engine used in Question 5 must exhaust 400 joules of energy, one of the laws of thermodynamics indicates that the engine can do no more than 600 joules of useful work. Which law of thermodynamics indicates this?

7. Does the following statement agree with the second law of thermodynamics? "No engine can transform its entire heat input into work."

8. Would it be possible to design a heat engine that produces no thermal pollution? Explain.

9. It is possible to float heat engines on the ocean and extract some of the internal energy of the water by extending a tube well beneath the ocean's surface. Why is it necessary for the heat engine to have this tube in order to satisfy the second law of thermodynamics?

10. A hurricane can be thought of as a heat engine that converts thermal energy from the ocean to the mechanical motion of its winds. Use this idea to explain why the wind speeds decrease as the hurricane moves away from the equator.

U.S. Environmental Science Services Administration

11. Many people have tried to build perpetual-motion machines. What restrictions does the first law of thermodynamics place on the possibility of building a perpetual-motion machine?

12. What restrictions does the second law of thermodynamics place on the possibility of building a perpetual-motion machine?

13. Explain how the following simplified statements of the first and second laws of thermodynamics are consistent with the versions given in this chapter. First: You cannot get ahead. Second: You cannot even break even.

14. A student proposes to run an automobile without using any fuel by building a windmill on top of the car. The car's motion will cause the windmill to rotate and generate electricity. The electricity will run a motor, maintaining the car's motion, which in turn causes the windmill to rotate. What, if anything, is wrong with this proposal?

15. What does it mean to say that the human body is a heat engine with an efficiency of 20%?

16. Why are nuclear power plants less efficient than coal-fired plants?

17. What happens to the efficiency of an ideal heat engine as its input temperature is increased while its exhaust temperature is held fixed?

18. If the input temperature of an ideal heat engine is fixed, what happens to its efficiency as its exhaust temperature is increased?

19. Heat engine A operates between 300°C and 20°C, whereas heat engine B operates between 300°C and 80°C. Which engine has the greater possible theoretical efficiency? Explain.

20. You are building a heat engine in which the temperature difference between the hot and cold regions is 100 K. Will it be more efficient to have your cold region as cold as possible or as hot as possible? Why?

21. An engineer claims that she could build a more efficient automobile engine if the materials science division could develop piston materials that could withstand higher operating temperatures. Why would this help?

22. A car company has just designed an ultra-fuel-efficient car, and they wish to advertise the best possible miles per gallon. If the engine can be thought of as a heat engine with a constant operating temperature, would it be better to run the trial on a hot day or a cold day? Why?

23. How is the following statement equivalent to the heat-engine form of the second law of thermodynamics? "The efficiency of a heat engine must be less than 1."

24. With his paddle-wheel apparatus, Joule determined that 4.2 joules of mechanical work are equivalent to 1 calorie of heat. Imagine that he had mistakenly used a heat engine instead and had measured the heat flowing into the engine and the work done by the engine to determine the conversion factor. Would this have produced a conversion factor for 1 calorie that was greater than, equal to, or less than 4.2 joules? Why?

25. You are installing a central air-conditioning system in which the main unit sits outside your home. For maximum cooling, should you locate the unit on the sunny or the shady side of the house? Why?

26. Bob moves into a new home that is heated with an electric heat pump. He decides that because no heat pump can be perfectly efficient, he will disconnect the heat pump and use the electricity to run a baseboard heater instead. Will Bob's energy bill increase, remain the same, or decrease? Why?

27. Would it be possible to keep a room cool by leaving the door of the refrigerator open? Why or why not?

28. An air-conditioner mechanic is testing a unit by running it on the workbench in an isolated room. What happens to the temperature of the room?

29. A salesperson tries to sell you a "new and improved" air conditioner that does not need a window opening. The unit just sits in the corner of the room and keeps it cool. Use the second law of thermodynamics to convince the salesperson that this will not work.

30. Imagine you are heating your home with a heat pump that uses a small amount of work to transfer heat from the cold outside air to the warm inside air. Your friend suggests that you set up a second heat engine using the air inside the house as the hot region and the outside air as the cold region to provide the necessary work to drive the heat pump. Which law or laws of thermodynamics does this money-saving scheme violate?

31. State the refrigerator form of the second law of thermodynamics in your own words.

32. Give an example that clearly illustrates the meaning of the refrigerator form of the second law of thermodynamics.

33. In what way is the following statement equivalent to the refrigerator form of the second law of thermodynamics? "The natural direction for the flow of heat is from hotter objects to colder objects."

34. Is the following statement equivalent to the refrigerator form of the second law of thermodynamics? "In moving energy from a cold region to a hot region, the energy delivered to the hot region must be the sum of the work performed plus the energy extracted from the cold region." Explain your reasoning.

35. The coefficient of performance for a refrigerator is defined as the ratio of the heat extracted from the colder system to the work required. Will this number be greater than, equal to, or less than 1 for a good refrigerator? Explain.

36. Why is the efficiency of a heat engine always less than 1, whereas the coefficient of performance for a heat pump is not so constrained?

37. You have two friends who always play the state lottery. Janice's strategy is to always select last week's winning number for this week's draw. In contrast, Jeremy has researched the winning number combinations for the last ten years and always selects a combination that has not yet won. Which strategy, if either, is more likely to be a winner? Explain.

38. You watch a friend flipping coins and notice that heads has come up four times in a row. Does this mean that it is more likely that tails will come up on the next throw? Explain.

39. What sums of two dice have the highest and lowest order?

40. What sums of three dice have the highest order?

41. Why does the sum of two dice equal 7 more often than any other number?

42. Your friend challenges that, given 25 chances to roll two dice, you cannot roll a sum of 7 at least three times. Should you accept the challenge? Why or why not?

43. Why do "boxcars" (a pair of sixes) occur so rarely when throwing two dice?

44. On average, how many times would you expect to roll "boxcars" (a pair of sixes) with two dice if you rolled the dice a total of 180 times?

45. State the entropy form of the second law of thermodynamics in your own words.

46. Give an example that clearly illustrates the meaning of the entropy form of the second law of thermodynamics.

47. One end of a steel bar is held over a flame until it is red hot. We know from Chapter 13 that when the bar is removed from the flame, the thermal energy will diffuse along the bar until the entire bar has the same equilibrium temperature. Use a microscopic model to explain why the bar's entropy (that is, its disorder) is increasing during this equilibration process.

48. What happens to the entropy of the universe as the orange liquid diffuses into the clear liquid?

© Cengage Learning/Charles D. Winters

49. What happens to the entropy of the universe as an ice cube melts in water? Explain.

50. A cold piece of metal is dropped into an insulated container of hot water. After the system has reached an equilibrium temperature, has the entropy of the universe increased or decreased? Explain.

51. Describe a system in which the entropy is decreasing. Is this system isolated from its surroundings?

52. What happens to the entropy of a human as it grows from childhood to adulthood? Is this consistent with the second law of thermodynamics? Explain.

Russell Streadbeck (both)

53. When water freezes to ice, does the order of the water molecules increase or decrease? What does this imply about the change in entropy in the rest of the universe?

54. A ringing bell is inserted into a large glass of water. The bell and the water are initially at the same temperature and are insulated from their surroundings. Eventually, the bell stops vibrating, and the water comes to rest.
    a. What happens to the mechanical energy of the bell?
    b. What happens to the temperature of the system?
    c. What happens to the entropy of the system?

55. If you slide a crate across the floor, kinetic energy is converted to thermal energy as it comes to rest. Why will adding thermal energy to a stationary crate not cause it to move?

56. Are Mexican jumping beans a violation of the second law of thermodynamics? Explain.

57. Imagine that you could film the motion of the gas molecules in the room. Would you be able to tell whether the film was running forward or backward? Would it make a difference if air were being released from a balloon? Explain.

58. You have an aquarium with a divider down the middle. One side is filled with hot water, and the other is filled with cold water. Imagine that as the divider is removed you can film the individual collisions between water molecules. When watching the film, how could you tell whether it was running forward or backward?

59. Which of the following statements explains why we are currently experiencing a worldwide energy crisis?
    a. The amount of energy in the world is decreasing rapidly.
    b. The entropy of the world is increasing rapidly.
    c. The entropy of the world is decreasing rapidly.

60. How does slowing the increase in entropy help solve the world's energy crisis?

61. Which results in the larger increase in the entropy of the universe: heating a liter of room-temperature water to boiling using natural gas or using electricity? Why?

62. Why is heating water on a gas stove more efficient than heating it on an electric stove?

63. It is the middle of winter, and you live in a house with electric baseboard heating. Your friend chides you for being wasteful for turning on the oven to 400°F for 45 minutes just to cook a single baked potato. How do you respond?

64. Since childhood we've been told to turn out the lights when we leave a room. Does this really reduce the electric bill during the winter for a house with electric heating? Why?

Exercises

65. What input energy is required if an engine performs 300 kJ of work and exhausts 400 kJ of heat?

66. How much work is performed by a heat engine that takes in 2000 J of heat and exhausts 800 J?

67. An engine takes in 9000 cal of heat and exhausts 4000 cal of heat each minute it is running. How much work does the engine perform each minute?

68. A heat engine requires an input of 10 kJ per minute to produce 3 kJ of work per minute. How much heat must the engine exhaust per minute?

69. What is the efficiency of a heat engine that does 50 J of work for every 200 J of heat it takes in?

70. An engine exhausts 1200 J of energy for every 3600 J of energy it takes in. What is its efficiency?

71. An engine has an efficiency of 40%. How much energy must be extracted to do 900 J of work?

72. An engine operates with an efficiency of 25%. If the engine does 600 J of work every minute, how many joules per minute are exhausted to the cold region?

73. An engine takes in 600 cal and exhausts 450 cal each second it is running. How much work does the engine perform each minute? What is the engine's efficiency?

74. How much work does an engine produce each second if it takes in 8000 cal and exhausts 5000 cal each second? What is the efficiency of the engine?

75. An engineer has designed a machine to produce electricity by using the difference in the temperature of ocean water at different depths. If the surface temperature is 20°C and the temperature at 50 m below the surface is 12°C, what is the maximum efficiency of this machine?

76. A heat engine takes in 1000 J of energy at 1000 K and exhausts 600 J at 500 K. What are the actual and maximum theoretical efficiencies of this heat engine?

77. An ideal heat engine has a theoretical efficiency of 60% and an exhaust temperature of 27°C. What is its input temperature?

78. What is the exhaust temperature of an ideal heat engine that has an efficiency of 50% and an input temperature of 400°C?

79. How much work is required by a refrigerator that takes in 1000 J from the cold region and exhausts 1500 J to the hot region?

80. A refrigerator uses 600 J of work to remove 2400 J of heat from a room. How much heat does it exhaust?

81. How much work per second (power) is required by a refrigerator that takes 700 J of thermal energy from a cold region each second and exhausts 1500 J to a hot region?

82. A heat pump requires 500 W of electrical power to deliver heat to your house at a rate of 2400 J per second. How many joules of energy are extracted from the cold air outside each second?

83. The coefficient of performance for a heat pump is defined as the ratio of the heat extracted from the colder system to the work required. If a heat pump requires an input of 400 W of electrical energy and has a coefficient of performance of 3, how much energy is delivered to the inside of the house each second?

84. If a refrigerator requires an input of 200 J of electrical energy each second and has a coefficient of performance of 5, how much heat energy is extracted from the refrigerator each second?

85. Show that four coins can be arranged in 16 different ways.

86. Show that the combination of four coins with the lowest order (two heads and two tails) is the one with the largest number of arrangements.

87. What is the probability of rolling a total of 6 with two dice?

88. What is the probability of rolling a sum of 10 with two dice?

89. The total number of possible states for three dice is $6 \times 6 \times 6 = 216$. What is the probability of throwing a sum equal to 5?

90. The total number of possible states for three dice is $6 \times 6 \times 6 = 216$. What is the probability of throwing a sum equal to 15?

The Big Picture

Waves—Something Else That Moves

Imagine standing near a busy highway trying to get the attention of a friend on the other side. How could you signal your friend? You might try shouting first. You could throw something across the highway, you could make a loud noise by banging two rocks together, you could shine a flashlight at your friend, and so on.

Signals can be sent by one of two methods. One method includes ways in which material moves from you to your friend—such as throwing a pebble. The other method includes ways in which energy moves across the highway without any accompanying material. This second method represents phenomena that we usually call waves.

The study of waves has greatly expanded the physics world view. Surprisingly, however, waves do not have a strong position in our commonsense world view. It's not that wave phenomena are uncommon, but rather that many times the wave nature of the phenomena is not recognized. Plucking a guitar string, for example, doesn't usually invoke images of waves traveling up and down the string. But that is what happens. The buzzing of a bee probably does not generate thoughts of waves either. We will discover interesting examples of waves in unexpected situations.

Waves are certainly common enough—we grow up playing with water waves and listening to sound waves—but

Raymond Gendreau/Stone/Getty

Sculling on Lake Powell creates interesting wave patterns.

most of us do not have a good intuitive understanding of the behavior of waves. Ask yourself a few questions about waves: Do they bounce off materials? When two waves meet, do they crash like billiard balls? Is it meaningful to speak of the speed of a wave? When speaking of material objects, the answers to such questions seem obvious, but when speaking of waves, the answers require closer examination.

We study waves for two reasons. First, because they are there; studying waves adds to our understanding of how the world works. The second reason is less obvious. As we delve deeper and deeper into the workings of the world, we reach limits beyond which we cannot observe phenomena directly. Even the best imaginable magnifying instrument is too weak to allow direct observation of the subatomic worlds. Our search to understand these worlds yields evidence only by indirect methods. We must use our

common experiences to model a world we cannot see. In many cases the modeling process can be reduced to asking whether the phenomenon acts like a wave or acts like a particle.

To answer the question of whether something acts like a wave or a particle, we must expand our commonsense world view to include waves. After you study such common waves as sound waves, we hope you will be ready to "hear" the harmony of the subatomic world.

It's not that wave phenomena are uncommon, but rather that many times the wave nature of the phenomena is not recognized.

© Cengage Learning/Charles D. Winters

15

Vibrations and Waves

▶Water drops falling onto the surface of water produce waves that move outward as expanding rings. But what is moving outward? Does the wave disturbance carry energy or momentum? What happens when two waves meet? How does wave motion differ from particle motion?

(See page 323 for the answer to this question.)

Don Bonsey/Stone/Getty

Circular waves are formed by falling water drops.

I F you stretch or compress a spring and let go, it vibrates. If you pull a pendulum off to one side and let it go, it oscillates back and forth. Such vibrations and oscillations are common motions in our everyday world. If these vibrations and oscillations affect surrounding objects or matter, a wave is often generated. Ripples on a pond, musical sounds, laser light, exploding stars, and even electrons all display some aspects of wave behavior.

Waves are responsible for many of our everyday experiences. Fortunately, nature has been kind; all waves have many of the same characteristics. Once you understand one type, you will know a great deal about the others.

We begin our study with simple vibrations and oscillations. We then examine common waves, such as waves on a rope, water waves, and sound waves, and later progress to more exotic examples, such as radio, television, light, and even "matter" waves.

▶ Extended presentation available in the *Problem Solving* supplement

Simple Vibrations

If you distort an object and release it, elastic forces restore the object to its original shape. In returning to its original shape, however, the inertia of the displaced portion of the object causes it to overshoot, creating a distortion in the opposite direction. Again, restoring forces attempt to return the object to its original shape and, again, the object overshoots. This back-and-forth motion is what we commonly call a **vibration,** or an **oscillation.** For all practical purposes, the labels are interchangeable.

A mass hanging on the end of a vertical spring exhibits a simple vibrational motion. Initially, the mass stretches the spring so that it hangs at the position where its weight is just balanced by the upward force of the spring, as shown in Figure 15-1. This position—called the **equilibrium position**—is analogous to the undistorted shape of an object. If you pull downward (or push upward) on the mass, you feel a force in the opposite direction. The size of this restoring force increases with the amount of stretch or compression you apply. If the applied force is not too large, the restoring force is proportional to the distance the mass is moved from its equilibrium position. If the force is too large, the spring will be permanently stretched and not return to its original length. In the discussion that follows, we assume that the stretch of the system is not too large. Many natural phenomena obey this condition, so little is lost by imposing this constraint.

Imagine pulling the mass down a short distance and releasing it as shown in Figure 15-2(a). Initially, a net upward force accelerates the mass upward. As the mass moves upward, the net force decreases in size (b), becoming zero when the mass reaches the equilibrium position (c). Because the mass has inertia, it overshoots the equilibrium position. The net force now acts downward (d) and slows the mass to zero speed (e). Then the mass gains speed in the downward direction (f). Again, the mass passes the equilibrium position (g). Now the net force is once again upward (h) and slows the mass until it reaches its lowest point (a). This sequence [Figure 15-2(a through a)] completes one **cycle.**

Actually, a cycle can begin at any position. It lasts until the mass returns to the original position *and* is moving in the same direction. For example, a cycle may begin when the mass passes through the equilibrium point on its way up (c) and end when it next passes through this point on the way up. Note that the cycle does not end when the mass passes through the equilibrium point on the way down (g). This motion is known as periodic motion, and the amount of time required for one cycle is known as the **period** *T.*

If we ignore frictional effects, energy conservation (Chapter 7) tells us that the mass travels the same distance above and below the equilibrium position.

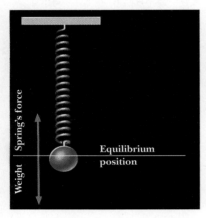

Figure 15-1 At the equilibrium position, the upward force due to the spring is equal to the weight of the mass.

period is the time to complete one cycle ▶

Figure 15-2 A time sequence showing one complete cycle for the vibration of a mass on a spring. The clocks show that equal time intervals separate the images.

This distance is marked in Figure 15-2 and is known as the **amplitude** of the vibration. In real situations the amplitude decreases and eventually the motion dies out because of the frictional effects that convert mechanical energy into thermal energy.

We can describe the time dependence of the vibration equally well by giving its **frequency** f, the number of cycles that occur during a unit of time. Frequency is often measured in cycles per second, or hertz (Hz). For example, concert A (the note that orchestras use for tuning) has a frequency of about 440 hertz, household electricity oscillates at 60 hertz, and your favorite FM station broadcasts radio waves near 100 million hertz.

There is a simple relationship between the frequency f and the period T—one is the reciprocal of the other:

$$f = \frac{1}{T}$$

◀ frequency $= \frac{1}{\text{period}}$

$$T = \frac{1}{f}$$

◀ period $= \frac{1}{\text{frequency}}$

To illustrate this relationship, let's calculate the period of a spring vibrating at a frequency of 4 hertz:

$$T = \frac{1}{f} = \frac{1}{4 \text{ Hz}} = \frac{1}{4 \text{ cycles/s}} = \frac{1}{4} \text{ s}$$

This calculation shows that a frequency of 4 cycles per second corresponds to a period of $\frac{1}{4}$ second. This makes sense because a spring vibrating four times per second should take $\frac{1}{4}$ of a second for each cycle. (When we state the period, we know it refers to one cycle and don't write "second per cycle.")

Q: What is the period of a mass that vibrates with a frequency of 10 times per second?

A: Because the period is the reciprocal of the frequency, we have

$$T = \frac{1}{f} = \frac{1}{10 \text{ Hz}} = 0.1 \text{ s}$$

We may guess that the time it takes to complete one cycle would change as the amplitude changes, but experiments show that the period remains essentially constant. It is fascinating that the amplitude of the motion does *not* affect the period and frequency. (Again, we have to be careful not to stretch the system "too much.") This means that a vibrating guitar string always plays the same frequency regardless of how hard the string is plucked.

Although the period for a mass vibrating on the end of a spring does not depend on the amplitude of the vibration, we may expect the period to change if we switch springs or masses. The stiffness of the spring and the size of the mass do change the rate of vibration.

The stiffness of a spring is characterized by how much force is needed to stretch it by a unit length. For moderate amounts of stretch or compression, this value is a constant known as the **spring constant** k. In SI units this constant is measured in newtons per meter. Larger values correspond to stiffer springs.

In trying to guess the relationship between the spring constant, mass, and period, we would expect the period to decrease as the spring constant increases because a stiffer spring means more force and therefore a quicker return to the equilibrium position. Furthermore, we would expect the period to increase as the mass increases because the inertia of a larger mass will slow the motion.

WORKING IT OUT *Period of a Mass on a Spring*

The mathematical relationship for the period of a mass on a spring can be obtained theoretically and is verified by experiment:

◀ period of a mass on a spring

$$T = 2\pi \sqrt{\frac{m}{k}}$$

where π is approximately 3.14.

As an example, consider a 0.2-kg mass hanging from a spring with a spring constant of 5 N/m:

$$T = 2\pi \sqrt{\frac{m}{k}} = 2\pi \sqrt{\frac{0.2 \text{ kg}}{5 \text{ N/m}}} = 6.28 \sqrt{\frac{1}{25} \text{ s}^2} = 1.26 \text{ s}$$

Therefore, this mass–spring combination vibrates with a period of 1.26 s, or a frequency of 0.793 Hz.

Q: What is the period of a 0.1-kg mass hanging from a spring with a spring constant of 0.9 N/m?

A: 2.09 s.

The Pendulum

The pendulum is another simple system that oscillates. Students are often surprised to learn (or to discover by experimenting) that the period of oscillation does not depend on the amplitude of the swing. To a very good approximation, large- and small-amplitude oscillations have the same period if we keep their amplitudes less than 30 degrees. This amazing property of pendula was first discovered by Galileo when he was a teenager sitting in church watching a swinging chandelier. (Clearly, he was not paying attention to the service.) Galileo tested his hypothesis by constructing two pendula of the same length and swinging them with different amplitudes. They swung together, verifying his hypothesis.

Let's consider the forces on a pendulum when it has been pulled to the right, as shown in Figure 15-3. The component of gravity acting along the string is balanced by the tension in the string. Therefore, the net force is the component of gravity at right angles to the string and directed toward the lower left. This restoring force causes the pendulum bob to accelerate toward the left. Although the restoring force on the bob is zero at the lowest point of the swing, the bob passes through this point (the equilibrium position) because of its inertia. The restoring force now points toward the right and slows the bob.

We found in free fall that objects with different masses fall with the same acceleration because the gravitational force is proportional to the mass. Therefore, we may expect that the motion of a pendulum would not depend on the mass of the bob. This prediction is true and can be verified easily by making two pendula of the same length with bobs of the same size made out of different materials so that they have different masses. The two pendula will swing side by side.

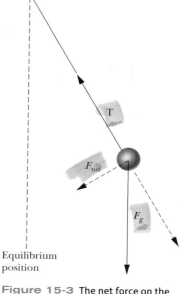

Figure 15-3 The net force on the pendulum bob accelerates it toward the equilibrium position.

Q: Why do we suggest using different materials?

A: If we use the same type of material, the size has to be different to get different masses. Different sizes may also affect the period. When doing an experiment, it is important to keep all but one factor constant.

We also know from our experiences with pendula that the period depends on the length of the pendulum; longer pendula have longer periods. Therefore, the length of the pendulum can be changed to adjust the period.

Because the restoring force for a pendulum is a component of the gravitational force, you may expect that the period depends on the strength of gravity, much as the period of the mass on the spring depends on the spring constant. This hunch is correct and can be verified by taking a pendulum to the Moon, where the acceleration due to gravity is only one-sixth as large as that on Earth.

Clocks

Keeping time is a process of counting the number of repetitions of a regular, recurring process, so it is reasonable that periodic motions have been important to timekeepers. Devising accurate methods for keeping time has kept many scientists, engineers, and inventors busy throughout history. The earliest methods for keeping time depended on the motions in the heavens. The day was determined by the length of time it took the Sun to make successive crossings of a north–south line and was monitored with a sundial. The month was

A strobe photograph of a pendulum taken at 20 flashes per second. Note that the pendulum bob moves the fastest at the bottom of the swing.

WORKING IT OUT *Period of a Pendulum*

The period of a pendulum is given by

period of a pendulum ▶

$$T = 2\pi\sqrt{\frac{L}{g}}$$

As an example, consider a pendulum with a length of 10 m:

$$T = 2\pi\sqrt{\frac{L}{g}} = 2\pi\sqrt{\frac{10 \text{ m}}{10 \text{ m/s}^2}} = 6.28\sqrt{1 \text{ s}^2} = 6.28 \text{ s}$$

Therefore, this pendulum would oscillate with a period of 6.28 s.

Are You On the Bus?

Q: What would you expect for the period of a 1.7-m pendulum on the Moon?

A: 6.28 s.

A replica of an early mechanical clock.

determined by the length of time it took the Moon to go through its phases. The year was the length of time it took to cycle through the seasons and was monitored with a calendar, a method of counting days.

As science and commerce advanced, the need grew for increasingly accurate methods of determining time. An early method for determining medium intervals of time was to monitor the flow of a substance such as sand in an hourglass or water in a water clock. Neither of these, however, was very accurate, and because they were not periodic, they had to be restarted for each time interval. It is interesting to note that Galileo kept time with a homemade water clock in many of his early studies of falling objects.

The next generation of clocks took on a different character, employing oscillations as their basic timekeeping mechanism. Galileo's determination that the period of a pendulum does not depend on the amplitude of its swing led to Christiaan Huygens's development of the pendulum clock in 1656, 14 years after Galileo's death. One of the difficulties Huygens encountered was to develop a mechanism for supplying energy to the pendulum to maintain its swing.

Seafarers spurred the development of clocks that would keep accurate time over long periods. To determine longitude requires measuring the positions of prominent stars and comparing these positions with their positions as seen from Greenwich, England, *at the same time.* Because pendulum clocks did not work on swaying ships, several cash prizes were offered for the design and construction of suitable clocks. Beginning in 1728, John Harrison, an English instrument maker, developed a series of clocks that met the criteria, but he was not able to collect his money until 1765. One of Harrison's clocks was accurate to a few seconds after 5 months at sea.

Any periodic vibration can be used to run clocks. Grandfather clocks use pendulums to regulate the hands and are powered by hanging weights. Mechanical watches have a balance wheel fastened to a spring. Electric clocks use 60-hertz alternating electric current. Digital clocks use the vibrations of quartz crystals or resonating electric circuits.

Modern time is kept with atomic clocks, which use the frequencies of atomic transitions (see Chapter 23) and are extremely insensitive to such changes in the clocks' environment as pressure and temperature. Atomic clocks are accurate to better than a second in 60 million years.

Resonance

We discovered with the mass on a spring and the pendulum that each system had a distinctive, natural frequency. The natural frequency of the pendulum is determined by its length and the acceleration due to gravity. Pulling the bob back and releasing it produces an oscillation at this particular frequency.

A child on a swing is an example of a life-size pendulum. If the child does not "pump" her legs and if no one pushes her, the amplitude of the swing continually decreases, and the child comes to rest. As every child knows, however, pumping or pushing greatly increases the amplitude. A less obvious fact is that the size of the effort—be it from pumping or pushing—is not important, but its timing is crucial. The inputs must be given at the natural frequency of the swing. If the child pumps at random times, the swinging dies out. This phenomenon of a large increase in the amplitude when a periodic force is applied to a system at its natural frequency is called **resonance.**

Resonance can also be achieved by using impulses at other special frequencies, but each of these has a definite relationship to the natural frequency. For example, if you push the child on the swing every other time, you are providing inputs at one-half of the natural frequency; every third time gives inputs at one-third of the natural frequency; and so on. Each of these frequencies causes resonance.

Modern time is kept by extremely accurate atomic clocks such as this F1 operated by the National Institute of Standards and Technology. It is accurate to 1 second in 80 million years.

Q: What happens to the amplitude of the swing if you push at twice the natural frequency?

A: In this case you would be pushing twice for each cycle. One of the pushes would negate the other, and the swing would stop.

More complex systems also have natural frequencies. If someone strikes a spoon on a table, you are not likely to mistake its sound for that of a tuning fork. The vibrations of the spoon produce sounds that are characteristic of the spoon. All objects have natural frequencies. The factors that determine these frequencies are rather complex. In general, the dominant factors are the stiffness of the material, the mass of the material, and the size of the object.

Resonance can have either good or bad effects. Although your radio receives signals from many stations simultaneously, it plays only one station at a time. The radio can be tuned so that its resonant frequency matches the broadcast frequency of your favorite station. Tuning puts the radio in resonance with one particular broadcast frequency and out of resonance with the frequencies of the competing stations. On the other hand, if the radio has an inferior speaker with one or two strong resonant frequencies, it will distort the sounds from the radio station by not giving all frequencies equal amplification.

Suppose you have a collection of pendula of different lengths, as shown in Figure 15-4. Notice that two of these pendula have the same length and thus the same natural frequency. The pendula are not independent because they are all tied to a common string. The motion of one of them is felt by all the others through pulls by the string. If you start the left-hand pendulum swinging, its back-and-forth motion creates a tug on the common string with a frequency equal to its natural, or resonant, frequency. Pendula with different frequencies jiggle a little bit but are not affected much. However, the pendulum with the same frequency resonates with the input frequency, drastically increasing its amplitude. In exactly the same way, objects resonate when input frequencies are the same as any of their natural frequencies.

If children pump at the right frequencies, they can increase the amplitudes of their motions.

Figure 15-4 Pendula with the same natural frequency resonate with each other.

Everyday Physics | *Tacoma Narrows Bridge*

Resonant effects can sometimes have disastrous consequences. In 1940 a new bridge across one of the arms of Puget Sound in the state of Washington was opened to traffic. It was a suspension bridge with a central span of 850 meters (2800 feet). Because the bridge was designed for two lanes, it had a width of only 12 meters (40 feet). Within a few months after it opened, early-morning winds in the Sound caused the bridge to oscillate in standing-wave patterns that were so large in amplitude that the bridge failed structurally and fell into the water below, as shown in the figure.

But why did this bridge fail when other suspension bridges are still standing (including the bridge that now spans the Sound at the location of the original)? The bridge was long, narrow, and particularly flexible. Motorists often complained about the vertical oscillations and nicknamed the bridge "Galloping Gertie." However, the amplitudes of the vertical oscillations were relatively small until that fateful morning. The wind was blowing along the arm of the Sound (perpendicular to the length of the bridge) at moderate to high velocities but was not near gale force. One may speculate that fluctuations in the wind speed matched the natural frequency of the bridge, causing it to resonate. However, the wind was reasonably steady, and wind fluctuations are normally quite random. Furthermore, the forces would be horizontal, and the oscillations were vertical.

The best explanation involves the formation and shedding of vortices in the wind blowing past the bridge. *Vortices* are the eddies that you get near the ends of the oars when you row a boat. Vortices rotate in opposite directions in the wind blowing over and under the bridge. As each vortex is shed, it exerts a vertical impulse on the bridge. Therefore, if the frequency of vortex formation and shedding is near the natural frequency of the bridge for vertical oscillations, a standing wave will form just like those on a guitar string. (The frequency does not have to match exactly; it only needs to be close. How close depends on the details of the bridge construction.)

The bridge would have been fine except for another unfortunate circumstance. Besides the vertical standing wave, there were also torsional, or twisting, standing waves on the bridge. Normally, the frequencies of the two standing waves are quite different. But for the Tacoma Narrows Bridge, the two frequencies were fairly close (eight per minute for the vertical motion compared with ten per minute for the twisting motion). This allowed some of the energy from the vertical motion to be transferred to the twisting motion that eventually led to the mechanical failure of the bridge.

1. What was the periodic driving force that caused the standing wave to be formed on the bridge?

2. What uncommon characteristic of the Tacoma Narrows Bridge ultimately led to its mechanical failure?

Special Collections Div. Univ. of Washington Libraries, photo by Farquharson (both)

The Tacoma Narrows Bridge collapsed when winds set up resonant vibrations.

Waves: Vibrations That Move

Most **waves** begin with a disturbance of some material. Some disturbances, such as the clapping of hands, are onetime, abrupt events, whereas others,

Figure 15-5 A wave pulse travels along a line of dominoes.

Figure 15-6 A wave disturbance can move along a chain of balls and springs.

such as the back-and-forth vibration of a guitar string, are periodic events. The simplest wave is a single pulse that moves outward as a result of a single disturbance.

Imagine a long row of dominoes lined up as shown in Figure 15-5. Once a domino is pushed over, it hits its neighbor, and its neighbor hits its neighbor, and so on, sending the disturbance along the line of dominoes. The key point is that "something" moves along the line of dominoes—from the beginning to the end—but it is not any individual domino.

Actually, the domino example is not completely analogous to what happens in most situations involving waves because there is no mechanism for restoring the dominoes in preparation for the next pulse. We can correct this omission by imagining a long chain of balls connected by identical springs, as shown in Figure 15-6. As the disturbance moves from left to right, individual balls are lifted up from their equilibrium positions and then returned to these equilibrium positions. The springs allow each ball to pull its neighbor away from equilibrium, just as the dominoes passed the disturbance from neighbor to neighbor by striking each other. After the pulse passes, the springs provide the restoring force that returns each ball to equilibrium. Notice that the pulse travels along the chain of balls without any of the balls moving in the direction of the pulse. Figure 15-6 shows the shape of the chain of balls and springs at the time the center ball reaches its maximum displacement.

In a similar manner, a pebble dropped into a pond depresses a small portion of the surface. Each vibrating portion of the surface generates disturbances in the surrounding water. As the process continues, the disturbance moves outward in circular patterns, such as those shown in Figure 15-7.

This type of disturbance, or pulse, occurs in a number of common, everyday events. A crowd transmits single pulses when a small group begins pushing. This push spreads outward through the crowd much as a ripple moves over a pond's surface. Similarly, the disturbance produced by a clap sends a single sound pulse through the air. Other examples of nonrecurrent waves include tidal bores, tidal waves, explosions, and light pulses emitted by supernovas (exploding stars).

Figure 15-7 Water drops produce disturbances that move outward in circular patterns.

Figure 15-8 In a transverse wave (a), the medium moves perpendicular to the direction of propagation of the wave, whereas in a longitudinal wave (b), the medium's motion is parallel to the direction of propagation.

(a)

(b)

Although a wave moves outward from the original disturbance, there is no overall motion of the material. As the wave travels through the medium, the particles of the material vibrate about their equilibrium positions. Although the wave travels down the chain, the individual balls of the chain return to their original positions. The wave transports energy rather than matter from one place to another. The energy of an undisturbed particle in front of the wave is increased as the wave passes by and then returns to its original value. In a real medium, however, some of the energy of the wave is left behind as thermal energy in the medium.

There are two basic wave types. A wave in which the vibration of the medium is perpendicular to the motion of the wave is called a **transverse wave.** Waves on a rope are transverse waves. A wave in which the vibration of the medium is along the same direction as the motion of the wave is called a **longitudinal wave.** Both types can exist in the chain of balls. If a ball is moved vertically, a transverse wave is generated [Figure 15-8(a)]. If the ball is moved horizontally, the wave is longitudinal [Figure 15-8(b)].

Transverse waves can move only through a material that has some rigidity; transverse waves cannot exist within a fluid because the molecules simply slip by each other. Longitudinal waves, on the other hand, can move through most materials because the materials can be compressed and have restoring forces.

Are You On the Bus?

Q: Is it possible to have transverse waves on the surface of water?

A: Transverse surface waves are possible because the force of gravity tends to restore the surface to its flat equilibrium shape. Actually, the motion of the individual water molecules is a combination of transverse motion and longitudinal motion; the water molecules follow elliptical paths.

One-Dimensional Waves

Because all waves have similar properties, we can look at waves that are easy to study and then make generalizations about other waves. Imagine a clothesline tied to a post, as in Figure 15-9. A flick of the wrist generates a single wave pulse that travels away from you. On an idealized rope, the wave pulse would maintain its shape and size. On a real rope, the wave pulse slowly spreads out. We will ignore this spreading in our discussion. The wave's speed can be calculated by dividing the distance the pulse travels by the time it takes.

The speed of the wave can be changed. If you pull harder on the rope, the pulse moves faster; the speed increases as the tension in the rope increases. The speed also depends on the mass of the rope; a rope with more mass per

Figure 15-9 While pieces of the rope vibrate up and down, the wave moves along the rope.

unit length has a slower wave speed. Surprisingly, the amplitude of the pulse does not have much effect on the speed.

These observations make sense if we consider the vibrations of a small portion of the rope. The piece of rope is initially at rest and moves as the leading edge of the pulse arrives. How fast the rope returns to its equilibrium position determines how the pulse passes through the region (and hence, the speed of the pulse). The more massive the rope, the more sluggishly it moves. Also, if the rope is under a larger tension, the restoring forces on the piece of rope are larger and cause it to return to its equilibrium position more quickly.

FLAWED REASONING

A physics teacher has offered his class a prize if they can send a transverse pulse down a long spring and then send a second pulse down the same spring in such a manner as to catch up with the first pulse. Three students have taken up the challenge.

Trever: "I will make the first pulse with a slow movement of my hand and then make a second pulse with a very quick jerk on the spring. That should send the second pulse down the spring at a quicker speed."

Lindzee: "I think the amplitude of the pulse is what matters, not how fast you move your hand. Send the first pulse down with a big amplitude and then send the second pulse down with a small amplitude."

Courtnee: "The textbook claims that pulse speeds don't depend on how the pulse was created but only on the tension and the mass density of the spring. We can't change the mass density after we send the first pulse, but we could tighten the spring. Send the first pulse and then pull the spring tighter before we send the second pulse."

All three students are wrong. **Find the flaws in their claims.**

ANSWER Courtnee correctly points out the flaws in her classmates' reasoning. The speed of a pulse down a stretched spring does not depend on the size or shape of the pulse or the manner in which it was created. Courtnee's suggestion of stretching the spring will indeed increase the speed of transverse pulses on the spring, but this will speed up *both* pulses. The physics teacher tricked them with an impossible challenge.

When a pulse hits the end that is attached to the post, it "bounces" off and heads back. This reflected pulse has the same shape as the incident pulse but is inverted, as shown in Figure 15-10. If the incident pulse is an "up" pulse (a **crest**), the reflected pulse is a "down" pulse (a **trough**). If the end of the rope is free to move up and down, the pulse still reflects, but no inversion takes place.

Figure 15-10 A wave pulse is inverted when it reflects from a fixed end. Note that the steep edge leads on the way in and on the way out.

Figure 15-13 This time sequence shows that the superposition of two wave pulses yields shapes that are the sum of the individual shapes.

Before

After

Figure 15-11 In contrast to a wave, the blue half of the ball leads on the way in and trails after reflection from the wall.

Figure 15-12 The two wave pulses on the rope pass through each other as if the other were not present.

What about the front and back of the wave pulse? To observe this you would generate a pulse that is not symmetric. The pulses shown in Figure 15-10 are steeper in front than in back. The steeper edge is away from you when the pulse moves down the rope and toward you when the reflected pulse returns. The leading edge continues to lead.

These inversions contrast with the behavior of a ball when it "reflects" from a wall. If the ball is not spinning, the top of the ball remains on top, but the leading edge is interchanged. Figure 15-11 shows that the blue half leads before the collision, whereas the green half leads afterward.

Superposition

Suppose you send a crest down the rope and, when it reflects as a trough, you send a second crest to meet it. An amazing thing happens when they meet. The waves pass through each other as if the other were not there. This is shown in Figure 15-12. Each pulse retains its own shape, clearly demonstrating that the pulses are not affected by the "collision." A similar thing happens when you throw two pebbles into a pond. Even though the wave patterns overlap, you can still see a set of circular patterns move outward from *each* splash.

In contrast, imagine what would happen if two particles—say, two Volkswagens—were to meet. Particles don't exhibit this special property of waves. It would certainly be a strange world if waves did not pass through each other. Two singers singing at the same time would garble each other's music, or the sounds from one might bounce off those from the other.

During the time the wave pulses pass through each other, the resulting disturbance is a combination of the individual ones; it is a **superposition** of the pulses. As shown in Figure 15-13, the distance of the medium from the equilibrium position, the **displacement,** is the algebraic sum of the displacements of the individual wave pulses. If we consider displacements above the equilibrium position as positive and those below as negative, we can obtain the shape of the resultant disturbance by adding these numbers at each location along the rope.

Everyday Physics | *Probing Earth*

Imagine drawing a circle to represent Earth. Further, imagine drawing a dot to show how far Earth's interior has been explored by direct drilling and sampling techniques. Where would you place the dot? The dot should be placed on the original circle. Earth is about 6400 kilometers (4000 miles) in radius, and we have drilled into its interior only about 12.2 kilometers, less than 0.2% of the distance to the center. Therefore, we must learn about Earth's interior using indirect means such as looking at signals from explosions and earthquakes.

Three kinds of wave are produced in an earthquake. One type travels along the surface, and the other two travel through Earth's interior; one of the interior waves is a longitudinal wave, and the other is a transverse wave. These waves move outward in all directions from the earthquake site and are received at numerous earthquake-monitoring sites around the world. The detection of these waves and their arrival times provides clues about Earth's interior.

Two major things happen to the waves: First, partial reflections occur at boundaries between distinctly different regions. Second, the waves change speed as the physical conditions—such as the elasticity and density—change. Changing a wave's speed usually results in the wave changing direction, a phenomenon known as refraction, which we will discuss in Chapter 18. As the waves go deeper into the interior, they speed up, causing them to change direction.

The longitudinal waves are called the primary waves, or P waves, and are created by the alternating expansion and compression of the rocks near the source of the shock. This push–pull vibration can be transmitted through solids, liquids, and gases. P waves move with the highest speeds and therefore are the first to arrive at a seismograph station. P waves move at about 5 kilometers per second (11,000 miles per hour!) near the surface and speed up to about 7 kilometers per second toward the base of the upper crust.

The secondary waves, or S waves, are transverse waves. In this case the rock movement is perpendicular to the direction the wave is traveling. S waves travel through solids but cannot propagate through liquids and gases because fluids lack rigidity.

We wouldn't be able to infer much about Earth's interior if only one signal arrived at each site. There would be a number of paths

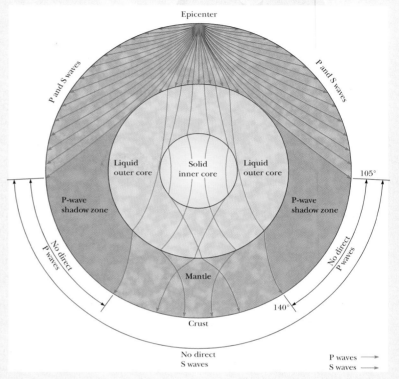

Cross section of Earth showing the paths of some waves produced by an earthquake.

that could account for the characteristics and timing of the signal. Fortunately, many sites receive multiple signals that allow large computers to piece the information together to form a model of Earth's interior. Information that is not received is also important. After an earthquake, many sites do not receive any transverse signals. This tells us that they are located in a shadow region behind a liquid core, as shown in the figure.

1. Would *transverse* or *longitudinal* waves be used to communicate with a submarine that was submerged in the ocean? Explain your choice.

2. What characteristic of all waves, transverse or longitudinal, explains the P wave shadow zone that is observed?

Source: H. Levin, *The Earth through Time* (Philadelphia: Saunders, 1992).

If two crests overlap, the disturbance is bigger than either one alone. A crest and a trough produce a smaller disturbance. If the crest and the trough are the same size and have symmetric shapes, they completely cancel at the instant of total overlap. A high-speed photograph taken at this instant yields a picture of a straight rope. This phenomenon is illustrated in Figure 15-12(d). This is not as strange as it may seem. If we take a high-speed photograph of a pendulum just as it swings through the equilibrium position, it would appear that the pendulum was not moving but simply hanging straight down. In either case, longer exposures would blur, showing the motion.

Periodic Waves

A rope moved up and down with a steady frequency and amplitude generates a train of wave pulses. All the pulses have the same size and shape as they travel down the rope. The drawing in Figure 15-14 shows a **periodic wave** moving to the right. New effects emerge when we examine periodic waves. For one thing, unlike the single pulse, periodic waves have a frequency. The frequency of the wave is the oscillation frequency of any piece of the medium.

An important property of a periodic wave is the distance between identical positions on adjacent wave pulses, called the **wavelength** of the periodic wave. This is the smallest distance for which the wave pattern repeats. It may be measured between two adjacent crests, or two adjacent troughs, or any two identical spots on adjacent pulses, as shown in Figure 15-15. The symbol used for wavelength is the Greek letter lambda λ.

The speed of the wave can be determined by measuring how far a particular crest travels in a certain time. In many situations, however, the speed is too fast, the wavelength too short, or the amplitude too small to allow us to follow the motion of a single crest. We then use an alternative procedure.

Suppose you take a number of photographs at the same frequency as the vertical vibration of any portion of the rope. You would find that all the pictures look the same. During the time the shutter of the camera was closed, each portion of the rope went through a complete cycle, ending in the position it had during the previous photograph. But this means that each crest moved from its original position to the position of the crest in front of it. That

Figure 15-14 A periodic wave on a rope can be generated by moving the end up and down with a constant frequency.

Figure 15-15 The wavelength of a periodic wave is the distance between any two identical spots on the wave.

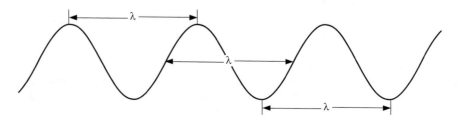

is, the crest moved a distance equal to the wavelength λ. Because the time between exposures is equal to the period T, the wave's speed v is

$$v = \frac{\lambda}{T}$$

Because the frequency is just the reciprocal of the period, we can change the equation to read

$$v = \lambda f$$

 ◀ speed = wavelength × frequency

Although we developed this relationship for waves on a rope, there is nothing special about these waves. This relationship holds for all periodic waves, such as radio waves, sound waves, and water waves.

WORKING IT OUT   *Speed of a Wave*

If you know any two of the three quantities in the wave equation, you can use this relationship to calculate the third. As an example, let's calculate the speed of a wave that has a frequency of 40 Hz and a wavelength of $\frac{3}{4}$ m. Multiplying the wavelength and the frequency gives us the speed:

$$v = \lambda f = \left(\tfrac{3}{4}\,\text{m}\right)(40\ \text{Hz}) = 30\ \text{m/s}$$

Q: If water waves have a frequency of 5 Hz and a wavelength of 8 cm, what is the wave speed?

A: $v = \lambda f = (8\ \text{cm})(5\ \text{Hz}) = 40\ \text{cm/s}$.

Are *You* On the Bus?

Standing Waves

When a periodic wave is confined, new effects emerge because of the superposition of the reflected waves with the original ones. Let's return to the example of a periodic wave moving down a rope toward a rigid post. When the periodic wave reflects from the post, it superimposes with the wave heading toward the post. The complete pattern results from the superposition of the original wave and reflections from both ends. In general, we get a complicated pattern with a small amplitude, but certain frequencies cause the rope to vibrate with a large amplitude. Figure 15-16 shows multiple images of a resonating rope. Although the superimposing waves move along the rope, they produce a resonant pattern that does not move along the rope. Because the pattern appears to stand still (in the horizontal direction), it is known as a **standing wave.**

It may seem strange that two identical waves traveling in opposite directions combine to produce a vibrational pattern that doesn't travel along the rope. We can see how this happens by using the superposition principle to find the results of combining the two traveling waves. Let's start at a time when the crests of the traveling wave moving to the right (the blue line in Figure 15-17) line up with the crests of the wave moving to the left (yellow line). (The blue and yellow lines lie on top of each other and are shown as a single green line.) Adding the displacements of the two traveling waves yields a wave that has the same basic shape, but twice the amplitude. This is shown by the black line in Figure 15-17(a).

FLAWED REASONING

The following question appears on the midterm exam: "A periodic wave is traveling to the right on a long, stretched rope. Two small pieces of yarn are tied to the rope, one at point A and the other at point B, as shown in the figure:

Draw an arrow for each piece of yarn, indicating the direction of its velocity when the picture was taken."

Brielle gives the following answer to this question: "Because the wave is moving to the right, the pieces of yarn must also be moving to the right. The wave is carried by the rope."

What is wrong with Brielle's reasoning, and what is the correct answer to the exam question?

ANSWER The wave is a transverse wave, meaning that the medium moves perpendicular to the direction of the wave. Therefore, the pieces of yarn can only move up or down. If we look at the wave at a slightly later time (when it has moved a little to the right), we see that the yarn at point A has moved upward and the yarn at point B has moved downward:

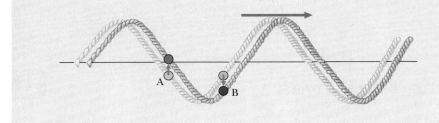

Figure 15-17(b) shows the situation a short time later. The blue wave has moved to the right, and the yellow wave has moved the same distance to the left. The superposition at this time produces a shape that still looks like one of the traveling waves but does not have as large an amplitude as before. A short time later, the crests of one wave line up with the troughs of the other. At this time the two waves cancel each other, and the rope is straight. Although the rope is straight at this instant, some parts of the rope are moving up while others are moving down. The remaining drawings in Figure 15-17 show how this pattern changes through the rest of the cycle as time progresses.

Notice that some portions of the rope do not move. Even though each traveling wave by itself would cause all pieces of the rope to move, the waves interfere to produce no motion at these points. Such locations are known as **nodes** and are located on the vertical lines indicated by N in Figure 15-17. The positions on the rope that have the largest amplitude are known as **antinodes** and are on the lines marked by A. Notice that the nodes and antinodes alternate and are equally spaced.

There is a relationship between the shape of the resonant pattern, or standing wave, and the moving periodic waves that superimpose to create it. The

Figure 15-16 A strobe drawing of a standing wave on a rope shows how the shape of the rope changes with time. The shape does not move to the left or right.

"wavelength" of the standing wave is equal to the wavelength of the underlying periodic wave. This can be seen in Figure 15-17.

Unlike the pendulum or the mass on a spring where there was only one resonant frequency, periodic waves that are confined have many different resonant frequencies. Strobe photographs of the standing wave with the lowest frequency, or **fundamental frequency,** show that the rope has shapes like those drawn in Figure 15-18. The images show how the shape of the rope changes during one-half cycle. At position 1 the rope has the largest possible crest. The displacement continually decreases until it becomes zero and the rope is straight (between positions 3 and 4). The rope's inertia causes it to overshoot and form a trough that grows in size. At the end of the half cycle, the rope is in position 6 and beginning its upward journey. The process then repeats.

This pattern has the lowest frequency and thus the longest wavelength of the resonant modes. Notice that one-half wavelength is equal to the length of the rope, or the wavelength is twice the length of the rope. Because this is also the wavelength of the traveling waves, the longest resonant wavelength on a rope with nodes at each end is twice the length of the rope.

If we slowly increase the frequency of the traveling wave, the amplitude quickly decreases. The vibrational patterns are rather indistinct until the next resonant frequency is reached. This new resonant frequency has twice the frequency of the fundamental and is known as the second **harmonic.** Six shapes of the rope for this standing-wave pattern are shown in Figure 15-19. When the rope is low on the left side, it is high on the right, and vice versa. Because the rope has the shape of a full wavelength, the wavelength of the traveling waves is equal to the length of the rope. Note that this pattern has one more node and one more antinode than the fundamental standing wave.

We can continue this line of reasoning. A third resonant frequency can be reached by again raising the frequency of the traveling wave. This third harmonic frequency is equal to three times the fundamental frequency. This standing wave has three antinodes and four nodes, including the two at the ends of the rope. Other resonant frequencies occur at whole-number multiples of the fundamental frequency.

Remember that the product of the frequency and the wavelength is a constant. This is a consequence of the fact that the speeds of all waves on this rope are the same. Therefore, if we increase the frequency by some multiple while keeping the speed the same, the wavelength must decrease by the same multiple. The fundamental wavelength is the largest, and its associated frequency is the smallest. As we march through higher and higher frequencies, we get shorter and shorter wavelengths. The wavelengths of the higher harmonics are obtained by dividing the fundamental wavelength by successive whole

Figure 15-17 This set of strobe drawings shows how two traveling waves (the blue and yellow lines) combine to form a standing wave (the black line). Only the black line would be visible in a photograph.

Q: How does the distance between adjacent nodes or antinodes compare with the wavelength?

A: Because one antinode is up when the adjacent ones are down, each antinodal region corresponds to a crest or a trough. Therefore, the distance between adjacent antinodes or adjacent nodes is one-half wavelength. The distance between adjacent nodes and antinodes is one-quarter wavelength.

Are You On the Bus?

Figure 15-18 The shapes of a rope oscillating as a standing wave of the lowest frequency.

Figure 15-19 The shapes of a rope for a standing wave with the second resonant frequency.

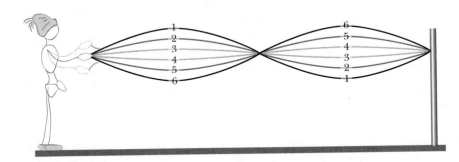

numbers; the wavelength of the second harmonic is one-half the wavelength of the fundamental wavelength.

Q: How does the wavelength of the third harmonic compare with the length of the rope?

A: The wavelength of the third harmonic is one-third the length of the fundamental wavelength. Because the fundamental wavelength is twice the length of the rope, the wavelength of the third harmonic would be two-thirds the length of the rope.

Interference

Standing waves on a rope are an example of the superposition, or **interference,** of waves in one dimension. If we use a two-dimensional medium—say, the surface of water—we can generate some new effects.

Suppose we use two wave generators to create periodic waves on the surface of water in a ripple tank like the one in Figure 15-20(a). Because the two waves travel in the same medium, they have the same speed. We also assume that the two sources have the same frequency and that they are **in phase;** that is, both sources produce crests at the same time, troughs at the same time, and so on. The superposition of these waves creates the interference pattern shown in the photograph and drawing of Figure 15-20. The bright regions are produced by the crests, whereas the dark regions are produced by the troughs.

In some places, crest meets crest to form a supercrest, and one-half period later, trough meets trough to form a supertrough. This meeting point is a region of large amplitude; the two waves form antinodal regions. In other places, crest and trough meet. Here, if the two waves have about the same amplitude, they cancel each other, resulting in little or no amplitude; the two waves form nodal regions.

Because of the periodic nature of the waves, the nodal and antinodal regions have fixed locations. These stationary interference patterns can be observed only if the two sources emit waves of the same frequency; otherwise, one wave continually falls behind the other, and the relationships between the two waves change. The two wave sources do not have to be in phase; there can be a time delay between the generation of crests by one source and the other as long as the time delay is constant. For simplicity we usually assume that this time delay is zero; that is, the two sources are in phase.

The regions of crests and troughs lie along lines. One such antinodal line lies along the perpendicular to the midpoint of the line joining the two sources. This is the vertical red line in Figure 15-20(c). This central antinodal line is the same distance from the two sources. Therefore, crests generated at the same time at the two sources arrive at the same time at this midpoint to form supercrests. Similarly, two troughs arrive together, creating a supertrough.

Consider a point *P* off to the right side of the central line, as shown in Figure 15-21. Although nothing changes at the sources, we get a different result. Crests from the two sources no longer arrive at the same time. Crests from the left-hand source must travel a greater distance and therefore take longer to get to *P*. The amount of delay depends on the difference in the two path lengths.

If the point *P* is chosen such that the distances to the sources differ by an amount equal to one-half wavelength, crests overlap with troughs and troughs overlap with crests at point *P*. The waves cancel. There are many points that have this path difference. They form nodal lines that lie along each side of the central line, as shown by the black lines in Figure 15-20(c). An antinodal line occurs when the path lengths differ by one wavelength; the next nodal line when the paths differ by $1\frac{1}{2}$ wavelengths, and so on.

The photograph in Figure 15-22 shows the interference pattern for water waves with a longer wavelength than those in Figure 15-20. The nodal lines are now more widely spaced for the same source separation. Therefore, longer wavelengths produce wider patterns. Actually, the width of the pattern depends on the relative size of the wavelength and the source separation. As the ratio of the wavelength to the separation gets bigger, the nodal lines spread out. If the wavelength is much larger than the separation, the pattern is essentially that of a single source, whereas if it is much smaller, the nodal lines are so close together they cannot be seen.

Diffraction

In the photographs in Figure 15-23, periodic water waves move toward a barrier. We see that the waves do not go straight through the opening in the barrier but spread out behind the barrier. This bending of a wave is called **diffraction** and is definitely not a property of particles. If a BB gun is fired repeatedly through an opening in a barrier, the pattern it produces is a precise "shadow" of the opening if we assume that no BBs bounce off the opening's edges.

(a)

(b)

PSSC PHYSICS, 2nd Edition, 1965; D.C. Heath & Co. and Education Development Center, Inc., Newton, MA.

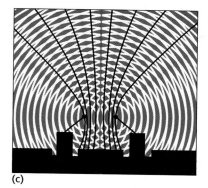
(c)

Figure 15-20 (a) A light bulb above a ripple tank produces light and dark lines on the floor due to the water waves. (b) The interference pattern produced by two point sources of the same wavelength and phase. (c) The locations of the nodal lines in this pattern are shown in black; the central antinodal line is shown in red.

Q: Would the central line still be an antinodal line if the two sources were completely out of phase—that is, if one source generates a crest at the same time as the other generates a trough?

A: The central line would now be a nodal line because crests and troughs would arrive at the same time.

Are *You* On the Bus?

Figure 15-21 Whether the region at *P* is a nodal or antinodal region depends on the difference in the path lengths from the two sources.

Path difference

Figure 15-22 The nodal lines (black) are more widely spaced for longer wavelengths. The central antinodal line (red) remains in the same location. Compare this with Figure 15-20.

(a) (b)

PSSC PHYSICS, 2nd Edition, 1965; D.C. Heath & Co. and Education Development Center, Inc., Newton, MA.

Figure 15-23 Ripple tank patterns of water moving upward and passing through a narrow barrier. Note that the amount of diffraction increases as the wavelength gets longer.

(a) (b) (c)

PSSC PHYSICS, 2nd Edition, 1965; D.C. Heath & Co. and Education Development Center, Inc., Newton, MA. (all)

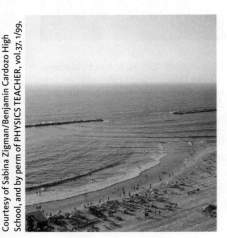

Courtesy of Sabina Zigman/Benjamin Cardozo High School, and by perm of PHYSICS TEACHER, vol.37, 1/99,

The waves are diffracted as they pass through the openings in the seawall, producing an interesting shape at the beach.

The amount of diffraction depends on the relative sizes of the wavelength and the opening. If the wavelength is much smaller than the opening, very little diffraction is evident. As the wavelength gets closer to the size of the opening, the amount of diffraction gets bigger. In Figure 15-23(c) the opening and the wavelength are approximately the same size, and the diffraction is evident.

Notice that diffraction produces nodal and antinodal lines similar to those observed in the interference patterns from two point sources. In this case there is a broad central antinodal region with nodal lines on each side, and the diffraction pattern is created by different portions of the wave interfering with themselves. The spacing of these lines is determined by the ratio of the wavelength and the width of the opening.

Summary

Vibrations and oscillations are described by the length of time required for one cycle, the period *T* (or its reciprocal, the frequency *f*), and the amplitude of the vibration, the maximum distance the object travels from the equilibrium point. When vibrations are small, the period is independent of the amplitude. The pendulum and a mass hanging on a spring are examples of systems that vibrate.

All systems have a distinctive set of natural frequencies. A simple system such as a pendulum has only one natural frequency, whereas more complex systems have many natural frequencies. When a system is excited at a natural frequency, it resonates with a large amplitude.

Waves are vibrations moving through a medium; it is the wave (energy) that moves through the medium, not the medium itself. Transverse waves vibrate perpendicular to the direction of the wave, whereas longitudinal waves vibrate parallel to the direction of the wave. The speed of a periodic wave is equal to the product of its frequency and its wavelength, $v = \lambda f$.

Waves pass through each other as if the other were not there. When they overlap, the shape is the algebraic sum of the displacements of the individual waves. When a periodic wave is confined, resonant patterns known as standing waves can be produced. Portions of the medium that do not move are called nodes, whereas portions with the largest amplitudes are known as antinodes. The fundamental standing wave has the lowest frequency and the longest wavelength.

Two identical periodic-wave sources with a constant phase difference produce an interference pattern consisting of large-amplitude antinodal regions and zero-amplitude nodal regions. The spacing of the interference pattern depends on the relative size of the wavelength and the source separation.

Waves do not go straight through openings or around barriers but spread out. This diffraction pattern contains nodal and antinodal regions and depends on the relative sizes of the wavelength and the opening.

CHAPTER 15 | *Revisited*

When waves move in a medium, the medium oscillates in place. No material is transported from one location to another; it is the disturbance that moves. Unlike with particles, when two waves pass through the same region at the same time, the individual disturbances are added together. Afterward, each wave retains its own identity.

Key Terms

amplitude   The maximum distance from the equilibrium position that occurs in periodic motion.

antinode   One of the positions in a standing-wave or interference pattern where there is maximum movement; that is, the amplitude is a maximum.

crest   The peak of a wave disturbance.

cycle   One complete repetition of a periodic motion. It may start anyplace in the motion.

diffraction   The spreading of waves passing through an opening or around a barrier.

displacement   In wave (or oscillatory) motion, the distance of the disturbance (or object) from its equilibrium position.

equilibrium position   A position where the net force is zero.

frequency   The number of times a periodic motion repeats in a unit of time. It is equal to the inverse of the period.

fundamental frequency   The lowest resonant frequency for an oscillating system.

harmonic   A frequency that is a whole-number multiple of the fundamental frequency.

in phase   Describes two or more waves with the same wavelength and frequency that have their crests lined up.

interference   The superposition of waves.

longitudinal wave   A wave in which the vibrations of the medium are parallel to the direction the wave is moving.

node   One of the positions in a standing-wave or interference pattern where there is no movement; that is, the amplitude is zero.

oscillation   A vibration about an equilibrium position or shape.

period   The shortest length of time it takes a periodic motion to repeat. It is equal to the inverse of the frequency.

periodic wave   A wave in which all the pulses have the same size and shape. The wave pattern repeats itself over a distance of one wavelength and over a time of one period.

resonance A large increase in the amplitude of a vibration when a force is applied at a natural frequency of the medium or object.

spring constant The amount of force required to stretch a spring by one unit of length. The spring constant is measured in newtons per meter.

standing wave The interference pattern produced by two waves of equal amplitude and frequency traveling in opposite directions. The pattern is characterized by alternating nodal and antinodal regions.

superposition The combining of two or more waves at a location in space.

transverse wave A wave in which the vibrations of the medium are perpendicular to the direction the wave is moving.

trough A valley of a wave disturbance.

vibration An oscillation about an equilibrium position or shape.

wave The movement of energy from one place to another without any accompanying matter.

wavelength The shortest repetition length for a periodic wave. For example, it is the distance from crest to crest or trough to trough.

Questions and exercises are paired so that most odd-numbered are followed by a similar even-numbered.

Blue-numbered questions and exercises are answered in Appendix B.

ʃ indicates more challenging questions and exercises.

WebAssign Many Conceptual Questions and Exercises for this chapter may be assigned online at WebAssign.

Conceptual Questions

1. If the net force on a mass oscillating at the end of a vertical spring is zero at the equilibrium point, why doesn't the mass stop there?

2. If the restoring force on a pendulum is zero when it is vertical, why doesn't it quit swinging at this point?

3. A mass is oscillating up and down on a vertical spring. When the mass is above the equilibrium point and moving downward, in what direction does the net force on the mass act? When the mass is above the equilibrium point and moving upward, what is the direction of the net force on the mass? Explain.

4. A mass is oscillating up and down on a vertical spring. When the mass is below the equilibrium point and moving downward, what is the direction of its acceleration? Is the mass speeding up or slowing down? Explain.

5. A mass is oscillating up and down on a vertical spring. If the mass is increased, will the period of oscillation increase, decrease, or stay the same? Will the frequency increase, decrease, or stay the same? Explain.

6. A grandfather clock (with a pendulum) keeps perfect time on Earth. If you were to transport this clock to the Moon, would its period of oscillation increase, decrease, or stay the same? Would its frequency increase, decrease, or stay the same? Explain.

7. You hang a 1-kilogram block from a spring and find that the spring stretches 15 centimeters. What mass would you need to stretch the spring 45 centimeters?

8. Which spring would you expect to have the greater spring constant, the one in the suspension of your Chevy or the one in your watch? Why?

9. Assume that you pull the mass on a spring 1 centimeter from the equilibrium position, let go, and measure the period of the oscillation. Would you expect the period to be larger, the same, or smaller if you pulled the mass 2 centimeters from the equilibrium position? Why?

10. The amplitude of a real pendulum decreases because of frictional forces. How does the period of this real pendulum change?

11. What is the period of the hand on a clock that measures the seconds? What is its frequency?

12. What is the period of the hand on a clock that measures the minutes? What is its frequency?

13. Suppose your grandfather clock runs too fast. If the mass on the pendulum can be moved up or down, which way would you move it to adjust the clock? Explain your reasoning.

ʃ 14. How does the natural frequency of a swing change when you move from sitting down to standing up?

© Cengage Learning/George Semple

---

15. You find that the exhaust system on your 1979 Chrysler Cordoba tends to rattle loudly when the tachometer, which measures the engine's frequency, reads 2000 rpm. It is relatively quiet at frequencies above or below 2000 rpm. Use the concept of resonance to explain this.

16. Why do soldiers "break step" before crossing a suspension bridge?

17. You hold one end of a spring in your hand and hang a block from the other end. After lifting the block up slightly and releasing it, you find that it oscillates up and down at a frequency of 2 hertz. At which of the following frequencies could you jiggle your hand up and down and produce resonance: 5 hertz, 4 hertz, 1.5 hertz, 1 hertz, or 0.5 hertz?

18. You stand to the side of the low point of a child's swing and always push the child in the same direction. Which of the following multiples of the fundamental frequency will not produce resonance: $\frac{1}{3}$, $\frac{1}{2}$, 1, or 2?

19. When you yell at your friend, are the air molecules that strike his ear the same ones that were in your lungs? Explain.

20. What is being transported along a clothesline when a wave moves from one end to the other?

21. Sonar devices use underwater sound to explore the ocean floor. Would you expect sonar to be a longitudinal or a transverse wave? Explain.

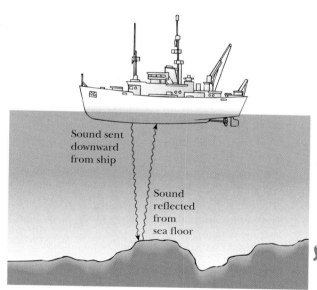

Sound sent downward from ship

Sound reflected from sea floor

22. You fasten one end of a long spring to the base of a wall and stretch it out along the floor, holding the other end in your hand. Describe how you would generate a transverse pulse on the spring. Describe how you would generate a longitudinal pulse on the spring.

23. Is it possible for a shout to overtake a whisper? Explain.

24. You generate a small transverse pulse on a long spring stretched between a doorknob and your hand. How could you generate a second pulse that would overtake the first pulse?

25. Which one or more of the following properties affect the speed of waves along a rope: amplitude of the pulse, shape of the pulse, tension in the rope, or the mass per unit length of the rope? Why?

26. You move your hand up and down to send a pulse along a long spring stretched between a doorknob and your hand. Which of the following would generate a slower-traveling pulse: Moving your hand the same distance as before, but more slowly; moving your hand a smaller distance at the same speed as before; or moving closer to the doorknob to decrease the tension in the spring?

27. You send a pulse of amplitude 5 centimeters down the right side of a spring. A moment later you send an identical pulse on the same side. The first pulse reflects from the fixed end and returns along the spring. When the reflected pulse meets the second pulse, will the resulting amplitude be less than, equal to, or greater than 5 centimeters? Explain your reasoning.

28. Imagine that the string in Figure 15-14 is tied to the pole with a loose loop such that the end is free to move up and down. A pulse of amplitude 10 centimeters is sent down the top of the string, and a moment later a second identical pulse is sent, also on the top. The first pulse reflects from the free boundary and returns along the string. When the reflected pulse meets the second pulse, will the resulting amplitude be less than, equal to, or greater than 10 centimeters? Explain your reasoning.

29. The pulse in the following figure is traveling on a string

to the right toward a fixed end. Draw the shape of the pulse after it reflects from the boundary.

30. A pulse in the shape of a crest is sent from left to right along a stretched rope. A trough travels in the opposite direction so that the pulses meet in the middle of the rope. Would you expect to observe a crest or a trough arrive at the right-hand end of the rope? Explain.

31. If shapes (a) and (b) in the following figure correspond to idealized wave pulses on a rope, what shape is produced when they completely overlap?

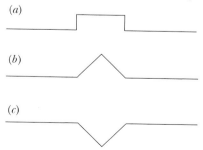

(a)

(b)

(c)

32. Repeat Question 31 for shapes (a) and (c).

33. Which of the following properties are meaningful for periodic waves but not for single pulses: frequency, wavelength, speed, amplitude?

34. In the following list of properties of periodic waves, which one is independent of the others: frequency, wavelength, speed, amplitude?

35. Two waves have the same speed but one has twice the frequency. Which wave has the longer wavelength? Explain.

36. If the frequency of a periodic wave is cut in half while the speed remains the same, what happens to the wavelength?

37. If the speed of a periodic wave doubles while the period remains the same, what happens to the wavelength?

38. What happens to the wavelength of a periodic wave if both the speed of the wave and the frequency are cut in half?

39. Travelers spaced 10 feet apart are all walking at 3 mph relative to a moving sidewalk. When the moving sidewalk ends, they continue to walk at 3 mph. An observer standing next to the moving sidewalk notes that the travelers are passing by at a frequency of 1 hertz. A second observer stands just beyond the end of the moving sidewalk and notes the frequency at which the travelers pass.

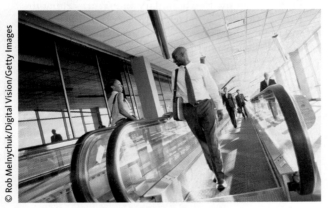

© Rob Melnychuk/Digital Vision/Getty Images

Would this frequency be greater than, equal to, or less than 1 hertz? Is the spacing between the travelers after leaving the moving sidewalk greater than, equal to, or less than 10 feet? Explain.

40. A waterproof electric buzzer has a membrane that vibrates at a constant frequency of 440 hertz. The buzzer is placed in a bucket of water. Knowing that the speed of sound is much greater in water than in air, will the frequency of the sound heard in the air be greater than, equal to, or less than 440 hertz? Will the wavelength of the sound in air be greater than, equal to, or less than what it was in the water? Explain. (*Hint:* Review Question 39 and think of the travelers as the wave crests.)

41. Draw a diagram to represent the standing-wave pattern for the third harmonic of a rope fixed at both ends. How many antinodes are there?

42. Draw a diagram to represent the standing-wave pattern for the fourth harmonic of a rope fixed at both ends. How many nodes are there?

43. How much higher is the frequency of the fifth harmonic on a rope than the fundamental frequency?

44. How much higher is the frequency of the sixth harmonic on a rope than that of the second?

45. How many antinodes are there when a rope fixed at both ends vibrates in its third harmonic?

46. How many nodes are there when a rope fixed at both ends vibrates in its fourth harmonic?

47. Standing waves can be established on a rope that is fixed on one end but free to slide up and down a pole on the other. The fixed end remains a node, while the free end must be an antinode. Draw diagrams to represent the standing-wave patterns for the two lowest frequencies.

48. How does the fundamental wavelength of standing waves on a string with one end fixed and the other free compare to the fundamental wavelength if the same string is held with both ends fixed?

49. How does the wavelength of the fourth harmonic on a rope with both ends fixed compare with the length of the rope?

50. How does the wavelength of the fourth harmonic on a rope with both ends fixed compare with that of the second harmonic?

51. A longitudinal standing wave can be established in a long aluminum rod by stroking it with rosin on your fingers. If the rod is held tightly at its midpoint, what is the wavelength of the fundamental standing wave? Assume that there are antinodes at each end of the rod and a node where the rod is held.

52. What is the wavelength of the fundamental standing wave for the rod in Question 51 if it is held midway between the center and one end? Will the resulting pitch be higher or lower than when the rod was held at its midpoint? Explain.

53. Two point sources produce waves of the same wavelength and are in phase. At a point midway between the sources, would you expect to find a node or an antinode? Explain.

54. Two point sources produce waves of the same wavelength and are completely out of phase (that is, one produces a crest at the same time as the other produces a trough). At a point midway between the sources, would you expect to find a node or an antinode? Why?

§ 55. What happens to the spacing of the antinodal lines in an interference pattern when the two sources are moved farther apart? Explain.

§ 56. As you increase the frequency, what happens to the spacing of the nodal lines in an interference pattern produced by two sources? Explain.

Courtesy of VWR Corporation

57. An interference pattern is produced in a ripple tank. As the two sources are brought closer together, does the

separation of the locations of maximum amplitude along the far edge of the tank decrease, increase, or remain the same? Why?

§ 58. As the frequency of the two sources forming an interference pattern in a ripple tank increases, does the separation of the locations of minimum amplitude along the far edge of the tank increase, decrease, or remain the same? Why?

§ 59. What happens to the spacing of the antinodal lines in an interference pattern when the two slits are moved farther apart? Explain.

60. As you increase the frequency, what happens to the spacing of the nodal lines in a diffraction pattern? Explain.

Exercises

61. If a mass on a spring takes 6 s to complete two cycles, what is its period?

62. If a mass on a spring has a frequency of 4 Hz, what is its period?

63. A Foucault pendulum with a length of 9 m has a period of 6 s. What is its frequency?

64. A mass on a spring bobs up and down over a distance of 30 cm from the top to the bottom of its path twice each second. What are its period and amplitude?

65. A spring hanging from the ceiling has an unstretched length of 80 cm. A mass is then suspended at rest from the spring, causing its length to increase to 89 cm. The mass is pulled down an additional 3 cm and released. What is the amplitude of the resulting oscillation?

66. A mass oscillates up and down on a vertical spring with an amplitude of 4 cm and a period of 2 s. What total distance does the mass travel in 10 s?

67. What is the period of a 0.4-kg mass suspended from a spring with a spring constant of 40 N/m?

68. A boy with a mass of 50 kg is hanging from a spring with a spring constant of 200 N/m. With what frequency does the boy bounce up and down?

69. By what factor would you have to increase the mass to double the period for a mass on a spring?

70. By what factor would you have to increase the spring constant to triple the frequency for a mass on a spring?

71. A pendulum has a length of 5 m. What is its period?

72. A girl with a mass of 40 kg is swinging from a rope with a length of 2.5 m. What is the frequency of her swinging?

§ 73. The highly idealized wave pulses shown in the figure below at a time equal to zero have the same amplitudes and travel at 1 cm/s. Draw the shape of the rope at 2, 4, 5, and 8 s.

§ 74. Work Exercise 73 but change the rectangular pulse from a crest to a trough.

75. A train consisting of identical 10-m boxcars passes you such that 25 boxcars pass you each minute. Find the speed of the train.

distance, cm

EXERCISE 73

76. You observe that 25 crests of a water wave pass you each minute. If the wavelength is 10 m, what is the speed of the wave?

77. A periodic wave on a string has a wavelength of 25 cm and a frequency of 3 Hz. What is the speed of the wave?

78. If the breakers at a beach are separated by 5 m and hit shore with a frequency of 0.3 Hz, at what speed are they traveling?

79. What is the distance between adjacent crests of ocean waves that have a frequency of 0.2 Hz if the waves have a speed of 3 m/s?

80. Sound waves in iron have a speed of about 5100 m/s. If the waves have a frequency of 400 Hz, what is their wavelength?

81. For sound waves, which travel at 343 m/s in air at room temperature, what frequency corresponds to a wavelength of 1 m?

82. What is the period of waves on a rope if their wavelength is 0.8 m and their speed is 2 m/s?

83. A rope is tied between two posts separated by 3 m. What possible wavelengths will produce standing waves on the rope?

84. A 3-m-long rope is tied to a thin string so that one end is essentially free. What possible wavelengths will produce standing waves on this rope?

85. What is the fundamental frequency on a 6-m rope that is tied at both ends if the speed of the waves is 18 m/s?

86. Tweety Bird hops up and down at a frequency of 0.5 Hz on a power line at the midpoint between the poles, which are separated by 20 m. Assuming Tweety is exciting the fundamental standing wave, find the speed of transverse waves on the power line.

20 Electricity

▶ We see electrical effects when sparks occur. On dry days, we may feel a shock when we touch a metal doorknob after walking across a carpet. If we look closely, we can see a spark jump between our hand and the doorknob. Although they may surprise us, the sparks do not hurt us. On the other hand, nature produces long and dangerous sparks during lightning storms. What determines the length of the sparks?

(See page 440 for the answer to this question.)

Jean-Loup Charmet/SPL/Photo Researchers, Inc.

A lightning storm produces very high voltages, creating large and dangerous electric currents.

THE early Greeks knew that amber—a fossilized tree sap currently used in jewelry—had the interesting ability to attract bits of fiber and hair after it was rubbed with fur. This was one way of recognizing an object that was electrified. In modern terminology we say the object is **charged.** This doesn't explain what **charge** is, but is a handy way of referring to this condition.

In 1600 English scientist William Gilbert published a pioneering work, *De Magnete*, in which he pointed out that this electrical effect was not an isolated property of amber but a much more general property of matter. Materials such as gems, glass, and sealing wax could also be charged. Rubbing together two objects made from different materials was the most common way of charging an object. In fact, both objects become charged.

After Gilbert, many experimenters joined in the activity of investigating electricity. However, the question of what happens when an object gains or loses this electrical property remained unanswered. A modern response may include words such as *electron* or *proton*. But we have to be careful that we are not simply substituting new words or phrases for old ones. To answer this question fully and thus expand our world view, we need to carefully examine our electrical world.

Electrical Properties

In an effort to explain electricity, Gilbert proposed the existence of an electric fluid in certain types of objects. He suggested that rubbing an object removed some of this fluid, leaving it in the region surrounding the object. Bits of fiber were attracted to the object by the "draft" of the fluid returning to the object. Although many other electrical phenomena could not be explained with this idea, it was the beginning of attempts to model the underlying, invisible processes that caused the electrical effects.

Little progress was made for more than a century. In the 1730s it was shown that charge from one object would be transferred to a distant object if metal wires connected them but not if silk threads connected them. Materials that can transfer charge are known as **conductors;** those that cannot are called nonconductors, or **insulators.** It was discovered that metals, human bodies, moisture, and a few other substances are conductors.

The discovery that moisture is a conductor explains why electrical effects vary from day to day. You normally experience bigger shocks in the winter when the humidity is naturally low. On more humid days, any charge that you get by scuffing your feet is quickly dissipated by the moisture in the air surrounding your body.

Many charged objects have to be suspended from insulators such as silk threads or plastic bases or they quickly lose their charge to Earth, or ground. In fact, we speak of **grounding** an object to ensure that it is not charged. Other objects hold their charges without being insulated. When we charge a plastic rod by rubbing it with a cloth, the charge stays on the rod even if we hold it. But our bodies are conductors. Why doesn't the charge flow to ground through our bodies? It stays on the rod because the rod is an insulator; charge generated at one end remains there. The charge can be removed by moving our hands along the charged end. As we touch the regions that are charged, the charges flow through our bodies to ground.

A metal rod cannot be charged by holding it in our hands and rubbing it with a cloth because metal conducts the charge to our hands. A metal rod can be charged if it is mounted on an insulating stand or if we hold it with an insulating glove; that is, the rod must be insulated from its surroundings.

Fuel trucks are grounded to remove the charge built up by the fuel flowing through the hoses.

© Cengage Learning/David Rogers

Even the flow of liquid through pipes is enough of a rubbing action to charge the liquid. Fuel trucks tend to build up charge as they dispense fuel. If this charge becomes large enough, sparking may occur and cause a fire or explosion. To avoid this danger, a conducting wire is connected from the truck to the ground to allow the charge to flow into the ground.

Two Kinds of Charge

It was also the early 1730s before there was any mention that charged objects could repel one another. Electricity, like gravity, was believed to be only attractive. It may seem strange to us now because both the attractive and repulsive aspects of electricity are easy to demonstrate. If you comb your hair, the comb becomes charged and can be used to attract small bits of paper. After contacting the comb, some of these bits are then repelled by the comb.

This phenomenon can be investigated more carefully using balloons and pieces of wool. If we rub a balloon with a piece of wool, the balloon becomes charged; it attracts small bits of paper and sticks to walls or ceilings. If we suspend this balloon by a thread and bring uncharged objects near it, the balloon is attracted to the objects [Figure 20-1(a)]. Everything seems to be an attractive effect.

Figure 20-1 A charged balloon is (a) attracted by uncharged objects, (b) repelled by an identically charged balloon, and (c) attracted by the wool used in charging it.

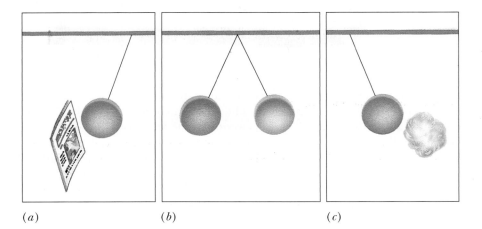

(a)   *(b)*   *(c)*

Charging another balloon in the same way demonstrates the new effect: the two balloons repel one another [Figure 20-1(b)]. Because we believe that any two objects prepared in the same way are charged in a like manner, we are led to the idea that like-charged objects repel one another.

Whenever we charge an object by rubbing it with another, both objects become charged. If we examine the pieces of wool, we find that they are also charged: they each attract bits of paper.

> **Q:** Will the two pieces of wool attract or repel each other?
>
> **A:** Because they have been charged in a like manner, they will repel each other.

Are *You* On the Bus?

The piece of wool and the balloon, however, attract each other after being rubbed together [Figure 20-1(c)]. If they had the same kind of charge, they would repel. We are therefore led to the idea that there must be two different kinds of charge and that the two kinds attract each other. These experiments can be summarized by stating the following:

> Like charges repel; unlike charges attract.

Conservation of Charge

Like Gilbert, Benjamin Franklin believed that electricity was a single fluid and that an excess of this fluid caused one kind of charged state, whereas a deficiency caused the other. Because he could not tell which was which, he arbitrarily named one kind of charge *positive* and the other kind *negative*. By convention the charge on a glass rod rubbed with silk or plastic film is positive (Figure 20-2), whereas that on an amber or rubber rod rubbed with wool or fur is negative.

At first glance these names seem to have no advantage over other possible choices, such as black and white or yin and yang. They were, however, more significant. Franklin's use of the fluid model led him to predict that charge should be conserved. The amount of electric fluid should remain the same; it is just transferred from one object to another. If you start with two uncharged objects and rub them together, the amount of excess fluid on one is equal to the deficiency on the other. In other words, the positive and negative charges are equal. Using his system of positive and negative numbers, we can add them and see that the total charge remains zero.

We no longer believe in Franklin's fluid model; the fluid model was abandoned as it could not account for later experimental observations. However, Franklin's idea about the **conservation of charge** has been verified to a high precision. It is one of the fundamental laws of physics. In its generalized form, it can be stated as follows:

> In an isolated system the total charge is conserved.

Figure 20-2 The charge on the glass rod is called positive. This is indicated by a few plus signs.

◀ conservation of charge

In our modern physics world view, all objects are composed of negatively charged electrons, positively charged protons, and uncharged neutrons. The electron's charge and the proton's charge have the same size. An object is uncharged (or neutral) because it has equal amounts of positive and negative charges, not because it contains no charges. For example, atoms are electrically neutral because they have equal numbers of electrons and protons.

Biosketch | *Franklin: The American Newton*

Benjamin Franklin

© The Concoran Gallery of Art/Corbis

When Benjamin Franklin (1706–1790) arrived in Paris as Revolutionary America's ambassador, he was known to the sympathetic French as a famous scientist who "stole the scepter from tyrants and lightning from the gods." Precisely because he was the "American Newton," he was the most effective representative the young nation-in-the-making could have sent.

Born in Boston, Franklin made his fortune and reputation in Philadelphia—then the largest English-speaking city outside of London. He was a one-man chamber of commerce who greeted newcomers, brought promising young people into his circle of friends, and helped organize everything from the first philosophical society in the country to a street-lighting district, a fire department, and a mental hospital. His publications on electricity brought him membership in the Royal Philosophical Society of London. (Newton had been president just 50 years earlier.) He invented lightning rods and used them on his own house to great effect.

Franklin believed that electricity was a fluid that pervaded all bodies in varying quantities. A body would seek to maintain electrical equilibrium. If there was a deficit, the body would be in a negative state, and electricity would flow into the body. If there was a surplus, electricity would flow out. He also believed that unlike gravitation, which is only attractive, electrified matter could repel as well as attract other electrified matter. It could also attract nonelectrified matter.

Like others of his era, he was also interested in heat, illumination, weather, and other aspects of science. Like Michael Faraday a bit later, Franklin was not a mathematician and so could not give definitive form to his arguments. Even so, this well-known patriot was the best known of early American scientists and a fascinating human being.

—Pierce C. Mullen, historian and author

Sources: Carl Van Doren, *Benjamin Franklin* (New York: Viking, 1938); *The Autobiography of Benjamin Franklin* (New York: Dover Thrift, 1966).

Positively charged objects may have an excess of positive charges or a deficiency of negative charges—that is, an excess of protons or a deficiency of electrons. We simply call them positively charged because the electrical effects are the same in both situations.

The modern view easily accounts for the conservation of charge when charging objects. The rubbing simply results in the transfer of electrons from one object to the other; whatever one object loses, the other gains.

Are You On the Bus?

Q: If you remove one electron from a neutral quarter and one electron from a neutral Buick, which (if either) has the greater net charge?

A: In either case, each proton is paired with an electron except for one. The net charge on either object would just be the charge of a single proton.

Induced Attractions

Attraction is more common than repulsion because charged objects can attract *uncharged* objects. How do we explain the observation that charged objects attract uncharged objects? Consider a positively charged rod and an uncharged metal ball. As the rod is brought near the ball, the rod's positive charges attract the negative charges and repel the positive charges in the ball. Because the charges in the ball are free to move, this results in an excess of negative charges on the near side and an excess of positive charges on the far side of the ball (Figure 20-3). Because charge is conserved, the excess negative charge on one side is equal to the excess positive on the other. Experi-

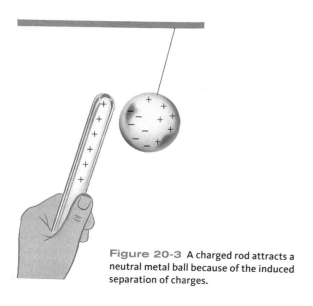

Figure 20-3 A charged rod attracts a neutral metal ball because of the induced separation of charges.

Figure 20-4 An insulating object is attracted because of the separation of charges within the molecules.

ments with balloons show that the electric force varies with distance; the force gets weaker as the balloons are moved farther apart. Thus, the ball's negative charge is attracted to the rod more than its positive charge is repelled. These induced charges result in a net attraction of the uncharged ball toward the rod.

Experiments show that if the ball is made of an insulating material, the attraction still occurs. In insulators the charges are not free to move across the object, but there can be motion on the molecular level. Although the molecules are uncharged, the presence of the charged rod may induce a separation of charge within the molecule. Other molecules are naturally more positive on one end and more negative on the other. The positive rod rotates these polar molecules so that their negative ends are closer to the rod, as shown in Figure 20-4. This once again results in a net attractive force.

A particularly graphic example of the attraction of an uncharged insulator is illustrated in Figure 20-5. A long two-by-four piece of lumber is balanced on

Figure 20-5 A standard two-by-four piece of lumber balanced on a curved dish can be rotated by the attractive force of a charged rod.

a watch glass so that the board is free to rotate. When a charged rod is brought near one end, the resulting attractive force produces a torque that rotates the board.

Q: Assuming that the two-by-four was attracted by a positively charged rod in the previous example, what direction would it rotate if a negatively charged rod was used?

A: It would rotate in the same direction. The interaction between any charged object and an uncharged object is always attractive.

We can now return to the charged comb that attracted and then repelled bits of paper. Initially, the bits of paper are uncharged. They are attracted to the comb by the induced charge. When the bits of paper touch the comb, they acquire some of the charge on the comb and are repelled because the comb and the paper have the same sign of charge and like charges repel one another.

FLAWED REASONING

Your teacher demonstrates that the south pole of a bar magnet is attracted to a charged glass rod. She then asks your class to predict what will happen when the charged glass rod is brought near the north pole of the magnet. The best student in the class answers, "The glass rod is positively charged, so the south pole of the magnet must be negatively charged. I know that a north pole will attract a south pole, so north poles of magnets must be positively charged. I predict that the glass rod will repel the north pole of the magnet."

Your teacher then performs the experiment and shows that the glass rod *attracts* the north pole. She then shows that a negatively charged rubber rod also attracts both poles of the magnet, demonstrating that the magnet must be electrically neutral. **Clearly the best student in the class made a mistake in his reasoning. What is it?**

ANSWER An electrical *attraction* between two objects only indicates that at least one of the objects is charged. The class already knew that the glass rod was charged, so the first demonstration teaches nothing about the charge (if any) on the magnet. If the charged rod had *repelled* one of the poles of the magnet, it would indicate that the magnet must be charged. However, bar magnets are normally not charged and are attracted by any charged rod.

The Electroscope

In most experiments we transfer so few charges that the total attraction or repulsion is small compared with the pull of gravity. Detecting that an object is charged becomes difficult unless it is very light—like bits of paper or a balloon. We get around this difficulty with a device called an *electroscope,* which gives easily observable results when it is charged. By bringing the object in question near the electroscope, we can deduce whether it is charged by the effect it has on the electroscope.

The essential features of an electroscope are a metal rod with a metal ball on top and two very light metal foils attached to the bottom. Figure 20-6 shows a homemade electroscope constructed from a chemical flask. The glass enclosure protects the very light foils from air currents and insulates the foils and rod from the surroundings.

Figure 20-6 An electroscope made from a flask, a metal rod, and two pieces of thin metal foil.

© Cengage Learning/David Rogers

If the electroscope isn't charged, the foils hang straight down under the influence of gravity. When we touch the ball of the electroscope with a charged rod, some of the rod's excess charge is shared with the electroscope. The charges flow from the rod onto the electroscope because the charges on the rod repel each other and therefore distribute themselves over the largest region possible, which includes the ball, the metal rod, and the foils. Because the foils have the same kind of charge, they repel each other and swing apart (Figure 20-7). The separation of the foils indicates that the electroscope is charged; the amount of separation is a rough indication of the amount of excess charge. This happens when the electroscope is touched with a positive rod as well as a negative one.

Figure 20-7 When an electroscope is charged by direct contact, it has the same charge as the rod.

© Cengage Learning/David Rogers

When a charged object is touched to an electroscope, the electroscope takes on the same kind of charge as the object. However, the electroscope only shows the presence of charge; it does not indicate the sign. We can determine the sign by slowly bringing a rod with a known positive or negative charge toward the ball of the charged electroscope and watching the motion of the foils.

For example, as a positive rod nears the electroscope, it induces a redistribution of charge in the electroscope. It repels the positive charge and attracts the negative charge, causing an increase in positive charge (or a decrease in negative charge) on the foils. If the foils were originally positive, the extra positive charge would cause them to move farther apart. If the foils were originally negative, the extra positive charge would cancel part of the negative charge and the foils would move closer together.

Are You On the Bus?

Q: When a negative rod is slowly brought near an electroscope that is initially charged, the foils move closer together. What is the charge on the electroscope?

A: The negative rod repels negative charges on the electroscope's ball, increasing their concentration on the foils. Because this negative charge reduces the charge on the foils, they must have been positive initially.

A charged rod will also separate the foils on an uncharged electroscope. As a positive rod is brought near an uncharged electroscope, some of the negative charges in the electroscope are attracted to the ball on top; the ball has a net negative charge, and the foils have a net positive charge, as shown in Figure 20-8(a). Note that the net charge on the electroscope is still zero because we did not touch it with the rod. We have only moved the charges around in the electroscope by bringing the charged rod near the ball. When the charged rod is removed, the charges redistribute themselves and the foils fall. An analogous phenomenon happens with a negative rod.

Imagine that while the positive rod is held near the electroscope you touch the ball with your finger, as shown in Figure 20-8(b). Negative charges are attracted by the large positive charge on the rod and travel from the ground through your body to the electroscope. If you first remove your finger and then the rod, the electroscope is now charged and the foils repel each other. This is known as charging the electroscope by induction. The electroscope acquires a charge *opposite* to that on the rod.

Figure 20-8 (a) The positively charged rod brought near the electroscope attracts negative charges to the ball, leaving the foils with an excess positive charge. (b) Touching the ball with a finger allows negative charges to flow into the electroscope, (c) leaving it with a net negative charge.

(a)   (b)   (c)

The Electric Force

Extended presentation available in the *Problem Solving* supplement

Simple observations of the attraction or repulsion of two charged objects indicate that the size of the electric force depends on distance. For instance, a charged object has more effect on an electroscope as it gets nearer. But we need to be more precise. How does the size of this force vary as the separation between two charged objects changes? And how does it vary as the amount of charge on the objects varies?

In 1785 French physicist Charles Coulomb measured the changes in the electric force as he varied the distance between two objects and the charges on them. He verified that if the distance between two charged objects is doubled (without changing the charges), the electric force on each object is reduced to one-fourth the initial value. If the distance is tripled, the force is reduced to one-ninth, and so on. This type of behavior is known as an **inverse-square relationship**—*inverse* because the force gets smaller as the distance gets larger, *square* because the force changes by the square of the factor by which the distance changes.

Coulomb also showed that reducing the charge on one of the objects by one-half reduced the electric force to one-half its original value. Reducing the charge on each by one-half reduced the force to one-fourth the original value. This means that the force is proportional to the product of the two charges.

Q: Coulomb could not measure charge directly. What technique could he have used to reduce the charge to one-half its original value?

A: He could have used identical conducting spheres and put the charged sphere into contact with a neutral one. By symmetry each sphere would have one-half the original charge.

Are You On the Bus?

These two effects are combined into a single relationship known as Coulomb's law:

$$F = k\frac{q_1 q_2}{r^2}$$

◀ Coulomb's law

In this equation, q_1 and q_2 represent the amount of charge on objects 1 and 2, r is the distance between their centers, and k is a constant (known as Coulomb's constant) whose value depends on the units chosen for force, charge, and distance.

Each object feels the force due to the other. The forces are vectors and act along the line between the centers of the two objects. The force on each object is directed toward the other if the charges have opposite signs and away from each other if the charges have the same sign (Figure 20-9). Because the two forces are due to the interaction between the two objects, the forces are an action–reaction pair. According to Newton's third law, the forces are equal in magnitude, point in opposite directions, and act on different objects.

Because the existence of an elementary, fundamental charge was not known until the 20th century, the unit of electric charge, the **coulomb** (C), was chosen for convenience in use with electric circuits. (We will formally define the coulomb later.) Using the coulomb as the unit of charge, the value of Coulomb's constant is determined by experiment to be

$$k = 9 \times 10^9 \; \frac{\text{N} \cdot \text{m}^2}{\text{C}^2}$$

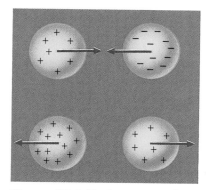

Figure 20-9 The forces on two charged objects are equal in size and opposite in direction in accordance with Newton's third law.

◀ Coulomb's constant

The coulomb is a tremendously large unit for the situations we have been discussing. For instance, the force between two spheres, each having 1 coulomb of charge and separated by 1 meter, is

$$F = k\frac{q_1 q_2}{r^2} = \left(9 \times 10^9 \ \frac{\text{N} \cdot \text{m}^2}{\text{C}^2}\right)\frac{(1 \ \text{C})(1 \ \text{C})}{(1 \ \text{m})^2} = 9 \times 10^9 \ \text{N}$$

This is a force of 1 million tons!

The charges that we have been discussing are much less. When you scuff your shoes on a carpet, the charges transferred are about a millionth of a coulomb. But even this charge is very large compared with the charge on a single electron or proton. The size of the charge on an electron or a proton is now known to be 1.6×10^{-19} coulomb. Conversely, it would take the charge on 6.25×10^{18} protons (or electrons) to equal a charge of 1 coulomb.

FLAWED REASONING

The following question appears on the final exam: "Two small, identical metal balls are hung next to each other from silk threads. Both balls are given the same net charge, such that they repel each other, as shown in Figure 20-10. If the net charge on ball 2 is reduced by half, will the angle that ball 1 makes with the vertical be smaller than, the same as, or larger than the angle that ball 2 makes with the vertical?"

Figure 20-10

Peter gives the following answer: "Ball 2 will still hang at the same angle as before. The charge on ball 1 has not changed, so it exerts the same force on ball 2. Ball 2 has half the charge, so it exerts half the force on ball 1. Therefore, ball 1 will hang at a smaller angle than ball 2." **Should Peter get full credit for this problem?**

ANSWER Coulomb's law automatically satisfies Newton's third law. The force exerted on ball 1 by ball 2 must be equal in magnitude (and opposite in direction) to the force exerted on ball 2 by ball 1. Reducing the charge on ball 2 by half will reduce the force felt by *each* ball by half. Both balls hang at the same, smaller angle.

Electricity and Gravity

The mathematical form of Coulomb's law is the same as that of the universal law of gravitation discussed in Chapter 5. Therefore, the drawing in Figure 5-2 illustrating the dependence of the inverse-square law on distance also holds for electricity if Earth is replaced by a charged object. There are, however, several important differences between gravity and electricity that govern the roles of these two forces in the universe.

There are two kinds of electric charge—positive and negative. Opposite charges attract each other and like charges repel each other. There is, however, only one kind of mass. Negative mass does *not* exist, and the gravitational force is never repulsive. Because of the induced charge distribution possible with two kinds of charge, conducting materials can shield a region from all

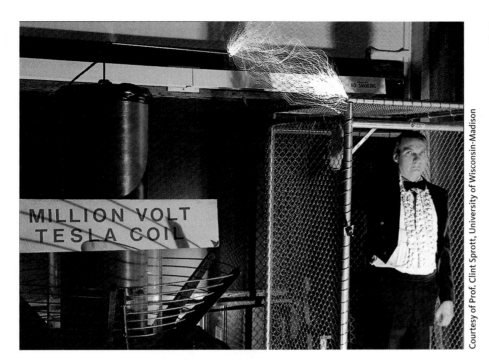

Professor Sprott is protected from the high-voltage discharge by the metallic cage.

Courtesy of Prof. Clint Sprott, University of Wisconsin-Madison

external electric forces. Suppose we were to build a large metal room. If this room were placed near a large electric charge, the charges in the metal walls would redistribute in such a way as to cancel the effect of the external charge for all locations in the room. The cancellation works only because there are two kinds of electric charge. This result is not possible for gravitational forces; a gravitation-free chamber cannot be created because there is no negative mass. Antigravity spaceships, although they make good material for science fiction writers, are not possible.

Another difference between electricity and gravity lies in the behavior of objects in the respective fields. The gravitational force is proportional to an object's mass, and therefore objects of all sizes and compositions have the same acceleration in a gravitational field. For example, all objects have the same free-fall acceleration near Earth's surface. In contrast, the electric force is proportional to the charge on an object and *not* to the object's mass. Therefore, charged objects have different accelerations in an electric field. For example, if we were to release a proton and an electron near a charged object, the electron would have a much larger acceleration. The forces on the electron and the proton are the same size, but the electron's mass is much smaller, resulting in a much larger acceleration. In this case, the accelerations would be in opposite directions because the charges have opposite signs.

Finally, all charges occur as integral multiples of one fixed size. Experiments have shown that the charges on electrons and protons are the same size. The charges on other ordinary subatomic particles are either this same size or integral multiples of this size; the charges can be 2 times or 3 times the charge on the proton but never 1.5 times this charge. Mass also occurs in lumps, but there are many different sizes—for instance, the masses of the electron, the proton, and the neutron. These can combine in many different ways, and the total masses are not multiples of a single unit mass.

A consequence of these differences is the different roles these two forces play in our lives. The electric force is the dominant force in the atomic world; it determines the properties of atoms and molecules. On the other hand, the gravitational force dominates on the macroscopic scale of people, planets, and

© Royalty-Free/Corbis

The electric force dominates on the atomic scale, such as in large crystals.

NASA

The gravitational force dominates in the solar system and beyond, such as in this galaxy.

galaxies. The reason electricity and gravity switch roles in these two domains is that macroscopic objects are essentially uncharged; that is, there are approximately equal numbers of positive and negative charges on most large objects. Even though the total attractive force between all the positive charges in Earth and all the negative charges in the Moon is tremendous, it is equal and opposite to the repulsive force between the negative charges in Earth and the negative charges in the Moon. The net electric force is essentially zero because of this cancellation. However, because there is only one kind of mass, there is nothing to cancel the much weaker gravitational force between the atoms in Earth and atoms in the Moon. So as we move from people to planets and eventually to galaxies, the gravitational force becomes the dominant force.

WORKING IT OUT *Gravitational and Electric Forces*

As a numerical example, let's calculate the sizes of the electric and gravitational forces between an electron and a proton in a hydrogen atom. We start by assuming that the electron and proton are separated by a distance of 5.29×10^{-11} m and use the known masses and charges of the electron and proton. (These are given on the back end pages of this text.)

$$F_g = G\frac{m_1 m_2}{r^2} = \left(6.67 \times 10^{-11} \frac{\text{N} \cdot \text{m}^2}{\text{kg}^2}\right) \frac{(9.11 \times 10^{-31}\,\text{kg})(1.673 \times 10^{-27}\,\text{kg})}{(5.29 \times 10^{-11}\,\text{m})^2}$$

$$= 3.63 \times 10^{-47}\,\text{N}$$

$$F_e = k\frac{q_1 q_2}{r^2} = \left(9 \times 10^9 \frac{\text{N} \cdot \text{m}^2}{\text{C}^2}\right) \frac{(1.6 \times 10^{-19}\,\text{C})^2}{(5.29 \times 10^{-11}\,\text{m})^2} = 8.23 \times 10^{-8}\,\text{N}$$

Therefore, the electric force is more than 10^{39} times as strong as the gravitational force. (That's 1000 trillion trillion trillion times as strong!) Because the two forces change in the same way as the separation of the electron and proton changes, the separation does not matter and the electric force is always this much stronger.

The Electric Field

Implicitly, we have assumed the force between two charges to be the result of some kind of direct interaction—sort of an action-at-a-distance interaction. This type of interaction is a little unsettling because there is no direct pushing or pulling mechanism in the intervening space. Electrical effects are evident even when there is a vacuum between the charges.

In many cases, it proves to be both conceptually and computationally simpler to separate the electrical interaction that one object feels as the result of another into two distinct steps. First, one of the objects generates, by virtue of its charge, an **electric field** at every point in space. Second, another object interacts, by virtue of its charge, with the electric field to experience the force. The electric field is created by electric charges and exerts forces on other electric charges. This method is analogous to what we did with the gravitational force near the end of Chapter 5.

If this were the only purpose of the field idea, it would play a minor role in our physics world view. In fact, it probably seems as if we are trading one unsettling idea for another. However, as we continue our studies, we will find that the electric field takes on an identity of its own. As we will learn in Chapter 22, electric and magnetic fields can travel through space as waves.

We define the electric field **E** at every point in space as the force exerted on a unit positive charge placed at the point. This is equivalent to the way that the gravitational field was defined, with the unit mass replaced by a unit positive charge.

◀ electric field = force on unit positive charge

Because force is a vector quantity, the electric field is a vector field; it has a size and a direction at each point in space. You could imagine the space around a positive charge as a "porcupine" of little arrows pointing outward, as shown in the diagrams in Figure 20-11. The arrows farther from the charge would be shorter to indicate that the force is weaker there.

Q: What does the electric field surrounding a negative charge look like?

A: An electric field surrounding a negative charge looks just like the electric field surrounding a positive charge except that all of the arrows are reversed.

The values for an actual electric field can be measured with a test charge. The unit of charge that we have been using is 1 coulomb. This is a very large amount of charge, and if we actually used 1 coulomb as our test charge, it would most likely move the charges that generated the field, thus disturbing the field. Therefore, we use a much smaller charge, such as 1 microcoulomb, and obtain the size of the field by dividing the measured force **F** by the size q of the test charge:

$$E = \frac{F}{q}$$

Notice that the units of electric field are newtons per coulomb (N/C).

If we know the sizes and locations of the charges creating the electric field, we can also calculate the value of the field at any point of interest by assuming that we place a 1-coulomb charge at the location and calculating the force on this charge using Coulomb's law. In doing this, we can take advantage of the fact that each charge acts independently; the effects simply add. This means that we calculate the contribution of each charge to the field and then add these contributions vectorially.

◀ electric field = $\dfrac{\text{force}}{\text{charge}}$

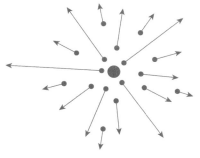

Figure 20-11 Sample electric field vectors around a positive source charge.

WORKING IT OUT *Electric Field*

What is the electric field at a distance of 2 m from a point charge of 3 μC?

By definition, the electric field at a point in space is equal to the force on a unit positive test charge placed at that point. In simple cases its magnitude can be obtained by calculating the force on a test charge q and then dividing by q.

$$E = \frac{F}{q} = \frac{1}{q}\,k\frac{Qq}{r^2} = k\frac{Q}{r^2}$$

We see that the value of the electric field depends on the source charge Q that creates the field, and is independent of the value of the test charge q that experiences the field. The magnitude of the electric field is

$$E = k\frac{Q}{r^2} = \left(9 \times 10^9 \, \frac{\text{N} \cdot \text{m}^2}{\text{C}^2}\right) \frac{3 \times 10^{-6} \, \text{C}}{(2 \, \text{m})^2} = 6750 \, \frac{\text{N}}{\text{C}}$$

Because the electric field is a vector, we must give its direction to completely specify it. The field points directly away from the positive source charge because that is the direction of the force on a unit positive test charge placed at the field point.

Once we know the value of the electric field at any point, we can calculate the force that any charge q would experience if placed at that point:

electric force = charge × electric field ▶

$$F = qE$$

This is read as, "The force on an object is equal to the net charge q on the object times the electric field **E** at the location of the object."

As an example, let's assume that we have generated a uniform electric field that points downward and has a size of 1000 newtons per coulomb. If we place an object in this field that has a positive charge of 1 microcoulomb, the object will experience a downward force of

$$F = qE = (10^{-6} \, \text{C})\left(10^3 \, \frac{\text{N}}{\text{C}}\right) = 10^3 \, \text{N}$$

If we change the charge on the object, calculating the new force is easy; we do not have to deal with the charges that produced the electric field.

Electric Field Lines

If we are only interested in what is going on at a single point, the electric field vector representation is helpful. However, it becomes cumbersome if we are interested in a region of space because each point in space may have a different electric field vector associated with it. To draw this, we would need to draw a different vector (possibly different in both magnitude and direction) at each point, and these would tend to overlap! To deal with this, we introduce an alternative representation using **electric field lines.**

Imagine a region containing charged particles that are fixed in place and create an electric field at every point in the region. We now draw an electric field line. Find the direction of the electric field at a starting point and take a small step in the direction of this vector. At the new point, again find the direction of the electric field and take a small step in this direction. Continuing this process creates a series of points that we connect with a smooth line. This line is an electric field line. The final step is to put a small arrow on the line to indicate the direction of travel. Starting at a new point in the region leads to a new electric field line. If we do this from enough different starting points, we can get a sample of lines throughout the region, as shown in Figure 20-12.

This sounds like a long, involved process—and it is. The good news is that we can often learn a great deal through qualitative sketches of electric field lines, which are much easier than doing the detailed calculations. We can use our intuition to draw the electric field lines surrounding an isolated positive source charge. A positive test charge would be repelled directly away from the positive source charge, so the electric field lines should start on the source charge and continue radially outward to infinity, as shown in Figure 20-13. The electric field vector at any point in space is then tangent to the electric field line that would pass through that point. Notice that this is consistent with the drawing in Figure 20-11 of the electric field vectors for a positive source charge. (If we are interested in a point that does not lie on one of the electric field lines that have been drawn, we can find the approximate direction of a line that passes through the point by looking at the surrounding lines.)

Notice that the electric field lines in Figure 20-13 start at locations equally spaced around the source charge. When the electric field lines are drawn this way, they give us visual information about the magnitude of the electric field vector at any point. Notice that the field lines are close together near the source charge, where the electric field is strong, and the field lines are far

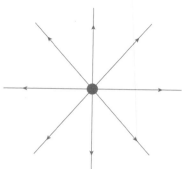

Figure 20-13 Electric field lines due to an isolated positive source charge.

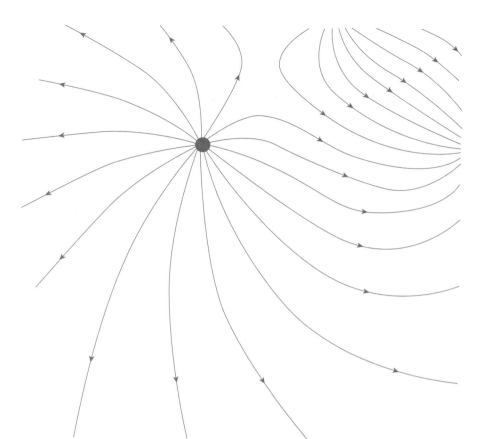

Figure 20-12 Electric field lines represent the total electric field in a region of space.

from each other at locations away from the source charge, where the electric field is weak. Indeed, any point in space that is 3 centimeters from the source charge should have the same size electric field. The spherical symmetry of our field lines ensures that the spacing between adjacent field lines is the same for all these points. In general, the strength of the electric field (that is, the length of the vector) is greater in regions where the electric field lines are closer together. Another way to say this is that the electric field is proportional to the density of electric field lines.

Q: How would the diagrams in Figure 20-13 change if the positive source charge were replaced by a negative source charge?

A: The arrows in the drawing would point in opposite directions.

When more than one source charge is present in a region, the field lines represent the total electric field in the region due to all the source charges. At locations very close to one of the source charges, the electric field lines should still be radially symmetric about that source charge (as its contribution will dominate). The number of field lines originating on a positive source charge or ending on a negative source charge should be proportional to the magnitude of the charge. This is because the electric field strength at a location 1 centimeter from a +2-coulomb source charge should be twice as big as the electric field strength 1 centimeter from a +1-coulomb source charge. Figure

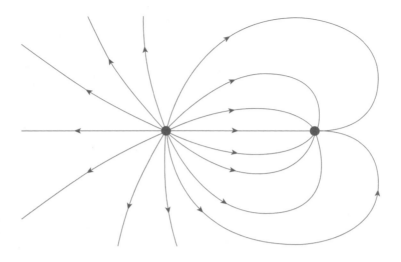

Figure 20-14 The electric field lines indicate that the magnitude of the positive charge is twice that of the negative charge.

20-14 shows the electric field lines around two source charges, one positive and one negative.

Are You On the Bus?

Q: If the positive source charge in Figure 20-14 has a charge of +6 microcoulombs, what is the magnitude of the charge on the negative source charge?

A: The number of electric field lines originating on a positive charge or ending on a negative charge is proportional to the size of the electric charge. Twelve field lines leave the positive source charge, but only six of those field lines end on the negative source charge. The negative charge must be half the size of the positive charge. It must have a charge of −3 microcoulombs.

Notice that electric field lines always begin on positive charges and end on negative charges. If the region you are considering contains more positive than negative charges, some lines will leave the region. If the region contains more negative charges, some lines will come in from outside the region.

Are You On the Bus?

Q: Why do electric field lines never cross?

A: The electric field lines represent the total electric field in a region due to all source charges. The electric field vector at any point in space is tangent to the electric field line through that point. If two electric field lines crossed at any point, the electric field at that point would have to be tangent to both electric field lines. It would have to point in two directions at a single location. This is not possible.

A common situation in which electric field lines provide physical insight is the case of charged parallel metal plates. If electrons are taken from one metal plate and placed on the other, both plates end up with the same amount of excess electric charge (positive on one plate and negative on the other). The charge will spread out on the facing surfaces of the plates until the charge density is uniform. Electric field lines originate on positive charge and end on negative charge, so the uniform charge distribution on the plates dictates that the electric field lines between the plates be parallel to each other, perpendicular to the plates, and uniformly spaced, as shown in Figure 20-15.

These electric field lines are very simple, and they represent an electric field that is uniform in strength everywhere between the two plates. The simplicity of the electric field lines predicts a result that is highly nonintuitive: if you place a small positive test charge at a location halfway between the plates, it experiences the same electric field (and hence the same electric force) as it would very close to the positive plate (or, in fact, anywhere else between the plates). It is possible to show that this is indeed the case mathematically, but it is much easier to use the concept of electric field lines.

Electric Potential

We must do work to lift a 1-kilogram block in Earth's gravitational field, and this work increases the gravitational potential energy of the block. When the block is released, it falls and the gravitational potential energy is changed to kinetic energy. Lifting a 5-kilogram block the same distance requires five times the work and the block gains five times the gravitational potential energy. In other words, knowing the energy that is gained by the 1-kilogram block allows us to calculate the energy that would be gained by any other mass moved between the same two points.

In the same way, work is required to move a charged particle in an electric field, and this work changes the **electric potential energy** of the particle. When we release the particle, this electric potential energy can be converted to kinetic energy. Therefore, we define the electric potential energy the same way we did for gravity. The electric potential energy of a charged object is equal to the work done in bringing the object from some zero reference location to the object's location.

As with gravitational potential energy, the value of the electric potential energy does not depend on the path, but it does depend on the reference location, the location of the object, and the charge on the object (Figure 20-16).

As with gravitational potential energy, the actual value of the electric potential energy is not important in physical problems; it is only the difference in energy between points that matters. If 10 joules of work is required to move a charged object from point A to point B, the electric potential energy of the object at point B is 10 joules higher than at point A. If point A is the zero reference point, the electric potential energy of the object at point B is 10 joules.

Because objects with different charges have different electric potential energies at a given point, it is often more convenient to talk about the energy available due to the electric field without reference to a specific charged object. The **electric potential** V at each point in an electric field is defined as the electric potential energy EPE divided by the object's charge q:

$$V = \frac{EPE}{q}$$

Notice that it doesn't matter which charged object we use to define the electric potential. This quantity is numerically equal to the work required to bring a positive test charge of 1 coulomb from the zero reference point to the specified point. The units for electric potential are joules per coulomb (J/C), a combination known as a **volt** (V). Because of this, we often speak of the electric potential as a *voltage*.

Defining the electric potential allows us to obtain the electric potential energy for any charged object by multiplying the potential by the charge. Once again, it is only the *potential difference* that matters. For instance, a 12-volt battery has an electric potential difference of 12 volts between its two terminals. This means that 1 coulomb of charge moving from one terminal to the other would gain or lose (1 coulomb)(12 volts) = 12 joules of energy.

Figure 20-15 The electric field lines between charged metal plates indicate that the electric field is everywhere uniform between the plates.

◀ electric potential energy = work from reference point

Figure 20-16 The work performed in taking a charge from A to B does not depend on the path.

◀ electric potential = $\dfrac{\text{electric PE}}{\text{charge}}$

Q: How much work is required to move 3 coulombs of positive charge from the negative terminal of a 12-volt battery to the positive terminal?

A: Each coulomb requires 12 joules of work, so 3 coulombs require 36 joules: $W = qV$ = (3 coulombs)(12 volts) = 36 joules.

If the potential difference, or voltage, between two points is high and the points are close together, the electric field can be strong enough to tear electrons from molecules in the air. The electrons are pulled one way, and the remaining positive ions are pulled the other way. Because electrons have much less mass than ions, they quickly accelerate to high velocities. As they accelerate they collide with other molecules, knocking additional electrons loose. This forms a cascade of electrons that we call a *spark*. Dry air breaks down in this fashion when the electric field reaches about 30,000 volts per centimeter.

Summary

Objects become electrically charged or uncharged by transferring charges. An uncharged object has equal numbers of positive and negative charges. Positively charged objects may have an excess of positive charges or a deficiency of negative charges. In an isolated system, the total charge is conserved. Charges can flow through conductors such as metal wires but not through insulators such as silk threads.

In electricity, unlike gravity, there are two different types of charge, and the direction of the force depends on the relative types of the charges. Like charges repel; unlike charges attract. A charged object can attract an uncharged object because of an induced separation of charges.

The electric force has the same mathematical form as the gravitational force. The electric force, however, can be repulsive as well as attractive. Electric charge comes in whole-number multiples of one fixed size. The electric charges on electrons and protons have the same size but are opposite in sign.

Electric charges are surrounded by an electric field that is equal to the force experienced by a unit positive charge. This electric field is a vector field with a size and a direction at each point in space. The units for electric field are newtons per coulomb. Electric field lines represent the electric field in a region of space. The electric field at any location points in a direction tangent to the field line at that location, and the strength of the electric field is proportional to the local density of electric field lines.

The electric potential energy of a charged object in an electric field is equal to the work done in bringing the object from some zero reference location. Only the differences in electric potential energy matter in physical situations. The electric potential equals the electric potential energy divided by the object's charge. This quantity is numerically equal to the work required to bring a positive test charge of 1 coulomb from the zero reference point. The units for electric potential are joules per coulomb, or volts.

CHAPTER 20 *Revisited*

Sparks occur when the electric field becomes strong enough to pull electrons from atoms. These electrons are accelerated by the electric field, obtaining sufficient kinetic energy to knock electrons from other atoms. This avalanche of electrons produces the visible sparks that we see. Electric fields of 30,000 volts per centimeter are strong enough to break down atoms in dry air. Because the electric field is proportional to the electric potential difference and inversely proportional to the distance, it takes large voltages to produce long sparks.

Everyday Physics | *Lightning*

Everybody has seen lightning (Figure A) and has probably been frightened by this spectacular display of energy at some time or another. Lightning is unpredictable and seems to occur randomly and instantaneously. Everything but the thunder is usually over in less than half a second. And yet the effects are not inconsequential: each year lightning causes about $80 million of damage in the United States. On the average, lightning kills 85 people a year and injures another 250 in the United States alone.

We still have a lot to learn about lightning. For instance, we don't know what causes the initial buildup of charge, what determines the paths the lightning takes, or what triggers the lightning bolts.

We do know that during a storm, clouds develop a separation of electric charge, with the tops of the clouds positively charged and the bottoms negatively charged, as shown in Figure B. The production of a lightning bolt begins when the negative charge on the bottom of the cloud gets large enough to overcome air's resistance to the flow of electricity (Figure C) and electrons begin flowing toward Earth along a zigzag, forked path at about 60 miles per second! As the electrons flow downward, they collide with air molecules and ionize them, producing more free electrons. Even though the current may be as large as 1000 amperes, this is not the lightning bolt we see.

Meanwhile, as the electrons approach the ground, the ground becomes more and more positively charged because of the repulsion of electrons in the ground. This positively charged region moves up through any conducting objects on the ground—houses, trees, and people—into the air (Figure D). When the downward-moving electrons meet the upward-flowing positive regions at an altitude of 100 meters or so, they form a complete circuit (Figure E), and the lightning begins. In less than 1 millisecond, up to 1 billion trillion electrons may reach the ground; the current may

Figure A Lightning bolts can generate currents up to 200,000 amperes.

© Photodisc Red/Getty Images

be as high as 200,000 amperes. Although the flow of charge is downward, the point of contact between the cloud charge and the ground charge is flowing *upward* at about 50,000 miles per second. (Observers report that the lightning bolt is moving downward because the pathway that is lit is the initial forked pathway coming from the cloud.) The upward surge heats the air to 50,000°F. One meter of the lightning's path may shine as bright as 1 million 100-watt lightbulbs (Figure F). This rapid heating of the air along the lightning's path also produces a shock wave that we hear as the thunder. What we see as a single lightning bolt is usually several bolts in rapid succession along the same path.

1. People near a lightning strike report that their hair begins to stand on end, warning of the impending strike. What causes this?

2. What process causes the loud noise that we call thunder?

(B)   (C)   (D)   (E)   (F)

Figures B–F The development of a lightning bolt.

Key Terms

charge   A property of elementary particles that determines the strength of its electric force with other particles possessing charge. Measured in coulombs, or in integral multiples of the charge on the proton.

charged   Possessing a net negative or positive charge.

conductor   A material that allows the passage of electric charge. Metals are good conductors.

conservation of charge   In an isolated system, the total charge is conserved.

coulomb   The SI unit of electric charge; the charge of 6.25×10^{18} protons.

electric field   The space surrounding a charged object, where each location is assigned a value equal to the force experienced by one unit of positive charge placed at that location. Electric field is measured in newtons per coulomb.

electric field lines   A representation of the electric field in a region of space. The electric field is tangent to the field line at any point, and its magnitude is proportional to the local density of field lines.

electric potential   The electric potential energy divided by the object's charge; the work done in bringing a positive test charge of 1 coulomb from the zero reference location to a particular point in space. Electric potential is measured in joules per coulomb.

electric potential energy   The work done in bringing a charged object from some zero reference location to a particular point in space.

grounding   Establishing an electrical connection to Earth in order to neutralize an object.

insulator   A material that does not allow the passage of electric charge. Ceramics are good insulators.

inverse-square relationship   A relationship in which a quantity is related to the reciprocal of the square of a second quantity; the electric force is proportional to the inverse square of the distance from the charge.

volt   The SI unit of electric potential, 1 joule per coulomb.

Questions and exercises are paired so that most odd-numbered are followed by a similar even-numbered.

Blue-numbered questions and exercises are answered in Appendix B.

❓ indicates more challenging questions and exercises.

WebAssign Many Conceptual Questions and Exercises for this chapter may be assigned online at WebAssign.

Conceptual Questions

1. A handheld glass rod can be charged by rubbing it with silk or a plastic bag while holding it in your hands. Would you conclude from this that glass is a conductor or an insulator? Why?

2. Inexperienced physics teachers often demonstrate the use of the electroscope by touching it with a charged glass rod at a single point. More experienced teachers typically drag the length of the rod across the top of the electroscope to increase the desired effect. Why does this help?

3. A neutral metal sphere is hanging from the ceiling by a thin insulating thread. Which of two equally charged rods, one metal and one rubber, would transfer the greater amount of charge to the metal sphere when touched to the sphere? Why?

4. Could you use fur to charge a metal rod that is held in your bare hand? Explain.

5. Why is it easier to charge a balloon on a dry day than on a humid day?

6. Why is it easier to demonstrate electrostatic phenomena in Fairbanks, Alaska, than in Honolulu, Hawaii?

7. Before an aircraft is fueled from a truck, a wire from the truck is attached to the aircraft. Why?

8. Why do you sometimes feel a shock when you touch a metal doorknob after crossing a carpeted floor?

9. If one rubber rod is charged with fur and another with plastic, they attract each other. How would you explain this?

10. You use a piece of silk to charge a glass rod. Does this leave the silk with a positive or a negative charge? Explain.

11. When you rub balloons with wool and place them near bits of paper, you find three categories of behavior. Balloons attract wool and bits of paper but repel other balloons. Pieces of wool attract balloons and bits of paper but repel other pieces of wool. Bits of paper are attracted to both the balloons and the pieces of wool but do not react with other bits of paper. When we have three distinct behaviors, why do we only need two types of charge in our model?

12. How would you know if you discovered a third kind of charge? What other objects would you expect the third kind of charge to attract and repel?

13. Your classmate claims that magnets are really just charged rods with opposite charges at the two ends. What experi-

---

ment could you perform to convince your classmate otherwise?

14. You find that the north pole of a bar magnet is attracted to a charged glass rod. You then discover that the south pole of the magnet is also attracted to the same rod. Will the north pole be attracted or repelled or experience no force if brought near a charged rubber rod? Explain.

15. You have three small balls, each hanging from an insulating thread. You find that balls 1 and 2 repel one another and that balls 2 and 3 repel one another. Will balls 1 and 3 attract or repel one another? Explain.

16. You have three small balls, each hanging from an insulating thread. You find that balls 1 and 2 attract one another and that balls 2 and 3 repel one another. Can any of the balls be neutral? If not, why not? If so, which one?

17. The nucleus of a helium atom contains two protons and two uncharged neutrons. If two electrons surround the nucleus, what is the total charge of the atom?

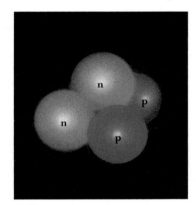

18. In the modern view of electricity, does an object with a negative charge have an excess of positive charge, an excess of negative charge, a deficiency of negative charge, or a deficiency of positive charge? Explain.

19. Describe how a charged balloon sticks to a wall.

20. When a charged comb is brought near bits of paper, the bits are first attracted to the comb and then repelled after the bits come in contact with the comb. Describe how the charge on the comb and on the bits of paper changes during this process.

21. Why are neutral objects attracted to both negatively and positively charged objects?

22. Describe how a negatively charged rod attracts (a) an uncharged conducting object and (b) an uncharged insulating object.

23. How could you use an electroscope of known positive charge to determine the sign of a charged rod?

24. How could you use a negatively charged rod to determine whether a charged electroscope has a negative or a positive charge without changing the charge on the electroscope?

25. Describe how you would use a positively charged rod to give an electroscope a negative charge.

26. Describe how you would use a positively charged rod to put a positive charge on an electroscope.

27. An electroscope initially has a net negative charge. Why do the foils come together and stay together when a human hand touches the electroscope?

28. An electroscope is initially given a net positive charge. You find that as you bring your hand near the top of the electroscope without touching it, the foils come slightly closer together. When you remove your hand, they move back apart. How do you account for this?

29. You have two metal spheres on insulating stands. You bring them each in turn near the ball of a positively charged electroscope. In each case the deflection of the leaves is reduced slightly. You then bring them each in turn near the ball of a negatively charged electroscope and find that sphere A causes the deflection to increase, whereas sphere B causes it to decrease. What can you say about the net charge on each of the two metal spheres?

30. You hold a charged rubber rod near the ball of an electroscope and then briefly touch the ball with your finger. After removing the rubber rod, you bring a charged glass rod near the ball. Will the deflection of the leaves increase, decrease, or stay the same? Why?

31. When Coulomb was developing his law, he did not have an instrument for measuring charge. And yet he was able to obtain spheres with $\frac{1}{2}, \frac{1}{3}, \frac{1}{4}$ of some original charge. How might he have used a set of identical spheres to do this?

32. You have three identical metal spheres on insulating stands. The spheres hold charges $Q_A = -2q$, $Q_B = -q$, and $Q_C = 4q$. First, sphere A is brought into contact with sphere C and separated. Second, sphere C is brought into contact with sphere B and separated. What is the resulting charge on sphere B?

33. Two uniformly charged insulating spheres are held atop fixed posts. The charge on one of the spheres is three times the charge on the other. Which diagram correctly represents the magnitude and direction of the electric forces on the two spheres? Explain your answer.

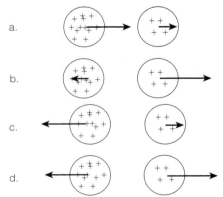

34. You have two positively charged objects. Object B has twice the charge of object A. When they are brought close together, object A experiences an electric force of 10 newtons. Is the electric force experienced by object B greater than, equal to, or less than 10 newtons? Explain.

35. Two charged objects are very far from any other charges. If the distance between them is cut to one-third its original value, what happens to the electric force between them?

36. Assume that you have two identically charged objects separated by a certain distance. How would the force change if the objects were twice as far away from each other?

37. Two charged objects are very far from any other charges. If the charge on each of them is tripled, what happens to the electric force between them?

38. Assume that you have two identically charged objects separated by a certain distance. How would the force change if one object had twice the charge and the other object kept the same charge?

39. Three identical charges are arranged as shown in the following figure. The distance from A to B is twice the distance from A to C. Which of the vectors best represents the force on charge A due to B and C? Justify your answer.

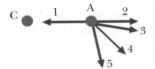

40. Two identical charges are fixed at the locations indicated in the following figure. A third charge, identical to the other two, is placed first at position A and then position B. For each case, determine the direction of the net electrical force on the third charge. If the force is zero, state that explicitly.

41. Although the formulas for electric and gravitational forces have the same form, there are differences in the two. What are some of these differences?

42. What are the similarities between electric and gravitational forces?

43. When we approach another person, we are not aware of the gravitational and electric forces between us. What are the reasons in each case?

44. Even though electric forces between electrons and protons are much stronger than gravitational forces, gravitational forces determine the motions in the Solar System. Why?

45. Why are the accelerations of all charged objects near a charged sphere not the same?

⌇46. In this chapter we compared the electric and gravitational forces between an electron and a proton. Why is the result valid for all separations?

47. A metal sphere with a charge of +2 coulombs experiences an electric force of 20 newtons directed to the left. If the charge on the sphere is increased to 6 coulombs, what force will it experience?

48. A metal sphere with a charge of +4 coulombs experiences an electric force of 50 newtons directed to the left. If the charge on the sphere is changed to −8 coulombs, what force will it experience?

49. The electric field lines between two parallel charged plates are everywhere perpendicular to the plates and parallel to each other, as shown in the following figure. A small positively charged particle placed at the midpoint between the plates experiences an electric force of magnitude F. If the charged particle were instead placed at location A, close to the positive plate, would the force on the particle be greater than, equal to, or less than F? Use the concept of field lines to justify your answer.

50. For the situation described in Question 49, the charged particle is released from rest at the midpoint and strikes the negative plate at speed v. If the particle had instead

been released from rest at point A, would it have struck the negative plate at a speed greater than, equal to, or less than v? Explain your reasoning.

51. An electron and a proton are released in a region of space where the electric field is vertically upward. How do the electric forces on the electron and proton compare?

52. How do the accelerations of the electron and proton in Question 51 compare?

53. A portion of an electric field is shown in the following figure. Four locations within the field are marked with **x**'s and labeled **a, b, c,** and **d.** Draw the vectors representing

the force on a positive test charge when placed at each of these locations. Explain your reasoning.

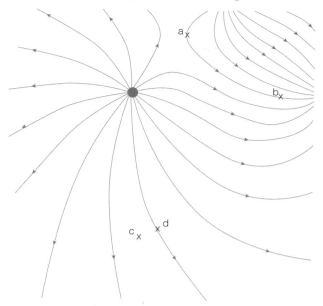

54. The field lines (solid lines) shown in the following figure represent the electric field in a certain region of space. A charged particle is released with an initial velocity and follows the trajectory from points 1 to 4, as shown by the dashed line. Draw vectors to represent the force on the charged particle at each location. What is the sign of the charged particle? Explain how you can tell.

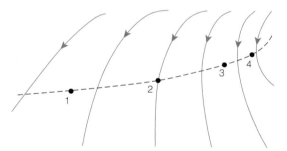

55. The following figure shows the electric field lines in a region of empty space. Is the electric field in the region between A and B greater than, equal to, or less than the electric field in the region between C and D? Explain.

Exercises

65. The nucleus of a certain type of uranium atom contains 92 protons and 143 neutrons. What is the total charge of the nucleus?

66. How many electrons does it take to have a total charge of -1 C?

67. What is the electric force of attraction between charges of $+3$ C and -6 C separated by a distance of 2 m?

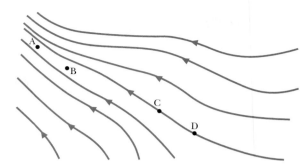

56. In the figure described in Question 55, points A and B and points C and D are the same distance apart. Is the potential difference between A and B greater than, equal to, or less than the potential difference between C and D? Explain.

57. A proton is observed moving at speed v_o at point A in space, where the electric potential is 750 volts. It moves to point B, where the electric potential is 950 volts. Is its speed at point B greater than, equal to, or less than v_o? Explain.

58. An electron is observed moving at speed v_o at point A in space, where the electric potential is 750 volts. It moves to point B, where the electric potential is 950 volts. Is its speed at point B greater than, equal to, or less than v_o? Explain.

59. How is the value of the electric potential at each point in space defined?

60. What is the definition of the value of the electric potential energy of a charged object at each point in space?

61. A proton is released from rest in a uniform electric field. Does the proton's electric potential energy increase or decrease? Does the proton move toward a location with a higher or lower electric potential? Explain.

62. An electron is released from rest in a uniform electric field. Does the electron's electric potential energy increase or decrease? Does the electron move toward a location with a higher or lower electric potential? Explain.

63. A chlorine ion contains 17 protons, 18 neutrons, and 18 electrons. What is the net charge of the ion?

64. What is the net charge of an iodine ion containing 53 protons, 74 neutrons, and 54 electrons?

68. By what distance must two charges of $+2$ C and $+8$ C be separated so that the repulsive force between them is 3.6 $\times 10^{10}$ N?

69. The nucleus of lithium contains three protons and three neutrons. When two electrons are removed from the neutral lithium atom, the remaining electron has an average distance from the nucleus of 0.018 nm. What is the force between the electron and the nucleus at this separation?

70. In an ionic solid such as ordinary table salt (NaCl), an electron is transferred from one atom to the other. If a distance of 0.1 nm separates the atoms, how strong is the electric force between them?

71. How much stronger is the electric force between two protons than the gravitational force between them?

72. Calculate the ratio of the electric force to the gravitational force between two electrons.

73. A 5-mC charge experiences a force of 2 N directed north. What is the electric field (magnitude and direction) at the location of the charge? Find the force on a −20-mC charge that replaces the 5-mC charge.

74. A −15-mC charge experiences a force of 25 N directed west. What is the electric field (magnitude and direction) at the location of the charge? If this charge is removed and not replaced, what is the electric field at this location?

75. What is the electric field at a distance of 3 cm from 4 mC of negative charge?

76. What is the electric field 2 m from 8 C of positive charge?

77. What is the electric field at a distance of 0.2 nm from a carbon nucleus containing six protons and six neutrons?

78. What is the electric field 5.3×10^{-11} m from a proton?

79. What is the electric field midway between charges of 2 C and 6 C separated by 2 m?

+2 C   +6 C

◄─────1m─────►◄─────1m─────►

80. What is the electric field midway between an electron and a proton separated by 0.2 nm?

81. What is the force on a proton located in an electric field of 3000 N/C? What is the proton's acceleration?

82. How would the values obtained in Exercise 81 change if the proton were replaced by an electron?

83. The electric potential energy of an object at point A is known to be 50 J. If it is released from rest at A, it gains 30 J of kinetic energy as it moves to point B. What is its potential energy at B?

84. If the object in Exercise 83 had a charge of +2 C, what would be the potential difference between points A and B?

85. How much work does a 12-V battery do in pushing 2 mC of charge through a circuit containing one lightbulb?

86. Points A and B each have an electric potential of +9 V. How much work is required to take 3 mC of charge from A to B?

87. What was your electric potential relative to a metal pipe if a spark jumped 1.3 cm (0.5 in.) through dry air from your finger to the pipe?

88. How far can a spark jump in dry air if the electric potential difference is 6×10^5 V? (This is why high-voltage sources are surrounded by a vacuum or an insulating fluid.)

21

Electric Current

▶ We live in an electrical world. It is hard to imagine a world without electricity. How does the movement of electrons produce the light that allows us to work and play indoors and at night?

(See page 462 for the answer to this question.)

© David H. Wells/Corbis

This night scene of Boathouse Row in Philadelphia is an example of the artistic use of lighting.

NEW phenomena often enter our world view accidentally, a result of someone with an open mind making an observation or conducting an experiment. Initially, these phenomena are often just curiosities, new challenges to our understanding of the world. Many of them give us new ways to do things and eventually find widespread use as our understanding of them increases. In some cases, however, a whole new technology is born. The electrical properties we studied in Chapter 20 were such curiosities. In the early days of investigation into these phenomena, little use was made of electricity other than building devices to amaze and shock friends. In the past 200 years, electricity has developed into a major technological presence in our society. Today, the United States, with 5% of the world's population, uses about one-fourth of the electricity generated in the world. This means that the average U.S. citizen uses five times the electricity the average world citizen does, including those in developed countries. When electricity usage of U.S. citizens is compared with that of citizens in third-world countries, the per capita usage is very much larger.

An important advance in understanding electricity came with the development of batteries. Batteries could make electric charges flow continuously as a current rather than in short bursts. This development eventually led to modern electric circuits that have had an obvious impact on our lives. Electric currents can produce heat and light, run motors and radios, operate sound systems and computers, and so on. The list is nearly endless.

An Accidental Discovery

Near the end of the 18th century, Italian anatomist Luigi Galvani announced that an excised frog's leg twitched when he touched it. It was well known from firsthand experiences with static electricity that twitches in humans could result from electric charges. But this was just a leg! Galvani was convinced that he had discovered the secret life force that he believed existed within all animals. Galvani found that a twitch occurred when he touched the frog's leg with a metal object but not when he touched it with an insulator. Furthermore, the metal had to be different from the metal hook holding the other end of the leg. It took two *different* metals to produce the twitch.

Another Italian scientist, Alessandro Volta, heard of these results and performed many experiments with sensitive electroscopes, searching for evidence that electric charge resides in animal tissue. He eventually convinced himself that the electricity was not in the frog's leg but was the result of touching it with the two metals.

Volta tried many different materials and discovered that an electric potential difference was produced whenever two different metals were joined. Some combinations of metals produced larger potential differences than others. One of his demonstrations was to place his tongue between pieces of silver and zinc that were held together at one end. The flow of electricity caused his tongue to tingle. You may have felt a similar sensation when a piece of metal or metal foil touched a metal filling in your tooth. A more detailed explanation of this process would take us beyond the scope of this book, but it is enough to realize that some chemical process moves charges from one metal to the other, creating a potential difference.

Batteries

Volta used his discovery that two dissimilar metals produce an electric potential difference to make the first battery. He made a single cell by putting a

A modern version of Volta's demonstration that zinc and copper rods inserted into a grapefruit produce a potential difference. (The cherry is for decorative purposes only.)

Q: How many $1\frac{1}{2}$-volt cells does it take to make a 9-volt battery?

A: Because the voltages of the cells add, it takes six cells.

piece of paper that had been soaked in a salt solution between pieces of silver and zinc. He then stacked these cells in a pile like the one shown in Figure 21-1. By doing this he was able to produce a larger potential difference. The potential difference of all the cells together was equal to the sum of the potential differences due to the individual cells.

A cell's potential difference depends on the choice of metals. Those commonly used in batteries have potential differences of $1\frac{1}{2}$ or 2 volts. Putting lead and lead oxide plates (known as electrodes) into dilute sulfuric acid (the electrolyte) makes a simple 2-volt cell. Connecting six of these cells together makes a 12-volt car battery. The positive electrode of one cell is connected to the negative electrode of the next. The connectors sticking out of the first and last cells (the terminals) are usually marked with a + or − sign to indicate their excess charges.

Batteries that can be recharged are called *storage batteries*. This recharging is usually accomplished with an electric charger with the negative terminals of the charger and battery connected to each other and the positive terminals connected to each other. The charger forces the current to run backward through the battery, reversing the chemical reactions and refreshing the battery. This electric energy is stored as chemical energy in the battery for later use.

Nonrechargeable flashlight batteries (sometimes called *dry cells*) are constructed with a carbon rod down the center, as shown in Figure 21-2. Carbon is a conductor and replaces one of the metals. A moist paste containing the electrolyte surrounds the rod. The other electrode is the zinc can surrounding the paste. The cell is then covered with an insulating material. These $1\frac{1}{2}$-volt cells are often used end to end (both pointing in the same direction) to provide the 3 volts used in many flashlights.

The voltage produced by an individual cell depends on the materials used and not on its size. The size determines the total amount of chemicals used and therefore the total amount of charge that can be transferred. We have seen that the voltage can be increased by placing cells end to end in a row, an arrangement called **series.** Cells (or batteries) can also be placed side by side, or **parallel,** as shown in Figure 21-3. The cells must all point in the same direction. This parallel arrangement does not increase the voltage but does increase the effective size of the battery. The larger amount of chemicals means that the batteries will last longer before they run down.

Figure 21-1 A schematic of Volta's battery composed of zinc and silver plates with moist paper between them.

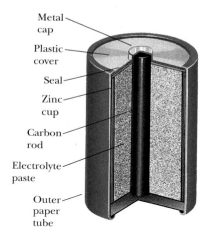

Figure 21-2 Schematic of a standard flashlight-type dry cell.

Q: When might you wish to connect batteries in parallel?

A: This is especially useful when the batteries are difficult to change, as in remote radio transmitters on mountaintops.

Household electricity has a number of differences from the electricity produced by batteries. The electricity in homes is supplied at a much higher voltage, typically 120 volts. (For historical reasons, household voltage is often referred to as "one-ten," but modern national standards in the United States and Canada specify that the nominal voltage be 120 volts, plus or minus 5%.) This electricity is supplied by huge electric generators in dams or power plants

Figure 21-3 Two cells in parallel have the same voltage as a single cell but last twice as long.

Courtesy of Vector Products, Inc.

A car battery can be recharged with a battery charger.

burning coal, oil, gas, or nuclear fuel. (We will discuss generators in the next chapter.) In household circuits the flow of charge (current) reverses direction 120 times a second. This is called *60-cycle alternating current* (ac) as opposed to the *direct current* (dc) supplied by batteries that flows in a single direction.

Because most of the uses of electricity we will discuss in this chapter depend only on the flow of charge and not on its direction, we will simplify our discussions by referring to flashlight batteries. Another advantage of using batteries is that they have low voltages, so you can experiment with some of the ideas mentioned here without fear of being shocked. **Household electricity**, on the other hand, **is dangerous and should not be used in experiments**.

Pathways

The invention of batteries provided experimenters with a way of producing a continuous flow of charge: a **current**. Although the properties of interacting electric charges are unchanged from the preceding chapter, new effects are observed when charges move through conducting pathways in a continuous fashion. The simplest way to learn about these is to experiment with a simple flashlight battery, some wire, and a few flashlight bulbs. The basic properties of household electricity are then simple extensions of these ideas.

Imagine that you have a battery, a wire, and a bulb. From your previous experiences, can you predict an arrangement that will light the bulb? A common response is to connect the bulb to the battery, as shown in Figure 21-4. The bulb doesn't light. It doesn't matter which end of the battery is used or which part of the bulb is touched. The charges don't flow from one end of the battery to the bulb and cause it to light. Suppose you hold the wire to one end of the battery and the metal tip of the bulb to the other end. Touching the free end of the wire to various parts of the bulb will eventually yield success. Two of the four possible arrangements for lighting the bulb are shown in Figure 21-5.

Comparing the methods for lighting the bulb with the ways that don't light it indicates that we must use two parts of each object: the two ends of the wire, the two ends of the battery, and the two metal parts of the bulb. The two parts of the bulb are the metal tip and the metal around the base (this is threaded on many bulbs). Whenever all six ends are connected in pairs, forming a continuous loop (no matter how you do it), the bulb lights.

Figure 21-4 An unsuccessful attempt to light a bulb.

Figure 21-5 Successful ways to light a bulb.

Q: Would the bulb still light if you turned the battery in the opposite direction in the arrangements shown in Figure 21-5?

A: Either end of the battery works because reversing the battery does not change the conducting path.

Figure 21-7 When a conducting material is placed between wires at the gap, the bulb lights.

Figure 21-6 Schematic drawing showing the continuous conducting path through a bulb.

The continuous conducting path (known as a **complete circuit**) allows the electric charges to flow from one end of the battery to the other end. If the bulb is to light, the conducting path must go through the bulb. You can verify this by examining a bulb with unfrosted glass, as illustrated in Figure 21-6. The part of the bulb that glows is a thin wire (called the *filament*) that is supported by two thicker wires coming out of the base. The bulb has to be taken apart to see that one of these support wires is connected to the metal tip and the other to the metal side of the bulb. The rest of the bulb is made from insulating materials so that there is a single conducting path through the filament within the bulb.

You can use the concept of a complete circuit to investigate the conductivity of different household materials. Connect a flashlight bulb to a flashlight battery, leaving a gap in the circuit, as shown in Figure 21-7. Test a variety of materials to see whether they are conductors by inserting them in the gap.

Combining the concept of a complete circuit with the idea of the conservation of charge leads to the conclusion that electricity flows out one end of the battery and back into the other end. The charge that leaves one end of the battery returns to the other end. Charge does not get lost or used up along the way.

Q: How does the information about pathways explain why the bulb in Figure 21-4 doesn't light?

A: Because there is no pathway for the charges to flow back to the other end of the battery, there is no complete circuit.

A Water Model

The flow of charge in a complete circuit is analogous to the flow of water in a closed system of pipes. The battery is analogous to a pump, the wires to the pipes, and the bulb (something that transforms energy in the flow to another form) to a paddle wheel that turns when the water is flowing (Figure 21-8). In a closed system of pipes, the water that leaves the pump returns to the intake side of the pump.

The water model is useful for clarifying the difference between current, charge, and voltage. The current is a measure of the amount of charge flowing past a given section of the circuit in a unit time. If we measure a charge ΔQ passing the point in an interval of time Δt, the current I in the circuit is

current = charge/time

$$I = \frac{\Delta Q}{\Delta t}$$

This expression is analogous to the rate at which water flows through a pipe. The flow of water is measured in units such as liters per second. Electric current is measured in **coulombs** per second, a unit known as the **ampere** (A). Flashlight batteries usually provide less than 1 ampere of current, typical household circuits are usually limited to a maximum of 20 amperes, and a car battery provides more than 100 amperes while starting the car. (The precise definition of the ampere will be described in the next chapter.)

The voltage between two points in a circuit is a measure of the change in electric potential between these two points. That is, voltage is the work done in moving one unit of electric charge between the two points. The work is equal to the distance the charge moves along the circuit multiplied by the force on the unit electric charge due to the electric field that exists in the circuit. You can also think of voltage as being analogous to pressure in our water model.

Although the water model is useful, it is important to realize its limitations. The electric circuit is always "filled" with charges; it does not need to be filled as you would a garden hose. A break in an electric circuit causes the electricity to stop flowing; it doesn't spill out the end of the wire as water would from a broken pipe. An electric circuit is not analogous to a sprinkler system used to water lawns. A valve blocks the flow of water, whereas a switch puts a gap in the conducting path to stop the flow of electricity.

Another difference is that one or both types of charge may be free to move in a conductor. In plasmas and ionized liquids and gases, both charges can move. Not until 1879 did experiments determine that only some of the negatively charged electrons are mobile in metals. For most macroscopic effects, positive charges moving in one direction are equivalent to negative charges moving in the opposite direction. For example, consider two neutral metal spheres. If you move 1 coulomb of positive charge from sphere A to sphere B, sphere

Figure 21-8 The flow of charge is analogous to the flow of water through a closed system.

Pump

B has an excess of 1 coulomb of positive charge. Sphere A has a deficiency of 1 coulomb of positive charge, or the equivalent, an excess of 1 coulomb of negative charge. The net charges on the two spheres would be identical if 1 coulomb of negative charge were moved from B to A. Because of this equivalence, as well as the historic origins of the subject, when discussing the macroscopic effect of current, we adopt the convention of assuming that the direction of the electric current is the same as the flow of the positive charges.

Resistance

Extended presentation available in the *Problem Solving* supplement

If a bulb is left connected to a battery, the battery runs down in less than a day. (The actual time depends on the particular bulb and battery used.) The bulb has the same brightness for most of this time, but near the end it dims and goes out. However, if you connect just a wire across the ends of the battery, the battery runs down in less than 1 hour, and the wire often gets too hot to touch. Because the battery is drained much faster and something is obviously happening in the wire, we infer that the current is larger through the wire than through the pathway with the bulb. The bulb offers more **resistance** to the flow of electricity. It is important to notice that the amount of current delivered by a battery depends on what is connected to the battery. Increasing the resistance in the circuit decreases the current through the battery, and decreasing the resistance in the circuit increases the current through the battery.

This notion of resistance also makes sense according to the water model. More water flows through wide pipes each minute than through narrow pipes. The bulb's filament is a thin section of wire and should offer more resistance. Our analogy also tells us that long pipes offer more resistance than short pipes. Therefore, we would expect the resistance of wires to increase with length and decrease with diameter. The other factor—not obvious from our water model—is that the resistance depends on the type of material used for the wires. These ideas are verified by experiments with wires of different sizes and materials. Table 21-1 compares the resistances of wires with the same length and diameter but made of different metals.

Resistance is a result of the interaction of the pathway with the flow of charge. The electrons in a wire feel a force due to the repulsion of the negative terminal and the attraction of the positive terminal and are accelerated; the electrons experience forces due to the electric field that exists in the wire. However, the electrons don't go very far before they bump into atoms, which causes them to lose speed and to be deflected in random directions. Although the average speed of the electrons due to their thermal motion is quite high, all of these collisions keep them from moving very fast along the wire; a typical average speed along the wire is on the order of millimeters per second. This impedance to the flow of charge determines the resistance of the wire.

As the electrons move through a wire, they undergo collisions that transfer kinetic energy to the atoms and cause the wire's temperature to increase. If enough energy is transferred, the wire gets hot enough to glow. Heating coils and elements in stoves, ovens, toasters, baseboard heaters, and lights do this. Resistors are used to control the voltages and currents in circuits used in such devices as radios and curling irons.

Resistance R is defined to be the voltage V across an object divided by the current I through the object:

Table 21-1	*The Resistance of Various 100-Meter-Long Wires with a Diameter of 1 Millimeter*
Metal	**Resistance (Ω)**
Silver	2.02
Copper	2.16
Gold	3.11
Aluminum	3.59
Tungsten	7.13
Iron	12.7
Nichrome	191

Resistors are used in electronic circuits.

$$R = \frac{V}{I}$$

resistance $= \dfrac{\text{voltage}}{\text{current}}$

Resistance is the number of volts across an object required to drive 1 ampere of current through the object, and is therefore measured in volts per ampere, a unit known as the **ohm** (Ω, uppercase Greek omega). This definition is always valid but is most useful when the resistance is constant or relatively constant. In this case this relationship is known as **Ohm's law**. The resistances of pieces of metal, carbon, and some other substances are approximately constant if they are maintained at a constant temperature. The resistance of the filament in a lightbulb increases as the filament heats up.

WORKING IT OUT *Ohm's Law*

Ohm's law provides us with a relationship between the resistance of, the potential difference (voltage) across, and the current through a circuit element. For example, suppose we have a 12-V battery and want to produce a current of 2 A in a particular circuit. What resistance should the circuit have?

$$R = \frac{V}{I} = \frac{12\ V}{2\ A} = 6\ \Omega$$

Q: If a lightbulb draws a current of 0.5 A when connected to a 120-V circuit, what is the resistance of its filament?

A: $R = V/R = (120\ V)/(0.5\ A) = 240\ \Omega$.

Ohm's law can be rearranged to find any of the three quantities when the other two are given. Suppose a heating element has a resistance of 9.6 Ω when hot. What current will it draw when connected to 120 V?

$$I = \frac{V}{R} = \frac{120\ V}{9.6\ \Omega} = 12.5\ A$$

Q: If a 3-V flashlight bulb has a resistance of 9 Ω, how much current will the bulb draw?

A: $I = V/R = (3\ V)/(9\ \Omega) = \frac{1}{3}\ A$.

The Danger of Electricity

Electricity is dangerous because our bodies are electric machines. Our muscles constrict when neurons "fire." This reaction is normally triggered by a complex chain of chemical reactions but can also occur when electric charges flow through the muscle. This process is the origin of twitches! This situation can become dangerous when the muscle is part of the heart or breathing system. In these cases the current can also affect the pacing signals, interrupting the normal processes. Currents as small as 50 milliamperes can interrupt breathing in some individuals.

Often people believe that the danger with electricity is the voltage. Although high voltages can be dangerous, it is the current through a body that is lethal. Ohm's law tells us that the current is equal to the voltage divided by the resistance of the path. In this case the body is part of the path. If the conditions are such that the total resistance is low, even a low voltage can cause a dangerously high current. The resistance of dry skin is high enough that you can place a finger across the terminals of a 9-volt battery without feeling anything. However, if you touch the terminals of the 9-volt battery to your tongue, you will feel it. The moisture of the tongue lowers the resistance.

One way of preventing high currents from passing through your body, especially through your heart, is to keep one hand in your pocket. Wearing rubber-soled shoes is also good.

A Model for Electric Current

We can use flashlight bulbs to develop a simple model of more complicated electric circuits. Each bulb serves as a visual indicator of the current through the bulb. Bulbs don't glow at all until the current exceeds a certain value; after that the brightness increases with increased current. Although the relationship between the current and the bulb's brightness is complicated, it is reasonable to assume that if one bulb is glowing more brightly than an identical bulb, it must have more current. We say that "more flow means more glow."

We begin by creating a "standard" to which we can refer; the brightness of a single bulb connected to a single battery will represent a standard current. We will also assume that all batteries and bulbs are identical.

Two bulbs can be connected to a battery so that there is a single path from the battery through one bulb, through the second bulb, and back to the other end of the battery. In this arrangement (Figure 21-9), the two bulbs are said to be in series with each other. We observe that the two bulbs have the same brightness and that they are dimmer than the single bulb in the standard circuit. This decrease in brightness indicates that there is less current in the series circuit than in our standard circuit. This conclusion is supported by the observation that the battery's lifetime in this series circuit is longer than the lifetime of the battery in the standard circuit. Therefore, the resistance of two bulbs in series is greater than that of a single bulb. (It is tempting to say that there is twice as much resistance in the series circuit because there are two bulbs. However, the resistance of the lightbulbs changes with brightness; consequently, the resistance does not double.)

Of special note is the observation that the two bulbs have the same brightness. This tells us that the current through each bulb is the same; whatever charge flows through the first bulb also flows through the second bulb. Electric charge is not being used up or lost along the pathway in agreement with the conservation of charge. The same amount of charge flows back into the battery as left it. It is the flow of charge through a bulb that causes the filament to heat up and glow. If the same amount of charge flows through each bulb each second, the two identical filaments will reach the same temperature and glow with the same brightness.

Figure 21-9 Two bulbs wired in series are equally bright but dimmer than the standard.

FLAWED REASONING

Alan connects a single bulb to a battery and notices that it is fairly bright. He then adds a second bulb in series with the first and finds that both bulbs are now dimmer and equally so. He makes the following conclusion: "The single bulb got all the current from the battery. The two bulbs must now *share* the current, each getting half. It makes sense that they should be dimmer."

Alan thinks he understands what is going on, but he has made a serious error in his reasoning. **Can you find it?**

ANSWER The word *share* has two meanings. We can share a book or we can share a pizza. If we share a book, you read it first, and then I read it. If we share a pizza, we first divide the pizza in two pieces and each take a piece. Alan believes that the current through the battery is always the same size and that the two bulbs are "sharing" this current as we would share a pizza. In fact, the bulbs share the current as we would share a book; first the current passes through one bulb (heating its filament), and then through the other (heating its filament). The two bulbs are dimmer than the single bulb because the increased resistance in the circuit reduces the current through the battery.

Figure 21-10 Two bulbs wired in parallel are equally bright and have the same brightness as the standard.

Figure 21-11 Bulbs B and C are in parallel, and the pair is in series with bulb A.

Another consequence of the equal brightness is that bulbs cannot be used to determine a direction for the flow of electricity. Everything can be explained equally well assuming a flow of negative charges in one direction (the actual situation in wires), a flow of positive charges in the other direction, or both charges flowing simultaneously in opposite directions (as happens in fluids). In fact, because only the motion of charges matters, one charge moving back and forth would also work. In household electricity the negative charges move back and forth with a frequency of 60 hertz.

Two bulbs can also be connected so that each bulb has its own path from one end of the battery to the other, as shown in Figure 21-10. In this arrangement the two bulbs are said to be wired in parallel. In contrast to the bulbs in series, the current in one bulb does not pass through the other, which can be seen by disconnecting either bulb and observing that the other is not affected. The two bulbs in parallel are equally bright, and each is as bright as our standard. Because each bulb has its own path, the battery in this circuit supplies twice as much current as the battery in the standard circuit. This can be verified experimentally; in this arrangement the battery runs down in one-half its normal lifetime.

Adding the extra bulb in parallel has increased the current through the battery, indicating that the resistance of the circuit must have decreased. Although a bulb can be correctly thought of as an obstacle to the flow of charge, the addition of one more bulb in a circuit can either increase or decrease the total resistance of the circuit, depending on how we add the bulb. When the new bulb is added in series (adding a new resistance on an existing line), the resistance of the circuit increases and the current through the battery decreases. When the new bulb is added in parallel (on a new path that did not exist before), the resistance of the circuit goes down and the current through the battery increases. Even though the new path contains resistance (the new bulb), it represents a new opportunity for flow that did not exist before.

Consider the following analogy. When a popular movie ends, people jostle to get out of the theater through the front door. The finite width of the door represents resistance to the flow of people. If the back door is also opened, the flow of people out of the building increases, even if the back door is narrow. It is a new opportunity for flow. Any path that is added in parallel lowers the total resistance of a circuit, regardless of how much resistance is contained on the new path.

Three or more bulbs can be connected in series or parallel or in combinations of series and parallel. The relative brightnesses of these bulbs can be predicted using the ideas that we have discussed. As an example, consider the combination of bulbs in Figure 21-11. Bulb A will be the brightest because all of the current must pass through it, but it will not be as bright as our standard. Bulbs B and C will be dimmer because the current splits—part going through bulb B and part through bulb C. The charge flowing into the junction J must be equal to that flowing out of the junction. This is just a consequence of the conservation of charge. Because the bulbs are identical, the resistances of the two paths are equal, and the current will split equally; bulbs B and C will be equally bright.

In general, when charge reaches the junction between two parallel branches of unequal resistance, more charge will flow through the easier branch. We say that "current favors the path of least resistance." This does not mean that all of the current takes the easier path; some current takes the more difficult path. If parallel path 1 has twice the resistance of path 2, path 1 will have one-half the current of path 2.

The German physicist Gustav Kirchhoff formalized two rules for analyzing the current in circuits. **Kirchhoff's junction rule** states:

Kirchhoff's junction rule ▶

The sum of the currents entering any junction in a circuit must equal the sum of the currents leaving that junction.

As we have seen, this is a consequence of the conservation of charge. We will discuss Kirchhoff's other rule in the next section.

If one of the paths in a circuit is a conducting wire without a lightbulb as shown in Figure 21-12, the path has very little resistance and virtually all of the current takes this path. This is known as a **short circuit**. When there is a short circuit, so little current flows through any bulbs in parallel to the short that the bulbs go out.

In general, parallel paths are not independent of each other. Making a change in one parallel path usually affects the current in the other path. The only exception to this rule is the case we started with in Figure 21-10, in which each path had direct connections to both sides of an ideal battery. These paths are said to be "parallel across the battery," and each path acts as if it has its own battery (for reasons that we will discuss later).

Figure 21-12 When the wire on the left is connected from one end of the battery to the other, producing a short circuit, the light goes out.

FLAWED REASONING

Three students are considering what will happen to the brightness of bulb B in Figure 21-11 when bulb C is removed from its socket.

Terry: "The current splits after passing through A. When C is removed, all the current must pass through B, so B will get brighter."

Judy: "Removing bulb C removes a pathway from the circuit, which will increase the resistance of the circuit. The current through the battery will go down, and B will get dimmer."

Kay: "You're both half right. The current through A will go down, but B will now get all the current instead of sharing it with C. It's like choosing between some of a large pizza or all of a small one. We need more information to make a decision."

Which student (if any) do you agree with?

ANSWER Kay correctly recognizes the conundrum. We know that the current through the battery (and hence through bulb A) must decrease when bulb C is removed, but we do not know by how much it will decrease. The model that we have developed so far does not allow us to make a prediction; we will need to add something to our model.

A Model for Voltage

If you leave your house and climb a nearby mountain, you gain a certain amount of gravitational potential energy. As you return to your house, you lose this same amount of gravitational potential energy, regardless of which path you choose to descend the mountain. It is the same with electric circuits. A 12-volt battery delivers 12 joules of electric potential energy to every coulomb of charge that passes through it. As that coulomb of charge travels through the circuit and returns to the battery, it must lose 12 joules of electric potential energy, regardless of which path it takes. This electric potential energy is delivered to the resistive elements in the circuit (such as bulbs) and converted to heat and light. This basic application of the conservation of energy is called **Kirchhoff's loop rule:**

> Along any path from the positive terminal to the negative terminal of a battery, the voltage drops across the resistive elements encountered must add up to the battery voltage.

◀ Kirchhoff's loop rule

This means that a single bulb connected to a battery will have a voltage drop equal to the voltage of the battery. We also know from experience that a bulb will glow more brightly when connected to a 12-volt battery than when it is connected to a 6-volt battery. This suggests that the bulb's brightness can be used to indicate the voltage drop across the bulb. If one bulb is brighter than another, it must have a larger voltage drop across it (assuming again that both bulbs are identical). This important idea, coupled with Kirchhoff's loop rule, allows us to answer many questions that are left unanswered by our present model for electric current.

Let's return to the conundrum presented in the last Flawed Reasoning feature. Our model for electric current allows us to predict that bulb A in Figure 21-11 will get dimmer when bulb C is removed from its socket. As Judy explains, removing bulb C removes a pathway from the circuit, increasing the resistance of the circuit. The current through the battery must decrease, and because all of the current passes through bulb A, bulb A must get dimmer. The current model, however, cannot predict what happens to bulb B. The total current in the circuit has decreased, but bulb B's share of that current has increased (from 50% to 100%). Which effect wins out? Our model for electric voltage answers the question without ambiguity. The sum of the voltage across bulb A and the voltage across bulb B must be equal to the battery voltage by Kirchhoff's loop rule. If bulb A gets dimmer when bulb C is removed, the voltage across bulb A must decrease. Because the voltage drop across bulbs A and B must be equal to the battery voltage, the voltage across bulb B must increase by the same amount and bulb B will therefore get brighter.

We are now in a position to understand the independence of branches that are parallel across an ideal battery—why each parallel branch acts as if it has its own battery. The voltage drops across each branch must equal the battery voltage by Kirchhoff's loop rule, regardless of what changes are made to the other branches. An ideal 6-volt battery is one that can maintain a potential difference of 6 volts across its terminals regardless of how much current is drawn by the circuit. In practice, real batteries have trouble delivering large currents because of limitations of the chemical processes inside the battery. This limitation manifests itself in the form of a drop in the battery voltage as the current increases. The addition of a second bulb in parallel to our standard bulb will actually dim the first bulb slightly if you are using an alkaline battery. The effect will not be noticeable with a rechargeable nickel-cadmium (NiCad) battery. The NiCad battery is said to be more ideal, and many branches can be connected across a NiCad battery before the effect appears. This assumes that each of those branches contains resistance and draws a small amount of current. If a parallel branch contains nothing but a conducting wire, it demands more current than the battery can supply and the voltage of the battery drops dramatically, causing the bulbs on the other branches to go out. This is called "shorting out the battery" and is similar to what happens on a hot summer day when Los Angeles has a "brownout" caused by the huge demand for current to run air conditioners.

The circuits in houses are wired in parallel so that electric devices can be turned on and off without affecting one another. As each new appliance is turned on, the electric company supplies more current. Each parallel circuit, however, is wired in series with a circuit breaker, as shown in Figure 21-13, to deliberately put a "weak link" in the circuit. If too many devices are plugged into one circuit, they will draw more current than the wires can safely carry. For example, the wires may heat up at a weak spot and start a fire. The circuit breaker interrupts the circuit, shutting everything off.

Some modern Christmas tree lights are wired in parallel. If one bulb burns out, it is the only bulb that goes out. The older style was wired in series; when one bulb went out, they all went out. That was rather inconvenient, because when one bulb burned out, all of them had to be tested until the defective one was found. The latest style of Christmas tree lights has the bulbs wired in series, but when a bulb burns out, a short circuit is formed across the burned-out bulb that allows the rest of the bulbs to stay lit.

Figure 21-13 Appliances are connected to household circuits in parallel. The circuit is protected from drawing too much current by a circuit breaker wired in series.

A circuit breaker protects household circuits from drawing too much current and starting a fire.

Over the years, a convention for drawing circuit elements has been developed. Like all symbols, the electrical symbols capture the functional essence of the element and omit nonessential characteristics. A lightbulb, for example, has no directional characteristic, but a battery does. Their symbols reflect this difference. Figure 21-14 gives the common symbols paired with a diagram like those we have been using.

Circuits are also drawn with sharper corners than exist in the actual circuits. This is simply a technique that has helped communication among experimenters.

Electric Power

When you buy electricity, it is energy that you are paying for. This electric energy is converted to heat, light, or motion. Why is it, then, that we talk about power? As we discussed in Chapter 7, **power** is the amount of energy used per unit time. But we need to be careful here; conservation of energy tells us that energy can be neither created nor destroyed. Formally, power is the amount of energy transformed from one form to another form divided by the elapsed time. In electricity, the energy the electrons gain from the electric field is converted to such forms as kinetic energy, heat, sound, and light. Most electric devices in our houses are rated by their power usage. Power is measured in **watts** (W). Many household lightbulbs are rated at 60, 75, or 100 watts. Electric heaters and hair dryers may use 1500 watts.

The electric meter connected between your house and the energy or power company records the energy you use, much as the odometer in your car records the miles you drive. Electric energy is usually billed at 8 to 25 cents per kilowatt-hour. A typical household with an electric range and an electric clothes dryer (but no electric heat) uses about 900 kilowatt-hours per month.

There are times when the energy losses in connecting wires become important. Sending electric energy long distances through wires from a power plant can result in significant energy losses. Two things can be done to minimize these losses. First, the wires should have as little resistance as possible, which means using large-diameter wires and low-resistance materials. Second, transformers (discussed in Chapter 22) allow the utility company to send the same energy through the wire by raising the voltage and simultaneously lowering the current. Lower current means less thermal loss in the transmission lines. Economic and engineering considerations dictate the extent to which each of these methods is used to minimize energy loss. Electricity is carried between energy companies and towns at very high voltages (up to 765,000 volts). The

Figure 21-14 Common symbols used in diagrams of electric circuits. (*Top to bottom:* resistor, battery, switch.)

This electric meter measures the energy used in a house or business. This meter reads 95,109 kilowatt-hours.

WORKING IT OUT *Electric Power*

We can obtain an expression for electric power by using the definition for electric energy we developed in Chapter 20 and the definition for current. Because energy is the charge multiplied by the voltage, we have

$$P = \frac{\Delta E}{\Delta t} = \frac{\Delta QV}{\Delta t} = \frac{\Delta Q}{\Delta t}V$$

This equation tells us that the power is equal to the current times the voltage:

▶ power = current × voltage

$$P = IV$$

Therefore, an electric appliance that draws a current of 10 A at a voltage of 120 V uses energy at a rate of

$$P = IV = (10 \text{ A})(120 \text{ V}) = 1200 \text{ W}$$

Are You On the Bus?

Q: What power is required to operate a clock radio if it draws 0.05 A from the household circuit?

A: $P = IV = (0.05 \text{ A})(120 \text{ V}) = 6 \text{ W}$.

By rearranging the expression for power, you can determine the current through a 100-W bulb:

$$I = \frac{P}{V} = \frac{100 \text{ W}}{120 \text{ V}} = 0.83 \text{ A}$$

We can obtain an alternative expression for electric power by using the relationship that the voltage is equal to the current multiplied by the resistance:

▶ power = current squared × resistance

$$P = IV = I(IR) = I^2R$$

In this form it is more obvious why a bulb in a circuit glows, yet the connecting wires do not. The current through the circuit is the same everywhere, but the section that has the highest resistance receives the most thermal energy per unit time. Because the filament is made of a fine wire of high-resistance material, it has the highest resistance; its temperature goes up, and it glows. The connecting wires also heat up, but they just don't get hot enough to glow or even feel warm.

voltage is lowered by transformers for distribution around towns and then lowered once more to the safer 120 volts used in homes.

Summary

Volta used the electric potential difference between two different metals to build the first battery. Batteries make charges flow continuously as a current, producing heat and light and running motors and other appliances.

The voltage produced by an individual cell depends on the materials used but not on its size. The size determines the amount of chemicals contained in the cell and therefore the total amount of charge that can be transferred. The

WORKING IT OUT *Cost of Electric Energy*

A device running at constant power uses an amount of energy equal to the power multiplied by the time; the longer we use it, the more energy it consumes. For example, a 60-W bulb burning for 10 h uses

$$\Delta E = P\Delta t = (60 \text{ W})(10 \text{ h}) = 600 \text{ Wh} = 0.6 \text{ kWh}$$

of electric energy. It is important to recognize that a kilowatt-hour (kWh) is a unit of energy, not power. Because 1 kW is equal to 1000 J/s and 1 h is the same as 3600 s, 1 kWh is 3.6 million J.

We can now calculate the monthly cost of electric energy for this typical household using 900 kWh per month. Let's assume that electric energy costs 10¢ per kWh:

$$\text{cost} = (900 \text{ kWh})(\$0.10/\text{kWh}) = \$90$$

Are You On the Bus?

Q: At this price, what would it cost to run a 1500-W heater continuously during an 8-h night?

A: $E = Pt = (1500 \text{ W})(8 \text{ h}) = 12,000 \text{ Wh} = 12 \text{ kWh}.$
Cost $= (12 \text{ kWh})(10¢/\text{kWh}) = \$1.20.$

voltages of cells in series are additive. Cells placed in parallel do not increase the voltage but increase the effective size of the battery.

The electricity in homes is usually supplied at 120 volts and as 60-hertz alternating current. The electricity from a battery is direct current and usually at a much lower voltage.

Electric charges can flow continuously only when there is a complete pathway, or circuit, made of conducting materials. The current is determined by the voltage and the total resistance of the circuit, $V = IR$. The voltage between two points in a circuit is equal to the change in electric potential between these two points, which is the work done in moving a unit of positive charge through the circuit. The most charge flows through the path with the least resistance. Conservation of charge requires that electricity flow out one end of the battery and back into the other. Charge does not get lost or used up along the way.

Conservation of energy requires that any electric potential energy gained by a charge as it passes through the battery must be lost by the charge as it passes through the circuit. The voltage drops across the resistive elements along any path between the terminals of a battery must sum to the voltage of the battery.

One or both types of charge may be free to move in a conductor. In plasmas and ionized liquids and gases, both charges can move, whereas only some of the negatively charged electrons are mobile in metals. For most macroscopic effects, positive charges moving in one direction are equivalent to negative charges moving in the opposite direction.

The resistance of wires increases with increasing length, decreases with increasing cross-sectional area, and depends on the type of material. Resistance R is the ratio of the voltage V across an object to the current I through the object, $R = V/I$. Resistance is measured in volts per ampere, or ohms (Ω).

A power company sells energy, not power. Power is the rate of using energy and is used to rate most electrical devices in our houses. Power is measured in watts (W) or kilowatts (kW), whereas household electric energy usage is measured in kilowatt-hours (kWh).

Everyday Physics | *The Real Cost of Electricity*

The real cost of electricity varies a great deal, depending on the source. We are most familiar with purchasing electric energy from the local energy company for use in our homes. But what is the cost of this electric energy? Examining your electric bill shows that you are charged something between 8¢ and 25¢ per kilowatt-hour. Let's use a representative value of 10¢ per kilowatt-hour. Knowing that 1 kilowatt-hour is equal to 3.6 million joules, we can divide these two numbers to learn that we can use about 36 million joules for each dollar we spend.

We can also buy electric energy packaged in a variety of batteries. In this case we pay for properties of the batteries that appeal to the manufacturer of the electric appliances. This may include the batteries' size, voltage, lifetime, or their ability to maintain current as they wear out. To discover what we are paying for the electric energy, we need to know the prices and the total electric energy that the batteries provide during their lifetimes. As shown in the text, the energy can be obtained by multiplying the total charge by the voltage rating of the battery. The total charge is obtained by multiplying the current rating by the battery's lifetime.

A standard flashlight battery—usually a carbon–zinc D cell—costs about 75¢ and has a rating of 1.5 volts. Its lifetime depends on the way it is used. Flashlight D cells are normally used for short intervals over a much longer time. Under these conditions the battery could supply 375 milliamperes for about 400 minutes, yielding a total charge of 9000 coulombs and a total energy of 13,500 joules. Therefore, you are getting 18,000 joules per dollar, 2000 times as expensive as household electricity. Alkaline batteries were a major improvement in performance over their carbon–zinc cousins, primarily in applications requiring fairly large, continuous currents, such as portable stereo radios. Alkaline batteries typically last about 50–75% longer than a standard battery, but under some conditions can last much longer. However, their costs are also higher. Are alkaline batteries a good buy? Only if their increase in cost is less than their corresponding increase in lifetime.

The development of mercury, silver, and lithium batteries allowed manufacturers to pack more energy in smaller volumes.

These devices have a much more stable current over their lifetimes, so they are especially useful in devices such as wristwatches and photocells in cameras, which require a particular value for the current. A watch battery delivers 1.3 volts and 10 millionths of an ampere for a year, giving a total energy of 410 joules. If these batteries sell for about $2, they have an energy–cost rating of 205 joules per dollar, or 175,000 times the cost of household electricity!

1. Your car battery is recharged by the alternator while the car is running. Is this electric energy cost-free?

2. A typical 6-volt lantern battery can supply a total charge of 10 ampere-hours, or about 36,000 coulombs, during its lifetime. If this battery cost $8.00, how much more expensive is this electric energy than household electricity?

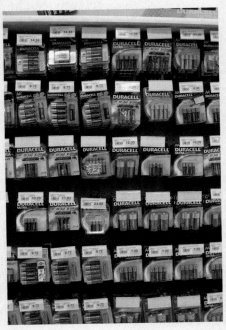

© Cengage Learning/Charles D. Winters

CHAPTER 21 | *Revisited*

When electrons move through a conductor, they collide with the atoms in the conductor, raising the average kinetic energy of the atoms and therefore the temperature of the conductor. The rise in temperature depends on the thermal properties of the material and its electrical resistance. Filaments in lightbulbs are made of thin wires of high-resistance materials to enhance the heating. If the rise in temperature is great enough, the wire glows, emitting light.

Key Terms

ampere   The SI unit of electric current, 1 coulomb per second.

complete circuit   A continuous conducting path from one end of a battery (or other source of electric potential) to the other end of the battery.

coulomb   The SI unit of electric charge. The amount of charge passing a given point in a conductor carrying a current of 1 ampere.

current   A flow of electric charge. Current is measured in amperes.

Kirchhoff's junction rule   The sum of the currents entering any junction in a circuit must equal the sum of the currents leaving that junction.

Kirchhoff's loop rule   Along any path from the positive terminal to the negative terminal of a battery, the voltage drops across the resistive elements encountered must add up to the battery voltage.

ohm   The SI unit of electric resistance. A current of 1 ampere will flow through a resistance of 1 ohm under 1 volt of potential difference.

Ohm's law   The resistance of an object is equal to the voltage across it divided by the current through it.

parallel   Two circuit elements are wired in parallel when the current can flow through one or the other but not both. Elements that are wired parallel to each other are directly connected to each other at both terminals.

power   The rate at which energy is converted from one form to another. In electric circuits the power is equal to the current times the voltage. Power is measured in joules per second, or watts.

resistance   The impedance to the flow of electric current. The resistance is equal to the voltage across the object divided by the current through it. Resistance is measured in volts per ampere, or ohms.

series   An arrangement of resistances (or batteries) on a single pathway so that the current flows through each element.

short circuit   A path in an electric circuit that has very little resistance.

watt   The SI unit of power, 1 joule per second.

Questions and exercises are paired so that most odd-numbered are followed by a similar even-numbered.

Blue-numbered questions and exercises are answered in Appendix B.

𝄼 indicates more challenging questions and exercises.

WebAssign   Many Conceptual Questions and Exercises for this chapter may be assigned online at WebAssign.

Conceptual Questions

1. Does the size of a cell affect its voltage, its lifetime, or the current it delivers?

2. When you stack three flashlight batteries in the same direction, you get a voltage of $3 \times 1\frac{1}{2}$ volts $= 4\frac{1}{2}$ volts. What voltage do you get if one of the batteries is turned to face in the opposite direction?

3. You have two 9-volt cells. What voltage would you get if you wired them in series?

4. If you have four 1.5-volt cells, what voltage would you get if you wired them in parallel?

5. What are the differences and similarities of the electricity supplied by a flashlight battery and a car battery?

6. What are the differences and similarities of the electricity supplied by a car battery and your local energy company?

7. Which bulbs in the following figure are the brightest? Which are the dimmest but still glow? Why?

8. Which bulbs in the following figure do not light? Explain.

Questions 7–10 Identical bulbs are connected to various combinations of identical batteries.

9. Which arrangements of batteries in the preceding figure will burn out most quickly? Which will last the longest? Why?

10. Which arrangements of batteries in the preceding figure will light the bulb for the longest time? Explain.

---

11. There are four different ways of connecting a battery, a bulb, and a wire so that the bulb lights. Two of these are shown in Figure 21-5. What are the other two?

12. Write a short statement describing how you would connect a battery, a bulb, and a single wire so that the bulb lights. Your statement should include all four ways without favoring any one way.

13. What is the difference, if any, in a circuit between a bulb burning out and removing the bulb from its socket?

14. Suppose that you ran speaker wires from the stereo in the family room to your bedroom, but you forgot to mark the wires. What could you do to tell which wires you should connect to which speaker?

🦶15. You are given a mystery box with six bolt heads on top, as shown in the following figure. You connect a battery and a bulb across each pair of bolt heads and find that the bulb lights for the following pairs: AB, CF, CE, and EF, but not for any of the other combinations. Draw diagrams showing all of the ways the box could be wired.

🦶16. A friend examines a box similar to the one in Question 15 and finds that the bulb lights for only the following combinations: AB, CD, and DE. Can your friend's observations be correct? Explain.

17. What is the difference between a volt and an ampere?

18. Car batteries are often rated in ampere-hours. What does this mean?

19. In the water model for electricity, what are the analogues of a battery, a switch, a wire, and a lightbulb?

20. In the water model for electricity, what are the analogues of charge, current, and voltage?

21. Which of the following affects the resistance of a wire: diameter, type of metal, length, or temperature? Explain.

22. Some power tools run poorly when connected to very long extension cords. Why? What could you do to improve the situation?

23. If the resistance connected to a battery is cut in half, what happens to the current through the battery?

24. A resistor has a current of 2 amperes when connected to a single battery. If a second identical battery is connected in series with the first, what is the current through the resistor? What is the current if the batteries are connected in parallel?

25. Why does a short circuit cause a battery to run down quickly?

26. Why is it dangerous to use a blow dryer while taking a bath?

27. If the only voltage you have is 120 volts, how could you light some 6-volt lightbulbs without burning them out?

28. In England the voltage supplied to homes is 240 volts. What would happen if you used an American lightbulb in England?

29. In a circuit with a single lightbulb, the current through the bulb is found to be 2 amperes. Is the current that flows through the battery greater than, equal to, or less than 2 amperes? Explain.

30. Two bulbs in series are connected to a battery. The current in the first bulb is found to be 1 ampere. What is the current through the battery?

31. If a second bulb is added in series to a circuit with a single bulb, does the resistance of the circuit increase, decrease, or remain the same? Does the current through the battery increase, decrease, or remain the same? Why?

32. When a bulb is added in parallel to a circuit with a single bulb, does the resistance of the circuit increase, decrease, or remain the same? Does current increase, decrease, or remain the same? Why?

33. If a string of five bulbs in series is added in parallel to a single bulb connected to a battery, you find that the brightness of the original bulb does not change. Has the current through the battery increased, decreased, or stayed the same? Has the resistance of the circuit increased, decreased, or stayed the same? Explain.

34. Which circuit offers the greater resistance to the battery, two bulbs in series or two bulbs in parallel? Why?

35. For the circuits in the following figure, is the current through the battery in circuit A greater than, equal to, or less than the current through the battery in circuit E? Explain your reasoning.

Questions 35–38 Identical batteries are connected to various combinations of identical bulbs.

36. For the circuits in the preceding figure, is the current through the battery in circuit A greater than, equal to, or less than the current through the battery in circuit B? Explain your reasoning.

61. Two bulbs have ratings of 60 watts and 120 watts. Which bulb carries the higher current? Explain.

62. Two bulbs have ratings of 60 watts and 120 watts. Which bulb has the higher resistance? Explain.

Exercises

63. What is the resistance of a lightbulb that draws 0.5 A when it is plugged into a 120-V outlet?

64. You plug a vacuum cleaner into a 120-V outlet with nothing else on the circuit. If the 15-A circuit breaker is tripped, what is the maximum possible resistance of the vacuum's motor?

65. The coils of a heater have a resistance of 10 Ω when hot. What current does the heater draw when plugged into a 120-V outlet?

66. You plug your new 9-Ω hair dryer into a 120-V outlet and trip the 10-A circuit breaker. If you plug the hair dryer into a line with a 15-A breaker, will it trip the breaker? Assume that nothing else is plugged into either circuit.

67. If a lightbulb has a resistance of 8 Ω and a current of 0.5 A, at what voltage is it operating?

68. A lightbulb has a resistance of 250 Ω. What voltage is required for the bulb to draw a current of 0.5 A?

69. Two $1\frac{1}{2}$-V batteries are connected in series to a 3-Ω resistor. How much current flows through each battery?

70. Two $1\frac{1}{2}$-V batteries are connected in parallel to a 3-Ω resistor. How much current flows through each battery?

71. Two 3-Ω resistors are connected in series to a 12-V battery. What single resistor, if connected to the battery alone (called the *equivalent resistance*), would draw this same current? What is the current through the battery?

72. A 2-Ω and a 4-Ω resistor are connected in series to a 12-V battery. What is the current through each resistor? Use Ohm's law to show that the voltage drops across the individual resistors add up to 12 V.

73. Two 4-Ω resistors are connected in parallel to a 12-V battery. Use the fact that the voltage across each of the resistors is 12 V to find the total current through the battery. What single resistor, if connected to the battery alone (called the *equivalent resistance*), would draw this same current?

74. A 3-Ω resistor is connected in parallel with a 12-Ω resistor, and the combination is connected to a 12-V battery. How much current does the battery supply?

75. A heating element is rated at 1400 W. How much current does it draw when it is connected to a 120-V line?

76. What is the power used by a toaster that draws a current of 8 A when connected to a 120-V line?

77. A DVD player has a maximum power rating of 24 W when plugged into a 120-V outlet. What is the maximum current that it requires?

78. If a clock draws a maximum current of 5 mA from a 120-V line, what is its maximum power consumption?

79. A 4-Ω resistor is connected to a 12-V battery. What is the current through the battery? What is the power dissipated by the resistor?

80. A 3-Ω resistor draws a current of 2 A when hooked up to a battery of unknown voltage. What is the battery's voltage? What is the power dissipated by the resistor?

81. What is the power rating of a heating coil with a resistance of 12 Ω that draws a current of 20 A?

82. A coffeemaker has a resistance of 12 Ω and draws a current of 10 A. What power does it use?

83. What is the resistance of a 60-W lightbulb in a 120-V circuit?

84. What is the resistance of the coil in a 1500-W heater?

85. If a hair dryer is rated at 1200 W, how much energy is used to operate the hair dryer for 8 min?

86. If a 60-W bulb is left on continuously in a secluded hallway, how much energy is used each month?

87. If electricity costs 15¢/kWh, how much does it cost to burn a 100-W bulb for one day?

88. A 1500-W heater for a sauna requires 40 min to heat the sauna to 190°F. What does this cost if electricity sells for 12¢/kWh?

22

Electro-magnetism

▶ During the 19th century, scientists discovered that electricity and magnetism were not separate phenomena but two different aspects of something called electromagnetism. It is electromagnetism that enables us to pop corn in our microwave ovens and brings television to us while we enjoy our popcorn. **What is this connection between electricity and magnetism?**

(See page 488 for the answer to this question.)

© Taxi/Getty Images

A radio telescope scanning the sky over Hawaii. The stars appear to move in circles as Earth spins on its axis.

A collection of permanent magnets.

MOST of us have played with magnets. Through play we learn that these little pieces of material attract and repel each other and that they attract some objects but have no effect on others. For instance, magnets stick to refrigerator doors, but not to the walls of a room; to paper clips, but not to paper; and so on.

Although magnetic properties are interesting, they, like electricity, appear to be rather isolated phenomena. Unlike electricity, however, magnetism does seem more permanent. And although there are similarities between these two phenomena, they seem to be separate properties of matter: a magnet with no net electric charge does not deflect the foils of an electroscope. So magnetism appears to be a different phenomenon from electricity. Yet some experimenters in the 17th century were more fascinated by the similarities than the differences between magnetism and electricity, and they searched for a connection.

Magnets

It has been known since at least the sixth century BC that lodestone, a naturally occurring mineral, attracts iron. It was also known that iron can be made magnetic by rubbing it with lodestone. The magnetized piece of iron then attracts other pieces of iron. The only elements that occur naturally in the magnetized state are iron, nickel, and cobalt. Many permanent magnets are made from alloys of these metals.

Imagine having three unmarked, magnetized rods. If we arbitrarily mark one end of one of the rods with an X, we find that it attracts one end of each of the other two magnets (Figure 22-1), and it repels their other ends. Let's mark these ends A (for attract) and R (for repel), respectively. We then find that the two A ends repel each other, the two R ends repel each other, and the A and R ends attract each other. Because the two A ends were determined in the same way using the first magnet, they should be *alike*. Therefore, the ends of the magnets (called **magnetic poles**) behave like the electric charges we studied in Chapter 20. We can summarize the behavior of the magnetic poles with the following simple statement:

magnets ▶

> Like poles repel; unlike poles attract.

Both electricity and magnetism seem to have two kinds of "charge." In electricity, however, it is possible to separate the two charges, or at least to put more of one kind of charge on one object than on the other. But in magnetism the poles always come in pairs that have the same strength.

Are *You* On the Bus?

Q: How might you search for a third kind of magnetic pole?

A: You could search for a magnetized material that attracts or repels both north poles *and* south poles. No such material has been discovered.

Suppose you try to separate the poles by breaking a magnet in half. You get two magnets, and each of these new magnets has two poles (Figure 22-2). If you break each of these in half, you get four magnets, and so on. Even if you continue dividing down to the atomic level, you always obtain magnets with two poles. It does not seem to be possible to isolate a single magnetic pole, a **magnetic monopole**. Many extensive searches for magnetic monopoles have been conducted. Although the existence of a magnetic monopole is not ruled out by the present, well-established theories, all searches have ended in failure and it does not seem likely that one will be discovered. The discovery of a magnetic monopole would most likely result in a Nobel Prize.

In naming the ends, or poles, of the magnets, we make use of the observation that a freely swinging magnet aligns itself along the north–south direction. The end that points north is called *north* and the other *south*. The compasses used for navigation are simply tiny magnets that are free to rotate. Magnets have been used as compasses since the 11th century, but not until 1600 did William Gilbert, an English physician and scientist, hypothesize that they work because Earth itself is a giant lodestone. Gilbert even made a spherically shaped piece of lodestone to show that a compass placed near it behaved as it did on Earth.

Magnets are surrounded by **magnetic fields** in the same way electric charges are surrounded by electric fields. These magnetic fields can be detected using a small compass. The direction of a magnetic field at a particular point is the direction in which the north pole of the compass points when placed at this point, and the torque that aligns the compass is a measure of the strength of the magnetic field. The photograph in Figure 22-3(a) was obtained by sprinkling iron filings on a piece of glass placed over a bar magnet. The iron filings line up along the direction of the magnetic field.

Figure 22-1 Magnetic poles can be identified by their attraction to and repulsion by known poles.

Q: How would the photograph in Figure 22-3 change if the poles of the magnet were reversed?

A: The photograph would not change because the iron filings do not tell us which way the field points.

(*a*)

(*b*)

Figure 22-2 Dividing a magnet results in two smaller magnets, each with both poles.

Figure 22-3 (a) Iron filings line up along the magnetic field lines surrounding a bar magnet. (b) The drawing indicates the direction of the magnetic field outside the magnet.

Figure 22-4 (a) Iron filings show that the magnetic field lines surrounding a long, straight, current-carrying wire are circles around the wire. The dark line at the left is the shadow of the wire. (b) The diagram shows the direction of the magnetic field for a current flowing away from you (into the page).

(a)

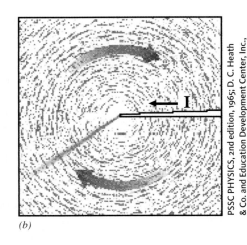

PSSC PHYSICS, 2nd edition, 1965; D. C. Heath & Co. and Education Development Center, Inc., Newton, MA.

(b)

Electric Currents and Magnetism

Although electricity and magnetism were well known for centuries, any connection between the two phenomena eluded experimenters until the 19th century. In 1820 Hans Christian Oersted, a Danish scientist, discovered a connection while performing a lecture demonstration: a compass needle experiences a force when it is brought near a current-carrying wire. This means that the current-carrying wire produces a magnetic field in the surrounding space. This discovery was fascinating and yet confusing. It was known that stationary charges did not produce magnetic fields, so the motion of the charges must produce the field. The photograph in Figure 22-4(a) shows that the magnetic field lines form circles about the wire.

The direction of the magnetic field is given by the *right-hand rule*: if you grasp the wire with the thumb of your right hand pointing in the direction of the positive current, your fingers encircle the wire in the direction of the magnetic field, as shown in Figure 22-5.

Bending the wire produces different field patterns. For instance, the magnetic field of a circular loop of wire is shown in Figure 22-6. The contributions to the magnetic field from each segment of the wire add together inside the loop to produce a field pointing into the page. Adding more loops to form the cylindrical structure, called a *solenoid* (Figure 22-7), produces a magnetic

Figure 22-5 The direction of the magnetic field is given by the right-hand rule. If your thumb points in the direction of the current, your fingers encircle the wire in the direction of the field.

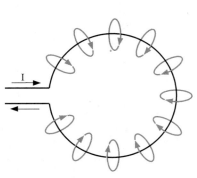

Figure 22-6 The magnetic field of this single loop of wire points into the page inside the loop and out of the page outside the loop.

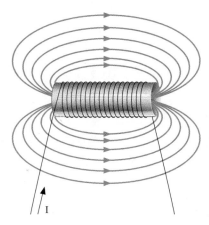

Figure 22-7 The magnetic field of a solenoid is like that of a bar magnet.

field like that of the bar magnet in Figure 22-3. The magnetic field inside the solenoid is relatively strong and rather uniform except near the ends.

Observations like these led French physicist André Ampère to suggest that *all* magnetic fields originate from current loops. We now believe that magnetism originates in current loops at the atomic level. In a simplified model of the atom, we visualize these currents arising from electrons orbiting atomic nuclei or spinning about their axes. The macroscopic magnetic properties of an object are determined by the superposition of these atomic magnetic fields. If their orientations are random, the object has no net magnetization; if they are aligned, the object is magnetized.

Making Magnets

Like electric charge, our universe is filled with magnetism. We do not normally detect the magnetism because of the random orientation of the atomic current loops. The atoms can be aligned, however, by the presence of a magnetic field. In most materials this alignment disappears when the magnetic field is removed. However, in some materials the atoms remain aligned and therefore retain their macroscopic magnetism.

A piece of iron can be magnetized by placing it in a strong magnetic field, by stroking it with a magnet, or by hitting it while it is in a magnetic field. Tapping an iron rod on the floor while holding it vertically will magnetize it because Earth's magnetic field has a vertical component. You can check for the resulting magnetization by holding a compass near the top and bottom of a metal object, such as a filing cabinet. The opening and closing of the file drawers provides the tapping; Earth's magnetic field does the rest. Figure 22-8 shows the orientation of a compass near the top and bottom of a file cabinet. On the other hand, an iron object that has been magnetized can lose its magnetism if it is dropped because the impact tends to randomize the alignment of the atoms. Heating the object can also randomize the atomic magnetic fields.

A versatile magnet can be constructed by wrapping wire around an iron core and connecting the ends of the wire to a battery, as illustrated in Figure 22-9. When current passes through the coil of wire, it generates a magnetic

The top of the cabinet is a magnetic south pole.

The bottom of the cabinet is a magnetic north pole.

© Cengage Learning/David Rogers (both)

Figure 22-8 The magnetic fields near the top and bottom of a metal file cabinet point in opposite directions.

Figure 22-9 A simple electromagnet can be constructed from a nail and some wire.

A large electromagnet is used to load cargo containers onto a ship.

Figure 22-10 A current-carrying wire produces a magnetic field surrounding the wire. Notice that the compasses above and below the wire point in opposite directions.

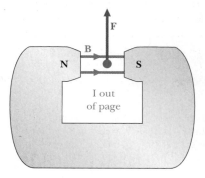

Figure 22-11 The magnetic field of the horseshoe magnet exerts a force on the current-carrying wire, causing it to jump out of the jaws of the magnet.

field along the axis of the coil. The magnetic field that is induced in the iron adds to that of the solenoid, increasing the strength of the magnetic field. These **electromagnets** are useful because they can be turned on and off, the strength of the field can be varied by varying the current, and they can produce large magnetic fields. Electromagnets can be used for such tasks as moving cars or sorting scrap iron from nonmagnetic metals.

The Ampere

A current-carrying wire exerts forces on compass needles (Figure 22-10). Therefore, by Newton's third law, a magnet should exert a force on the wire. This effect was quickly verified by experiment. A wire between the jaws of a large horseshoe magnet jumps out of the gap when the current is turned on (Figure 22-11). This force is biggest when the wire is perpendicular to the magnetic field. The force on the wire is always perpendicular to the wire and to the magnetic field.

These experiences indicate that two current-carrying wires should exert forces on each other. They do. The magnetic field produced by each current exerts a force on the other. If the currents are in the same direction, the wires attract each other (Figure 22-12); if the currents are in opposite directions, they repel.

This effect is used to define the unit of current—the basic electrical unit in the metric system. Consider two long parallel wires separated by 1 meter and carrying the same current. If the force between these wires is 2×10^{-7} newton on each meter of wire, the current is defined as 1 **ampere** (A). The *coulomb* (C) is then defined as the amount of charge passing a given point in one of these wires during 1 second. This is the same as the charge on 6.25×10^{18} protons.

The source of the force on each wire is the magnetic field produced by the other wire. We can also use the interacting wires to define a field strength for magnetism. The strength of the magnetic field at a distance of 1 meter from a long straight wire carrying a current of 1 ampere is 2×10^{-7} **tesla** (T). Another unit for magnetic field strength, the **gauss** (G), is often used instead of tesla, where 1 tesla = 10,000 gauss.

(a)

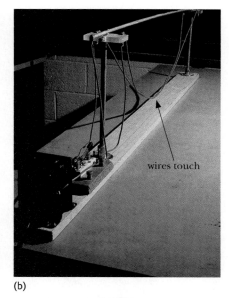

(b)

Figure 22-12 (a) When the switch is open, there is no attractive force between the wires. (b) When the switch is closed, the currents are in the same direction, and the wires are attracted to each other.

Everyday Physics | *Superconductivity*

Under normal conditions all conductors exhibit resistance to the passage of electric current. This makes sense if you visualize the conduction electrons bumping their way through a metal consisting of an array of vibrating atoms. The electrons are accelerated along the wire by the electric potential difference provided by a battery and then lose their acquired speeds in collisions. The surprise is that the electrical resistance of some materials goes to zero at very low temperatures; they become *superconductors*.

This knowledge is not a new breakthrough; superconductivity was first discovered by Dutch physicist Heike Kamerlingh Onnes in 1911. Three years earlier Onnes had developed a process for liquefying helium and was therefore able to study the properties of materials at very low temperatures. The first superconductor was solid mercury, which lost its electrical resistance at a critical temperature of 4.15 K. Experimenters searched for metals or alloys that exhibited superconductivity, and new superconductors were found with critical temperatures as high as 23 K.

In 1972 three American physicists, John Bardeen, Leon Cooper, and J. Robert Schrieffer, received the Nobel Prize for explaining superconductivity in 1957. Their theory—called the BCS theory—showed how electrons pair up to create a resonance effect that allows the electron pairs to travel effortlessly through the material. The electrical resistance does not simply drop to a very low value; it drops to zero! Once a current has been established in a superconducting material, it will persist without any applied voltage. Such supercurrents have been observed to last for years, which can have great practical applications in the generation of magnetic fields. Large magnetic fields require very large currents. In ordinary materials this means generating a lot of thermal energy at high cost and diverting the heat to avoid melting the magnets. In contrast, once the large currents have been established in superconducting magnets, they can be disconnected from the power supply; no further electric energy is needed.

Theoretical models for superconductivity predicted an upper limit for the critical temperature of about 30 K. Had this upper limit been true, superconductivity would have remained in the domain of very low temperatures. The only gases with boiling points this low are helium (which is expensive) and hydrogen (which is explosive). In 1986, nearly 80 years after the discovery of the first superconductor, a major new class of superconductors was discovered with much higher critical temperatures. These new superconductors

Argonne National Laboratory and the U.S. Department of Energy

A permanent magnet levitates above a superconductor because its magnetic field cannot penetrate the superconducting material.

are *ceramics* made from copper oxides mixed with such rare-earth elements as lanthanum and yttrium and have critical temperatures as high as 125 K. This new limit is significant because this temperature is higher than the boiling point of nitrogen, an abundant gas that is relatively inexpensive to liquefy and safe to use, allowing liquid nitrogen to be used to keep a material superconducting. A tremendous increase in the use of superconductors will occur if a superconducting material can be made with a critical temperature above room temperature (or at least above those obtained by ordinary refrigeration).

One important potential application of superconductivity is in the transportation of electric energy. With conventional transmission lines, much of the electricity produced at a distant power plant is lost to resistive heating effects in the wires that carry this energy to our homes. A superconducting transmission line would eliminate these losses. The problem, of course, is that we need to cool these conduits in order to make them superconducting. However, if we can use ordinary refrigeration, the cooling costs will be much less than the costs of resistive losses.

Superconductors have a second property that is important. They expel magnetic fields when they become superconducting. This also means that magnetic fields cannot penetrate the superconductors, and they will therefore repel magnets. This is dramatically shown by floating a permanent magnet above a superconductor, as shown in the photo above. This effect may find applications in improving the magnetic levitation of trains for high-speed transportation. A prototype train has been constructed in Japan using helium-cooled superconducting magnets to levitate the train as well as to propel it at speeds up to 300 mph.

Many other applications have been proposed. The switching properties of superconducting materials could have a large impact on the field of computer electronics. It may also be possible to construct superconducting generators and motors. These new superconducting materials may also find applications in the field of medical imaging. However, many of these applications will require major technological breakthroughs in fabrication (for instance, ceramics are brittle and therefore difficult to form into wires) and in finding materials that will carry larger currents.

1. The boiling point of nitrogen is 77.36 K. Why is it important that superconductors have a critical temperature higher than this value?

2. List three possible applications of high-temperature superconducting materials.

Figure 22-13 Earth is a giant magnet with its south magnetic pole in the Northern Hemisphere. This early view of Earth's magnetism has been replaced by one in which the magnetic field is generated by electric currents in Earth's core.

The magnets that are used to hold notes or pictures on refrigerator doors produce fields on the order of 0.3 tesla, whereas large laboratory magnets produce fields of 2.5 teslas. The theoretical limit for a permanent magnetic field is 5 teslas. Electromagnets made with ordinary wires have produced steady fields of 34 teslas, whereas those made with superconducting wires have not yet exceeded 22 teslas. Combinations of these two types of magnets have reached 45 teslas. At the upper extreme, pulsed magnets have produced fields of 70 teslas. Even higher fields approaching 1000 teslas have been generated for extremely short times, but the forces are so large that the magnets destroy themselves.

The Magnetic Earth

As mentioned earlier, a simple compass shows that Earth acts as if it had a huge magnet in its core, as illustrated by Figure 22-13. The strength of Earth's magnetic field at Earth's surface is typically 5×10^{-5} tesla (0.5 gauss). In the United States its horizontal component points generally northward and an even larger component points downward.

Measurements of the magnetic field show that one of Earth's magnetic poles is located in northeastern Canada just north of Hudson Bay, about 1300 kilometers from the geographic North Pole. Earth's other magnetic pole is almost directly on the other side of Earth. A line through Earth connecting these two poles is tilted about 12 degrees from Earth's rotational axis, which passes through Earth's geographic North and South Poles.

Q: Is the magnetic pole located in northern Canada a north or a south magnetic pole?

A: Because this pole attracts the north pole of a compass and because opposite poles attract, this magnetic pole must actually be a south magnetic pole. However, the geographical location of the pole is still known as the magnetic North Pole.

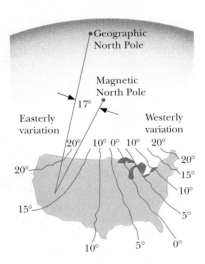

Figure 22-14 The magnetic variation between the directions to the geographic and magnetic North Poles changes across the United States.

The tilt of Earth's magnetic axis from its rotational axis makes finding true north with a compass a bit complicated. The only time your compass will point to true north is when the magnetic pole is between you and the geographic pole. Because of local variations in the magnetic field, this situation occurs along an irregular line running from Florida to the Great Lakes, as shown in Figure 22-14. If you are east of this line, your compass will point to the west of true north. If you are west of this line, your compass points to the east of true north. The difference in the directions to the magnetic and geographic poles is known as the *magnetic variation*. As an example, the magnetic variation in Bozeman, Montana, is currently 15.5 degrees east.

Airplane pilots measure magnetic headings and refer to their aeronautical maps to find the true headings. The large numbers painted on the ends of runways correspond to the magnetic directions of the runways divided by 10 and rounded.

At one time scientists imagined that Earth's magnetic field was caused by magnetized solid iron in Earth's interior, but they now believe this cannot be true. Earth's interior is known to be hot enough that the iron and nickel are in a liquid state. In a liquid state, the atomic magnetic fields do not remain aligned but rather take on random orientations, thus eliminating a macroscopic magnetic effect.

The best model is that large electric currents circulating in the molten interior cause Earth's magnetic field. Such currents could easily produce the field that we observe on Earth, as well as the general features of the magnetic fields of the other planets. However, there are some difficulties with this theory. No one understands the details of the mechanisms for producing and maintaining the currents.

The most puzzling aspect of Earth's magnetic field is its reversals, when the North and South Poles switch locations. There is strong evidence that Earth's magnetic field has reversed directions 171 times in the last 17 million years. This evidence comes from the rocks on both sides of the mid-Atlantic rift. As molten rock emerges from the rift, it cools and solidifies, and at that moment the direction of Earth's magnetization is locked into the rocks and preserved. Samples from the ocean's floor show that the directions of the magnetic fields in the rocks alternate as we approach the rift. Although we know the reversals have occurred, no one has been able to propose a satisfactory mechanism for these reversals or account for the source of the vast amount of energy required to reverse the field.

Charged Particles in Magnetic Fields

◀ Extended presentation available in the *Problem Solving* supplement

Recall that a magnet has no effect on a charged object other than the normal electrostatic attraction between neutral and charged objects. On the other hand, a current-carrying wire in a magnetic field does experience a force. Because an electric current is a stream of charged particles, the motion of the charges must be important. As bizarre as it seems, the magnetic force on a charged particle is zero unless the charge is moving!

In addition, the direction of the magnetic force is probably not the direction you would predict. Recall that the current (the motion of charged particles) is along the wire in Figure 22-11 but that the wire experiences a force in a direction perpendicular to the wire and perpendicular to the magnetic field. Thus, we expect a charged particle moving in a magnetic field to experience a force that is at right angles to its velocity and to the magnetic field.

The strength of the force depends on the charge q, the strength of the magnetic field B, the speed v, and the angle between the field and the velocity. The magnetic force is maximum when the field and velocity are perpendicular:

$$F_{max} = qvB$$

◀ maximum magnetic force = charge × speed × magnetic field

When the velocity and the magnetic field are parallel, the force is zero.

The magnetic force produces some interesting phenomena. First, because the force is always perpendicular to the particle's velocity, it never changes the particle's speed, only its direction. If the particle's velocity is parallel to the magnetic field, the particle does not experience a force, and it moves in a straight line. On the other hand, if the velocity is perpendicular to the field, the particle experiences a centripetal force that causes it to move in a circular path [Figure 22-15(a)]. If the velocity has components parallel and perpendicular to the magnetic field, these two motions will superimpose, and the particle will follow a helical path [Figure 22-15(b)] along the magnetic field.

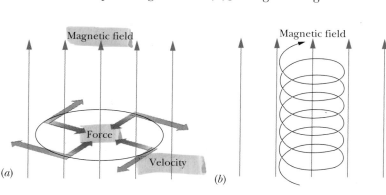

Figure 22-15 (a) The charged particle follows a circular path if its velocity is perpendicular to the magnetic field. (b) The path is helical if its velocity has components both perpendicular and parallel to the field.

The aurora borealis is caused by cosmic rays following magnetic field lines toward the magnetic North Pole.

The magnetic force on charged particles is the cause of dramatic effects known as the *aurora borealis* (northern lights) and the *aurora australis* (southern lights). As charged particles, mostly from the Sun but also from outer space, approach Earth, they interact with Earth's magnetic field. This interaction causes the charged particles to follow helical paths along Earth's magnetic field so that they strike Earth's atmosphere in the regions of the magnetic poles. The collisions of these cosmic rays with the oxygen and nitrogen molecules in the atmosphere produce the spectacular evening light shows high over the north and south magnetic poles.

The magnetic forces on charged particles are important in many scientific and technologic devices as varied as containment vessels used in developing the future technology of fusion reactors, particle accelerators used in research, and television sets in our homes. We will discuss many of these in the remaining chapters.

WORKING IT OUT   *Magnetic Force*

If a proton is moving perpendicular to a uniform magnetic field, it will experience a magnetic force that will pull the proton out of its straight-line path and cause it to travel in a circle of radius r_c, known as the "cyclotron radius." Find the cyclotron radius for a proton moving at 3×10^6 m/s perpendicular to a magnetic field of 1.5 T.

The maximum magnetic force on the proton (for motion perpendicular to the field) is given by

$$F = qvB = (1.6 \times 10^{-19}\,\text{kg})(3 \times 10^6\,\text{m/s})(1.5\,\text{T}) = 7.2 \times 10^{-13}\,\text{N}$$

This very small force causes a large acceleration because it is acting on a small mass (the proton).

$$a = \frac{F}{m} = \frac{7.2 \times 10^{-13}\,\text{N}}{1.67 \times 10^{-27}\,\text{kg}} = 4.3 \times 10^{14}\,\text{m/s}^2$$

This acceleration is centripetal, so we can now solve for the cyclotron radius:

$$a = \frac{v^2}{r_c}$$

$$r_c = \frac{v^2}{a} = \frac{(3 \times 10^6\,\text{m/s})^2}{4.3 \times 10^{14}\,\text{m/s}^2} = 0.021\,\text{m} = 2.1\,\text{cm}$$

Magnetism and Electric Currents

In the evolution of the physics world view, a small number of basic themes have emerged that reflect our biases but also fuel our searches. One such theme is symmetry. A number of times, progress has been made while searching for symmetric effects. Early in the development of our understanding of electricity and magnetism, the questions of symmetry haunted some experimenters. In particular, they asked: given that an electric current produces a magnetic field, does a magnetic field produce an electric current? To investigate this, you might try wrapping some wire around a magnet and connecting the wire to an *ammeter*, an instrument for measuring current. Although similar experiments were performed with large magnets and sensitive instruments, no effect was found.

British scientist Michael Faraday discovered the connection in 1831. He found that motion is the key to producing an electric current with a magnet. If

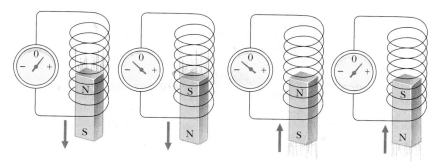

Figure 22-16 A magnet inserted into or removed from a coil of wire produces a current. (When the needle is centered, there is no current.)

a wire is moved through a magnetic field (but not parallel to the field), a current is produced in the wire. This current is due to the motion of the wire in the magnetic field; there are no batteries. The current is largest if the motion is perpendicular to the field and increases with the speed of the wire. Having the benefit of hindsight, we see that this is understandable because of the magnetic force on the charges in the wire.

Because the principle of relativity (Chapter 9) must be valid for electricity and magnetism, we know that it is the *relative* motion of the wire and the magnet that is important. It doesn't matter which is moving relative to the laboratory and which is stationary. Therefore, we should be able to generate a current in a stationary wire by moving the magnet. This principle can be demonstrated with a simple experiment. If a bar magnet is placed in a coil of wire and quickly withdrawn, a current is generated that is easily detected. Quickly inserting the magnet into the coil generates a current in the opposite direction. Reversing the direction of the bar magnet also reverses the direction of the current. The four possibilities are shown in Figure 22-16.

Faraday also discovered that motion is not the only way of producing a current with a magnetic field. A current is produced if the strength of the magnetic field varies with time, even when there is no relative motion of the wire and the magnet. An increasing field produces a current in one direction; a decreasing field produces a current in the opposite direction.

After a long series of experiments conducted over several years, Faraday was able to generalize his results in terms of magnetic field lines. These lines are visualized as pointing along the direction of the magnetic field at all points in space. The number of lines in a given region of space represents the strength of the magnetic field. The lines are closer together where the magnetic field is stronger.

Faraday showed that if the number of magnetic field lines passing through a loop of wire changed *for any reason*, a current was produced in the loop (Figure 22-17). The voltage (and hence the current) generated in the loop depends

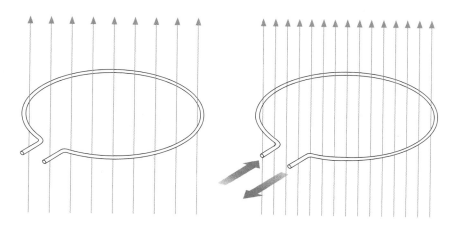

Figure 22-17 Changing the number of magnetic field lines passing through a loop of wire produces a current. The direction of the current is given by Lenz's law.

on how fast the number of field lines passing through the loop changes—the faster the change, the larger the voltage. This explanation of these phenomena is now known as *Faraday's law*.

Faraday's law ▶

The direction of the current induced in the coil by the changing number of magnetic field lines is given by *Lenz's law*, which states that the current always produces a magnetic field to oppose the change that created it. For example, in Figure 22-17 the number of lines in the upward direction is increasing. Therefore, the induced current will produce a magnetic field that points downward to try to cancel out the increase. Notice that the current shown in Figure 22-17 does this. If the number of lines had been decreasing, the current would be induced in the opposite direction to produce a magnetic field in the upward direction to try to maintain the original field.

Lenz's law ▶

Are You On the Bus?

Q: What would you see on the ammeter if you dropped a magnet through the coil in Figure 22-16?

A: As the magnet enters the coil, the needle will swing to one side because of the increase in the number of field lines passing through the coil. As the magnet exits, the needle will swing to the other side because the number of lines decreases.

FLAWED REASONING

The following question appears on the midterm: "A copper loop lies flat on a table. If the north pole of a bar magnet is lowered quickly toward the loop, as shown in Figure 22-18, a current is induced in the loop in the direction given by Lenz's law. Find this direction and explain how you used Lenz's law."

Figure 22-18

Jose gives the following answer: "Magnetic field lines leave the north pole of a bar magnet, so the field lines through the loop are directed downward. Lenz's law states that an induced magnetic field will be generated in the opposite direction—that is, upward. By the right-hand rule, the induced current must be counterclockwise to produce this field."

The induced current is counterclockwise, as Jose claims, but not for the reason given. **Explain Lenz's law to Jose.**

ANSWER There are two magnetic fields in the loop: the *external* field of the bar magnet and the *induced* field due to the current. Jose believes that the induced field opposes the external field, but this is not what Lenz's law asserts. The induced field opposes any *change* in the external field, not the field itself. Jose gets the correct direction by luck. If the bar magnet is moving upward, Jose's reasoning yields the wrong answer. The external field in the loop is downward and growing weaker. Jose predicts that the induced field is upward to oppose the downward external field. Lenz tells us that the induced field is downward to oppose the weakening of the external field.

Transformers

These discoveries about magnetism—especially its connections with electricity—have many practical uses. For example, they are fundamental to the operation of transformers used to change the voltage of alternating-current electricity. The same amount of electric power can be delivered through wires at low voltage and high current as at high voltage and low current (Chapter 21). For instance, 10 amperes at 12 volts supplies the same power as 1 ampere at 120 volts. The particular choice of what amperage and voltage are used depends on the circumstances. The large cylindrical containers on power poles are transformers for reducing the voltage to 120 volts before it enters our homes and businesses.

The schematic diagram in Figure 22-19 shows the essential features of a transformer. One coil, called the *primary coil*, is connected to a source of alternating current. The alternating current in the primary coil produces an alternating magnetic field that is transmitted to the *secondary coil* by the iron core. This alternating magnetic field produces an alternating voltage in the secondary coil.

The size of the voltage produced in the secondary coil depends on the ratio of the number of loops in the two coils. A transformer designed to reduce the voltage by a factor of 2 would have one-half as many loops in the secondary as in the primary coil.

Neighborhood transformers reduce the voltage for household use.

© Cengage Learning/George Semple

Q: Why won't a transformer work with direct-current electricity?

A: Direct current will not produce the varying magnetic field necessary to induce a current in the secondary coil.

Generators and Motors

The electric generator is another application of Faraday's discovery. If we rotate the loop of wire in Figure 22-17 by 90 degrees, the number of magnetic field lines passing through the loop drops to zero. The change in the number of field lines passing through the loop produces a voltage around the loop, which produces a current. The loop is usually rotated by a steam turbine or falling water in a hydroelectric facility. The steam can be produced by many means: by burning wood, coal, petroleum, or natural gas or by using heat from the Sun or nuclear reactors.

Figure 22-20 contains a series of drawings of the magnetic field and a wire loop in an electric generator that illustrates how the rotating loop produces electric voltages (and currents). Figure 22-20(a) shows the plane of the loop parallel to the magnetic field lines. The number of lines passing through the loop in this orientation is zero. As the loop rotates at a constant speed, the number of lines initially increases rapidly, producing a large voltage. As it continues to rotate (b), the number of lines passing through the loop continues to increase but at a slower rate. The voltage drops to zero when the plane of the loop is perpendicular to the field lines (c) and the number of lines through the loop is a maximum. The number of lines through the loop now decreases (d), and the voltage increases in the opposite direction. It increases to a maximum (e) and then returns to zero when the plane of the loop is perpendicular to the field lines once again. The voltage increases to a maximum in the opposite direction, and the entire cycle repeats. This generator produces the alternating voltage shown in Figure 22-21(a).

A simple change in the way the voltage is carried to the external circuit converts the generator to produce pulsating direct current [Figure 22-21(b)].

Figure 22-19 The essential features of a transformer.

Figure 22-20 The number of magnetic field lines passing through a loop of wire changes as it rotates. This produces the alternating voltage shown in Figure 22-21.

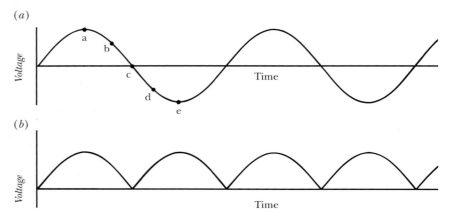

Figure 22-21 (a) A loop rotating in a magnetic field produces an alternating voltage. The letters show the voltage produced when the loop has the positions shown in Figure 22-20. (b) The pulsating direct current produced by using a commutator.

Figure 22-22 Schematic of a direct-current generator.

A connector (called a *commutator*), shown in Figure 22-22, reverses the connections from the loop to the outside circuit each half turn. This pulsating current can then be electronically smoothed to produce a constant direct current like that from a battery.

A direct-current motor is basically a direct-current generator run backward. In fact, the first such motor was produced during an exhibition in 1873 when a technician setting up a demonstration of generators hooked one up the wrong way and "discovered" a motor! In a generator we rotate the loop in a magnetic field, which produces a voltage that moves electric charges. In a motor the sequence is reversed; we apply a voltage, causing the charges to move. This current in a magnetic field produces a force that rotates the loop.

Figure 22-20 can be used to illustrate the operation of a motor. When a voltage is applied to the loop when it is in the position shown in Figure 22-20(a), the magnetic field exerts a torque on the current-carrying loop. The forces on the long sides of the loop are in opposite directions because the currents are in opposite directions. The torque decreases as the loop rotates and becomes zero when the plane of the loop is vertical. If nothing changes as the loop coasts through this position, it will oscillate and eventually come to a stop. However, the commutator reverses the direction of the current so that the torque continues to act in the same direction.

The similarity of a motor and a generator is important in some electric-powered vehicles. While the vehicle is speeding up or cruising, the engine acts as an electric motor. However, when the brakes are applied, the connections

Everyday Physics | *"Wireless" Battery Charger*

An electric toothbrush has a base designed to hold the tooth-brush handle when not in use. As shown in the following figure, the handle has a plastic cylindrical hole that fits loosely over a matching plastic cylinder on the base. How does the toothbrush get charged when there are no metal contacts? When the handle is placed on the base, a changing current in a solenoid inside the base cylinder (the primary coil) induces a current in a coil inside the handle (the secondary coil). This induced current charges the battery in the handle. The size of the induced current is amplified by inserting a core of ferromagnetic material, such as iron, into the primary coil. The primary coil then magnetizes the iron core such that its poles reverse 120 times each second (for 60-cycle current).

This causes a greater change in the magnetic field inside the secondary coil.

Wireless charging is used in a number of other "cordless" devices. One significant example is the inductive charging that is used by some electric car manufacturers that avoids direct metal-to-metal contact between the car and the charger.

1. Why is N_H, the number of coils in the handle, less than N_B, the number of coils in the base unit?

2. Find the ratio N_B/N_H if the handle contains a 9-volt recharge-able battery.

This electric toothbrush uses magnetic induction to charge its battery. The primary coil (coil 1) of N_B turns is in the base unit and the secondary coil (coil 2) of N_H turns is in the handle.

Gilette

to the engine are changed so that it acts as a generator to recharge its batteries. The current exerts a torque on the engine, which in turn exerts a torque on the wheels to stop the vehicle. This is not a perpetual-motion machine because only part of the energy is recovered. However, regenerative braking can extend the range of the vehicle.

A Question of Symmetry

The connection between electricity and magnetism appears to be complete— a *changing* magnetic field produces an electric current, and an electric current produces a magnetic field. Notice, however, that the situation is not quite symmetric. It requires a changing magnetic field to produce the electric current, but the electric current produces a steady magnetic field.

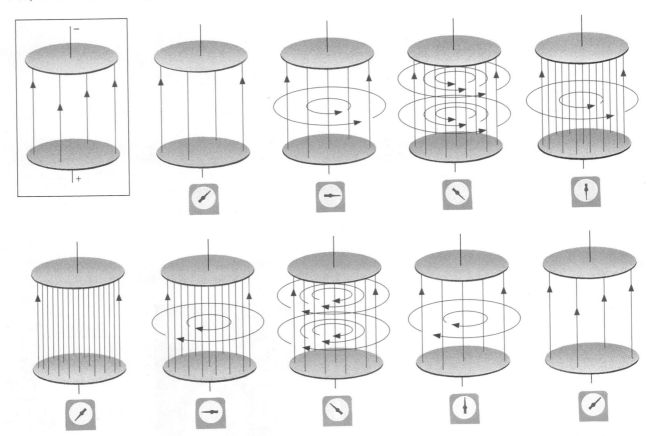

Figure 22-23 A steady electric field between the parallel plates shown in the boxed diagram at the left does not produce a magnetic field. An increasing electric field produces a magnetic field in one direction, whereas a decreasing field produces a magnetic field in the opposite direction.

Actually, there is more to this connection that becomes apparent only when everything is expressed in terms of the electric and magnetic fields. Consider two parallel plates connected to a battery, as shown in Figure 22-23. As current from the battery builds up charges on the plates, an electric field is generated in the region between the plates. Even though no charges are flowing between the plates, a magnetic field is produced in the region surrounding the plates that matches the magnetic field surrounding the wires. Therefore, the origin of the new field cannot be the charges. Furthermore, the magnetic field disappears when the current stops. The electric field between the plates continually increases as long as charge is flowing to the plates and remains constant when the charge stops flowing. Therefore, the magnetic field is produced by a *changing* electric field. As the charges leave the plates, the electric field decreases and a magnetic field in the opposite direction is produced.

We can make the analogy closer by looking at the region between the poles of an electromagnet. As the magnetic field between the poles increases, an electric field is produced in the surrounding region (Figure 22-24). When the magnetic field reaches its maximum strength and is no longer changing, the electric field disappears. As the magnetic field decreases, an electric field in the opposite direction is produced.

It is important to realize that a changing magnetic field produces an electric field in *empty* space. There is no need for a wire. If a wire is present, the electric field exerts forces on the charges within the wire and produces a current. But the important point is that even in the absence of the wire, the electric field is present.

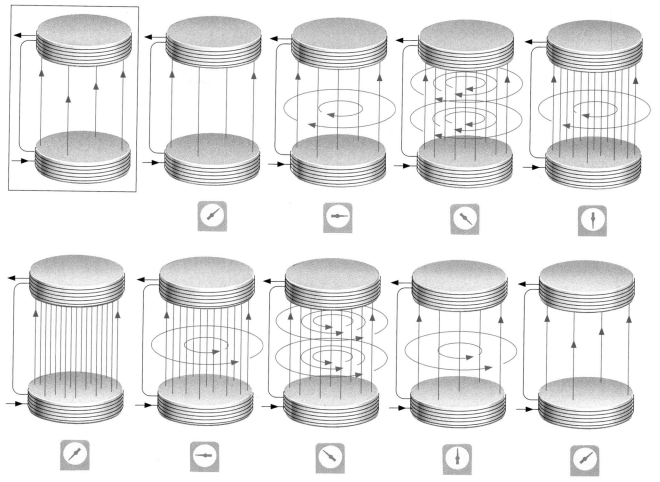

Figure 22-24 A steady magnetic field between the poles of the electromagnet shown in the boxed diagram at the left does not produce an electric field. An increasing magnetic field produces an electric field in one direction, whereas a decreasing magnetic field produces an electric field in the opposite direction.

If we use these results and focus on the fields rather than the currents, the situation is completely symmetric: a *changing* magnetic field generates an electric field, and a *changing* electric field generates a magnetic field. There is an intimate relationship between electricity and magnetism.

Electromagnetic Waves

If the magnetic field changes at a constant rate (that is, if it changes by the same amount each second), the electric field that is produced is constant. A rapidly changing magnetic field produces a large electric field, and a slowly changing magnetic field produces a smaller one. However, if the magnetic field starts out changing slowly and then increases its rate of change, the electric field starts out small and grows larger. Thus, it is possible for a changing magnetic field to produce a changing electric field.

The parallel plates in Figure 22-23 show the symmetric effect; a changing electric field produces a magnetic field. The rate of change of the electric field determines the size of the magnetic field. Therefore, a changing electric field can produce a changing magnetic field.

We have discussed a lot of "rates of change," and this can be quite confusing, but the payoff for understanding the process is worth the effort. We have discovered a sequential chain of field productions: one changing field produces another changing field, and then this new changing field produces the first kind of field again, and so on. The two fields generate each other in empty space.

We have only argued that this is possible, but the process was rigorously deduced in the 1860s by James Clerk Maxwell. Maxwell showed that this was a consequence of a set of four equations that he and others had developed to describe electricity and magnetism and the many connections between them. The equations, called *Maxwell's equations* in honor of his contributions, summarize all of electricity and magnetism.

Maxwell combined these equations into a single equation that had the same form as the equations that describe periodic waves, whether they are waves on a rope, water waves, or sound waves. Oscillating electric and magnetic fields can combine to produce waves that travel through space (Figure 22-25). As calculated by Maxwell, these **electromagnetic waves** take the form of oscillating electric and magnetic fields that travel with a speed equal to that of light. In 1887 German physicist Heinrich Hertz was able to produce electromagnetic waves on one side of a room and detect them on the other. The existence of electromagnetic waves affirmed Faraday's belief that these fields had their own identities.

Maxwell | *A Man for All Seasons*

James Clerk Maxwell was born on June 13, 1831, into a prominent Edinburgh family. *Clerk* (pronounced "Clark") was his original family name before the Clerks and the Maxwells intermarried. His "hyphenated" name was taken as a result of legal manipulations that prevented the family from holding extensive property in one unit.

Maxwell was brought up in a thoroughly Scottish environment, with kilts, Gaelic, and a burr. He was a curious and gifted child. His father, who had serious scientific interests, encouraged this bent. He attended Edinburgh Academy and in 1847 entered Edinburgh University, where a distinguished faculty welcomed him. One of his professors was famous for his textbooks in physics—some of which were used in the United States until after World War II. In 1850 Maxwell went to Cambridge University, first to Peterhouse College and then to Newton's old haunt, Trinity College. He was a brilliant student of mathematics and began a lifelong study of mathematical models of Saturn's rings. He won the most prestigious mathematical award in British school mathematics and began what would be a wonderfully productive review of the pioneering electromagnetic experiments of Michael Faraday.

Maxwell returned as a professor to Aberdeen, Scotland, and then to King's College in London. He married but had no children. He retired to his Galloway estate to write the monumental *Treatise*

on Electricity and Magnetism. His work in this field generated important advances in the electrical industry, in technology, and in every branch of physical science.

Although retired and in ill health, Maxwell later designed and founded the Cavendish Laboratory and was its first director. This was the first major experimental laboratory explicitly designed with shielding from outside radiation and vibration.

James Clerk Maxwell

He also developed an equally powerful kinetic theory of gases, which led in turn to further breakthrough work in thermodynamics. He continued the Saturn work and contributed to astrophysics, color vision, optics, photoelasticity, servomechanics, and, in the long run, color photography. He left a remarkable legacy. In tribute, Albert Einstein always displayed Maxwell's portrait alongside that of Isaac Newton. He endures as one of the greatest scientists in history.

—Pierce C. Mullen, historian and author

Source: C. W. Francis Everitt, *James Clerk Maxwell: Physicist and Natural Philosopher* (New York: Scribner, 1975).

North Wind Picture Archives

FLAWED REASONING

All of us struggle with misconceptions as we build our world views. Even Isaac Newton made mistakes in his reasoning. He believed that "luminiferous ether" must fill all space to provide a medium for the propagation of light waves. If we jiggle gelatin on one side of a bowl, the jiggle moves across the bowl. Newton felt it was impossible for the jiggle to move across an empty bowl. **Explain how an electromagnetic wave can propagate through empty space.**

ANSWER The electromagnetic wave propagates through empty space by re-creating itself. A changing electric field creates a changing magnetic field—it waves the magnetic field. This changing magnetic field in turn creates a changing electric field—it waves the electric field. And then the process starts over again. It is somewhat like the popular preschool story of *Harold and the Purple Crayon* by Crockett Johnson. In this story, Harold gets to where he wants to go by drawing himself a set of stairs with his purple crayon and using his imagination.

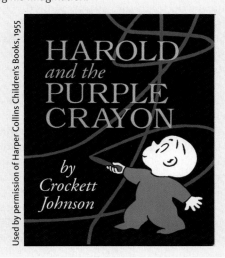

Used by permission of Harper Collins Children's Books, 1955

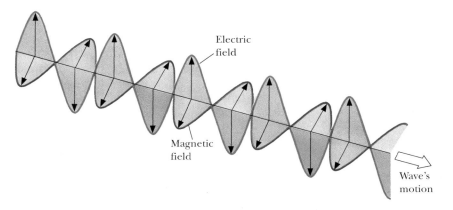

Electric field

Magnetic field

Wave's motion

Figure 22-25 An electromagnetic wave propagating through space. The electric and magnetic fields are perpendicular to each other and to the direction the wave is moving.

Electromagnetic waves are produced whenever electric charges are accelerated. If the charges have a periodic oscillatory motion, the wave will have a fixed frequency and therefore a fixed wavelength according to the relationship we developed in Chapter 15 between speed, wavelength, and frequency:

$$v = \lambda f$$

◀ speed = wavelength × frequency

Maxwell's equations require that the speed be that of light, but place no restrictions on the frequency, as shown in the diagram of the electromagnetic

Figure 22-26 The electromagnetic spectrum.

An infrared photo of the Washington, D.C., area looking north, showing the Potomac River in blue and vegetation in red. This view from a space shuttle covers 68 square miles.

A receiver dish for satellite television signals.

spectrum in Figure 22-26. Although these waves all have the same basic nature, the different ranges in frequency are produced (and detected) by different means. The boundaries between the various named regions are not distinct; they overlap quite a bit.

The lowest frequencies and the longest wavelengths belong to the radio waves. These are produced by large devices such as broadcast radio antennas. Microwaves are also produced electronically, but the devices are smaller, ranging in size from a few millimeters to a few meters. These devices are used in microwave ovens, radar, and long-distance transmission of telephone calls.

The frequencies of visible light extend from 4.0 to 7.5×10^{14} hertz. Although visible light occupies only a small region of the complete spectrum, it is obviously very important to us. This region is bounded on the low side by infrared (IR) radiation, waves whose wavelengths are too long (beyond the red) to be seen by the human eye. Infrared radiation is most noticeable when given off by hot objects, especially those that are red hot. This is the radiation you feel across the room from a fire or a heating element. The radiation that lies beyond violet is known as ultraviolet (UV) waves. This is the component of sunlight that causes suntans (and sunburns if it is excessive). Visible light and its neighbors are produced at the atomic level, and their properties are valuable clues about the structure of matter at the atomic level.

We are all familiar with X rays from visits to the doctor or dentist. X rays have high frequencies and are very penetrating. They are produced by the rapid acceleration of electrons in X-ray machines and are emitted by atoms. Gamma rays are an even higher-frequency radiation that originate in the nuclei of atoms. We will study these in more detail in later sections of this textbook.

Radio and TV

Radio is a means of coding electromagnetic waves with the information in sound waves so that they can be transmitted through space, intercepted, and converted back into sound. Television is the same sort of process with the addition of the video information. Sound waves are changed to an electric signal by a microphone. In one version the sound waves cause a coil to vibrate in a magnetic field. This produces a current in the coil that can then be amplified. Audio frequencies are in the range of 20 hertz to 20 kilohertz.

If audio frequencies were broadcast directly, there could only be one radio station in any geographic region. Instead, the audio signal is combined with a broadcast signal. A station broadcasting at "1450 on your dial" sends out waves with a frequency of 1450 kilohertz (often called *kilocycles*). This is the *carrier frequency*. The audio signal is used to vary, or *modulate*, the carrier signal. Two modulation methods are *amplitude modulation* (AM) and *frequency modulation*

(FM). The frequency of the sound determines the frequency of the modulation, and the loudness of the sound determines the amplitude of the modulation. In AM radio the audio signal causes the amplitude of the carrier signal to vary; in FM radio it causes the frequency of the carrier signal to vary (Figure 22-27).

Q: Although modern police and fire sirens produce sound waves rather than radio waves, they are modulated. Is this modulation AM or FM?

A: Because these sirens have oscillating frequencies, they must be FM.

Are *You* On the Bus?

(*a*)

Figure 22-27 (a) AM and (b) FM radio waves.

(*b*)

In either case, the final signal is then amplified and sent to the antenna, where it causes electrons to move up and down the antenna wire. The accelerations of these electrons produce electromagnetic waves that are broadcast. As these waves hit the antenna in your radio, they cause the electrons in the antenna to move back and forth, producing oscillating currents. Although all nearby stations are received simultaneously, the radio is tuned so that it resonates with only one of them at a time. Your radio amplifies this station's signal.

The radio contains electric circuits to filter out the carrier frequency and retain only the electrical version of the sound information. These signals are then sent to a speaker for reconversion to sound waves. In one version of a speaker, the electric signal generates a magnetic field that interacts with a magnet to move a diaphragm. This diaphragm then moves air to generate sound.

The allocation of the possible ranges of carrier frequencies to various types of broadcast is a governmental responsibility that is complicated by the history of the development of such advances in broadcasting techniques as FM stereo and TV. AM radio stations broadcast between 550 and 1500 kilohertz, FM radio between 88 and 108 megahertz, and TV in three regions between 54 and 890 megahertz. Other regions are assigned to citizens-band receivers, ships, airplanes, police, amateur radio operators, and satellite communication.

Summary

Magnetic poles behave similarly to electric charges: like poles repel and unlike poles attract. However, magnetic poles always occur in pairs. Magnets attract some objects but have no effect on others.

Magnets have no effect on stationary charges and do not deflect the foils of an electroscope. On the other hand, a current-carrying wire produces a magnetic field and is attracted or repelled by other magnets or current-carrying wires. The force on the wire is always perpendicular to the wire and to the

WORKING IT OUT *FM Wavelength*

You are driving through Bozeman, Montana, listening to FM radio station 103.5 (KZMY). What is the wavelength of the carrier signal for this station?

The reading on the radio dial is the carrier frequency, expressed in megahertz. The carrier frequency is 103.5×10^6 Hz, or 1.035×10^8 Hz. Radio waves, like all other waves, obey the wave equation, $v = \lambda f$. Because radio waves are electromagnetic waves, they travel at the speed of light, $v = c = 3 \times 10^8$ m/s. We can easily find the wavelength of the KZMY carrier signal:

$$\lambda = \frac{c}{f} = \frac{3 \times 10^8 \text{ m/s}}{1.035 \times 10^8 \text{ Hz}} = 2.899 \text{ m}$$

or, about the length of your car.

magnetic field. Two current-carrying wires attract each other if the currents are in the same direction; they repel if the currents are in opposite directions. The field strength for magnetism is defined as 2×10^{-7} tesla at a distance of 1 meter from a wire carrying a current of 1 ampere.

All magnetic fields originate from current loops. Naturally occurring magnetism originates in current loops at the atomic level. Earth's magnetism has a strength at the surface of about 5×10^{-5} tesla (0.5 gauss) and is caused by large electric currents circulating in Earth's molten interior.

A charged particle moving in a magnetic field experiences a force at right angles to its velocity and to the magnetic field. The strength of the force depends on the angle between the field and the particle's motion. It is maximum when they are perpendicular and zero when they are parallel.

If a wire and magnetic field move relative to each other, a current is produced in the wire, provided the motion is not parallel to either the wire or the field. This current is largest if the motion is perpendicular to the field and increases with the relative speed. A current also occurs in a loop of wire if the magnetic field inside the loop varies with time. An increasing field produces a current in one direction; a decreasing field produces a current in the opposite direction.

Field lines can be used to represent magnetic fields. The magnetic field is strongest in regions where the field lines are close together. If the number of magnetic field lines passing through a loop of wire changes *for any reason*, a current is produced in the loop. The voltage (and hence the current) generated in the loop depends on the rate of change—the faster the change, the larger the voltage.

There is a close connection between changing electric and magnetic fields. A changing magnetic field can generate a changing electric field, and a changing electric field can generate a changing magnetic field, creating electromagnetic waves that travel through empty space. These waves are produced whenever electric charges are accelerated. When the charges have a periodic oscillatory motion, the wave has a fixed frequency and wavelength. The spectrum of these waves ranges from low-frequency radio waves, to visible light, to high-frequency X rays and gamma rays.

CHAPTER 22 *Revisited*

The connection between your favorite TV show and your microwave oven lies in the electromagnetic waves that are created by combining electric and magnetic effects. These waves can resonate with molecules in a microwave oven to raise the temperature of the popcorn and can be beamed through space to bring us television (and radio) programming.

Key Terms

ampere The SI unit of electric current. The current in each of two parallel wires when the magnetic force per unit length between them is 2×10^{-7} newton per meter.

electromagnet A magnet constructed by wrapping wire around an iron core. An electromagnet can be turned on and off by turning the current in the wire on and off.

electromagnetic wave A wave consisting of oscillating electric and magnetic fields. In a vacuum, electromagnetic waves travel at the speed of light.

gauss A unit of magnetic field strength; 10^{-4} tesla.

magnetic field The space surrounding a magnetic object, where each location is assigned a value determined by the torque on a compass placed at that location. The direction of the field is in the direction of the north pole of the compass.

magnetic monopole A hypothetical, isolated magnetic pole.

magnetic pole One end of a magnet; analogous to an electric charge.

tesla The SI unit of magnetic field.

Questions and exercises are paired so that most odd-numbered are followed by a similar even-numbered.

Blue-numbered questions and exercises are answered in Appendix B.

💲 indicates more challenging questions and exercises.

WebAssign Many Conceptual Questions and Exercises for this chapter may be assigned online at WebAssign.

Conceptual Questions

1. If you are given three iron rods, how could you use them to find the one that is not magnetized?

2. If the labels on a magnet are missing, how would you determine which pole is the north pole?

3. You have three iron bars, each of which may or may not be a permanent magnet. Each rod is painted green on one end and yellow on the other. You perform three experiments and find the following: the green end of bar A attracts the green end of bar B; the yellow end of bar A repels the green end of bar B; and the green end of bar B repels the yellow end of bar C. Which of these three results indicates that bar A must be a permanent magnet? Explain.

4. Consider the experiments described in Question 3. Would the green end of bar A attract, repel, or have no interaction with the green end of bar C? Explain.

5. You have three iron bars, each of which may or may not be a permanent magnet. Each rod is painted black on one end and white on the other. You perform three experiments and find the following: the black end of bar A repels the white end of bar B; the white end of bar A attracts the white end of bar C; and the white end of bar B attracts the white end of bar C. Is bar C a permanent magnet? Explain.

6. Consider the experiments described in Question 5. Would the black end of bar A attract, repel, or have no interaction with the white end of bar C? Explain.

7. If a bar magnet is broken into two pieces, how many magnetic poles are there?

8. We label magnetic poles such that like poles repel and unlike poles attract. Would it have been possible to use a labeling convention in which like poles attract and unlike poles repel? Why or why not?

9. How is the direction of the magnetic field defined at each point in space?

10. Your classmate claims: "To find the direction of the magnetic field at any point, you simply hold a compass in your hand and see what direction the needle points. Because the needle always points parallel to the ground, the magnetic field must also be parallel to the ground." This conclusion is inconsistent with the statement in the text that Earth's field points strongly downward in the United States. What is the flaw in your classmate's reasoning?

11. Magnetic field lines in a region are shown in the following figure. A freely rotating compass needle is released in the orientation shown. Will the needle rotate clockwise, counterclockwise, or not at all? Explain.

12. Oersted found that the magnetic field lines surrounding a current-carrying wire circle around and close on themselves. Because the field lines for a bar magnet also close on themselves, what does this imply about the direction of the field lines inside the magnet? Do the field lines point from north to south or from south to north?

13. When you look down the axis of a solenoid so that the current circulates in the counterclockwise direction, are you looking in the direction of the magnetic field or opposite of it? Explain.

14. How would the photograph in Figure 22-4 change if the current were in the opposite direction?

---

15. In the following figure, you are looking at the end of a long straight wire. The current in the wire is directed out of the page. At the point labeled A, is the direction of the magnetic field left, right, up, down, into the page, or out of the page? What is the direction at point B? Explain.

A   ⊙   B

I (out of page)

16. A long straight wire carries a current in the direction shown in the following figure. At the point labeled A, is the direction of the magnetic field left, right, up, down, into the page, or out of the page? What is the direction of the magnetic field at the point labeled B? Explain.

17. The following figure shows the top view of a compass next to a long straight wire. An arrow indicates the north pole of the compass. When a switch is closed, a current is established in the wire, and the compass needle initially rotates as indicated. Is the direction of the current in the wire into the page or out of the page? Explain.

Top view

○—○→ ○   Current-carrying
              wire

18. The following figure shows the top view of a compass next to a long straight wire. The north pole of the compass is indicated by an arrow. When a switch is closed, the current is directed out of the page as indicated. Will the compass needle initially rotate clockwise or counterclockwise? Explain.

Top view

○——→ ⊙   Current carried
              out of the page

19. In electric circuits and telephone lines, two wires carrying currents in opposite directions are twisted together. How does this reduce the magnetic fields surrounding the wires?

20. Can you use a compass to locate the electric wires located inside the walls of your house? Explain.

21. Two long straight wires carry identical currents in opposite directions, as shown in the following figure. At the points labeled A and B, is the direction of the magnetic field left, right, up, down, into the page, or out of the page? Explain.

22. Two wires carry current into the page, as shown in the following figure. One wire carries a current I, and the other carries a current 2I. Which of the arrows best represents the direction of the magnetic field at point A, which is the same distance from both wires? Explain.

23. In Melville's *Moby Dick*, Captain Ahab regained the confidence of his crew when he fixed the compass that had been damaged by an electrical storm so that it pointed north. How might he have done this?

24. Lodestone (magnetite) is an igneous rock, one that forms from molten material. How do you suppose it became magnetized?

25. What do you expect will happen to the strength of the magnetism of a bar magnet that is dropped on a hard floor?

26. Would you expect the head of a steel hammer to be magnetized? Explain.

27. Two long straight wires carry currents in opposite directions. The current in wire 1 is twice the current in wire 2. If the net magnetic force per unit length on wire 1 is 4 newtons per meter, what is the net magnetic force per unit length on wire 2?

28. Three long straight wires each carry identical current in the directions shown in the following figure. What is the direction of the net force on the wire at the right?

29. How is the unit of current defined?

30. How is the unit of charge defined?

31. One of Earth's magnetic poles is located in Antarctica. Is it a magnetic north pole or a magnetic south pole?

32. Why might it be more proper to call a north magnetic pole a "north-seeking" magnetic pole?

33. If you want to walk toward the geographic North Pole while in New York City, what compass heading would you follow?

34. How far does a compass in Portland, Oregon, point away from the geographic North Pole?

35. Why are there more cosmic rays in Antarctica than in Hawaii?

36. We can model Earth's magnetic field as due to a single large current-carrying loop that runs around the equator just below the surface. In this model, is the direction of the current east to west or west to east? Why?

37. If a charged particle travels in a straight line, can you say that there is no magnetic field in that region of space? Explain.

38. Can you accelerate a stationary charged particle with a magnetic field? An electric field? Explain.

39. A proton and an electron with the same velocity enter a bending magnet with a magnetic field perpendicular to their velocity. Compare the motions of the proton and electron.

40. A magnet produces a magnetic field that points vertically upward. In what direction does the force on a proton act if it enters this region with a horizontal velocity toward the east?

41. A conducting loop is lying flat on the ground. The north pole of a bar magnet is brought down toward the loop. As the magnet approaches the loop, will the magnetic field created by the induced current point up or down? What is the direction of the current in the loop? Explain.

42. The magnet in Question 41 is now lifted upward. Which, if either, of your answers will change? Explain.

43. Consider the case in which the north pole of a bar magnet is being moved toward a conducting copper ring. Do the field lines created by the induced current point toward the bar magnet or away? Will this induced field pull on the magnet or push against it? Explain.

44. Consider the case in which the north pole of a bar magnet is being moved away from a conducting copper ring. Do the field lines created by the induced current point toward the bar magnet or away? Will this induced field pull back on the magnet or push it away? Explain.

45. A copper ring is oriented perpendicular to a uniform magnetic field, as shown in the following figure. If the ring is suddenly moved in the direction opposite the field lines, will the magnitude of the net magnetic field in the center of the loop (the uniform field plus the induced

field) be greater than, equal to, or less than the magnitude of the uniform field? Explain.

Questions 45 and 46.

46. A copper ring is oriented perpendicular to a uniform magnetic field, as shown in the preceding figure. The ring is quickly stretched such that its radius doubles over a short time. As the ring is being stretched, is the magnitude of the net magnetic field in the center of the loop (the uniform field plus the induced field) greater than, equal to, or less than the magnitude of the uniform field? Explain.

47. Quickly inserting the north pole of a bar magnet into a coil of wire causes the needle of a meter to deflect to the right. Describe two actions that will cause the needle to deflect to the left.

48. How could you produce a current in a solenoid by rotating a small magnet inside the coil?

49. When a transformer is plugged into a wall socket, it produces 9-volt alternating-current electricity for a portable tape player. Is the coil with the larger or smaller number of turns connected to the wall socket? Why?

50. If you had an old car that needed a 6-volt battery to drive the starter motor, could you use a 12-volt battery with a transformer to get the necessary output voltage? How?

51. What is the purpose of using a commutator in a motor?

52. What effect does a commutator have on the electrical output of a generator?

53. Describe an electromagnetic wave propagating through empty space.

54. How are electromagnetic waves generated?

55. Which of the following is *not* an electromagnetic wave: radio, television, blue light, infrared light, or sound?

56. Which of the following electromagnetic waves has the lowest frequency: radio, microwaves, visible light, ultraviolet light, or X rays? Which has the highest frequency?

57. What is the difference between X rays and gamma rays?

58. How fast do X rays travel through a vacuum?

59. How is sound encoded by an AM radio station?

60. How is sound transmitted by an FM radio station?

61. At what frequency does radio station FM 102.1 broadcast?

62. What does it mean when your local disc jockey says that you are "listening to radio 1380"?

Exercises

63. The record for a steady magnetic field is 45 T. What is this field expressed in gauss?

64. What is the magnetic field (expressed in gauss) of a refrigerator magnet with a magnetic field of 0.3 T?

65. The magnetic field at the equator of Jupiter has been measured to be 4.3 G. What is this field expressed in teslas?

66. The magnetic fields associated with sunspots are on the order of 1500 G. How many teslas is this?

67. An electron has a velocity of 5×10^6 m/s perpendicular to a magnetic field of 1.5 T. What force and acceleration does the electron experience?

68. What force and acceleration would a proton experience if it has a velocity of 5×10^6 m/s perpendicular to a magnetic field of 1.5 T?

69. A metal ball with a mass of 3 g, a charge of 1 μC, and a speed of 40 m/s enters a magnetic field of 20 T. What are the maximum force and acceleration of the ball?

70. A very small ball with a mass of 0.1 g has a charge of 10 μC. If it enters a magnetic field of 10 T with a speed of 30 m/s, what are the maximum force and acceleration the ball experiences?

71. If you wanted the maximum magnetic force on the ball in Exercise 69 to equal the gravitational force on the ball, what charge would you need to give it?

72. What speed would be needed in Exercise 70 for the maximum magnetic force to be equal to the gravitational force?

73. An ink drop with charge $q = 3 \times 10^{-9}$ C is moving in a region containing both an electric field and a magnetic field. The strength of the electric field is 3×10^5 N/C, and the strength of the magnetic field is 0.2 T. At what speed must the particle be moving perpendicular to the magnetic field so that the magnitudes of the electric and magnetic forces are equal?

74. How would your answer to Exercise 73 change if the charge on the ink drop were doubled?

75. A particle with charge $q = 5$ μC and mass $m = 6 \times 10^{-5}$ kg is moving parallel to Earth's surface at a speed of 1000 m/s. What minimum strength of magnetic field would be required to balance the gravitational force on the particle?

76. What minimum strength of magnetic field is required to balance the gravitational force on a proton moving at speed 5×10^6 m/s?

77. A transformer is used to convert 120-V household electricity to 6 V for use in a portable CD player. If the primary coil connected to the outlet has 400 loops, how many loops does the secondary coil have?

78. The voltage in the lines that carry electric power to homes is typically 2000 V. What is the required ratio of the loops in the primary and secondary coils of the transformer to drop the voltage to 120 V?

79. A transformer is used to step down the voltage from 120 V to 6 V for use with an electric razor. If the razor draws a current of 0.5 A, what current is drawn from the 120-V lines? What is the ratio of the loops in the primary and secondary coils of the transformer?

80. Your 120-V outlet is protected by a 15-A circuit breaker. What is the maximum current that you can provide to an appliance using a transformer with one-tenth as many loops in the secondary as in the primary?

81. You are using a transformer with 800 loops in the primary and 80 loops in the secondary. If the 120-V input is supplying 2 A, what is the current in the appliance connected to the transformer?

82. You are using a transformer with 800 loops in the primary and 80 loops in the secondary. If the transformer is plugged into a 120-V outlet protected by a 15-A circuit breaker, what is the maximum power rating of an appliance that could be used with this transformer?

83. How long does it take a radio signal from Earth to reach the Moon when it is 384,000 km away?

84. The communications satellites that carry telephone messages between New York City and London have orbits above Earth's equator at an altitude of 13,500 km. Approximately how long does it take for each message to travel between these two cities via the satellite?

85. Many microwave ovens use microwaves with a frequency of 2.45×10^9 Hz. What is the wavelength of this radiation, and how does it compare with the size of a typical oven?

86. An X-ray machine used for radiation therapy produces X rays with a maximum frequency of 2.4×10^{20} Hz. What is the wavelength of these X rays?

87. The ultraviolet light that causes suntans (and sunburns!) has a typical wavelength of 300 nm. What is the frequency of these rays?

88. The radioactive isotope most commonly used in radiation therapy is cobalt-60. It gives off two gamma rays with wavelengths of 1.06×10^{-12} m and 9.33×10^{-13} m. What are the frequencies of these gamma rays?

Martin Dohrn/Science Photo Library/Photo Researchers, Inc.

89. What is the range of wavelengths for AM radio?

90. What is the range of wavelengths for FM radio?

91. What is the wavelength of the carrier wave for an AM radio station located at 1090 on the dial?

92. What is the longest wavelength used for TV broadcasts?

The Story of the Quantum

In 1887 German physicist Heinrich Hertz generated sparks in an electrostatic machine and successfully caused a spark to jump across the gap of an isolated wire loop on the other side of his laboratory. Hertz had transmitted the first radio signal a distance of a few meters. Hertz was not interested in the societal implications of sending signals; he was testing James Clerk Maxwell's prediction of the existence of electromagnetic waves. Hertz also mentioned the curious fact that when ultraviolet light illuminated his loop, more sparks were produced.

The 19th century had been the age of steam engines and heat. It is no wonder that tremendous strides were made in the study of thermodynamics—the behavior of gases under different conditions of pressure and temperature. By the 1890s entirely new areas

© Floyd Dean/Taxi/Getty Images

The shape of a snowflake is determined by the quantum-mechanical interactions of water molecules with each other.

of confusing phenomena provided huge intellectual challenges to scientists. In 1895 Wilhelm Roentgen discovered a new form of radiation emanating from an electromagnetic device known as a cathode ray tube. The following year Henri Becquerel observed that a lump of uranium emitted some sort of energy that radiated through the wood of his desk and fogged his photographic plates. In 1900 Max Planck puzzled over unusual results when he calculated the spectrum of radiation from heated objects.

Some of the greatest minds of the 20th century grappled with these issues of radiation and matter, experimental fact and theory. They continually received information that was unbelievable. It was a crisis in the process of discovery itself, born in turmoil and confusion. What were they to believe? A philosopher of that era put it this way: "The senses do not lie. They just do not tell us the truth."

The only possible resolution was a revolution in the physics world view. Ideas that had been taken for granted were discarded, and the unbelievable became believable. The core of the revolution seemed innocent enough: Energy exists in "chunks," or quantum units. At first glance quantization may not seem that radical. The paintings of the French Impressionists look normal when viewed from afar, and our weight seems continuous even though it is really the sum of the weights of discrete atoms. But it is not just a matter of adding the many tiny components of this new discreteness to get the whole; it is the tiny components that control the character of the whole.

The quantum nature of energy was the first of three changes leading to this new physics. New insights into the nature of light and, finally, new insights into the nature of matter followed. Together these changes revealed the character of the universe, from the shape of snowflakes to the existence of neutron stars.

© Bettmann/Corbis

Heinrich Hertz

In the remaining chapters, we will follow the development of these ideas. Although it may seem easier to simply learn "the answers," they are unbelievable without knowing the experiments, the conflicts, the debates, and the eventual resolutions that led to them. As you track these developments, keep in mind Sir Hermann Bondi's comment: "We should be surprised that the gas molecules behave so much like billiard balls and not surprised that electrons don't."

23

The Early Atom

> ▶ We get much of our information about atoms from the light they emit. In fact, the pretty arrays of color emitted by atoms—known as atomic fingerprints—distinguish one type of atom from the others and hold many clues about atomic structure. What does the light tell us about atoms?
>
> (See page 516 for the answer to this question.)

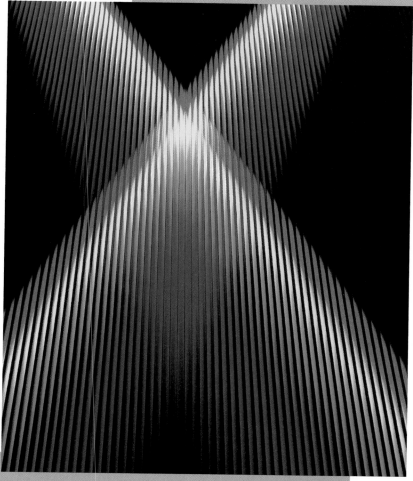

David Parker/SPL/Photo Researchers, Inc.

Diffraction gratings separate the different colors of light. These gratings can be used to probe the structure of atoms.

NEW world views don't emerge in full bloom—they start as seeds. Although a change may begin because of a single individual, it usually evolves from the collective efforts of many. Ideas are proposed. As they undergo the scrutiny of the scientific community, some are discarded and others survive, although almost always in a modified form. These survivors then become the beginnings of new ways of looking at the world.

During the early years of the last century, a number of phenomena were known or discovered that eventually led to several models for the atom. After exploring the phenomena, we will examine several atomic models that emerged in response to the experiments.

Periodic Properties

Figure 23-1 Mendeleev's arrangement of the elements.

Once the values of the atomic weights of the chemical elements were determined (Chapter 11), it was tempting to arrange them in numerical order. When this was done, a pattern emerged; the properties of one element corresponded closely to those of an element farther down the list and to another one even farther down the list. A periodicity of characteristics became apparent.

Russian chemist Dmitri Mendeleev wrote the symbols for the elements on cards and spent a great deal of time arranging and rearranging them. For some reason he omitted hydrogen, the lightest element, and began with the next lightest one known at that time, lithium (Li). He laid the cards in a row. When he reached the 8th element, sodium (Na), he noted that it had properties similar to lithium and therefore started a new row. As he progressed along this new row, each element had properties similar to the one directly above it. The 15th element, potassium (K), is similar to lithium and sodium, and so Mendeleev began a third row. His arrangement of the first 16 elements looked like Figure 23-1.

The next known element was titanium, but it did not have the same properties as boron (B) and aluminum (Al); it was similar to carbon (C) and silicon (Si). Mendeleev boldly claimed that there was an element that had not yet been discovered. He went on to predict two other elements, even predicting their properties from those of the elements in the same column. Later, these elements were discovered and had the properties that Mendeleev had predicted. Mendeleev's work was the beginning of the modern periodic table of the elements. At that time no theory could explain why the elements showed these periodic patterns. However, the existence of the periodicity was a clue to scientists that there might be an underlying structure to atoms.

Atomic Spectra

Another thing fascinated and confused scientists around the beginning of the 20th century. When an electric current passes through a gas, the gas gives off a characteristic color. A modern application of this phenomenon is the neon sign. Although each gas emits light of a particular color, it isn't necessarily unique. However, if you pass the light through a prism or a grating to spread out the colors, the light splits up into a pattern of discrete colored lines (Figure 23-2), which is always the same for a given gas but different from the patterns produced by other gases.

Are You On the Bus?

Q: If a particular gas glows with a yellow color when a current is passed through it, will its spectrum necessarily contain yellow light?

A: The spectrum will not necessarily contain yellow light because the yellow sensation could result from a combination of red and green spectral lines. In fact, red and green light from two different helium–neon lasers produces a beautiful yellow even though each laser has a narrow range of frequencies.

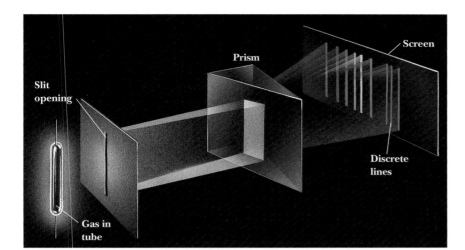

Figure 23-2 The emission spectrum of a gas consists of a small number of distinctly colored lines.

The various colors of light emitted by neon-type signs are due to different gases.

©Cengage Learning/Charles D. Winters

Each gas has its own particular set of spectral lines, collectively known as its **emission spectrum**. The emission spectra for a few gases are shown in the schematic drawings in Figure 23-3. Because these lines are unique—sort of an atomic fingerprint—they identify the gases even if they are in a mixture. Vaporized elements that are not normally gases also emit spectral lines, and these can be used to identify them.

Spectral lines also appear when white light passes through a cool gas (Figure 23-4). In this case, however, they do not appear as bright lines but as dark lines in the continuous rainbow spectrum of the white light; that is, light is missing in certain narrow lines. These particular wavelengths are absorbed or scattered by the gas to form this **absorption spectrum**.

All the lines in the absorption spectrum correspond to the lines in the emission spectrum for the same element. However, the emission spectrum has many lines that do not appear in the absorption spectrum, as illustrated in Figure 23-5.

Because the lines that appear in both spectra are the same, the absorption spectrum can also be used to identify elements. In fact, this is how helium was discovered. Spectral lines were seen in the absorption spectrum of sunlight that did not correspond to any known elements. These lines were attributed to

Figure 23-3 Schematic drawings of the emission spectra for sodium, mercury, helium, and hydrogen near the visible region.

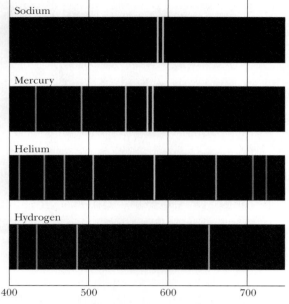

Sodium

Mercury

Helium

Hydrogen

400   500   600   700

Wavelength in nanometers (nm)

Figure 23-4 The absorption spectrum consists of a rainbow spectrum with a number of dark lines.

Figure 23-5 The absorption spectrum for an element has fewer lines than its emission spectrum.

Emission spectrum

Absorption spectrum

a new element that was named *helium* after the Greek name for the Sun, *helios.* Helium was later discovered on Earth.

Spectral lines are a valuable tool for identifying elements. Although this is of great practical use, the scientific challenge was to understand the origin of the lines. Clearly, the lines carry fundamental information about the structure of atoms. The lines pose a number of questions. The most important question concerns their origin: what do spectral lines have to do with atoms? Also, why are certain lines in the emission spectra missing in the absorption spectra?

Cathode Rays

During the latter half of the 19th century, two new technologies contributed to the next advance in our understanding of atoms: vacuum pumps and high-voltage devices. It was soon discovered that a high voltage applied across two electrodes in a partially evacuated glass tube produces an electric current and light in the tube. Furthermore, the character of the phenomenon changes radically as more and more of the air is pumped from the tube. Initially, after a certain amount of the air is removed, sparks jump between the electrodes, followed by a glow throughout the tubes like that seen in modern neon signs. As more air is pumped out, the glow begins to break up and finally disappears. When almost all of the air has been removed, only a yellowish-green glow comes from one end of the glass tube.

These discoveries led to a flurry of activity. Experimenters made tubes with a variety of shapes and electrodes from a variety of materials. The glow always appeared on the end of the glass tube opposite the electrode connected to the negative terminal of the high-voltage supply. When the wires were exchanged, the glow appeared at the other end, showing that its location depended on which electrode was negative. A metal plate placed in a tube cast a shadow (Figure 23-6). These effects indicated that rays were coming from the negative electrode. Because this electrode was known as the *cathode* (the other electrode is the *anode*), the rays were given the name **cathode rays**.

The path of the cathode rays could be traced by placing a fluorescent screen along the length of the tube. The screen glowed wherever it was struck by the cathode rays. A magnet was used to see whether the rays were particles or waves. We saw in the previous chapter that a magnetic field exerts a force on a moving charged particle, but this does not occur for a beam of light. A magnet placed near the tube deflected the cathode rays, indicating that they were indeed charged particles.

Q: Based on the information given, would you expect the cathode rays to be positively or negatively charged?

A: Because they come from the cathode and the cathode is negatively charged, we expect the cathode rays to also have a negative charge. The directions of the magnetic deflections were also consistent with this expectation.

Figure 23-6 Cathode rays cast a shadow of the cross onto the end of the tube.

Figure 23-7 The cathode rays are deflected by an electric field.

This being the case, cathode rays should also be deflected by an electric field. Early experiments showed no observable effect until British scientist J. J. Thomson used the tube sketched in Figure 23-7. By placing the electric field plates inside the tube and by using a much better vacuum, he could observe electric deflection of the cathode rays. The direction of the deflection was that expected for negatively charged particles.

J. J. Thomson

▶ Extended presentation available in the *Problem Solving* supplement

The Discovery of the Electron

Thomson went beyond these qualitative observations. He compared the amount of deflection using a known electric field with that of a known magnetic field. Using these values, he obtained a value for the ratio of the charge to the mass for the cathode particles. Furthermore, Thomson showed that the *charge-to-mass ratio* was always the same regardless of the gas in the tube or the cathode material. This was the first strong clue that these particles were universal to all matter and therefore an important part of the structure of matter.

This ratio could be compared with known elements. The charge-to-mass ratio for positively charged hydrogen atoms (called hydrogen **ions**) had previously been determined; Thomson's number was about 1800 times as large. If Thomson assumed that the charges on the particles were all the same, the mass of the hydrogen ion was 1800 times that of the cathode particle. Conversely, if he assumed the masses to be the same, the charge on the cathode ray was 1800 times as large. (Of course, any combination in between was also possible.)

Thomson believed that the first option—assuming that all charges are equal—was the correct one. (Other data supported this assumption.) This assumption meant that the cathode particles had a much smaller mass than a hydrogen atom, the lightest atom. At this point, he realized that the atom had probably been split!

This was a special moment in the building of the physics world view. The discussion that had originated at least as far back as the early Greeks was centered on the question of the existence of atoms, never on their structure. It had always been assumed that these particles were indivisible. Thomson's work showed that the indivisible building blocks of matter are divisible; they have internal structures.

Around 1910 American experimentalist Robert Millikan devised a way of measuring the size of the electric charge, thus testing Thomson's assumption.

Millikan produced varying charges on droplets of oil between a pair of electrically charged plates (Figure 23-8). By measuring the motions of individual droplets in the electric field, he determined the charge on each droplet and confirmed the widely held belief that electric charge comes in identical chunks, or **quanta**. A particle may have one of these elemental units of charge, or two, or three, but never two and a half. This smallest unit of charge is 1.6×10^{-19} coulomb, an extremely small charge; a charged rod would have more than billions upon billions of these charges.

◄ elementary charge

The mass of the cathode particles could now be calculated from the value of the charge and the charge-to-mass ratio. The modern value for the mass of the cathode particle is 9.11×10^{-31} kilogram. This particle—a component part of all atoms with a mass $\frac{1}{1800}$ the size of the hydrogen atom and carrying the smallest unit of electric charge—is known as an **electron**. The electron is so small that it would take a trillion trillion of them in the palm of your hand before you would notice their mass.

◄ mass of the electron

Thomson's Model

Thomson's work showing that atoms have structure was an exciting breakthrough. A natural question emerged immediately: what does an atom look like? One of its parts is the electron, but what about the rest? Thomson proposed a model for this structure, picturesquely known as the "plum-pudding" model, with the atom consisting of some sort of positively charged material with electrons stuck in it much like plums in a pudding (Figure 23-9). Because atoms were normally electrically neutral, he hypothesized that the amount of positive charge was equal to the negative charge and that atoms become charged ions by gaining or losing electrons.

Figure 23-8 The Millikan oil-drop experiment showed that electric charges are quantized.

Recall that gases were known to emit distinct spectral lines. Thomson suggested that forces disturbing the atom caused the electrons to jiggle in the positive material. These oscillating charges would then generate the electromagnetic waves observed in the emission spectrum. However, the model could not explain why the light was emitted only as definite spectral lines, nor why each element had its own unique set of lines.

Thomson also attempted to explain the repetitive characteristics in the periodic table. He suggested that each element in the periodic table had a different number of electrons. He hoped that with different numbers of electrons, different arrangements would occur that might show a periodicity and thus a clue to the periodicity shown by the elements. For example, three electrons might arrange themselves in a triangle, four in a pyramid, and so on. Perhaps there was a periodicity to the shapes. Patterns did occur in his model, but the periodicity did not match that of the elements.

Figure 23-9 Thomson's model of the atom.

Rutherford's Model

Ernest Rutherford, a student of Thomson, did an experiment that radically changed his teacher's model of the atom. A key tool in Rutherford's work was the newly discovered radioactivity, the spontaneous emission of fast-moving atomic particles from certain elements. Rutherford had previously shown that one of these radioactive products is a helium atom without its electrons. These particles have two units of positive charge and are known as **alpha particles**.

Rutherford bombarded materials with alpha particles and observed how the alpha particles recoiled from collisions with the atoms in the material. This procedure was much like determining the shape of an object hidden under a table by rolling rubber balls at it (Figure 23-10). Rutherford used the alpha

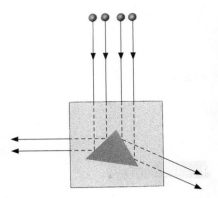

Figure 23-10 The shape of a hidden object can be determined by rolling rubber balls and seeing how they are scattered.

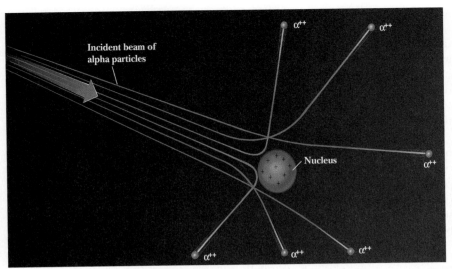

Figure 23-11 The positively charged alpha particles are repelled by the positive charges in the nucleus.

particles as his probe because they were known to be about the same size as atoms. (One doesn't probe the structure of snowflakes with a sledgehammer.) The target needed to be as thin as possible to keep the number of collisions by a single alpha particle as small as possible, ideally only one. Rutherford chose gold because it could be made into a thin foil.

The alpha particles are about 4 times the mass of hydrogen atoms and 7300 times the mass of electrons. Thomson's model predicted that the deflection of alpha particles by each atom (and hence by the foil) would be extremely small. The positive material of the atom was too spread out to be an effective scatterer. Consider the following analogy. Use 100 kilograms of newspapers to make a wall one sheet thick. Glue table-tennis balls on the wall to represent electrons. If you were to throw bowling balls at the wall, you would clearly expect them to go straight through with no noticeable deflection.

Initially, Rutherford's results matched Thomson's predictions. However, when Rutherford suggested that one of his students look for alpha particles in the backward direction, he was shocked with the results. A small number of the alpha particles were actually scattered in the backward direction! Rutherford later described his feelings by saying, "It was almost as incredible as if you fired a 19-inch shell at a piece of tissue paper and it came back and hit you!"

On the basis of his experiment, Rutherford proposed a new model in which the positive charge was not spread out but was concentrated in a very, very tiny spot at the center of the atom, which he called the **nucleus**. The scattering of alpha particles is shown in Figure 23-11. Returning to our analogy, imagine what would happen if all the newspapers were crushed into a small region of space: most of the bowling balls would miss the paper and pass straight through, but occasionally a ball would score a direct hit on the 100-kilogram ball of newspaper and recoil.

Rutherford determined the speed of the alpha particles by measuring their deflections in a known magnetic field. Knowing their speeds, he could calculate their kinetic energies. Because he knew the charges on the alpha particle and the nucleus of the gold atom, he could also calculate the electric potential energy between the two. These calculations allowed him to calculate the closest distance an alpha particle could approach the nucleus of the gold atom before its kinetic energy had been entirely converted to potential energy. These experiments gave him an upper limit on the size of the gold nucleus. The nucleus has a diameter approximately $\frac{1}{100,000}$ the size of its atom. This means that atoms are almost entirely empty space.

Q: If the nucleus were the size of a basketball, what would be the diameter of the atom?

A: Assuming that a basketball has a diameter of about 24 centimeters, the diameter of the atom would be (100,000)(0.24 meter) = 24 kilometers (15 miles) across.

Our work in electricity tells us that the negative electrons are attracted to the positive nucleus. Rutherford proposed that they don't crash together at the center of the atom because the electrons orbit the nucleus much as the planets in our solar system orbit the Sun (Figure 23-12). The force that holds the orbiting electrons to the nucleus is the electric force. (The gravitational force is negligible for most considerations at the atomic level.)

Q: If atoms are mostly empty space, why can't we simply pass through each other?

A: The atom's electrons in the outer regions keep us apart. When two atoms come near each other, the electrons repel each other strongly.

Rutherford's work led to a new explanation for the emission of light from atoms. The orbiting electrons are accelerating (Chapter 4), and accelerating charges radiate electromagnetic waves (Chapter 22). A calculation showed that the orbital frequencies of the electron correspond to the region of visible light. But there is a problem. If light is radiated, the energy of the electron must be reduced. This loss would cause the electron to spiral into the nucleus. In other words, this model predicts that the atom is not stable and should collapse. Calculations showed that this should occur in about a billionth of a second! Because we know that most atoms are stable for billions of years, there is an obvious defect in the model.

Rutherford's "solar system" model of the atom also failed to predict spectral lines and to account for the periodic properties of the elements. The resolution of these problems was accomplished by atomic models proposed during the two decades following Rutherford's work.

Figure 23-12 Rutherford's model of the atom. (The size of the nucleus is greatly exaggerated, and the electrons are known to be even smaller than the nucleus.)

Radiating Objects

The Thomson and Rutherford models of the atom could explain some of the experimental observations, but they had serious deficiencies. Clearly, a new model was needed. Interestingly, the germination of the next model began in yet another area of physics, one having to do with the properties of the radiation emitted by a hot object.

All objects glow. At normal temperatures the glow is invisible; our eyes are not sensitive to it. As an object gets warmer, we feel this glow—we call it heat. At even higher temperatures, we can see the glow. The element of an electric stove or a space heater, for example, has a red-orange glow. In all these cases, the object is emitting electromagnetic radiation. The frequencies, or colors, of the electromagnetic radiation are not equally bright, as can be seen in the photographs of Figure 23-13. Furthermore, the color spectrum shifts toward the blue as the temperature of the object increases.

All solid objects give off continuous spectra that are similar, but not identical. However, the radiation is the same for all materials at the same temperature if the radiation comes from a small hole in the side of a hollow block. Typical distributions of the brightness of the electromagnetic radiation versus wavelength are shown in the graph of Figure 23-14 for several temperatures. When the temperature changes, the position of the hump in the curve

Rutherford | *At the Crest of the Wave*

Well, I made the wave, didn't I?
—Ernest Rutherford, first Baron Rutherford of Nelson

Ernest Rutherford (1871–1937) grew up on a small farm near Nelson, New Zealand, at a time when that island nation was new and very much part of the expansive British Empire. He received a sound education in Nelson and won a scholarship to the University of New Zealand. His diligence and brilliance there earned him an imperial scholarship to Cambridge University, where J. J. Thomson, director of the Cavendish Laboratory, put him to work measuring electromagnetic waves. Teacher and student then moved on to analyze X-ray ionization in gases and aspects of the photoelectric effect. Henri Becquerel's discovery of radioactivity created a vast new arena for physicists to explore.

Rutherford had just begun to examine radium ionization because he had discovered that two types of radiation in this strange gas, alpha and beta, had different penetrating properties. The former was stopped by a thin layer of common air; the latter passed through a thin aluminum foil. Rutherford made a series of great discoveries involving alpha rays. He left England in 1898 and continued his work as a new professor at McGill University in Montreal, Quebec, Canada.

Rutherford's students and colleagues at McGill, the most notable of whom was Frederick Soddy, continued to work on explanatory mechanisms for radioactivity and published the first substantial study of that complex subject. In 1908 Rutherford received the Nobel Prize in chemistry. His address was a model of scientific reporting. He explained the importance of alpha radiations, their identity as helium atoms, and their use in analyzing atomic numbers. In 1909 Rutherford accepted a chair of physics

at Manchester University, where he continued to work on the cutting edge of atomic energy questions and in 1912 attracted perhaps his most outstanding student, Niels Bohr. Another brilliant young physicist, Henry Moseley, arrived in 1913. Moseley's death on the Gallipoli battlefront in World War I led to Rutherford's futile request that scientists be exempted from frontline service. That piece of advice was taken by the United States in World War II. Rutherford also

Lord Ernest Rutherford

contributed to wartime research in a number of areas, most notably submarine detection.

The New Zealander was knighted in 1914 and returned to Cambridge in 1919 when Thomson retired. He was made a baron in 1931, which gave him a seat in the House of Lords. Rutherford, always gifted with a splendid clarity of vision and expression, lived out his life full of honors and continued scientific challenges. The most complex problem for him was a matter of resolving the key components into a simpler form so that they could be tackled. The great revolutions in physics—relativity, atomic physics, and quantum theory—all saw him at the center of ongoing work. His comment quoted previously was true: he had indeed been at the crest of the wave that swept over this heroic age of physics.

—Pierce C. Mullen, historian and author

Sources: David Wilson, *Rutherford: Simple Genius* (Cambridge, Mass.: MIT Press, 1983); Arthur S. Eve, *Rutherford* (New York: Macmillan, 1939).

Figure 23-13 The color of the hot filament and its continuous spectrum change as the temperature is increased.

changes; the wavelength with the maximum intensity varies inversely with the temperature—that is, as the temperature goes up, the highest point of the curve moves to smaller wavelengths.

Q: What happens to the frequency corresponding to the maximum intensity as the temperature increases?

A: Because the product of the frequency and the wavelength must equal the speed of light, the frequency must increase as the wavelength decreases. Therefore, the frequency increases as the temperature increases.

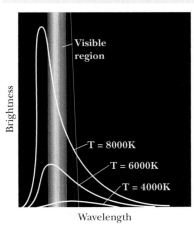

Figure 23-14 This graph shows the intensity of the light emitted by a hot object at various wavelengths for three different temperatures.

Figure 23-15 The theoretical ideas and the experimental data for the intensity of the light emitted by a hot body differed greatly in the low-wavelength (high-frequency) region.

The fact that the same distribution is obtained for cavities made in any material—gold, copper, salt, or whatever—puzzled and excited the experimenters. However, all attempts to develop a theory that would explain the shape of this curve failed. Most attempts assumed that the material contained some sort of charged atomic oscillators that generated light of the same frequency as their vibrations. Although the theories accounted for the shape of the large-wavelength end of the curve, none could account for the small contribution at the shorter wavelengths (Figure 23-15). Nobody knew why the atomic oscillators did not vibrate as well at the higher frequencies.

A solution was discovered in 1900 by Max Planck, a German physicist. He succeeded by borrowing a new technique from another area of physics. The technique called for temporarily pretending that a continuous property was discrete in order to make particular calculations. After finishing the calculation, the technique called for returning to the real situation—the continuous case—by letting the size of the discrete packets become infinitesimally small.

Planck was attempting to add up the contributions from the various atomic oscillators. Unable to get the right answer, he temporarily assumed that energy existed in tiny chunks. Planck discovered that if he didn't take the last step of letting the chunk size become very small, he got the right results! Planck announced his success in cautious tones—almost as if he were apologizing for using a trick on nature.

Planck's "incomplete calculation" was in effect an assumption that the atomic oscillators could vibrate only with certain discrete energies. Because the energy could have only certain discrete values, it would be radiated in little bundles, or quanta (singular, **quantum**). The energy E of each bundle was given by

$$E = hf$$

◄ energy quantum = Planck's constant × frequency

Planck's constant ▶ where *h* is a constant now known as *Planck's constant* and *f* is the frequency of the atomic oscillator. An oscillator could have an energy equal to 1, 2, 3, or more of these bundles, but not $1\frac{5}{8}$ or $2\frac{1}{2}$. The value of Planck's constant, 6.63 $\times 10^{-34}$ joule-second, was obtained by matching the theory to the radiation spectrum.

WORKING IT OUT *Energy of a Quantum*

We can get a feeling for the size of the energies of the atomic oscillators from our knowledge that red light has a frequency around 4.6×10^{14} Hz. Substituting these numbers into our relationship for the energy yields

$$E = hf = (6.63 \times 10^{-34}\,\text{J} \cdot \text{s})(4.6 \times 10^{14}\,\text{Hz}) = 3.05 \times 10^{-19}\,\text{J}$$

Are **You** On the Bus?

Q: What is the size of the energy quantum for violet light with a frequency of 7×10^{14} Hz?

A: 4.64×10^{-19} J.

Planck's idea explained why the contribution of the shorter wavelengths was smaller than predicted by earlier theories. Shorter wavelengths mean higher frequencies and therefore more energetic quanta. Because there is only a certain amount of energy in the macroscopic object and this energy must be shared among the atomic oscillators, the high-energy quanta are less likely to occur.

This discreteness of energy was in stark disagreement with the universally held view that energy was a continuous quantity and any value was possible. At first thought you may want to challenge Planck's result. If this were true, it would seem that our everyday experiences would have demonstrated it. For example, the kinetic energy of a ball rolling down a hill appears to be continuous. As the ball gains kinetic energy, its speed appears to increase continuously rather than jerk from one allowed energy state to the next. There doesn't seem to be any restriction on the values of the kinetic energy.

We are, however, talking about very small packets of energy. A baseball falling off a table has energies in the neighborhood of 10^{19} times those of the quanta. These energy packets are so small that we don't notice their size in our everyday experiences. On our normal scale of events, energy seems continuous.

The Photoelectric Effect

Shortly before 1900, another phenomenon was discovered that connected electricity, light, and the structure of atoms. When light is shined on certain clean metallic objects, electrons are ejected from their surfaces (Figure 23-16). This effect is known as the **photoelectric effect**. The electrons come off with a range of energies up to a maximum energy, depending on the color of the light. If the light's intensity is changed, the rate of ejected electrons changes, but the maximum energy stays fixed.

Are **You** On the Bus?

Q: Does the name *photoelectric effect* make sense for this phenomenon?

A: Yes. The root *photo* is often associated with light; the most obvious example is photography. The *electric* root refers to the emitted electrons.

Planck : *Founder of Quantum Mechanics*

Max Planck personified science in the early 20th century. He fought a difficult battle against a technical problem in physics, and his victory ushered in a revolution that he neither wanted nor welcomed. He originated quantum mechanics; that is, he laid the foundation for modern physics and destroyed commonsense physical philosophy. He was awarded the Nobel Prize in physics, and his renown in the field was second only to that of Einstein.

Although he did not support Hitler, Planck stayed in Germany under the Hitler regime to provide an institutional basis for the rebirth of German science after the Third Reich. He lost a son in World War I, and a second was executed for plotting to assassinate Hitler. Late in the war, a bombing raid destroyed his priceless laboratory, correspondence, and manuscript collections. His was a noble and tragic life.

Planck's family background was filled with culture and learning. His father was a law professor at Kiel when Max, his sixth child, was born in 1858. He was gifted in many areas: mathematics, music, languages, history, and science. In his youth most friends would have predicted that he would become a humanist or a musician. He became a physicist because he wanted to solve problems that were intractable.

In his work with the radiation from heated bodies, he discovered that Maxwell's laws predicted a host of phenomena but not the equilibrium distribution Planck needed. His work in thermodynamics resulted in two constants (and constants are rare indeed in science), with *h* the most commonly encountered. As he expressed

it in 1910, "The introduction of the quantum of action *h* into the theory should be done as conservatively as possible; that is, alterations should only be made that have shown themselves to be absolutely necessary."

As the major figure in German physical science, Planck presided over great institutions and expenditures. He was a professor who led two dozen students through their doctorates. He served on an astounding 650 doctoral examination committees, and he championed, as only a powerful figure could, the careers of some great women and Jewish colleagues.

Max Planck

Max Planck lived long enough to see new science born from the ashes of the Third Reich. In 1946 the first institute to bear his name appeared in the British zone of occupation. At 89 he was the sole German to be invited to the 300th (belated) anniversary of Isaac Newton's birth. Planck died full of years, toil, pain, and honor in October 1947.

—Pierce C. Mullen, historian and author

Source: John L. Heilbron, *The Dilemmas of an Upright Man: Max Planck as Spokesman for German Science* (Berkeley: University of California Press, 1986).

The ideas about atoms and light prevalent at the end of the 19th century accounted for the possibility of such an effect. Because Planck's oscillators produced light, presumably light could jiggle the atoms. Light is energy, and when it is shined on a surface, some of this energy can be transferred to the electrons within the metal. If the electrons accumulate enough kinetic energy, some should be able to escape from the metal. But no theory could account for the details of the photoelectric process. For instance, if the light source is weak, the old theories predict that it should take an average of several hours before the first electron acquires enough energy to escape. On the other hand, experiments show that the average time is much less than 1 second.

A range of kinetic energies made sense in the old theories—electrons bounced about within the metal accumulating energy until they approached the surface and escaped. There were bound to be differences in their energies. But a maximum value for the kinetic energy of the electrons was puzzling. The wave theory of light said that increasing the intensity of the light shining on the metal increases the kinetic energy of the electrons leaving the metal. However, experiments show that the intensity has no effect on the maximum kinetic energy. When the intensity is increased, the number of electrons leaving the metal increases, but their distribution of energy is unchanged.

One thing does change the maximum kinetic energy. If the frequency of the incident light increases, the maximum kinetic energy of the emitted electrons

Figure 23-16 The photoelectric effect. Light shining on a clean metal surface causes electrons to be given off from the surface.

increases. This result was totally unexpected. If light is a wave, the amplitude of the wave should govern the amount of energy transferred to the electron. Frequency (color) should have no effect.

The established ideas about light were inadequate to explain the photo-electric effect. The conflict was resolved by Albert Einstein. In yet another outstanding paper written in 1905, he claimed that the photoelectric effect could be explained by extending the work of Planck. Einstein said that Planck didn't go far enough: not only are the atomic oscillators quantized, but the experimental results could be explained only if light itself is also quantized! Light exists as bundles of energy called **photons**. The energy of a photon is given by Planck's relationship:

energy of a photon ▶

$$E = hf$$

Although it was universally agreed that light was a wave phenomenon, Einstein said that photons acted like particles, little bundles of energy.

Einstein was able to explain all the observations of the photoelectric effect. The ejection of an electron occurs when a photon hits an electron. A photon gives up its *entire* energy to a single electron. The electrons that escape have different amounts of energy, depending on how much they lose in the material before reaching the surface. Also, the fact that the first photon to strike the surface can eject an electron easily accounts for the short time between turning on the light and the appearance of the first electron. A maximum kinetic energy for the ejected electrons is now easy to explain. A photon cannot give an electron any more energy than it has, and the chances of an electron getting hit twice before it leaves the metal are extremely small.

Are You On the Bus?

Q: How does this model explain the change in the maximum electron energy when the color of the light changes?

A: This change occurs because the photon's color and energy depend on the frequency: the higher the frequency, the more energy the photon can give to the electron.

Einstein's model suggests a different interpretation of intensity. Because all photons of a given frequency have the same energy, an increase in the intensity of the light is due to an increase in the number of photons, not the energy of each one. Therefore, the model correctly predicts that more electrons will be ejected by the increased number of photons but their maximum energy will be the same.

Einstein's success introduced a severe conflict into the physics world view. Diffraction and interference effects (Chapter 19) required that light behave as a wave; the photoelectric effect required that light behave as particles. How can light be both? These two ideas are mutually contradictory. This is not like saying that someone is fat and tall, but rather like saying that someone is short and tall. The two descriptions are incompatible. This conflict took some time to resolve. We will examine this important development in the next chapter.

Meanwhile, Einstein's work inspired a new, improved model of the atom.

Bohr's Model

The model of the atom proposed by Rutherford had several severe problems. It didn't account for the periodicity of the elements, it was unable to give any clue about the origins of the spectral lines, and it implied that atoms were extremely unstable. In 1913 Danish theoretician Niels Bohr proposed a new model incorporating Planck's discrete energies and Einstein's photon into Rutherford's model.

Scientists are always using models when attempting to understand nature. The trick is to understand the limitations of the models. Bohr challenged the Rutherford model, stating that it was a mistake to assume that the atom is just a scaled-down solar system with the same rules; electrons do not behave like miniature planets. Bohr proposed a new model for the hydrogen atom based on three assumptions, or postulates.

First, he proposed that the angular momentum of the electron is quantized. Only angular momenta L equal to whole-number multiples of a smallest angular momentum are allowed. This smallest angular momentum is equal to Planck's constant h divided by 2π. Bohr's first postulate can be written

◀ Bohr's first postulate

$$L_n = n\frac{h}{2\pi}$$

◀ allowed angular momenta

where n is a positive integer (whole number) known as a **quantum number**.

In classical physics the angular momentum L of a particle of mass m traveling at a speed v in a circular orbit of radius r is given by $L = mvr$ (Chapter 8). Therefore, the restriction on the possible values of the angular momentum puts restrictions on the possible radii and speeds. Bohr showed that this restriction on the electron's angular momentum means that the only possible orbits are those for which the radii obey the relationship

$$r_n = n^2 r_1$$

◀ allowed radii

where r_1 is the smallest radius and n is the same integer that appears in the first postulate. This means that the electrons cannot occupy orbits of arbitrary size but only a certain discrete set of allowable orbits, as illustrated in Figure 23-17. The numerical value of r_1 is 5.29×10^{-11} meter.

There is also a definite speed associated with each possible orbit. This means that the kinetic energies are also quantized. Because the value of the electric potential energy depends on distance, the quantized radii mean that the potential energy also has discrete values. Therefore, there is a discrete set of allowable energies for the electron.

Figure 23-17 Electrons can only occupy certain orbits. These are designated by the quantum number n.

Q: Would you expect the energy of the electron to increase or decrease for larger orbits?

A: Because the electric force is analogous to the gravitational force, we can think about putting satellites into orbit around Earth. Higher orbits require bigger rockets and therefore more energy. Therefore, larger orbits should have more energy.

Bohr's second postulate states that an electron does not radiate when it is in one of the allowed orbits. This statement is contrary to the observation that accelerated charges radiate energy at a frequency equal to their frequency of vibration or revolution (Chapter 22). Bohr challenged the assumption that this property is also true in the atomic domain. Radically breaking away from what was accepted, Bohr said that an electron has a constant energy when it is in an allowed orbit.

◀ Bohr's second postulate

But atoms produce light. If orbiting electrons don't radiate light, how does an atom emit light? Bohr's third postulate answered this question: a single photon is emitted whenever an electron jumps down from one orbit to another. Energy conservation demands that the photon have an energy equal to the

◀ Bohr's third postulate

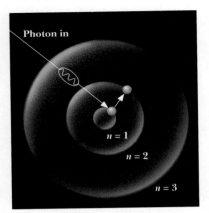

Figure 23-18 An electron can absorb a photon and jump to a higher-energy orbit.

Figure 23-19 When an electron jumps to a lower-energy orbit, it emits a photon.

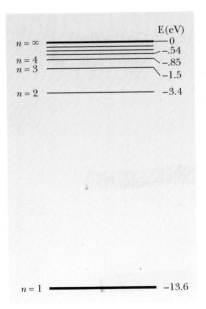

Figure 23-20 The energy-level diagram for the hydrogen atom.

difference in the energies of the two levels. It further demands that jumps up to higher levels can occur only when photons are absorbed. The electron is normally in the smallest orbit, the one with the lowest energy. If the atom absorbs a photon, the energy of the photon goes into raising the electron into a higher orbit (Figure 23-18). Furthermore, it is not possible for the atom to absorb part of the energy of the photon. It is all or nothing.

The electron in the higher orbit is unstable and eventually returns to the innermost orbit, or ground state. If the electron returns to the ground state in a single jump, it emits a new photon with an energy equal to that of the original photon (Figure 23-19). Again, this is different from our common experience. When a ball loses its mechanical energy, its temperature rises—the energy is transferred to its internal structure. As far as we know, an electron has no internal structure. Therefore, it loses its energy by creating a photon.

Occasionally, the phrase "a quantum leap" is used to imply a big jump. But the jump needn't be big. There is nothing big about the quantum jumps we are discussing. A quantum leap simply refers to a change from one discrete value to another.

Ordinarily, an amount of energy is stated in joules, the standard energy unit. However, this unit is extremely large for work on the atomic level, and a smaller unit is customarily used. The **electron volt** (eV) is equal to the kinetic energy acquired by an electron falling through a potential difference of 1 volt. One electron volt is equal to 1.6×10^{-19} joule. It requires 13.6 electron volts to remove an electron from the ground state of the hydrogen atom.

Atomic Spectra Explained

Bohr's explanation accounted for the existence of spectral lines, and his numbers even agreed with the wavelengths observed in the hydrogen spectrum. To illustrate Bohr's model, we draw energy-level diagrams like the one in Figure 23-20. The ground state has the lowest energy and appears in the lowest position. Higher-energy states appear above the ground state and are spaced to indicate the relative energy differences between the states.

To see how Bohr's model gives the spectral lines, consider the energy-level diagram for the hypothetical atom shown in Figure 23-21(a). Suppose the electron has been excited into the $n = 4$ energy level. There are several ways that it can return to the ground state. If it jumps directly to the ground state, it emits the largest-energy photon. We will assume it appears in the blue part of the emission spectrum in Figure 23-21(b). This line is marked $4 \rightarrow 1$ on the spectrum, indicating that the jump was from the $n = 4$ level to the $n = 1$ level.

The electron could also return to the ground state by making a set of smaller jumps. It could go from the $n = 4$ level to the $n = 2$ level and then from the $n = 2$ level to the ground level. These spectral lines are marked $4 \rightarrow 2$ and $2 \rightarrow 1$. The energies of these photons are smaller, so their lines occur at the red end of the spectrum. The lines corresponding to the other possible jumps are also shown in Figure 23-21(b).

Are *You* On the Bus?

Q: How does the total energy of the photons that produce the $4 \rightarrow 2$ and $2 \rightarrow 1$ lines compare with the photon that produces the $4 \rightarrow 1$ line?

A: Conservation of energy requires that the two photons have the same total energy as the single photon.

A single photon is not energetic enough for us to see. In an actual situation, a gas contains a huge number of excited atoms. Each electron takes only

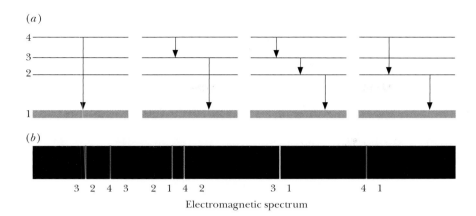

Figure 23-21 (a) Four possible ways for the electron in a hypothetical atom to drop from the n = 4 level to the n = 1 level. (b) The emission spectrum produced by these jumps.

one of the paths we have described in returning to the ground state. The total effect of all the individual jumps is the atomic emission spectrum we observe.

Bohr's scheme also explains the absorption spectrum. As we discussed earlier, the absorption spectrum is obtained when white light passes through a cool gas before a prism or a diffraction grating disperses it. In this case, selected photons from the beam of white light are absorbed and later reemitted. Because the reemitted photons are distributed in all directions, the intensities of these photons are reduced in the original direction of the beam. This removal of the photons from the beam leaves dark lines in the continuous spectrum.

The solution to the mystery of the missing lines in the absorption spectrum is easy using Bohr's model. For an absorption line to occur, there must be electrons in the lower level. Then they can be kicked up one or more levels by absorbing photons of the correct energy. Because almost all electrons in the atoms of the gas occupy the ground state, only the lines that correspond to jumps from this state show up in the spectrum. Jumps up from higher levels are extremely unlikely because electrons excited to a higher level typically remain there for less than a millionth of a second. Thus, there are fewer lines in the absorption spectrum (Figure 23-22) than in the emission spectrum, in agreement with the observations.

FLAWED REASONING

A large glass jar filled with helium gas is placed on the table and a hot, glowing chunk of steel is placed behind it. (*Caution:* This should only be done by professionals.) Joshua and Darcy are discussing what they would see if they looked through a diffraction grating at the light passing through the glass jar as the steel cools:

Joshua: "We should see the absorption spectrum for helium. The helium in the jar isn't changing, so the spectrum should remain constant as the steel cools."

Darcy: "But we learned that the spectrum for hot, glowing objects depends on temperature. As the steel cools, the entire spectrum, including the absorption lines, should move to longer wavelengths, a shift from blue toward red."

What will they see?

ANSWER There are two pieces to this puzzle. If they look directly at the piece of hot steel through the grating, they see a colorful rainbow spectrum. As the steel cools, the light will get dimmer, with the blue end of the spectrum fading the fastest. When the light passes through the jar of helium, certain wavelengths are removed by absorption, leaving dark lines in the rainbow pattern. The wavelengths that are absorbed by the helium do not change, so the absorption spectrum remains constant as the steel cools. However, the lines disappear as the colors dim because the photons of the required energy are no longer present in the light.

Figure 23-22 The absorption spectrum for the hypothetical atom in Figure 23-21.

Electromagnetic spectrum

"The Periodic Table"

The Periodic Table

Bohr's model helped unravel the mystery of the periodic properties of the chemical elements, a major unresolved problem for several decades. The table displaying the repeating chemical properties had grown considerably since Mendeleev's discovery and looked much like the modern one shown on the inside front cover of this text. About the time Bohr proposed his model, it became widely accepted that if the elements were numbered in the order of increasing atomic masses, their numbers would correspond to the number of electrons in the neutral atoms.

Bohr's theory gave a fairly complete picture of hydrogen (H): one electron orbiting a nucleus with one unit of positive charge (Figure 23-23). However, it could not explain the details of the spectrum of the next element, helium (He), because of the effects of the mutual repulsion of the two electrons. But the theory was successful for the helium atom with one electron removed (the He$^+$ ion). Bohr correctly assumed that the helium atom has two electrons in the lowest energy level. Because the electrons presumably have separate orbits, the collection of orbits of the same size was called a **shell**.

The next element, lithium (Li), has spectral lines similar to those of hydrogen. This similarity could be explained if lithium had one electron orbiting a nucleus with three units of positive charge and an inner shell of two electrons. The two inner electrons partially shield the nucleus from the outer electron, so the outer electron "sees" a net charge of approximately +1.

How many electrons can be in the second shell? The clue is found with the next element that has properties like lithium—sodium (Na), element 11. Sodium must have 1 electron outside two shells. Because the first shell holds 2, the second shell holds 8. The next lithium-like element is potassium (K), with 19 electrons. This time we have 1 electron orbiting 18 inner electrons: one shell of 2 and two shells of 8. And so on for succeeding elements.

Are You On the Bus?

Q: How many electrons are in the fourth shell?

A: The next lithium-like element is rubidium (Rb), with 37 electrons. Therefore, the 36 inner electrons must be arranged in shells with 2, 8, 8, and 18 electrons.

Figure 23-23 The electron shell structure for hydrogen, helium, lithium, and sodium.

Hydrogen   Helium   Lithium   Sodium

Bohr : *Creating the Atomic World*

On January 16, 1939, the *Drottingholm* carried Niels Bohr into New York harbor. His arrival was eagerly awaited by a small group of physicists, among them several who had fled Hitler's spreading power. Bohr's hosts knew that he would bring the latest news from a Europe rapidly drifting toward war. Most important, he would relay the latest information on German science. He would tell them about work on the atom.

Niels Bohr is one of the best-known physicists of the 20th century. His pioneering work on atomic structure earned the Nobel Prize in 1922—the year after his friend Einstein received his. Bohr was comfortable with the commonsense violations of quantum physics and uncertainty. He believed in an underlying order, but it was a good deal messier than that which Einstein could accept. For Bohr, God could play dice because probability ruled the universe. Atomic physics, however, was no gamble. It was the future of humankind.

Niels Hendrik Bohr was born in the heart of Copenhagen in 1885. His father was a physiologist, and Bohr was reared in the embrace of a close professional family. He received his doctorate from the University of Copenhagen and then traveled to Cambridge to study with J. J. Thomson—the first Cavendish Professor. Bohr then moved on to the industrial city of Manchester to study with Ernest Rutherford, a pioneer in radiation studies. Armed with considerable laboratory and theoretical experience, Bohr returned to the University of Copenhagen as professor of physics. His work received a high honor when the Carlsberg Breweries foundation constructed a new building for theoretical physics (1921) and a House of Honor (1932) for the most distinguished Dane—Niels Bohr.

Bohr was naturally gregarious, and on his accession to the university professorship in Copenhagen he recognized that his small country would be a natural gathering place for scientists who had been so alienated from one another in World War I. Young American men and women of science also joined this cosmopolitan and vibrant society as American science came of age. But an old friend, a Jewish physicist of considerable talent named Lise Meitner, visited Bohr on her way to exile in Sweden and gave him the information on the Berlin experiments of Christmastide 1938: the uranium atom had released fission by-products. Great energy had been released. This was the portentous news Bohr brought to New York that cold January day in 1939.

Bohr remained in Copenhagen until it became too dangerous, and then he and his wife fled in September 1943 to Sweden, and a month later to England. British and American scientists of the highest rank wanted access to him and his knowledge of the German nuclear effort. Bohr worked assiduously, counseling British and American officials on postwar nuclear policy. He was convinced that there was no secret to nuclear energy and that developed industrial societies could acquire that capability if they desired. His efforts were recognized in the Ford Motor Company's Atoms

for Peace Award—a part of President Eisenhower's policy on atomic energy.

Niels Bohr was first and foremost a scientist, but he was a committed and engaged human being. More clearly than most, he saw that science has consequences for society and that scientists should accept the task of educating their publics about those consequences. Neither President Franklin Roosevelt nor Prime Minister Winston Churchill could fathom Bohr's concern; later world leaders did recognize his prescience and his humanity.

Niels Bohr

Niels Bohr Archive, Courtesy of AIP Emilio Segre Visual Archives

Bohr continues to fascinate us. In 1998 British playwright and author Michael Frayn presented a popular and controversial drama, *Copenhagen*. In 2000 it was a hit on the New York stage. The title encapsulates a hypothetical meeting in Copenhagen between Bohr and his former student Werner Heisenberg in 1941. Bohr knew that an Allied program to develop atomic energy for military purposes was under way in Britain and the United States. Many leading scientists in that program were former associates and students. Bohr also recognized Heisenberg's leadership in the nascent Nazi bomb project. The two had been close, almost like father and son—certainly like mentor and pupil. Germany's destruction of European Jewry was also known. It was a meeting fraught with tension, memories, and cutting-edge science. The drama in *Copenhagen* takes place with these two protagonists and a third, more anti-German participant: Bohr's wife, Margrethe.

Did Heisenberg come to renew this old friendship with Bohr simply to spy on the Allied program? Did he come to assure his old friend that he would do his best as head of the atomic research program in Germany to ensure that the program would fail? Did he come simply to secure a blessing from the most civilized man in the world—the man whose word could confer moral comfort? In a series of flashbacks and references to scientific work in physics over the previous two decades, the actors build their characters and enlarge the audience's understanding of the colossal stakes in this nuclear game. "Bohr, I have to know! I'm the one who has to decide! If the Allies are building a bomb, what am I choosing for my country?" Does a physicist have a moral right to create a weapon of such vast destructive power? Neither man can answer sufficiently to satisfy the other. Margrethe passes the final judgment: "He is one of them." Scientists have become the creator and destroyer of worlds.

—Pierce C. Mullen, historian and author

Sources: Ruth Moore, *Niels Bohr* (New York: Knopf, 1966; MIT paperback edition, 1985); *Physics Today* (October 1985); R. Harre, *Scientific Thought 1900–1960: A Selective Survey* (Oxford: Oxford University Press, 1969); Michael Frayn, *Copenhagen* (New York: Random House–Anchor Books, 2000).

In this scheme the chemical properties of atoms are determined by the electrons in their outermost shells. This idea can be extended to other vertical columns in the periodic table. The outer shells of elements in the right-hand column (group 8) are completely filled. These elements do not react chemically with other elements and rarely form molecules with themselves. They are inert. They are also called the *noble elements* because they don't usually mix with the common elements. The properties of group-8 elements led to the idea that the most stable atoms are those with filled outer shells.

The elements in group 7 are missing one electron in the outer shell and, like the elements in group 1, are chemically very active. They become much more stable by gaining an electron to fill this shell. Therefore, they often appear as negative ions. On the other hand, the elements in group 1 are more stable if they give up the single electron in the outer shell to become positive ions. Therefore, group-1 atoms readily combine with group-7 atoms. The atom from group 1 gives up an electron to that from group 7, and these two charged ions attract each other. Common table salt—sodium chloride—is such a combination. As soon as the two atoms combine, their individual properties—reactive metal and poisonous gas—disappear because the outer structure has changed.

Are You On the Bus?

Q: What other salts besides sodium chloride would you expect to find in nature?

A: Most of the combinations of group-1 and group-7 elements should occur in nature. Therefore, we expect to find such salts as sodium fluoride, sodium bromide, potassium fluoride, potassium chloride, and rubidium chloride.

Atoms can also share electrons. Oxygen (O) is missing two electrons in the outer shell, and hydrogen (H) either is missing one or has an extra. They can both have "filled" outer shells if they share electrons. Two hydrogens combine with one oxygen to form the stable water molecule H_2O.

These ideas were a great accomplishment for the Bohr model of the atom. Finally, there was a simple scheme for explaining the large collection of properties of the elements. The model provided a *physical* theory for understanding the *chemical* properties of the elements.

X Rays

The cathode-ray experiments that we discussed earlier in this chapter led to another important discovery that had an impact on the new atomic theory. In 1895 German scientist Wilhelm Roentgen saw a glow coming from a piece of paper on a bench near a working cathode-ray tube. The paper had been coated with a fluorescent substance. Fluorescence was known to result from exposure to ultraviolet light, but no ultraviolet light was present. The glow persisted even when the paper was several meters from the tube, and the tube was covered with black paper!

Roentgen conducted numerous experiments demonstrating that although these new rays were associated with cathode rays, they were different. Cathode rays could penetrate no more than a few centimeters of air, but the new rays could penetrate several meters of air or even thin samples of wood, rubber, or metal. Using electric and magnetic fields, Roentgen could not deflect these rays, but he did find that if he deflected the cathode rays, the rays came from the new spot where the cathode rays struck the glass walls. The new rays were also shown to be emitted by many different materials. Roentgen named them **X rays** because of their unknown nature.

X ray of a hand hit by pellets.

The discovery of X rays quickly created public excitement because people could see their bones and internal organs. This discovery was quickly put to use by the medical profession. In 1901 Roentgen received the first Nobel Prize in physics. We now know that X rays are high-frequency electromagnetic radiation created when the electrons in the cathode-ray tube are slowed down as they crash into the metal target inside the tube.

A typical X-ray spectrum (Figure 23-24) consists of two parts: a continuous spectrum produced by the rapid slowing of the bombarding electrons and a set of discrete spikes. These spikes are from X-ray photons emitted by the atoms that electrons are crashing into and therefore carry information about the element. They occur when a bombarding electron knocks an atomic electron from a lower energy level completely out of the atom. An electron in a higher energy level fills the vacancy. When this electron falls into the vacated level, an X-ray photon is emitted. The energies of these X rays can be used to calculate the energy differences of the atom's inner electron orbits, which are characteristic of the element.

Shortly after Bohr proposed his model of the atom, Henry Moseley, a British physicist, used Bohr's model and X rays to resolve some discrepancies in the periodic table. In several cases, the order of the elements determined by their atomic masses disagreed with their order determined by their chemical properties. For instance, according to atomic masses, cobalt should come after nickel in the periodic table, yet its properties suggested that it should precede nickel.

Bohr's model predicted that the energies of these characteristic X rays would increase as the number of electrons in the atom increased. Moseley bombarded various elements with electrons and studied the X rays that were given off. His experimental results agreed with Bohr's theoretical prediction. In almost all cases, the results matched the order obtained from the atomic masses; however, they showed that cobalt should come before nickel. As a result of this work, we now know that the correct ordering of the elements is by **atomic number** (the number of electrons in a neutral atom) and not by atomic mass.

Figure 23-24 A typical X-ray spectrum contains a continuous background and characteristic lines from the inner atomic levels.

Summary

When an electric current passes through a gas, it gives off a characteristic emission spectrum. The absorption spectrum is obtained by passing white light through a cool gas, producing dark lines in the continuous, rainbow spectrum. The emission spectrum has all the lines in the absorption spectrum plus many additional ones.

Cathode rays are electrons with a charge-to-mass ratio about 1800 times that for hydrogen ions. Electrons have an electric charge of 1.6×10^{-19} coulomb and a mass of 9.11×10^{-31} kilogram. The cathode-ray experiments also led to the discovery of X rays, a form of high-frequency electromagnetic radiation.

Rutherford used alpha particles to show that the positive charge was concentrated in a nucleus with a diameter $\frac{1}{100,000}$ the size of the atom. Thus, Rutherford's "solar system" model of the atom is almost entirely empty space. His model failed to explain the chemical periodicity and details of the spectral lines.

All objects emit electromagnetic radiation. The spectrum of radiation emitted through a small hole in a heated, hollow block is the same for all materials at the same temperature. The frequencies are not equally bright, and the maximum of the spectrum shifts toward the blue as the temperature of the cavity increases. The spectrum drops to zero on both the high-frequency (small-wavelength) and low-frequency (long-wavelength) sides.

Planck's analysis of the electromagnetic spectrum emitted by hot cavities provided the first clues that phenomena on the atomic level are quantized.

The energy E of the quanta depends on Planck's constant h and the frequency f: $E = hf$.

Planck's idea was expanded to include the electromagnetic radiation by Einstein's efforts to explain the photoelectric effect. When light is shined on certain clean metallic surfaces, electrons are ejected with a range of kinetic energies up to a maximum energy, depending on the color of the light. If the light's intensity is changed, the rate of ejecting electrons changes, but the maximum kinetic energy stays fixed. This is explained by assuming that light is also quantized, existing as bundles of energy called photons. The energy of a photon is given by Planck's relationship, $E = hf$. A photon gives its entire energy to one electron in the metal.

Bohr's model for the hydrogen atom incorporated Planck's discrete energies and Einstein's photons into Rutherford's model. It was based on three postulates: (1) Allowed electron orbits are those in which the angular momentum is an integral multiple of a smallest possible value. (2) An electron in an allowed orbit does not radiate. (3) When an electron jumps from one orbit to another, it emits or absorbs a photon whose energy is equal to the difference in the energies of the two orbits.

The Bohr model was successful in accounting for many atomic observations, especially the emission and absorption spectra for hydrogen and the ordering of the chemical elements.

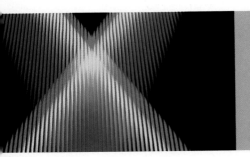

CHAPTER 23 *Revisited*

The light emitted by an atom is created when an atomic electron moves from one energy level to a lower one. The size of the jump determines the energy of the photon and—via Planck's relationship—the color of the emitted light. Therefore, the colors provide us with information about the energy levels of the atom's electronic structure.

Key Terms

absorption spectrum The collection of wavelengths missing from a continuous distribution of wavelength because of the absorption of certain wavelengths by the atoms or molecules in a gas.

alpha particle The nucleus of the helium atom.

atomic number The number of electrons in the neutral atom of an element. This number also gives the order of the elements in the periodic table.

cathode ray An electron emitted from the negative electrode in an evacuated tube.

electron The basic constituent of atoms that has a negative charge.

electron volt A unit of energy equal to the kinetic energy acquired by an electron or proton falling through an electric potential difference of 1 volt. The electron volt is equal to 1.6×10^{-19} joule.

emission spectrum The collection of discrete wavelengths emitted by atoms that have been excited by heating or electric currents.

ion An atom with missing or extra electrons.

nucleus The central part of an atom containing the positive charges.

photoelectric effect The ejection of electrons from metallic surfaces by illuminating light.

photon A particle of light. The energy of a photon is given by the relationship $E = hf$, where f is the frequency of the light and h is Planck's constant.

quantum (pl., *quanta*) The smallest unit of a discrete property. For instance, the quantum of charge is the charge on the proton.

quantum number A number giving the value of a quantized quantity. For instance, a quantum number specifies the angular momentum of an electron in an atom.

shell A collection of electrons in an atom that have approximately the same energy.

X ray A high-energy photon with a range of frequencies in the electromagnetic spectrum lying between the ultraviolet and the gamma rays.

Questions and exercises are paired so that most odd-numbered are followed by a similar even-numbered.

Blue-numbered questions and exercises are answered in Appendix B.

❦ indicates more challenging questions and exercises.

WebAssign Many Conceptual Questions and Exercises for this chapter may be assigned online at WebAssign.

Conceptual Questions

1. Today, we think of the periodic table as being arranged in order of increasing proton (or electron) number. Why did Mendeleev not use this approach?

2. If Mendeleev was ordering the elements according to their masses, why did he not produce a table with only one row?

3. What element in Mendeleev's periodic table is most similar to silicon (Si)?

4. What elements would you expect to have chemical properties similar to chlorine (Cl)?

5. Why are the spectral lines for elements sometimes called "atomic fingerprints"?

6. How could you determine whether there is oxygen in the Sun?

7. Would you obtain an emission or an absorption spectrum at location A in the setup shown in the following figure?

8. Would you obtain an emission or an absorption spectrum at location B in the setup shown in the preceding figure?

9. The emission spectra shown in the following figure were all obtained with the same apparatus. What elements can you identify in sample (a)? Are there any that you cannot identify?

Questions 9 and 10.

10. What element(s) can you identify in sample (b) of the preceding figure? Are there any that you cannot identify?

11. How does the number of lines in the absorption spectrum for an element compare with the number in the emission spectrum?

12. What are the differences between an emission spectrum and an absorption spectrum?

13. When your authors were students, their textbooks did not have color. Could the spectra in Figure 23-3 still be used to identify atoms if they were in black and white? Explain.

14. Two graphs of brightness versus wavelength are shown in the following figure. Identify which is an absorption spectrum and which is an emission spectrum.

15. To which of the brightness-versus-wavelength graphs in Question 14 does the line spectrum in the figure correspond?

16. Sketch the brightness-versus-wavelength graph for the hydrogen spectrum shown in Figure 23-3.

17. Suppose you were a 19th-century scientist who had just discovered a new phenomenon known as *zeta rays* (yes, we're making this up). What experiment could you

perform to determine whether zeta rays were charged particles or an electromagnetic wave? Could this experiment distinguish between neutral particles and an electromagnetic wave?

18. Imagine that you determined that the zeta rays from Question 17 were charged particles. How would you determine the sign of the charge?

19. In the Millikan oil-drop apparatus shown in Figure 23-8, an electric field provides a force that balances the gravitational force on charged oil drops. Millikan found that he needed the electric field to point down toward the floor. Was the net charge on the oil drops a result of an excess or a deficit of electrons?

20. Why was it not necessary for Millikan to use oil drops with an excess of only one electron to determine the charge of a single electron?

21. Millikan's oil-drop experiment was used to determine the charge on a single electron. Why should Millikan also get credit for determining the electron's mass?

22. What do we mean when we say that a physical quantity is quantized?

23. Why did Thomson feel that electrons were pieces of atoms rather than atoms of a new element?

24. How does Rutherford's model of an atom explain why most alpha particles pass right through a thin gold foil? How does it account for why some alpha particles are scattered backward?

25. Rutherford's model predicted that atoms should be unstable; the electrons should spiral into the nucleus in extremely short times. What caused this instability in Rutherford's model?

26. Rutherford's model provided an explanation for the emission of light from atoms. What was this mechanism and why was it unsatisfactory?

27. Why can the curves for the intensities of the colors emitted by hot solid objects not serve as "atomic fingerprints" of the materials?

28. If all objects emit radiation, why don't we see most of them in the dark?

29. As you move to the right along the horizontal axis of Figure 23-14, is the frequency increasing or decreasing? Explain your reasoning.

30. For an object at a temperature of 8000 K, use Figure 23-14 to determine whether the light intensity is greater for light in the ultraviolet or in the infrared.

31. You measure the brightness of two different hot objects, first with a blue filter and then with a red filter. You find that object A has a brightness of 25 in the blue and 20 in the red. Object B has a brightness of 12 in the blue and 3 in the red. The brightness units are arbitrary but the same for all measurements. Which is the hotter of the two objects?

32. The curves in Figure 23-14 show the intensities of the various wavelengths emitted by an object at three different temperatures. The region corresponding to visible light is indicated. How would the color of the object at 8000 K compare to the color of the object at 4000 K?

33. Why do astronomers often use the terms *color* and *temperature* interchangeably when referring to stars?

Jerry Schad/Photo Researchers, Inc.

Questions 33 and 34.

34. Why are blue stars thought to be hotter than red stars?

35. What assumptions did Planck make that enabled him to obtain the correct curve for the spectrum of light emitted by a hot object?

36. What assumptions did Einstein make that enabled him to account for the experimental observations of the photoelectric effect?

37. What property of the emitted photoelectrons depends on the intensity of the incident light?

38. What property of the emitted photoelectrons depends on the frequency of the incident light?

39. If a metal surface is illuminated by light at a single frequency, why don't all the photoelectrons have the same kinetic energy when they leave the metal's surface?

40. How is it possible that ultraviolet light can cause sunburn but no amount of visible light will?

41. You find that if you shine ultraviolet light on a negatively charged electroscope, the electroscope discharges even if the intensity of the light is low. Red light, however, will not discharge the electroscope even at high intensities. How do you account for this?

42. You find that if you shine ultraviolet light on a negatively charged electroscope, the electroscope discharges. Can you discharge a positively charged electroscope the same way? Why or why not?

43. What are the three assumptions of Bohr's model of the atom?

44. Why did Bohr assume that the electrons do not radiate when they are in the allowed orbits?

45. An electron in the $n = 3$ energy level can drop to the ground state by emitting a single photon or a pair of photons. How does the total energy of the pair compare with the energy of the single photon?

46. If electrons in hydrogen atoms are excited to the fourth Bohr orbit, how many different frequencies of light may be emitted?

47. How can the spectrum of hydrogen contain so many spectral lines when the hydrogen atom only has one electron?

48. What determines the frequency of a photon emitted by an atom?

49. Why does the spectrum of lithium (element 3) resemble that of hydrogen?

50. How does Bohr's model explain that there are more lines in the emission spectrum than in the absorption spectrum?

51. How many electrons would you expect to find in each shell for chlorine (Cl)?

52. Radon (element 86) is a gas. Would you expect the molecules of radon to consist of a single atom or a pair of atoms? Why?

53. Sodium does not naturally occur as a free element. Why?

54. What effective charge do the outer electrons in aluminum "see"? (Aluminum is element 13.)

55. What type of electromagnetic wave has a wavelength about the size of an atom? (The electromagnetic spectrum is given in Figure 22-26.)

56. Are X rays deflected by electric or magnetic fields? Explain.

57. How does an X ray differ from a photon of visible light?

58. Why would you not expect an X-ray photon to be emitted every time an inner electron is removed from an atom?

Exercises

59. What is the charge-to-mass ratio for a cathode ray?

60. What is the charge-to-mass ratio for a hydrogen ion (an isolated proton)?

61. Given that the radius of a hydrogen atom is 5.29×10^{-11} m and that its mass is 1.682×10^{-27} kg, what is the average density of a hydrogen atom? How does it compare with the density of water?

62. What is the average density of the hydrogen ion (an isolated proton) given that its radius is 1.2×10^{-15} m and that its mass is 1.673×10^{-27} kg? It is interesting to note that such densities also occur in neutron stars.

63. If you were helping your younger brother build a scale model of an atom for a science fair and wanted it to fit in a box 1 m on each side, how big would the nucleus be?

64. A student decides to build a physical model of an atom. If the nucleus is a rubber ball with a diameter of 1 cm, how far away would the outer electrons be?

65. What is the energy of a photon of orange light with a frequency of 5.0×10^{14} Hz?

66. What is the energy of the most energetic photon of visible light?

67. A photon of green light has energy 3.6×10^{-19} J. What is its frequency?

68. An X-ray photon has energy 1.5×10^{-15} J. What is its frequency?

69. A microwave photon has an energy of 2×10^{-23} J. What is its wavelength?

70. A photon of yellow light has a wavelength of 6.0×10^{-7} m. What is its energy?

71. What is the angular momentum of an electron in the ground state of hydrogen?

72. What is the angular momentum of an electron in the $n = 4$ level of hydrogen?

73. What is the radius of the $n = 4$ level of hydrogen?

74. What is the quantum number of the orbit in the hydrogen atom that has 36 times the radius of the smallest orbit?

75. What is the frequency of a photon of energy 3 eV?

76. What is the energy, in electron volts, of a yellow photon of wavelength 6.0×10^{-7} m?

77. According to Figure 23-20, it requires a photon with an energy of 10.2 eV to excite an electron from the $n = 1$ energy level to the $n = 2$ energy level. What is the frequency of this photon? Does it lie in, above, or below the visible range?

78. When a proton captures an electron, a photon with an energy of 13.6 eV is emitted. What is the frequency of this photon? Does it lie in, above, or below the visible range?

79. What is the ratio of the volumes of the hydrogen atom in the $n = 1$ state compared with those in the $n = 2$ state?

80. The diameter of the hydrogen atom is 10^{-10} m. In Bohr's model this means that the electron travels a distance of about 3×10^{-10} m in orbiting the atom once. If the orbital frequency is 7×10^{15} Hz, what is the speed of the electron? How does this speed compare with that of light?

81. What difference in energy between two atomic levels is required to produce an X ray with a frequency of 2×10^{18} Hz?

82. What is the frequency of the X ray that is emitted when an electron drops down to the ground state from an excited state with 1000 eV more energy?

The Modern Atom

▶Modern lasers have allowed us to take truly three-dimensional pictures (holography), perform surgery inside our bodies, weld retinas in our eyes, cut steel plates for industry and fabric for blue jeans, and perform myriad other amazing tasks. What makes laser light so special?

(See page 541 for the answer to this question.)

SuperStock

A laser being used to perform eye surgery.

BOHR'S model of the atom lacked "beauty"; the way the theory blended classical and quantum ideas together in a seemingly contrived way to account for the experimental results was troublesome. Bohr used the ideas of Newtonian mechanics and Maxwell's equations until he ran into trouble; then he abruptly switched to the quantum ideas of Planck and Einstein. For example, the electron orbits were like Newton's orbits for the planets. The force between the electron and the nucleus was the electrostatic force. According to Maxwell's equations, these electrons should have continuously radiated electromagnetic waves. Because this obviously didn't happen, Bohr postulated discrete orbits with photons being emitted only when an electron jumped from a higher orbit to a lower one. There was no satisfactory reason for the existence of the discrete orbits other than that they gave the correct spectral lines.

Bohr's model of the atom was remarkably successful in some areas, but it failed in others. Even though other models have replaced it, it was nevertheless important in advancing our understanding of matter at the atomic level.

Successes and Failures

Accounting for the stability of atoms by postulating orbits in which the electron did not radiate was a mild success of the Bohr theory. The problem was that nobody could give a fundamental reason for why this was so. Why *don't* the electrons radiate? They are accelerating, and according to the classical theory of electromagnetism, they should radiate.

The Bohr model successfully provided the numerical values for the wavelengths in the spectra of hydrogen and hydrogen-like ions. It provided qualitative agreement for the elements with a single electron orbiting a filled shell, but it could not predict the spectral lines when there was more than one electron in the outer shell.

But all along new data were being collected. Close examination of the spectral lines, especially when the atoms were in a magnetic field, revealed that the lines were split into two or more closely spaced lines. The Bohr theory could not account for this. Also, the different spectral lines for a particular element were not of the same brightness. Apparently, some jumps were more likely than others. The Bohr theory provided no clue to why this was so.

Although Bohr's model successfully described the general features of the periodic table, it could not explain why the shells had a certain capacity for electrons. Why did the first shell contain two electrons? Why not just one, or maybe three? And what's so special about the next level that it accepts eight electrons?

There were other loose ends. Einstein's theory of relativity was established, and it was generally accepted that the speeds of the electrons were fast enough and the measurements precise enough that relativistic effects should be included. But Bohr's model was nonrelativistic. Clearly, this was a transitional model.

De Broglie's Waves

Extended presentation available in the *Problem Solving* supplement

In 1923 a French graduate student proposed a revolutionary idea to explain the discrete orbits. Louis de Broglie's idea is easy to state but hard to accept: *electrons behave like waves.* De Broglie reasoned that if light could behave like particles, then particles should exhibit wave properties. He viewed all atomic objects as having this dualism; each exhibits wave properties *and* particle properties.

◄ electrons behave like waves

Louis de Broglie

De Broglie's view of the hydrogen atom had electrons forming standing-wave patterns about the nucleus like the standing-wave patterns on a guitar string (Chapter 16). Not every wavelength will form a standing wave on the string; only those that have certain relationships with the length of the string will do so.

Imagine forming a wire into a circle. The traveling waves travel around the wire in both directions and overlap. When a whole number of wavelengths fits along the circumference of the circle, a crest that travels around the circle arrives back at the starting place at the same time another crest is being generated. In this case the waves reinforce each other, creating a standing wave. This process is illustrated in Figure 24-1 for a wire and in Figure 24-2 for an atom.

De Broglie postulated that the wavelength λ of the electron is given by the expression

$$\lambda = \frac{h}{mv}$$

Figure 24-1 The shaft holding the wire executes simple harmonic motion that creates waves traveling around the circular wire in both directions. Standing waves form when the circumference is equal to a whole number of wavelengths.

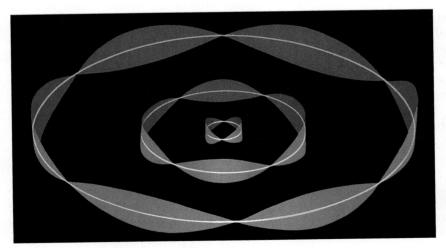

Figure 24-2 The electrons form standing waves about the nucleus.

where h is Planck's constant, m is the electron's mass, and v is the electron's speed. With this expression de Broglie could reproduce the numerical results for the orbits of hydrogen obtained by Bohr. The requirement that the waves form standing-wave patterns automatically restricted the possible wavelengths and therefore the possible energy levels.

De Broglie's idea put the faculty at his university in an uncomfortable position. If his idea proved to have no merit, they would look foolish awarding a Ph.D. for a crazy idea. The work was very speculative. Although he could account for the energy levels in hydrogen, de Broglie had no other experimental evidence to support his hypothesis. To make matters worse, de Broglie's educational background was in the humanities. He had only recently converted to the study of physics and was working in an area of physics that was not well understood by the faculty. Further, de Broglie came from an influential noble family that had played a leading role in French history. He carried the title of prince. His thesis supervisor resolved the problem by showing the work to Einstein, who indicated that the idea was basically sound.

Although de Broglie's idea was appealing, it was still speculative. It was important to find experimental confirmation by observing some definitive wave behavior, such as interference effects, for electrons. An electron accelerated through a potential difference of 100 volts has a speed that yields a de Broglie wavelength comparable to the size of an atom. Therefore, to see interference effects produced by electrons, we need atom-sized slits. Because X rays have atom-sized wavelengths, they can be used to search for possible setups. Photographs of X rays scattered from randomly oriented crystals, such as the one shown in Figure 24-3, show that the regular alignment of the atoms in crystals can produce interference patterns, such as those produced by multiple slits. Therefore, we might look for the interference of electrons from crystals.

Experimental confirmation of de Broglie's idea came quickly. Two American scientists, C. J. Davisson and L. H. Germer, obtained confusing data from a seemingly unrelated experiment involving the scattering of low-energy electrons from nickel crystals. The scattered electrons had strange valleys and peaks in their distribution. While they were showing these data at a conference at Oxford, it was suggested to Davisson that this anomaly might be a diffraction effect. Further analysis of the data showed that electrons do exhibit wave behavior. The photograph in Figure 24-4 shows an interference pattern produced by electrons. De Broglie was the first of only two physicists to receive a Nobel Prize for their thesis work.

Nothing in de Broglie's idea suggests that it should apply only to electrons. This relationship for the wavelength should also apply to all material particles. Although it may not apply to macroscopic objects, assume for the moment that it does. A baseball (mass = 0.14 kilogram) thrown at 45 meters per second (100 mph) would have a wavelength of 10^{-34} meter! This is an incredibly small distance. It is 1 trillion-trillionth the size of a single atom. This wavelength is smaller when compared with an atom than an atom is when compared with the solar system. Even a mosquito flying at 1 meter per second would have a wavelength that is only 10^{-27} meter.

You shouldn't spend much time worrying that your automobile will diffract off the road the next time you drive through a tunnel. Ordinary-sized objects traveling at ordinary speeds have negligible wavelengths.

Figure 24-3 An interference pattern produced by light.

Figure 24-4 An interference pattern produced by electrons scattering from randomly oriented aluminum crystallites.

Q: Why would you not expect baseballs thrown through a porthole to produce a measurable diffraction pattern?

A: For diffraction to be appreciable, the wavelength must be about the same size as the opening.

WORKING IT OUT *Electron Microscope*

Electron microscopes take advantage of the wave nature of particles. Electrons are accelerated to high speeds, giving them a very short de Broglie wavelength. Diffraction effects require that the wavelength of our probe be small compared to the size of the object being observed. How fast would electrons in our microscope need to be traveling to observe a gold atom that is 0.288 nm across?

To avoid diffraction effects, the wavelength of our electrons must be smaller than the size of the gold atom. Let's assume a wavelength that is smaller by one order of magnitude, or $\lambda = 0.0288$ nm $= 2.88 \times 10^{-11}$ m. The formula for the de Broglie wavelength

$$\lambda = \frac{h}{mv}$$

can then be rewritten to solve for the needed electron speed:

$$v = \frac{h}{m\lambda} = \frac{6.63 \times 10^{-34}\,\text{J} \cdot \text{s}}{(9.11 \times 10^{-31}\,\text{kg})(2.88 \times 10^{-11}\,\text{m})} = 2.53 \times 10^{7}\,\text{m/s}$$

nearly 10% of the speed of light!

Everyday Physics | *Seeing Atoms*

Atoms are so small that they cannot be seen with the most powerful optical microscopes. This limitation is due to the wave properties of light that we discussed in Chapter 19. The relatively large wavelength of the light scatters from the atoms without yielding any information about the individual atoms. Although electron microscopes have much higher resolution (electrons can be accelerated to high speeds by electric fields, giving them wavelengths less than $\frac{1}{100}$ as long as wavelengths of visible light), even the most powerful electron microscope does not allow us to view individual atoms.

A new type of microscope, developed in the early 1980s, provides views of atomic surfaces with a resolution thousands of times greater than is possible with light. The developers of the *scanning tunneling microscope,* or STM, were awarded the Nobel Prize in 1986 (along with the inventor of the electron microscope).

The STM uses a sharp probe (several atoms across at the tip) placed close to the surface (about 1 ten-billionth of a meter) to "view" the surface under investigation. As the probe is moved back and forth over the surface, electrons are pulled from the atoms to the probe, creating a current that is monitored by a computer. The current is greatest when the probe is directly over an atom and quite small when the probe is between atoms. By carefully mapping the surface, the operator can plot the structure of the surface with a resolution smaller than the atomic dimensions. Computer displays like the one shown here clearly indicate the orderly arrangement of the atoms on the surface of the material. These measurements

A scanning tunneling microscope image of iodine atoms absorbed on a platinum surface. Note the orderly arrangement of the atoms and that one of the atoms is missing.

Courtesy of VEECO Instruments and Purdue University

can also show the location of individual atoms that are deposited onto the surface. Such studies are important in understanding the physics of surfaces as well as of the catalytic processes used in manufacturing.

1. What principle of physics limits the resolution of an optical microscope or an electron microscope?

2. Why is the STM not limited by the same resolution issues that apply to optical and electron microscopes?

Waves and Particles

Imagine the controversy de Broglie caused. Interference of light is believable because parts of the wave pass through each slit and interfere in the overlap region (Chapter 19). But electrons don't split in half. Every experiment designed to detect electrons has found complete electrons, not half an electron. So how can electrons produce interference patterns?

The following series of thought experiments was proposed by physicist Richard Feynman to summarize the many experiments that have been conducted to resolve the wave–particle dilemma. Although they are idealized, this sequence of experiments gets at crucial factors in the issue.

We will imagine passing various things through two slits and discuss the patterns that would be produced on a screen behind the slits. In each case there are three pieces of equipment: a source, two slits, and a detecting screen arranged as in Figure 24-5.

In the first situation, we shoot indestructible bullets at two narrow slits in a steel plate. Assume that the gun wobbles so that the bullets are fired randomly at the slits. Our detecting screen is a sandbox. We simply count the bullets in certain regions in the sandbox to determine the pattern.

Imagine that the experiment is conducted with the right-hand slit closed. After 1 hour we sift through the sand and make a graph of the distribution of bullets. This graph is shown in Figure 24-6(a). The curve is labeled N_L to indicate that the left-hand slit was open. Repeating the experiment with only the right-hand slit open yields the similar curve N_R shown in Figure 24-6(b).

Figure 24-5 The experimental setup for the thought experiment concerning the wave–particle dilemma.

Q: What curve would you expect with both slits open?

A: The curve with both slits open should be the sum of the first two curves.

When both slits are opened, the number of bullets hitting each region during 1 hour is just the sum of the numbers in the previous experiments—that is, $N_{LR} = N_L + N_R$.

◀ bullets add

This is just what we expect for bullets or any other particles. This graph is shown in Figure 24-6(c).

Now imagine repeating the experiment using water waves. The source is now an oscillating bar that generates straight waves. The detecting screen is a collection of devices that measure the energy (that is, the intensity) of the wave arriving at each region. Because these are waves, the detector does not detect the energy arriving in chunks but rather in a smooth, continuous manner. The graphs in Figure 24-7 show the average intensity of the waves at each position across the screen for the same three trials.

Once again, the three trials yield no surprises. We expect an interference pattern when both slits are open. We see that the intensity with two slits open is not equal to the sum of the two cases with only one slit open, $I_{LR} \neq I_L + I_R$, but this is what we expect for water waves or for any other waves.

◀ water waves interfere

Both sets of experiments make sense, in part, because we used materials from the macroscopic world—bullets and water waves. However, a new reality emerges when we repeat these experiments with particles from the atomic world.

This time we use electrons as our "bullets." An electron gun shoots electrons randomly toward two narrow slits. Our screen consists of an array of devices that can detect electrons. Initially, the results are similar to those obtained in the bullet experiment: electrons are detected as whole particles, not fractions of particles. The patterns produced with either slit open are the expected ones

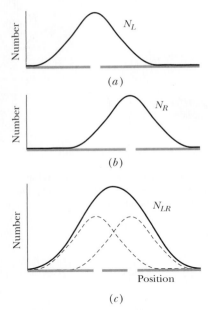

Figure 24-6 Distribution of bullets with (a) the left-hand slit open (N_L), (b) the right-hand slit open (N_R), and (c) both slits open (N_{LR}).

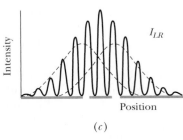

Figure 24-7 Intensity of water waves with (a) the left-hand slit open (I_L), (b) the right-hand slit open (I_R), and (c) both slits open (I_{RL}).

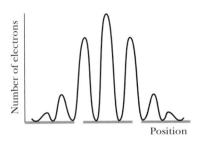

Figure 24-8 The distribution of electrons with two slits open.

electrons interfere ▶

and are the same as those in Figure 24-6(a) and (b). The surprise comes when we look at the pattern produced with both slits open: we get an interference pattern like the one for waves (Figure 24-8)!

Note what this means: if we look at a spot on the screen that has a minimum number of electrons and close one slit, we get an *increase* in the number of counts at that spot. Closing one slit yields more electrons! This is not the behavior expected of particles.

One possibility that was suggested to explain these results is that the electrons are somehow affecting each other. We can test this by lowering the rate at which the source emits electrons so that only one electron passes through the setup at a time. In this case we may expect the interference pattern to disappear. How can an electron possibly interfere with itself? Each individual electron should pass through one slit or the other. How can it even know that the other slit is open? But the interference pattern doesn't disappear. Even though there is only one electron in the apparatus at a time, the same interference effects are observed after a large number of electrons are measured.

The set of experiments can be repeated with photons. The results are the same. The detectors at the screen see complete photons, not half photons. But the two-slit pattern is an interference pattern. Photons behave like electrons. Photons and electrons exhibit a duality of particle and wave behavior. Table 24-1 summarizes the results of these experiments.

Table 24-1	*Summary of Experiments with Two Slits*	
Source	**Detection**	**Pattern**
Bullets	Chunks	No interference
Water waves	Continuous	Interference
Electrons or photons	Chunks	Interference

FLAWED REASONING

Your friend argues: "The interference pattern for electrons passing through two slits isn't that mysterious. We know that two electrons interact via the electric force, and so an electron passing through one slit is going to affect the motion of an electron passing through the other slit."

What experimental evidence could you use to persuade your friend that the interference pattern is not caused by the interaction between electrons?

ANSWER Even if the beam is turned down so low that only one electron passes through the slits at a time, the same interference pattern is produced.

Probability Waves

Nature has an underlying rule governing this strange behavior. We need to unravel the mystery of why electrons (or photons, for that matter) are always detected as single particles and yet collectively they produce wavelike distributions. Something is adding together to give the interference patterns. To see this we once again look at water waves.

There is something about water waves that does add when the two slits are open—the displacements (or heights) of the individual waves at any instant of time. Their sum gives the displacement at all points of the interference pattern—that is, $h_{12} = h_1 + h_2$. The maximum displacement at each point is the amplitude of the resultant wave at that point.

With light and sound waves we observe the intensities of the waves, not their amplitudes. Intensity is proportional to the square of the amplitude. It is important to note that the resultant, or total, intensity is *not* the sum of the individual intensities. The total intensity is calculated by adding the individual displacements to obtain the resulting amplitude of the combined waves and then squaring this amplitude. Thus, the intensity depends on the amplitude of each wave *and* the relative phase of the waves. As you can see in Figure 24-9, the intensity of the combined waves is larger than the sum of the individual intensities when the displacements are in the same direction and smaller when they are in opposite directions.

There is an analogous situation with electrons. The definition for a **matter–wave amplitude** (called a *wave function*) is analogous to that for water waves.

(*a*)

(*b*)

(*c*)

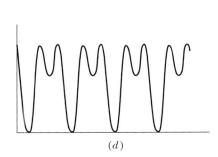

(*d*)

Figure 24-9 The two individual waves (a and b) are combined (c) by adding the displacements before squaring. Note the difference when the individual displacements are squared before adding (d).

(This matter–wave amplitude is usually represented by the Greek letter ψ, and pronounced "sigh".) The square of the matter–wave amplitude is analogous to the intensity of a wave. In this case, however, the "intensity" represents the likelihood, or probability, of finding an electron at that location and time. There is one important difference between the two cases. We can physically measure the amplitude of an ordinary mechanical wave, but there is *no* way to measure the amplitude of a matter wave. We can only measure the value of this amplitude squared.

These ideas led to the development of a new view of physics known as **quantum mechanics,** which are the rules for the behavior of particles at the atomic and subatomic levels. These rules replace Newton's and Maxwell's rules. A quantum-mechanical equation, called *Schrödinger's equation* after Austrian physicist Erwin Schrödinger, is a wave equation that provides all possible information about atomic particles.

A Particle in a Box

It is instructive to look at another simple, though somewhat artificial, situation. Imagine you have an atomic particle that is confined to a box. Further imagine that the box has perfectly hard walls so that no energy is lost in collisions with the walls and that the particle moves in only one dimension, as shown in Figure 24-10.

From a Newtonian point of view, there are no restrictions on the motion of the particle; it could be at rest or bouncing back and forth with any speed. Because we assume that the walls are perfectly hard, the sizes of the particle's momentum and kinetic energy remain constant. In this classical situation, things happen much as we would expect from our commonsense world view. The particle is like an ideal Super Ball in zero gravity. It follows a definite path; we can predict when and where it will be at any time in the future.

Until the early 1900s, it was assumed that an electron would behave the same way. We now know that atomic particles have a matter–wave character and their properties—position, momentum, and kinetic energy—are governed by Schrödinger's equation. When these particles are confined, their wave nature guarantees that their properties are quantized. For example, the solutions of Schrödinger's equation for the particle in a box are a set of standing waves. In fact, the solutions are identical to those for the guitar string that we examined in Chapter 16. There is a discrete set of wavelengths that fits the conditions of confinement.

Figure 24-10 A particle in a box has quantized values for its de Broglie wavelength.

Are You On the Bus?

Q: How would you expect the wavelength of the fundamental standing wave to compare with the length of the box?

A: As with the guitar string analogy, we would expect the wavelength to be twice the length of the box.

The fundamental standing wave corresponds to the lowest energy state (the ground state), and one-half of a wavelength fits into the box, as shown in Figure 24-11(a). The probability of finding the particle at some location depends on the square of this wave amplitude. These probability values [Figure 24-11(b)] help locate the particle. You can see that the most likely place for finding the atomic particle is near the middle of the box because ψ^2 is large there. The probability of finding it near either end is quite small.

A bizarre situation arises with the higher energy levels. The wave amplitude and its square for the next higher energy level are shown in Figure 24-12. Now the most likely places to find the particle are midway between the center and

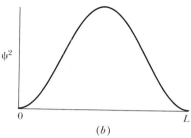

Figure 24-11 The (a) wave amplitude and (b) probability distribution for the lowest energy state for a particle in a box.

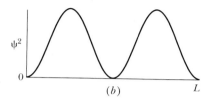

Figure 24-12 The (a) wave amplitude and (b) probability distribution for the first excited state for a particle in a box.

either end of the box. The least likely places are near an end or near the center. It is tempting to ask how the atomic particle gets from one region of high probability to the other. How does it cross the center where the probability of locating it is zero? This type of question, however, does not make sense in quantum mechanics. Particles do not have well-defined paths; they have only probabilities of existing throughout the space in question.

Q: Where are the most likely places to find the particle in the third energy level?

A: The square of the third standing wave will have maxima at the center and between the center and each side of the box (see Figure 16-6).

Are *You* On the Bus?

Suppose you try to find the particle. You would find it in one place, not spread out over the length of the box. However, you cannot predict where it will be. All you can do is predict the probability of finding it in a particular region. In addition, even if you know where it is at one time, you still cannot predict where it will be at a later time. Atomic particles do not follow well-defined paths as classical objects do.

The existence of quantized wavelengths means that other quantities are also quantized. Because de Broglie's relationship tells us that the momentum is inversely proportional to the wavelength, momentum is quantized. The kinetic energy is proportional to the square of the momentum. Therefore, kinetic energy must also be quantized. Thus, the particle in a box has quantized energy levels that can be labeled with a quantum number to distinguish one level from another as we did for the Bohr atom.

The Quantum-Mechanical Atom

The properties of the atom are calculated from the Schrödinger equation just as those for the particle in a box. Quantum mechanics works; it not only accounts for all the properties of the Bohr model but also corrects most of the deficiencies of that earlier attempt.

But this success has come at a price. We no longer have an atom that is easily visualized; we can no longer imagine the electron as being a little billiard ball moving in a well-defined orbit. It is meaningless to ask "particle" questions such as how the electron gets from one place to another or how fast it will be going after 2 minutes. The electron orbits are replaced by standing waves that represent probability distributions. The best we can do is visualize the atom as an electron cloud surrounding the nucleus. This is no ordinary cloud; the density of the cloud gives the probability of locating an electron at a given point in space. The probability is highest where the cloud is the most dense and lowest where it is the least dense. An artist's version of these three-dimensional clouds is shown in Figure 24-13. (Although each drawing is

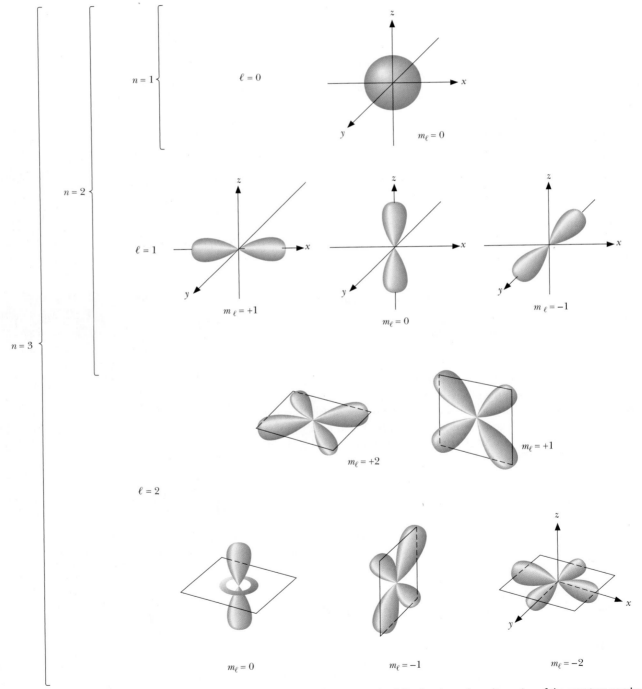

Figure 24-13 Probability clouds for a variety of electron states. See the following pages for a discussion of the quantum numbers n, ℓ, and m_{ℓ}.

Everyday Physics | *Psychedelic Colors*

In Chapter 17 we discussed why objects have colors. The object's color depends on the colors reflected from the illuminating light, but there is an exception to this. When certain materials are illuminated with ultraviolet (UV) light, they glow brightly in a variety of colors. This process is called *fluorescence* and is responsible for the colors seen on the "psychedelic" posters that were popular in the 1960s.

The "black lights" used to illuminate these posters radiate mostly in the UV region, which is invisible to our eyes. Although the light gives off a little purple, there are certainly no reds, yellows, or greens present. And yet these bright colors appear in the posters. The atomic model explains their presence. UV photons are very energetic and can kick electrons to higher energy levels. If the electron returns by several jumps as shown in Figure A, it gives off several photons, each of which has a lower energy. When one or more of these photons lie in the visible region of the electromagnetic spectrum, the material fluoresces.

A laundry detergent manufacturer once claimed that its soap could make clothes "whiter than white" by pointing out that the soap contained fluorescent material. This makes sense scientifically because some of the UV light is converted to visible light, making the shirt give off more light in the visible region. You may wish to observe various types of clothing and laundry soap under UV light to see whether they fluoresce. Do not look directly at the black light because UV light is dangerous to your eyes.

This process is also at work in fluorescent lights. The gas atoms inside the tube are excited by an electric discharge and emit UV photons. These photons cause atoms in the coating on the inside of the tube to fluoresce, giving off visible photons (Figure B).

A related phenomenon, *phosphorescence,* occurs with some materials. These materials continue to glow after the ultraviolet light is turned off. Some luminous watch dials are phosphorescent. In these materials the excited electrons remain much longer in upper energy states. In effect, phosphorescence is really a delayed fluorescence.

1. If the excited electron returns to its original state in a single jump, no fluorescence occurs. Why?

2. What is the difference between phosphorescence and fluorescence?

(a)

Figure A Fluorescence may occur when an electron makes several jumps in returning to its original state.

(b)

General Electric Company (both)

Figure B In a fluorescent lamp, a gas discharge produces ultraviolet light that is converted into visible light by a phosphor coating on the inside of the tube. (a) The two-component phosphor of praseodymium and yttrium fluorides produces the visible soft violet-pink light flooding the face of the researcher (b).

UV

confined to a finite space, the probability cloud extends to infinity, getting rapidly thinner the farther away you go from the center.)

The loss of orbiting electrons also means, however, the loss of the idea of an accelerating charge continuously radiating energy. The theory agrees with nature: atoms are stable.

As with the particle in a box, the simple act of confining the electron to the volume of the atom results in the quantization of its properties. If we allow the particle in the box to move in all three dimensions, we have standing waves in three dimensions and therefore three independent quantum numbers, one for each dimension. Similarly, the three-dimensionality of the atom yields three quantum numbers.

The particular form that these numbers take depends on the symmetry of the forces involved. In the case of atoms, the force is spherically symmetric, and the three quantum numbers are associated with the energy n, the size ℓ of the angular momentum, and its direction m_ℓ. However, these three quantum numbers were not adequate to explain all the features of the atomic spectra. The additional features could be explained by assuming that the electron spins on its axis. A fourth quantum number m_s was added that gave the orientation of the electron spin. There are only two possible spin orientations, usually called "spin up" and "spin down." Although the classical idea of the electron spinning like a toy top does not carry over into quantum mechanics, the effects analogous to those of a spinning electron are accounted for with this additional quantum number. The quantum number is retained, and, for convenience, we use the Newtonian language of electron spin.

The most recent model of the atom combines relativity and the quantum mechanics of electrons and photons in a theory known as *quantum electrodynamics* (QED), which is even more abstract than the quantum-mechanical model. In this theory, however, the concept of electron spin is no longer just an add-on but is a natural result of the combination of quantum mechanics and relativity.

Wolfgang Pauli

The Exclusion Principle and the Periodic Table

Let's return to the periodicity of the chemical elements that we discussed in Chapter 23. The periodicity required that we think of electrons existing in shells, but the theory did not tell us how many electrons could occupy each shell. The introduction of the quantum numbers said that electrons could exist only in certain discrete states and that these states formed shells, but it did not say how many electrons could exist in each state.

exclusion principle ▶

In 1924 Wolfgang Pauli suggested that no two electrons can be in the same state; that is, no two electrons can have the same set of quantum numbers. This statement is now known as the Pauli **exclusion principle.** When this principle is applied to the quantum numbers obtained from Schrödinger's equation and the electron spin, the periodicity of the elements is explained, as we will demonstrate.

Table 24-2 gives the first two quantum numbers for the first 30 elements. The values of the angular momentum quantum number ℓ are restricted by Schrödinger's equation to be integers in the range from 0 up to $n - 1$, and the values for the direction of the angular momentum m_ℓ are all integers from $-\ell$ to $+\ell$. Unlike the Bohr model, the angular momentum of the lowest energy state is zero, further supporting the notion that we cannot expect these atomic particles to act classically. For each value of n, ℓ, and m_ℓ, two spin states are available: spin up and spin down.

The good news is that the relationships between the quantum numbers explain why the orbital shells have different capacities, and this in turn explains the properties of the elements in the periodic table shown on the inside front cover of this

Table 24-2	Ground State Quantum Numbers for the First 30 Elements							
Atomic Number	**Element**	$n = 1$ $\ell = 0$	2 0	2 1	3 0	3 1	3 2	4 0
1	hydrogen (H)	1						
2	helium (He)	2						
3	lithium (Li)	2	1					
4	beryllium (Be)	2	2					
5	boron (B)	2	2	1				
6	carbon (C)	2	2	2				
7	nitrogen (N)	2	2	3				
8	oxygen (O)	2	2	4				
9	fluorine (F)	2	2	5				
10	neon (Ne)	2	2	6				
11	sodium (Na)	2	2	6	1			
12	magnesium (Mg)	2	2	6	2			
13	aluminum (Al)	2	2	6	2	1		
14	silicon (Si)	2	2	6	2	2		
15	phosphorus (P)	2	2	6	2	3		
16	sulfur (S)	2	2	6	2	4		
17	chlorine (Cl)	2	2	6	2	5		
18	argon (Ar)	2	2	6	2	6		
19	potassium (K)	2	2	6	2	6	0	1
20	calcium (Ca)	2	2	6	2	6	0	2
21	scandium (Sc)	2	2	6	2	6	1	2
22	titanium (Ti)	2	2	6	2	6	2	2
23	vanadium (V)	2	2	6	2	6	3	2
24	chromium (Cr)	2	2	6	2	6	5	1
25	manganese (Mn)	2	2	6	2	6	5	2
26	iron (Fe)	2	2	6	2	6	6	2
27	cobalt (Co)	2	2	6	2	6	7	2
28	nickel (Ni)	2	2	6	2	6	8	2
29	copper (Cu)	2	2	6	2	6	10	1
30	zinc (Zn)	2	2	6	2	6	10	2

text. The $n = 1$ state has only one angular momentum state and two spin states, so its maximum capacity is two electrons, both with $n = 1$ and $\ell = 0$, but with different spins. The first element, hydrogen, has one electron in the lowest energy state, whereas helium, the second element, has both of these states occupied. Because there are no more $n = 1$ states available and the Pauli exclusion principle does not allow two electrons to fill any one state, this completes the first shell.

The next two electrons go into the states with $n = 2$ and $\ell = 0$ because these states are slightly lower in energy than the $n = 2$ and $\ell = 1$ states. This takes care of lithium (element 3) and beryllium (4). There are six states with $n = 2$ and $\ell = 1$ because m_ℓ can take on three values (–1, 0, 1), and each of these can be occupied by two electrons, one with spin up and the other with spin down. These six states correspond to the next six elements in the periodic table, ending with neon (element 10). This completes the second shell, and the completed shell accounts for neon being a noble gas.

The electrons in the next eight elements occupy the states with $n = 3$ and $\ell = 0$ or $\ell = 1$. This completes the third shell, ending with the noble gas argon (18). The two states with $n = 4$ and $\ell = 0$ are lower in energy than the rest of the $n = 3$ states and are filled next. These correspond to potassium (19) and calcium (20). Then the remaining 10 states corresponding to $n = 3$ and $\ell = 2$ are filled, yielding the transition elements scandium (21) through zinc (30). Thus, the quantum-mechanical picture of the atom accounts for the observed periodicity of the elements.

The Uncertainty Principle

Courtesy of University of Hamburg

Werner Heisenberg

The interpretation that atomic particles are governed by probability left many scientists dissatisfied and hoping for some ingenious thinker to rescue them from this foolish predicament. German physicist Werner Heisenberg showed that there was no rescue. He argued that there is a fundamental limit to our knowledge of the atomic world.

Heisenberg's idea—that there is an indeterminacy of knowledge—is often misinterpreted. The uncertainty is not due to a lack of familiarity with the topic, nor is it due to an inability to collect the required data, such as the data needed to predict the outcome of a throw of dice or the Kentucky Derby. Heisenberg was proposing a more fundamental uncertainty—one that results from the wave–particle duality.

Imagine the following thought experiment. Suppose you try to locate an electron in a room void of other particles. To locate the electron, you need something to carry information from the electron to your eyes. Suppose you use photons from a dim lightbulb. Because this is a thought experiment, we can also assume that you have a microscope so sensitive that you will see a single photon that bounces off the electron and enters the microscope (Figure 24-14).

You begin by using a bulb that emits low-energy photons. These have low frequencies and long wavelengths. The low energy means that the photon will not disturb the electron much when it bounces off it. However, the long wavelength means that there will be lots of diffraction when the photon scatters from the electron (Chapter 19). Therefore, you won't be able to determine the location of the electron precisely.

To improve your ability to locate the electron, you now choose a bulb that emits more energetic photons. The shorter wavelength allows you to determine the electron's position relatively well. But the photon kicks the electron so hard that you don't know where the electron is going next. The smaller the wavelength, the better you can locate the electron but the more the photon alters the electron's path.

Figure 24-14 Heisenberg's thought experiment to determine the location of an electron.

Heisenberg argued that we cannot make any measurements on a system of atomic entities without affecting the system in this way. The more precise our measurements, the more we disturb the system. Furthermore, he argued, the measured and disturbed quantities come in pairs. The more precisely we determine one half of the pair, the more we disturb the other. In other words, the more *certain* we are about the value of one, the more *uncertain* we are about the value of the other. This is the essence of the uncertainty principle.

Two of these paired quantities are the position and momentum along a given direction. (Recall from Chapter 6 that the momentum for a particle is equal to its mass multiplied by its velocity.) As we saw in the thought experiment just described, the more certain your knowledge of the position, the more uncertain your knowledge of the momentum. The converse is also true.

This idea is now known as Heisenberg's **uncertainty principle.** Mathematically, it says that the product of the uncertainties of these pairs has a lower limit equal to Planck's constant. For example, the uncertainty of the position along the vertical direction Δy multiplied by the uncertainty of the component of the momentum along the vertical direction Δp_y must always be greater than Planck's constant h:

$$\Delta p_y \Delta y > h$$

◀ uncertainty principle

This principle holds for the position and component of momentum along the same direction. It does not place any restrictions on simultaneous knowledge of the vertical position and a horizontal component of momentum.

Another pair of variables that is connected by the uncertainty principle is energy and time, $\Delta E \Delta t > h$. This mathematical statement tells us that the longer the time we take to determine the energy of a given state, the better we can know its value. If we must make a quick measurement, we cannot determine the energy with arbitrarily small uncertainty. Stated in another way, the energy of a stable state that lasts for a long time is well determined. However, if the state is unstable and exists for only a short time, its energy must have some range of possible values given by the uncertainty principle.

WORKING IT OUT   *Uncertainty*

The speed of a proton is measured to be 5.00×10^4 m/s $\pm$ 0.003%. What is the minimum uncertainty in the position of the proton along the direction of its velocity?

We begin by finding the uncertainty in the proton's momentum. The momentum of the proton is:

$$p = mv = (1.67 \times 10^{-27} \text{ kg})(5.00 \times 10^4 \text{ m/s}) = 8.35 \times 10^{-23} \text{ kg} \cdot \text{m/s}$$

Because the uncertainty is 0.003% of this value, we have

$$\Delta p = 0.00003p = 2.51 \times 10^{-27} \text{ kg} \cdot \text{m/s}$$

We can now use Heisenberg's uncertainty principle to find the minimum uncertainty in the proton's position:

$$\Delta x > \frac{h}{\Delta p} = \frac{6.63 \times 10^{-34} \text{ J} \cdot \text{s}}{2.51 \times 10^{-27} \text{ kg} \cdot \text{m/s}} = 2.64 \times 10^{-7} \text{ m}$$

FLAWED REASONING

Two students are discussing the interpretation of Heisenberg's uncertainty principle:

Kristjana: "A particle cannot have a well-defined position and a well-defined momentum at the same time, which really goes against our common sense."

Matthew: "There's nothing really strange about the uncertainty principle—we learn in lab that all measurements have some degree of uncertainty."

Which of these students really understands the uncertainty principle?

ANSWER Matthew is mistakenly thinking of the uncertainty principle as putting a limit on our ability to measure properties that, at least in principle, have well-defined values. Kristjana understands that it is more fundamental than that and really does challenge our common sense.

Q: What does the uncertainty principle say about the energy of the photons emitted when electrons in the $n = 2$ state of hydrogen atoms drop down to the ground state?

A: Because the electrons spend a finite time in the $n = 2$ state, the energy of that state must have a spread in energy. Therefore, the photons have a spread in energy that shows up as a nonzero width of the spectral line.

"BUT YOU CAN'T GO THROUGH LIFE APPLYING HEISENBERG'S UNCERTAINTY PRINCIPLE TO EVERYTHING."

© Sidney Harris. Used by permission. ScienceCartoonsPlus.com.

The Complementarity Principle

Frustrating questions emerged as physicists built a world view of nature on the atomic scale. Is there *any* underlying order? How can one contemplate things that have mutually contradictory attributes? Can we understand the dual nature of electrons and light, which sometimes behave as particles and other times as waves?

There was no known way out. Physicists had to learn to live with this wave–particle duality. A complete description of an electron or a photon requires both aspects. This idea was first stated by Bohr and is known as the **complementarity principle.**

complementarity principle ▶

The complementarity principle is closely related to the uncertainty principle. As a consequence of the uncertainty principle, we discover that being completely certain about particle aspects means that we have no knowledge about the wave aspects. For example, if we are completely certain about the position and time for a particle, the wave aspects (wavelength and frequency) have infinite uncertainties. Therefore, wave and particle aspects do not occur at the same time.

Q: If you could determine which slit the electron goes through, would this have any effect on the two-slit interference pattern?

A: It would destroy the interference pattern. Knowledge of the particle properties (that is, the path of the electron) precludes the wave properties.

The idea that opposites are components of a whole is not new. The ancient Eastern cultures incorporated this notion as part of their world view. The most common example is the yin-yang symbol of tai chi tu. Later in life, Bohr was so attracted to this idea that he wrote many essays on the existence of complementarity in many modes of life. In 1947, when he was knighted for his work in physics, he chose the yin-yang symbol for his coat of arms.

Determinism

Classical Newtonian mechanics and the newer quantum mechanics have been sources of much debate about the role of cause and effect in the natural world. With Newton's laws of motion came the idea that specifying the position and momentum of a particle and the forces acting on it allowed the calculation of its future motion. Everything was determined. It was as if the universe was an enormous machine. This idea was known as the *mechanistic view*. In the 17th century, René Descartes stated, "I do not recognize any difference between the machines that artisans make and the different bodies that nature alone composes."

These ideas were so successful in explaining the motions in nature that they were extended into other areas. Because the universe is made of particles whose futures are predetermined, it was suggested that the motion of the entire universe must be predetermined. This notion was even extended to living organisms. Although the flight of a bumblebee seems random, its choices of which flowers to visit are determined by the motion of the particles that make up the bee. These generalizations caused severe problems with the idea of free will—that humans had something to say about the future course of events.

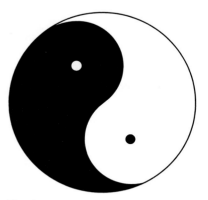

The yin-yang symbol reflects the complementarity of many things.

Is the flight of the bumblebee predetermined or a result of free will?

Einstein felt that the probabilistic nature of modern physics did not reflect the true reality of nature.

There were, however, some practical problems with actually predicting the future in classical physics. Because measurements could not be made with absolute precision, the position and momentum of an object could not be known exactly. The uncertainties in these measurements would lead to uncertainties about the calculations of future motions. However, at least in principle, certainty was possible. It was also impossible to measure the positions and momenta of all atomic particles in a small sample of gas, let alone all those in the universe. However, the motion of each atom was predetermined, and therefore the properties of gases were predetermined. Even though humans could not determine the paths, nature knew them. The future was predetermined.

With the advent of quantum mechanics, the future became a statistical issue. The uncertainty principle stated that it was impossible *even in principle* to measure simultaneously the position and momentum of a particle. The mechanistic laws of motion were replaced by an equation for calculating the matter waves of a system that gave only *probabilities* about future events. Even if we know the state of a system at some time, the laws of quantum mechanics do not permit the calculation of a future, only the probabilities for each of many possible futures. The future is no longer considered to be predetermined but is left to chance.

One of the main opponents of this probabilistic interpretation was Albert Einstein. His objections did not arise out of a lack of understanding. He understood quantum mechanics very well and even contributed to its interpretation. He believed that the path of an electron was governed by some (hidden) deterministic set of rules (like an atomic version of Newton's laws), not by some unmeasurable probability wave. The new physics didn't fit into his philosophy of the natural world. Einstein's famous rebellious quote is, "I, at any rate, am convinced that [God] is not playing at dice."

But nobody has ever found those hidden rules, and quantum mechanics continues to work better than anything else that has been proposed. The point is that hoping doesn't change the physics world view. Einstein spent a lot of time trying to disprove the very theory that he helped begin. He did not succeed. New work has shown that quantum mechanics is a complete theory, proving that there are no hidden variables.

Lasers

The understanding of the quantized energy levels in atoms and the realization that transitions between these levels involved the absorption and emission of photons led to the development of a new device that produced a special beam of light. Assume that we have a gas of excited atoms. Further assume that an electron drops from the $n = 3$ to the $n = 2$ level in the energy diagram in Figure 24-15(a). A photon is emitted in some random direction. It could escape the gas, or it could interact with another atom with an electron in one of two ways. It could be absorbed by an atom with an electron in the $n = 2$ level, exciting the electron to the $n = 3$ level. This excited atom would then emit another photon at a random time in a random direction. On the other hand, the original photon could stimulate an electron in the $n = 3$ level to drop to the $n = 2$ level, causing the emission of another photon [Figure 24-15(c)]. This latter process is known as **stimulated emission.** Moreover, this new photon does not come out randomly. It has the same energy, the same direction, and the same phase as the incident photon; that is, the two photons are coherent. These photons can then stimulate the emission of further photons, producing a coherent beam of light.

The device that produces coherent beams of light is called a **laser,** which is the acronym derived from *l*ight *a*mplification by *s*timulated *e*mission of *r*adiation. A laser produces a beam of light that is very different from that emitted

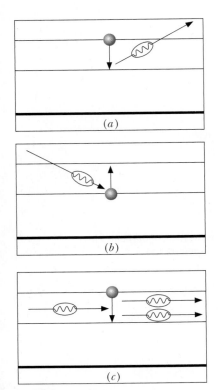

Figure 24-15 (a) Spontaneous emission. (b) Absorption. (c) Stimulated emission.

Spontaneous emission
random directions

Laser output

Mirror 1   Mirror 2

Energy input

Figure 24-16 A schematic of a simple laser.

by an ordinary light source such as a flashlight. The laser beam has a very narrow range of wavelengths, is highly directional, and can be quite powerful. The laser beam is a single color because all the photons have the same energy. For instance, the helium–neon laser usually has a wavelength of 632.8 nanometers. The beam is highly directional because the stimulated photons move in the same direction as those doing the stimulating. The fact that all the photons have the same phase means that the amplitude of the resulting electromagnetic wave is very large. The intensity of a collection of coherent photons is obtained by adding the amplitudes and then squaring the sum. The intensity of a collection of incoherent photons is obtained by squaring the amplitudes and then adding these squares. The difference can be illustrated by considering a collection of five photons. In the laser beam, we have $(2 + 2 + 2 + 2 + 2)^2 = 100$, whereas in the ordinary light beam we have $(2^2 + 2^2 + 2^2 + 2^2 + 2^2) = 20$. The effect is even more drastic for the large number of photons in a laser beam. These factors combine to allow us to clearly see a 1-milliwatt laser beam shining on the surface of a 100-watt lightbulb.

Making a working laser was more complicated than the preceding description implies. A method had to be found for "building" a beam of many photons. This was done by putting mirrors at each end of the laser tube to amplify the beam by passing it back and forth through the gas of excited atoms. One of the mirrors was only partially silvered, so that a small part of the beam was allowed to escape (Figure 24-16).

The construction of a laser was difficult because a photon is just as likely to be absorbed by an atom with an electron in the lower level as it is to cause stimulated emission of an electron in an excited level. Usually, most of the atoms are in the lower energy state, so most of the photons are absorbed and only a few cause stimulated emission. Therefore, building a laser depends on developing a *population inversion,* a situation in which there are many more electrons in the excited state than in the lower energy state. This is usually done by exciting the atom's electrons into a *metastable state* that decays into an unpopulated energy state. A metastable state is one in which the electrons remain for a long time. The electrons can be excited by using a flash of light, an electric discharge, or collisions with other atoms.

Lasers have a wide range of uses, including surveying and surgery. In surveying, the light beam defines a straight line, and by pulsing the beam, surveyors can use the time for a round-trip to measure distances. This same method is used on a gigantic scale to determine the distance to the Moon to an accuracy of a few centimeters. A very short pulse (less than 1 nanosecond long) of laser light is sent through a telescope toward the Moon's surface. The beam is so well collimated that it spreads out over an area only a few kilometers in diameter. The Apollo astronauts left a panel of retroreflectors (Chapter 17) on the surface that reflects the light back to the telescope, allowing the round-trip

A laser-surveying instrument used to ensure proper placement of underground sewer pipes.

Light from an argon laser is carried by a fiber-optic cable to perform surgery deep within an ear.

time to be measured. These measurements serve as a test of the validity of the general theory of relativity because the theory predicts the detailed orbit of the Moon.

Laser "knives" are used in optometry and surgery. Two leading causes of blindness are glaucoma and diabetes. In treating glaucoma, a laser beam "drills" a small hole to relieve the high pressure that builds up in the eye. One of the complications of diabetes is the weakening of the walls of blood vessels in the eye to the point where they leak. The laser can be shined into the eye to cause coagulation to stop the bleeding. The energy in laser beams can also be used to weld detached retinas to the back of the eye. In laser surgery, the beam coagulates the blood as it slices through the tissue, greatly reducing bleeding. Laser beams can also be directed by fiber optics through tiny incisions to locations that would otherwise require major incisions and long healing times.

Lasers also have widespread use in the marketplace. Laser beams read the audio and video information stored in the pits on CDs and DVDs. Likewise, the lasers at store checkout counters read the bar codes on the product identification labels, allowing the computer to print out a short description of the object and its current price. This has greatly reduced billing errors as well as the time required to check out.

Lasers have made practical holography possible (Chapter 19) and opened up a whole new research tool in holographic measurements.

Summary

The Bohr model was successful in accounting for many atomic observations, especially the emission and absorption spectra for hydrogen and the ordering of the chemical elements. However, it failed to explain the details of some processes and could not give quantitative results for multielectron atoms. It described the general features of the periodic table but could not provide the details of the shells, nor explain how many electrons could occupy each shell. Primary among the Bohr model's failures was the lack of intuitive reasons for its postulates. Finally, it was nonrelativistic.

The replacement of Bohr's model of the atom began with de Broglie's revolutionary idea that electrons behave like waves. The de Broglie wavelength of the electron is inversely proportional to its momentum. The primary consequence of this wave behavior is that confined atomic particles form standing-wave patterns that quantize their properties.

Although successful, this new understanding led to a wave–particle dilemma for electrons and other atomic particles: they are always detected as single particles, and yet collectively they produce wavelike distributions. The behavior of these particles is governed by a quantum-mechanical wave equation that provides all possible information about the particles. For example, the probability of finding the particle at a given location is determined by the square of its matter–wave amplitude. As a consequence, atomic particles do not follow well-defined paths as classical particles do. This quantum-mechanical view of physics replaced the classical world view containing Newton's laws and Maxwell's equations.

Four quantum numbers are associated with an electron bound in an atom: the energy n, the size ℓ and direction m_ℓ of the angular momentum, and the orientation m_s of the spin. The periodicity of the chemical elements is explained by the existence of shells of varying capacity determined by the constraints of quantum mechanics and the Pauli exclusion principle. This principle states that no two electrons can have the same set of quantum numbers.

According to the complementarity principle, all atomic entities require both wave and particle aspects for a complete description, but these cannot appear at the same time. These ideas also led to Heisenberg's uncertainty principle, a statement that there is an indeterminacy of knowledge that results from the wave–particle duality. Mathematically, the product of the uncertainties of paired quantities has a lower limit equal to Planck's constant.

The understanding of the quantified energy levels in atoms led to the development of lasers, which produce coherent beams of light through the stimulated emission of radiation by electrons in excited, metastable states.

CHAPTER 24 : *Revisited*

Laser light is special for a number of reasons. In many of the applications mentioned in the opening question, the laser light's key attribute is that it is (nearly) unidirectional, allowing for a very concentrated beam. With proper optical focusing, this beam can be concentrated to a very small spot, drastically increasing the intensity—the amount of energy per unit area—striking a surface. In holography the most important attribute is its coherence, allowing a beam of light to be split and recombined while preserving the phase information.

Key Terms

complementarity principle   A complete description of an atomic entity such as an electron or a photon requires both a particle description and a wave description, but not at the same time.

exclusion principle   No two electrons can have the same set of quantum numbers.

laser   A device that uses stimulated emission to produce a coherent beam of electromagnetic radiation. *Laser* is the acronym from *l*ight *a*mplification by *s*timulated *e*mission of *r*adiation.

matter–wave amplitude   The wave solution to Schrödinger's equation for atomic and subatomic particles. The square of the matter–wave amplitude gives the probability of finding the particle at a particular location.

quantum mechanics   The rules for the behavior of particles at the atomic and subatomic levels.

stimulated emission   The emission of a photon from an atom because of the presence of an incident photon. The emitted photon has the same energy, direction, and phase as the incident photon.

uncertainty principle   The product of the uncertainty in the position of a particle along a certain direction and the uncertainty in the momentum along this same direction must be greater than Planck's constant: $\Delta p_y \Delta y > h$. A similar relationship applies to the uncertainties in energy and time.

Questions and exercises are paired so that most odd-numbered are followed by a similar even-numbered.

Blue-numbered questions and exercises are answered in Appendix B.

❨ indicates more challenging questions and exercises.

WebAssign Many Conceptual Questions and Exercises for this chapter may be assigned online at WebAssign.

Conceptual Questions

1. Make a list summarizing the successes and failures of the Bohr theory.

2. The theory of special relativity requires that time, distance, and energy be treated in a new way as a particle approaches the speed of light. Because these corrections were understood at the time that Bohr developed his model of the hydrogen atom, could he have included

them and achieved better agreement with experimental results? Why or why not?

3. You find that the lowest frequency at which you can set up a standing wave in a wire loop, as shown in Figure 24-1, is 10 hertz. When you increase the driving frequency slightly above 10 hertz, the resonance goes away. What is

---

the next frequency at which resonance will again appear? Explain.

4. A 48-centimeter-long wire loop is used to demonstrate standing waves, as shown in Figure 24-1. What are the three longest wavelengths that will produce standing waves?

5. One may be tempted to interpret the de Broglie wave of an electron as a modified orbital path around the nucleus in which the electron deviates up and down from a circular path as it orbits. Does this model overcome the difficulties of Bohr's model? Explain.

6. For standing waves on a guitar string, adjacent antinodes are always moving in opposite directions. Use this principle to explain why a standing-wave pattern with three antinodes cannot exist on a wire loop.

7. Your friend claims that light is a wave. What experimental evidence could you cite to demonstrate that light behaves like a particle?

8. Your friend claims that electrons are particles. What experimental evidence could you cite to demonstrate that electrons behave like waves?

9. In Chapter 10 we found that an infinite amount of energy is required to accelerate a massive particle to the speed of light. What does this imply about the mass of a photon?

10. De Broglie argued that his relationship between wavelength and momentum should apply to photons as well as to electrons. For massless particles, the relationship takes the form $p = h/\lambda$. Which has more momentum, a red photon or a blue photon? Explain.

11. When applied to photons, the de Broglie relationship $p = h/\lambda$ shows that mass is not required for a particle to have momentum, which disagrees with our classical definition for particles. What other quantity that classically depends on mass can also be attributed to a photon?

12. Which of the following technical terms can be used to describe both an electron and a photon: wavelength, velocity, mass, energy, or momentum? Explain.

13. Why is it not correct to say that an electron is a particle that sometimes behaves like a wave and light is a wave that sometimes acts like a particle?

14. Why do you think that the particle nature of the electron was discovered before its wave nature?

15. The wavelength of red light is 600 nanometers. An electron with a speed of 1.2 kilometers per second has the same wavelength. Will the electron look red? Explain.

16. An electron and a proton have the same speeds. Which has the longer wavelength? Why?

17. Bohr could never really explain why an electron was limited to certain orbits. How did de Broglie explain this?

18. What do standing waves have to do with atoms?

19. Why is the wave behavior of bowling balls not observed?

20. Why doesn't a sports car diffract off the road when it is driven through a tunnel?

21. When we perform the two-slit experiment with electrons, do the electrons behave like particles, waves, or both? What if we perform the experiment with photons?

22. Two students are discussing what happens when you turn down the rate at which electrons are fired at two slits. Tyson claims, "Because you still get an interference pattern even with only one electron at a time, each electron must interfere with itself. As weird as it sounds, each electron must be going through both slits." Ulricht counters, "That's crazy. I can't be at class and on the ski slope at the same time. Each electron must pass through only one slit." Which student is correct? Explain.

23. What are the differences between an electron and a photon?

24. What are the similarities between an electron and a photon?

25. If the two-slit experiment is performed with a beam of electrons so weak that only one electron passes through the apparatus at a time, what kind of pattern would you expect to obtain on the detecting screen?

26. In the two-slit experiment with photons, what type of pattern do you expect to obtain if you turn the light source down so low that only one photon is in the apparatus at a time?

27. Two lightbulbs shine on a distant wall. To obtain the brightness of the light, do we add the displacements or the intensities? Explain.

28. Two coherent sources of light shine on a distant screen. If we want to calculate the intensity of the light at positions on the screen, do we add the displacements of the two waves and then square the result, or do we square the displacements and then add them? Explain.

Blue-numbered **answered in Appendix B** ♪ = more challenging questions

29. What meaning do we give to the square of the matter–wave amplitude?

30. The waves that we studied in Chapters 15 and 16 could be classified as either longitudinal or transverse. Why does this classification scheme have no meaning when applied to de Broglie's matter waves?

31. Where would you most likely find an electron in the lowest energy state for a one-dimensional box, as shown in Figure 24-11?

32. Where would you most likely find an electron in the first excited state for a one-dimensional box, as shown in Figure 24-12?

33. In discussing the first excited state of a particle in a one-dimensional box, your classmate claims, "The electron has to get from one region of high probability to the other without spending much time in between. It must be accelerating, so it must radiate energy." What is wrong with this reasoning?

34. Why doesn't the quantum-mechanical model of the atom have the problem of accelerating charges emitting electromagnetic radiation?

35. Where would you most likely find the electron if it is in a quantum state with $n = 2$, $\ell = 1$, and $m_\ell = -1$, as shown in Figure 24-13?

36. Where would you most likely find the electron if it is in a quantum state with $n = 3$, $\ell = 2$, and $m_\ell = -2$, as shown in Figure 24-13?

37. How many electrons can have the quantum numbers $n = 5$ and $\ell = 1$?

38. How many electrons can have the quantum numbers $n = 5$ and $\ell = 4$?

39. Make a list showing the quantum numbers for each of the four electrons in the beryllium atom when it is in its lowest energy state.

40. Make a list showing the quantum numbers for each of the 10 electrons in the neon atom when it is in its lowest energy state.

41. Your friend argues: "The Heisenberg uncertainty principle shows that we can never be certain about anything. If this is true, we shouldn't believe anything that science teaches." How do you respond?

42. Like light, electrons exhibit diffraction when passed through a single slit. Use the Heisenberg uncertainty principle to explain why narrowing the slit (that is, improving the knowledge of the electron's position in a direction perpendicular to the beam) causes the diffraction pattern to get wider.

43. Explain why the Heisenberg uncertainty principle does not put any restrictions on our simultaneous knowledge of a particle's momentum and its energy.

44. According to the uncertainty principle, why does a tennis ball appear to have a definite position and velocity but an electron does not?

45. De Broglie's relationship gives the wavelength for an electron of given momentum. If we have some idea where the electron is, what does the uncertainty principle tell us about the electron's wavelength?

46. In principle you can balance a pencil with its center of mass directly above its point, even if this point is perfectly sharp. Explain why the uncertainty principle makes this feat impossible.

47. Explain why Bohr's model of the atom is not compatible with the uncertainty principle.

48. What does the uncertainty principle say about the energy of an excited state of hydrogen that exists only for a short time? How will this affect the emission spectrum?

49. Einstein once complained, "The quantum mechanics is very imposing. But an inner voice tells me that it is still not the final truth. The theory yields much, but it hardly brings us nearer to the secret of the Old One. In any case, I am convinced that He does not throw dice." What about quantum mechanics was troubling him?

50. If the Bohr model of the atom has been replaced by the newer quantum-mechanical models, why do we still teach the Bohr model?

51. What quantum-mechanical variable is complementary to time?

52. What quantum-mechanical variable is complementary to position?

53. If we are certain about the energy of a particle, what does the Heisenberg uncertainty principle say about the uncertainty in its frequency?

54. If we are certain about the position of a particle, what does the Heisenberg uncertainty principle say about the uncertainty in its wavelength?

55. Why do some minerals glow when they are illuminated with ultraviolet light?

56. When ultraviolet light from a "black light" shines on crayons, visible light is emitted. How can you account for this?

Aaron Haupt/Photo Researchers, Inc.

57. Phosphorescent materials continue to glow after the lights are turned off. How can you use the model of the atom to explain this?

58. Can infrared rays cause fluorescence? Why or why not?

59. What is stimulated emission?

60. How does light from a laser differ from light emitted by an ordinary lightbulb?

Exercises

61. What is the de Broglie wavelength of a Volkswagen (mass = 1000 kg) traveling at 30 m/s (67 mph)?

62. A bullet for a .30-06 rifle has a mass of 10 g and a muzzle velocity of 900 m/s. What is its wavelength?

63. Nitrogen molecules (mass = 4.6×10^{-26} kg) in room-temperature air have an average speed of about 500 m/s. What is a typical wavelength for these nitrogen molecules?

64. What is the de Broglie wavelength for an electron traveling at 50 m/s?

65. What is the wavelength for a photon with energy 2 eV?

66. What is the wavelength for an electron with energy 2 eV?

67. What speed would an electron need to have a wavelength equal to the diameter of a hydrogen atom (10^{-10} m)?

68. What is the speed of a proton with a wavelength of 2 nm?

69. What is the size of the momentum for an electron that is in the lowest energy state for a one-dimensional box whose length is 2 nm?

70. What is the size of the momentum for an electron that is in the first excited energy state for a one-dimensional box whose length is 2 nm?

71. A child runs straight through a door with a width of 0.75 m. What is the uncertainty in the momentum of the child perpendicular to the child's path?

72. A Honda Civic passes through a tunnel with a width of 10 m. What uncertainty is introduced in the Civic's momentum perpendicular to the highway?

73. A proton passes through a slit that has a width of 10^{-10} m. What uncertainty does this introduce in the momentum of the proton at right angles to the slit?

74. Repeat Exercise 73 for an electron.

75. What is the minimum uncertainty in the position along the highway of a Ford Escort (mass = 1000 kg) traveling at 20 m/s (45 mph)? Assume that the uncertainty in the momentum is equal to 1% of the momentum.

76. What is the uncertainty in the momentum of an Acura NSX with a mass of 1200 kg parked by the curb? Assume that you know the location of the car with an uncertainty of 1 cm.

© American Honda Motor Co., Inc.

77. What is the uncertainty in the location of a proton along its path when it has a speed equal to 0.1% the speed of light? Assume that the uncertainty in the momentum is 1% of the momentum.

78. What is the uncertainty in each component of the momentum of an electron confined to a box approximately the size of a hydrogen atom, say, 0.1 nm on a side?

79. Consider an electron confined to a diameter of 0.1 nm, about the size of a hydrogen atom. If the electron's speed is on the order of the uncertainty in its speed, approximately how fast is it traveling? Assuming that the electron can still be treated without making relativistic corrections, find its kinetic energy (in electron volts).

80. Repeat Exercise 79 for a proton confined to a diameter of 10^{-14} m, about the size of a nucleus.

81. Electrons in an excited state decay to the ground state with the release of a photon. The uncertainty in the time that an electron spends in the excited state can be estimated by the average time electrons spend in that state. What is the spread in energy, in electron volts, of the photons from a state with a lifetime of 2×10^{-8} s?

82. The lifetime of an excited nuclear state is 5×10^{-12} s. What will be the spread in energy, in electron volts, of the photons from this state?

83. If the photons from an excited atomic state show a spread in energy of 2×10^{-4} eV, what is the lifetime (see Exercise 81) of the state?

84. The uncertainty principle allows for "violation" of conservation of energy for times less than the associated uncertainty. For how long could an electron increase its energy by 10 eV without violating conservation of energy?

The Big Picture

The Subatomic World

Kamioka Observatory, ICRR (Institute for Cosmic Ray Research), The University of Tokyo

The quantum-mechanical model of the atom has been very successful. From a large collection of seemingly unrelated phenomena has come a rather complete picture of the atom. This development has created a very profound change in our physics world view.

Our study of the physical world now goes one "layer" deeper, to that of the atomic nucleus. The search for the structure of the nucleus led to the discovery of new forces in nature, forces much more complicated than those of gravity and electromagnetism. It is not possible to write simple expressions for the nuclear forces like the one we wrote for gravity; the nuclear forces depend on much more than distance and mass. This discovery radically changed our basic concepts about forces. The models for the interactions between particles that began

The giant Super Kamiokande neutrino detector is buried 1 kilometer under a mountain near Tokyo, Japan. Physicists are cleaning the faces of the photomultipliers before they are submerged.

with Newton's action at a distance and progressed to the field concept in Maxwell's time now involve the exchange of fundamental particles.

Because the nuclear force is so strong, the energy associated with it is very large. The power of nuclear energy became evident with the explosion of nuclear bombs over Japan in World War II. The release of this energy in the form of bombs or nuclear power plants poses serious questions not only for scientists and engineers but also for our entire society.

In our search for the ultimate structure of all matter, we have discovered that atoms are not the most basic building blocks in nature. They are composed of electrons and very small, very dense nuclei. The nuclei are in turn composed of particles. These were originally believed to be electrons and protons. This notion was appealing because it required only three elemental building blocks: the electron, the proton, and the photon. Later, the neutron replaced the electron in the nucleus, but the overall picture for the fundamental structure was still appealing.

But as scientists delved deeper and deeper into the subatomic world, new particles emerged. The number of "elementary" particles grew to more than 300! In current theories, most of these are composed of a small number of more basic particles called quarks.

The search for the elusive elementary particles is analogous to a child playing with a set of nested eggs. Each egg is opened only to reveal another egg. However, the child eventually reaches the end of the eggs. In one version the smallest egg contains a bunny. Is there an end to the search for the elementary particles? Do quarks represent the bunny?

Gerald F. Wheeler

Are the layers of matter like the nested dolls, or is there a final set of elementary particles?

25

The Nucleus

▶ As our knowledge of nuclei has grown, whole new technologies have been developed to make use of this new knowledge. The new technologies are as varied as obtaining medical images of our internal organs and determining the age of mummies or artifacts. How can nuclear techniques be used to determine the age of an artifact?

(See page 570 for the answer to this question.)

Scott Camazine/Photo Researchers, Inc.

Magnetic resonance image of a normal human brain.

THE **nucleus** of an atom is unbelievably small. If you could line up a trillion (10^{12}) of them, they would stretch only a distance equal to the size of the period at the end of this sentence. Another way of visualizing these sizes is to imagine expanding things to more human sizes and then comparing relative sizes. If we imagine a baseball as big as Earth, the baseball's atoms would be approximately the size of grapes. Even at this scale, the nucleus would be invisible! To "see" the nucleus, we need to expand one of these grape-sized atoms until it is as big as the Superdome in New Orleans. The nucleus would be in the middle and would be about the size of a grape.

These analogies are a little risky because the quantum-mechanical view does not consider atomic entities to be classical particles. The images do, however, demonstrate the enormous amount of ingenuity required to study this submicroscopic realm of the universe. You don't just pick up a nucleus and take it apart.

We learn about the nucleus by examining what pieces come out—through either naturally occurring radioactivity or artificially induced nuclear reactions. Studying the nucleus is more difficult than studying the atom because we already understood the electric force that held the atom together. There, the task was to find a set of rules that governs the behavior of the atom. These rules were provided by quantum mechanics. In nuclear physics the situation was turned around; quantum mechanics provided the rules, but at first there wasn't a good understanding of the forces.

The Discovery of Radioactivity

Discoveries in science are often not anticipated. Radioactivity is such a case. Radioactivity is a nuclear effect that was discovered 15 years before the discovery of the nucleus itself. Although the production of X rays is not a nuclear effect, their study led to the discovery of radioactivity. Four months after Roentgen's discovery of X rays (Chapter 23), a French scientist, Henri Becquerel, was looking for a possible symmetry in the X-ray phenomenon. Roentgen's experiment had shown that X rays striking certain salts produced visible light. Becquerel wondered whether the phenomenon might be symmetric—whether visible light shining on these fluorescing salts might produce X rays.

His experiment was simple. He completely covered a photographic plate with paper so that no visible light could expose it. Then he placed a fluorescing mineral on this package and took the arrangement outside into the sunlight (Figure 25-1). He reasoned that the sunlight might activate the atoms and cause them to emit X rays. The X rays would easily penetrate the paper and expose the photographic plate.

The initial results were encouraging; his plates were exposed as expected. However, during one trial, the sky was clouded over for several days. Convinced that his photographic plate would be slightly exposed, he decided to start over. But, being meticulous, he developed the plate anyway. Much to his surprise, he found that it was completely exposed. The fluorescing material—a uranium salt—apparently exposed the photographic plate without the aid of sunlight. Becquerel soon discovered that all uranium salts exposed the plates—even those that did not fluoresce with X rays.

Figure 25-1 Becquerel's experimental setup to test whether X rays would penetrate the paper and expose the photographic plate.

Are You On the Bus?

Q: What hypothesis does Becquerel's observation suggest about the origin of the radiation?

A: Because the radiation came from all uranium salts, it is likely that the phenomenon is a property of uranium and not of the particular salt.

This new phenomenon appeared to be a special case of Roentgen's X rays. They penetrated materials, exposed photographic plates, and ionized air molecules; but unlike X rays, the new rays occurred naturally. There was no need for a cathode-ray tube, a high voltage, or even the Sun!

Becquerel did a series of experiments that showed that the strength of the radiation depended only on the amount of uranium present—either in pure form or combined with other elements to form uranium salts. Later it was realized that this was the first clue that radioactivity was a nuclear effect rather than an atomic one. Atomic properties change when elements undergo chemical reactions; this new phenomenon did not. The nuclear origin of this radiation was further supported by observations that the exposure of the photographic plates did not depend on outside physical conditions such as strong electric and magnetic fields or extreme pressures and temperatures.

Pierre Curie, a colleague of Becquerel, and Marie Sklodowska Curie developed a way of quantitatively measuring the amount of radioactivity in a sample of material. Marie Curie then discovered that the element thorium was also radioactive. As with uranium, the amount of radiation depended on the amount of thorium and not on the particular thorium compound. It was also not affected by external physical conditions.

The Curies found that the amount of radioactivity present in pitchblende, an ore containing a large percentage of uranium oxide, was much larger than expected from the amount of uranium present in the ore. They then attempted to isolate the source of this intense radiation, a task that turned out to be difficult and time-consuming. Although they were initially unable to isolate the new substance, they did obtain a sample that was highly concentrated. The new element was named polonium after Marie Curie's native Poland. Further work led to the discovery of another highly radioactive element, radium. A gram of radium emits more than a million times the radiation of a gram of uranium. A sample of radium generates energy at a rate that keeps it hotter than its surroundings. This mysterious energy source momentarily created doubt about the validity of the law of conservation of energy.

FLAWED REASONING

A question on the midterm exam asks: "If radium (which is radioactive) and chlorine (which is not radioactive) combine to form radium chloride, is the compound radioactive?"

Shannon gives the following answer: "Not necessarily. We found that the reactive metal sodium combines with the poisonous gas chlorine to form common table salt. A compound can have very different properties than the elements that form it."

Shannon's answer sounds reasonable, but she is missing an important concept. **What is it?**

ANSWER The *chemical* properties of elements are governed by the structure of their outer electron shells. These properties can change when compounds are formed because the outer shell structure is changed. Radioactivity, on the other hand, is independent of conditions outside the nucleus. Radium chloride is radioactive because it contains radium nuclei.

Types of Radiation

These radioactive elements were emitting two types of radiation. One type could not even penetrate a piece of paper; it was called **alpha (α) radiation** after the first letter in the Greek alphabet. The second type could travel through a meter of air or even thin metal foils; it was named **beta (β) radiation** after the second letter in the Greek alphabet.

Curie | *Eight Tons of Ore*

Nothing in life is to be feared. It is only to be understood.
—Marie Sklodowska Curie

Maria Sklodowska (1867–1934) was born in Warsaw, the daughter of a rather rigid and demanding father and a brilliant, artistic mother. She was well tutored at home, so it was natural for her to want a university education. Warsaw was under Russian rule, and higher education for women was impossible in Poland. Her sister Bronya shared her thirst for learning, and the two made a pact. Maria would work to support her older sister in the pursuit of a medical degree in Paris. With her new economic security, Bronya would then support Maria, who had long dreamed of studying at the Sorbonne.

Maria taught briefly in a Warsaw school and served as a governess in eastern Poland. When she was 24, she eagerly accepted her sister and new brother-in-law's invitation to come and live with them in Paris. Because their apartment was two hours away from the Sorbonne and to save tram fare, Marie—the French version of her name—took an attic room in an ancient and unheated building. She endured bitter cold but rejoiced in her freedom to learn and to work: "It was like a new world to me, the world of science." In 1893 she took her first university degrees, receiving a first in physics and a runner-up in mathematics. Her intention was to obtain a teaching diploma and return to Poland. Fate intervened and she met a charming, shy, and handsome young professor, Pierre Curie; they were married in 1896. Pierre had no sooner finished his final degree than he persuaded his new wife to pursue the same high goal.

She began research under the supervision of a distinguished committee—two of whom would be awarded Nobel Prizes. She sought to explore thoroughly the most critical aspects of the new phenomenon of radioactivity that was discovered by Henri Becquerel. After an exhaustive analysis of radiation intensity and the elements and compounds that produced it, Marie came to a brilliant conclusion: this strange energy came from the interior of the atoms.

With Pierre's assistance she began to assay possible sources of radioactive materials. They located a considerable amount of ore in an old and well-known mine at Joachimstal in Bohemia. The Curies had tons of the ore shipped to Paris, where they dug, boiled, and refined this material in order to isolate one-tenth of a gram of the most intense radiation source—a radium salt. When she presented her thesis research in 1903, her committee noted that this was the most significant work ever submitted by a student at the Sorbonne. Later that year Henri Becquerel and Pierre and Marie

Curie shared the Nobel Prize in physics for their work with radioactivity—a word she was the first to use scientifically.

Marie Curie

The Nobel Prize, the new and mysterious subject of radiation, and the love story of the two young scientists combined to promote the importance of the Nobel Prizes and to focus an intense journalistic attention on the Curies. From that time onward, Marie—Pierre died tragically in 1906 when he was struck by a heavy, horse-drawn freight wagon—was the subject of popular interest to an extent unknown to scientists previously. Marie concentrated her incredible work ethic on rearing her two young daughters and discovering the ultimate element from which the intense radiation in pitchblende emanated. Her reclusive and reticent style later made Albert Einstein a better subject for story-hungry writers.

Despite taboos against women, Marie Curie's outstanding record secured her a position as the first female professor at the Sorbonne. She raised her two young daughters, Irene and Eve, in a circle of elite children of similar age. They were to be instructed only by close friends and university professors, and the standards of their scientific training were extraordinary. Irene Joliot-Curie won the Nobel Prize in chemistry in 1935—and her husband was also a Nobel laureate. Marie was the first of three scientists to win a second Nobel Prize—this time in chemistry for her isolation of pure radium and polonium. Curium, element 96, was named in honor of Marie and Pierre Curie.

World War I shattered the European scientific community. Marie took her tremendous drive and dedication to the trenches. She set up mobile X-ray units and trained a generation of radiologists. Like Florence Nightingale a generation earlier, she took on any military authority who stood in the way of medical treatment.

She did not live quite long enough to see her daughter and son-in-law receive their Nobel Prizes. She died of leukemia on July 4, 1934. She was eventually interred in the great Pantheon. Poland produced her, France nourished her, and the world inherits her great work.

—Pierce C. Mullen, historian and author

Sources: Rosalynd Pflaum, *Grand Obsession: Marie Curie and Her Work* (New York: Doubleday, 1989); Susan Quinn, *Marie Curie: A Life* (New York: Simon & Schuster, 1995); Eve Curie, *Madame Curie* (New York: Doubleday, 1938).

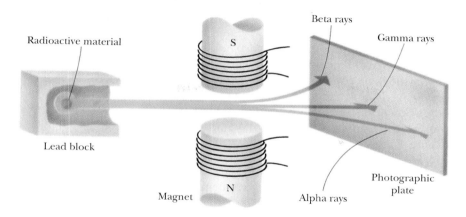

By 1900 several experimentalists had shown that beta radiation could be deflected by a magnetic field, demonstrating that it was a charged particle. These particles had negative charges and the same charge-to-mass ratio as the newly discovered electrons. Later it was concluded that these **beta particles** were in fact electrons that were emitted by nuclei.

In 1903 Rutherford showed that alpha radiation could also be deflected by magnetic fields and that these particles had a charge of +2. (The unit of charge is assumed to be the elementary charge—that is, the magnitude of the charge on the electron or proton.) Because they had a larger charge than electrons and yet were much more difficult to deflect, it was concluded that they were much more massive than electrons. Six years later Rutherford showed that **alpha particles** are the nuclei of helium atoms.

The third type of radiation was discovered in 1900. Naturally, it was named after the third letter of the Greek alphabet and is known as **gamma (γ) radiation.** This radiation has the highest penetrating power; it can travel through many meters of air or even through thick walls. Unlike the other two types of radiation, gamma radiation was unaffected by electric and magnetic fields (Figure 25-2). Gamma rays, like X rays, are now known to be very high-energy photons. Although the energy ranges associated with these labels overlap, gamma-ray photons usually have more energy than X-ray photons, which in turn have more energy than photons of visible light.

The Nucleus

The recognition that radioactivity is a nuclear phenomenon indicated that nuclei have an internal structure. Because nuclei emitted particles, it was natural to assume that nuclei were composed of particles. It was hoped there would be only a small number of different kinds of particles and that all nuclei would be combinations of these.

Rutherford's early work with alpha particles showed that he could change certain elements into others. While bombarding nitrogen with alpha particles, Rutherford discovered that his sample contained oxygen. Repeated experimentation convinced him that the oxygen was being *created* during the experiment. In addition to heralding the beginning of artificially induced transmutations of elements, this experiment led to the discovery of the proton.

Rutherford noticed that occasionally a fast-moving particle emerged from the nitrogen. These particles were much lighter than the bombarding alpha particles and were identified as **protons.** Rutherford and James Chadwick bombarded many of the light elements with alpha particles and found 10 cases in which protons were emitted. By 1919 they realized that the proton was a fundamental particle of nuclei.

The hydrogen atom has the simplest and lightest nucleus, consisting of a single proton. The next lightest element is helium. Its nucleus has approximately four times the mass of hydrogen. Because helium has two electrons, it must have two units of positive charge in its nucleus. If we assume that the helium nucleus contains two protons, we obtain the correct charge but the wrong mass—it would only be twice that of hydrogen. What accounts for the other two atomic mass units? One possibility is that the nucleus also contains two electron–proton pairs. Each pair would be neutral in charge and contribute 1 atomic mass unit. Thus, the helium nucleus might consist of four protons and two electrons.

This scheme had some appeal. It accounted for the electrons emitted from nuclei, and it could be used to "build" nuclei. However, it had several serious defects that caused it eventually to be discarded. For example, the spins of the electron and proton cause each of them to act as miniature magnets, and therefore the magnetic properties of a nucleus should be a combination of those of its protons and electrons. These magnetic values did not agree with the experimental results. The existence of electrons in the nucleus was also incompatible with the uncertainty principle (Chapter 24). If the location of the electron were known well enough to say that it was definitely in the nucleus, its momentum would be large enough to easily escape the nucleus. This left the "extra" mass unexplained.

The Discovery of Neutrons

By 1924 Rutherford and Chadwick began to suspect there was another nuclear particle. The **neutron** would have a mass about the same as the proton's but no electric charge. Chadwick received a Nobel Prize (1935) for the experimental verification of the neutron's existence in 1932. The "extra" mass in nuclei was now explained. Nuclei are combinations of protons and neutrons, as illustrated in Figure 25-3. Helium nuclei are made of 2 protons and 2 neutrons, oxygen nuclei have 8 protons and 8 neutrons, gold nuclei have 79 protons and 118 neutrons, and so on.

When it is not necessary to distinguish between neutrons and protons, they are often called **nucleons.** The number of nucleons in the nucleus essentially determines the atomic mass of the atom because the electrons' contribution to the mass is negligible. The scale of relative atomic masses has been chosen so that the atomic mass of a single atom in atomic mass units is nearly equal to the number of nucleons in the nucleus. This equivalence is attained by setting the mass of the neutral carbon atom with 12 nucleons equal to 12.0000 atomic mass units (Chapter 11). The masses of the electron, proton, and neutron are

Figure 25-3 The helium nucleus (a) consists of two protons and two neutrons, and the oxygen nucleus (b) has eight protons and eight neutrons.

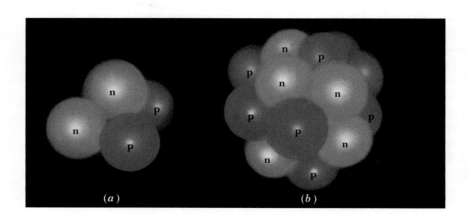

(a)   (b)

Table 25-1	*Masses of Electrons, Protons, and Neutrons in Kilograms, Atomic Mass Units, and Units in Which the Mass of the Electron = 1*		
Particle	**kg**	**amu**	**$m_e 1$**
Electron	9.109×10^{-31}	0.000 549	1
Proton	1.6726×10^{-27}	1.007 276	1836
Neutron	1.6750×10^{-27}	1.008 665	1839

given in Table 25-1 for several different mass units. The masses of nuclei vary from 1 to about 260 atomic mass units and the corresponding radii vary from 1.2 to 7.7×10^{-15} meter.

Isotopes

While studying the electric and magnetic deflection of ionized atoms, J. J. Thomson discovered that the nuclei of neon atoms are not all the same. The neon atoms did not all bend by the same amount. Because each ion had the same charge, some neon atoms must be more massive than others. It was shown that the lighter ones had masses of about 20 atomic mass units, whereas the heavier ones had masses of about 22 atomic mass units. This discovery destroyed the concept that all atoms of a given element were identical, an idea that had been accepted for centuries. We now know that all elements have nuclei that have different masses. These different nuclei are known as **isotopes** of the given element. Isotopes have the same number of protons but differ in the number of neutrons.

Q: Given that neon is the 10th element, how many neutrons and protons would each of its isotopes have?

A: Because each nucleon contributes about 1 atomic mass unit and neon must have 10 protons, the lighter isotope with 20 atomic mass units must have 10 neutrons and the heavier one with 22 atomic mass units must have 12 neutrons.

Although each isotope of an element has its own nuclear properties, the chemical characteristics of isotopes do not differ. Adding a neutron to a nucleus does not change the electric charge; therefore, the atom's electronic structure is essentially unchanged. Because the electrons govern the chemical behavior of elements, the chemistry of all isotopes of a particular element is virtually the same. Some differences, however, can be detected. There are slight changes in the electronic energy levels that show up in the atomic spectra. The extra mass of the heavier isotopes also slows the rates at which chemical and physical reactions occur.

Hydrogen has three isotopes (Figure 25-4). The most common one contains a single proton. The next most common has one proton and one

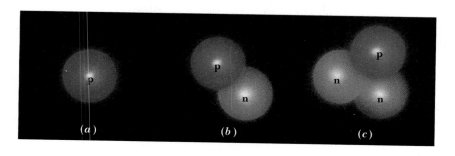

Figure 25-4 Hydrogen has three isotopes: (a) hydrogen, (b) deuterium, and (c) tritium.

neutron; it has about twice the mass of ordinary hydrogen. This "heavy" hydrogen, known as *deuterium,* is stable and occurs naturally as 1 atom out of approximately every 6000 hydrogen atoms. The third isotope has one proton and two neutrons. This "heavy, heavy" hydrogen, known as *tritium,* is radioactive.

Isotopes of other elements do not have separate names. The symbolic way of distinguishing between isotopes is to write the chemical symbol for the element with two added numbers to the left of the symbol. A subscript gives the number of protons, and a superscript gives the number of nucleons—that is, the total number of neutrons and protons. The most common isotope of carbon has six protons and six neutrons and is written $^{12}_{6}$C. Because there is some redundancy here (the number of protons is already specified by the chemical symbol), the isotope is sometimes written ^{12}C and read "carbon-12."

The Alchemists' Dream

The discovery of radioactivity changed another long-held belief about atoms. If the particle emitted during radioactive decay is charged, the resulting nucleus (commonly called the **daughter nucleus**) does not have the same charge as the original (the **parent nucleus**). The parent and daughter are not the same element because the charge on the nucleus determines the number of electrons in the neutral atom, and this number determines the atom's chemical properties. The belief that atoms are permanent was wrong; they can change into other elements. We see that nature has succeeded where the alchemists failed.

Although we can't control the process of nuclear transformation, we can bombard materials as Rutherford did and change one element into another. But what kinds of reactions are possible? Can we, perhaps, change lead into gold? As scientists examined various processes that might change nuclei, they observed that the conservation laws are obeyed. These laws include the conservation of mass–energy, linear momentum, angular momentum, and charge, which we studied earlier. In addition, a new conservation law was discovered, ◄ conservation of nucleons the conservation of nucleons. Although the nucleons may be rearranged or a neutron changed into a proton, the total number of nucleons after the process is the same as before.

The reaction in which Rutherford produced oxygen (O) and a proton (p) by bombarding nitrogen (N) with alpha particles (α) can be written in the form

$$\,^4_2\alpha + \,^{14}_7\text{N} \rightarrow \,^{17}_8\text{O} + \,^1_1\text{p}$$

where the arrow separates the initial nuclei on the left from the final nuclei on the right. Notice that the conservation of charge is obeyed; there are $2 + 7 = 9$ protons on the left-hand side of the arrow and $8 + 1 = 9$ protons on the right-hand side. The number of nucleons must also be conserved. There are $4 + 14 = 18$ nucleons on the left and $17 + 1 = 18$ nucleons on the right.

Neutrons are effective at producing nuclear transformations because they do not have a positive charge and can penetrate to the nuclei, even with low energies. An example that occurs naturally in the atmosphere is the conversion of nitrogen to carbon via neutron (n) bombardment:

$$\,^1_0\text{n} + \,^{14}_7\text{N} \rightarrow \,^{14}_6\text{C} + \,^1_1\text{p}$$

Let's now look at changes in the nucleus that occur naturally. For example, when a nucleus emits an alpha particle, it loses two neutrons and two protons. Therefore, the parent nucleus changes into a nucleus that is two elements lower in the periodic chart. This daughter nucleus has a nucleon number that is four less. For instance, when $^{226}_{88}\text{Ra}$ decays by alpha decay, the daughter nucleus has $88 - 2 = 86$ protons and $226 - 4 = 222$ nucleons. The periodic table printed on the inside front cover of this text tells us that the element with 86 protons is radon, which has the symbol Rn. Therefore, the daughter nucleus is $^{222}_{86}\text{Rn}$. This process can be written in symbolic form as

$$^{226}_{88}\text{Ra} \rightarrow {}^{222}_{86}\text{Rn} + {}^{4}_{2}\alpha$$

◄ alpha decay

Are You On the Bus?

Q: What daughter results from the alpha decay of $^{232}_{90}\text{Th}$?

A: The daughter will have $232 - 4 = 228$ nucleons and $90 - 2 = 88$ protons, so it must be $^{228}_{88}\text{Ra}$.

FLAWED REASONING

Two students are discussing alpha decay:

Brielle: "Our teacher claims that the alchemists were right; atoms can be changed from one element to another. She said that radium becomes radon when it emits an alpha particle, but that doesn't make sense. When radium gives up two protons and two neutrons, it still has the same number of electrons. Therefore, it should still react chemically like radium."

Hyrum: "Well, maybe the radium atom loses two electrons after the alpha decay, so it becomes electrically neutral. I think the atom is still radium until the electrons leave."

What important principle are these students misunderstanding?

ANSWER It is the number of protons, not electrons, in an atom that defines its chemical identity. A chlorine atom with an extra electron is a chlorine ion, not an argon atom. As soon as the alpha particle is ejected from the nucleus of the radium atom, the energy levels change to those of the radon atom.

The electron that is emitted in beta (β) decay is not one of those around the nucleus, nor is it one that already existed in the nucleus. The electron is ejected from the nucleus when a neutron decays into a proton. (We will study the details of this process in Chapter 27.) A new element is produced because the number of protons increases by one. The number of neutrons decreases by one, but it is the change in the proton number that causes the change in element. The daughter nucleus is one element higher in the periodic chart and has the same number of nucleons. For example,

$$^{24}_{11}\text{Na} \rightarrow {}^{24}_{12}\text{Mg} + {}^{0}_{-1}\beta^{-}$$

◄ beta minus decay

Q: What is the daughter of $^{228}_{89}\text{Ac}$ if it undergoes beta decay?

A: The daughter must have the same number of nucleons and one more proton. Therefore, it is $^{228}_{90}\text{Th}$.

Are You On the Bus?

An inverse beta-decay process has also been observed. In this process an atomic electron (e) in an inner shell is captured by one of the protons in the nucleus to become a neutron. This **electron capture** decreases the number of protons by one and increases the number of neutrons by one. Therefore, the daughter nucleus belongs to the element that is one lower in the periodic table and has the same number of nucleons as the parent. For example,

electron capture ▶

$$^{7}_{4}\text{Be} + {}^{0}_{-1}\text{e} \rightarrow {}^{7}_{3}\text{Li}$$

This process is detected by observing the atomic X rays given off when the outer electrons drop down to fill the energy level vacated by the captured electron. The evidence for this transformation is very convincing; the spectral lines are characteristic of the daughter nucleus and not the parent nucleus.

Q: What daughter is produced when $^{234}_{93}\text{Np}$ undergoes electron capture?

A: Once again, the number of nucleons does not change, only this time the number of protons decreases by one to yield $^{234}_{92}\text{U}$.

A third type of beta decay was observed in 1932. In this case the emitted particle has one unit of positive charge and a mass equal to that of an electron. This positive electron (**positron**) is the antiparticle of the electron, a concept that will be described further in Chapter 27. The emission of a positron results from the decay of a proton into a neutron. This process is called *beta plus decay* to distinguish it from *beta minus decay,* the process that produces a negative electron. Like electron capture, this decreases the number of protons by one and increases the number of neutrons by one. For example,

beta plus decay ▶

$$^{15}_{8}\text{O} \rightarrow {}^{15}_{7}\text{N} + {}^{0}_{1}\beta^{+}$$

Q: What is the result of $^{11}_{6}\text{C}$ decaying via beta plus decay?

A: Because a proton turns into a neutron, we have $^{11}_{5}\text{B}$.

Nuclei that emit gamma rays do not change their identities, because gamma rays are high-energy photons and do not carry charge. The nucleus has discrete energy levels analogous to those in the atom. If the nucleus is not in the ground state, it may change to a lower energy state with the emission of the gamma ray. This process often happens after a nucleus has undergone one of the other types of decay to become an excited state of the daughter nucleus. Table 25-2 summarizes the changes produced by each type of decay.

▶ Extended presentation available in the *Problem Solving* supplement

Radioactive Decay

Equal amounts of different radioactive materials do not give off radiation at the same rate. For example, 1 gram of radium emits 20 million times as much

Table 25-2 *Changes in the Number of Protons, Neutrons, and Nucleons for Each Type of Radioactive Decay*

Decay	Changes in the Number of		
	Protons	Neutrons	Nucleons
α	−2	−2	−4
β^{-}	+1	−1	0
Electron capture	−1	+1	0
$\beta+$	−1	+1	0
γ	0	0	0

radiation per unit time as 1 gram of uranium. The **activity** of a radioactive sample is a measure of the number of decays that take place in a certain time. The unit of activity is named the **curie** (Ci) after Marie Curie and has a value of 3.7×10^{10} decays per second, which is the approximate activity of 1 gram of radium.

◀ curie

Two factors determine the activity of a sample of material. First, the activity is directly proportional to the number of radioactive atoms in the sample. Two grams of radium will have twice the activity of 1 gram. Second, the activity varies with the type of nucleus. Some nuclei are quite stable, whereas others decay in a matter of seconds and still others in millionths of a second.

Both factors can be illustrated with an analogy that emphasizes the random nature of radioactive decay. Imagine that you have 36 dice. Throwing the dice can simulate the radioactive decay process. Each throw represents a certain elapsed time. Assume that any die whose number 1 is face up represents an atom that has decayed during the last period.

Let's look at the "activity" of the sample of dice. How many dice would you expect to have 1s up after the first throw? Because each die has six faces that are equally likely to appear as the top face, there is a 1-in-6 chance of having 1s up. Because there are 36 dice in the sample, we expect that, on the average, 6 (that is, $\frac{1}{6}$ of 36) will "decay" on the first throw. This number represents the activity of the sample.

On the average, 6 of the 36 dice will have 1s up.

If you double the number of dice in your sample, how many of the 72 dice should show 1s up after the first throw? On average, 12 ($\frac{1}{6}$ of 72) will have 1s up. From this analogy we see that the level of radioactivity is directly proportional to the number of radioactive atoms in a sample.

In a sample of radioactive material, the number of nuclei of the original isotope continually decreases, which decreases its activity. To illustrate this, imagine that we once again have 36 dice in our sample. After each throw of the dice, we remove the dice with 1s up. They are no longer part of the sample because they have changed to a new element. After the first throw there will be, on the average, 6 dice that decay. Removing them reduces the sample size to 30.

Q: On average, how many of these 30 dice do you expect to have 1s up on the next throw?

A: The number of 1s up will probably be $\frac{1}{6} \times 30 = 5$.

Are You On the Bus?

Removing the 5 dice that are expected to have 1s up on the second throw leaves a sample size of 25. Each successive throw of the dice yields a smaller sample size and consequently a lower level of activity on the next throw. The same is true of the radioactive sample. As with the dice, when a nucleus decays, it is no longer part of the sample. The activity of the radioactive sample decreases with time.

This change in the activity with time has a simple behavior, due to the probabilistic nature of radioactivity. The graph in Figure 25-5 shows this behavior. Notice that the time it takes the activity to drop to one-half its value is constant throughout the decay process. This time is called the **half-life** of the sample. In one half-life the activity of the sample decreases by a factor of 2. During the next half-life, it decreases by another factor of 2, to $\frac{1}{4}$ of its initial value. After each additional half-life, the activity would be $\frac{1}{8}, \frac{1}{16}, \frac{1}{32}, \ldots$, that of the original value. It does not matter when you begin counting. If you wait 1 half-life, your counting rate drops by one-half.

Figure 25-5 A graph of the activity of a radioactive material versus time. Notice that the time needed for the activity to decrease by a factor of 2 is always the same.

Twelve-sided Arabian dice have a smaller probability for a given number to be on top.

This decay law also applies to the number of radioactive nuclei remaining in the sample. The time it takes to reduce the population of radioactive nuclei by a factor of 2 is also equal to the half-life. The important point is that you do not get rid of the radioactive material in 2 half-lives. Let's say that you initially have 1 kilogram of a radioactive element. One half-life later you will have $\frac{1}{2}$ kilogram. After 2 half-lives, you will have $\frac{1}{4}$ kilogram, and so on. In other words, one-half of the remaining radioactive nuclei decays during the next half-life.

The second factor that determines the activity of a sample depends on the character of the nuclei. As we saw in the previous chapter, quantum mechanics determines the probability that something will occur. Therefore, the dice analogy is especially appropriate for nuclear decays. The probability in our dice analogy depends on the number of faces on each die. Six faces means there is a probability of 1 in 6 that a given face will be up on each throw of the dice. Imagine that instead of using 6-sided dice, we use Arabian dice, which have 12 sides. In this case there are more alternatives, and the probability of 1s up is lower. Each die has a 1-in-12 chance of decaying. On the first throw of 36 dice, we would expect an average of 3 dice with 1s up. The half-life of the Arabian dice is twice as long. This variation also occurs with radioactive nuclei, but with a much larger range of probabilities. Nuclear half-lives range from microseconds to trillions of years.

This process is all based on the probability of random events. With dice it is the randomness of the throwing process; with radioactive nuclei it is the quantum-mechanical randomness of the behavior of the nucleons within the nucleus. As with all probabilities, any prediction is based on the mathematics of statistics. Each time we throw 36 dice, we should not expect exactly 6 dice to have 1s up. Sometimes there will be only 5; other times there may be 7 or 8. In fact, the number can vary from 0 to 36! However, statistics tells us that as the total number of dice increases, our ability to predict also increases. Because there are more than 10^{15} nuclei in even a small sample of radioactive material, our predictions of the number of nuclei that will decay in a half-life are very reliable.

Although our predictions get better with large numbers, we must remember the probabilistic nature of dice and radioactive nuclei; we cannot predict the behavior of an individual nucleus. One particular nucleus may last a million years, whereas an "identical" one may last only a millionth of a second.

Radioactive Clocks

The decay rate of a radioactive sample is unaffected by physical and chemical conditions ordinarily found on Earth (excluding the extreme conditions found in nuclear reactors or bombs). Normal conditions involve energies that can rearrange electrons around atoms but cannot cause changes in nuclei. The nucleus is protected by its electron cloud and the large energies required to change the nuclear structure, which means that radioactive samples are good time probes into our history. Knowing the half-life of a particular isotope and the products into which it decays, we can determine the relative amount of parent and daughter atoms and calculate how long the isotope has been decaying. In effect, we have a radioactive "clock" for dating events in the past.

One of the first such clocks gave us the age of Earth. Before this, many different estimates were made by different groups. A 17th-century Irish bishop traced the family histories in the Old Testament and obtained a figure of 6000 years. Another estimate was obtained by calculating the length of time required for all the rivers of the world to bring initially freshwater oceans up to their present salinity. Another value was obtained by assuming that the Sun's energy output was due to its slow collapse under the influence of the gravitational force. These latter two methods yielded estimates on the order of 100,000 years. Charles Darwin suggested that Earth was much older, as his theory of evolution indicated that more time was required to produce the observed biodiversity.

The best value was obtained through the radioactivity of uranium. Uranium and the other heavy elements are formed only during supernova explosions that occur near the end of some stars' lives. Therefore, any uranium found on Earth was present in the gas and dust from which the solar system formed. Uranium-238 decays with a half-life of 4.5 billion years into a stable isotope of lead ($^{206}_{82}$Pb) through a long chain of decays. Suppose we find a piece of igneous rock and determine that one-half of the uranium-238 has decayed into lead-206. We then know that the rock was formed at a time in the past equal to 1 half-life. The oldest rocks are found in Greenland and have a ratio of lead-206 to uranium-238 of a little less than 1, meaning that a little less than one-half of the original uranium-238 has decayed. This indicates that the rocks are about 4 billion years old. Incidentally, the rocks brought back from the Moon have similar ratios, establishing that Earth and Moon were formed about the same time.

Another radioactive clock is used to date organic materials. Living organisms contain carbon, which has two isotopes of interest—the stable isotope $^{12}_{6}$C and the radioactive isotope $^{14}_{6}$C, which has a half-life of 5700 years. Cosmic rays bombarding our atmosphere continually produce carbon-14 to replace those that decay. Because of this replacement, the ratio of carbon-12 to carbon-14 in the atmosphere is relatively constant. While the plant or animal is living, it continually exchanges carbon with its environment and therefore maintains the same ratio of the two isotopes in the tissues that exists in the atmosphere. As soon as it dies, however, the exchange ceases and the amount of carbon-14 decreases because of the radioactive decay. By examining the ratio of the amount of carbon-14 to that of carbon-12 in the plant or animal, we can learn how long ago death occurred.

This is one of the rocks brought back from the Moon by the Apollo astronauts.

WORKING IT OUT *Radioactive Dating*

As an example of radioactive dating, suppose a piece of charred wood is found in a primitive campsite. To find out how long ago the campsite was occupied, one examines the ratio of carbon-14 to carbon-12 in the wood. Suppose that the ratio is only one-eighth of the atmospheric value. Roughly how long has it been since the piece of wood was put into the fire?

As long as the piece of wood was part of a living tree, the ratio of carbon-14 to carbon-12 in the wood is the same as the ratio in the atmosphere. Once the wood is cut from the tree, carbon-14 in the wood begins to decay and is not replenished from the atmosphere. Because $\frac{1}{2} \times \frac{1}{2} \times \frac{1}{2} = \frac{1}{8}$, the wood was removed from the living tree 3 half-lives ago. This gives an approximate age for the campsite of 3×5700 years = 17,100 years.

Because the Dead Sea Scrolls were written on parchment, their age was determined by this technique. Similar studies indicate that the first human beings may have appeared on the North American continent about 27,000 years ago.

Radioactive dating with carbon-14 is not useful beyond 40,000 years (seven half-lives), as the amount of carbon-14 becomes very small. Other radioactive clocks can be used to go beyond this limit. These clocks indicate that humans, or at least prehumans, may have been around for more than 3.5 million years, mammals about 200 million years, and life for 3–4 billion years.

Radiocarbon dating of this skeleton found in the Tyrolean Alps has helped estimate that the man lived around 5000 years ago.

Paul Hanny/Eyedea Presse

Radiation and Matter

How would you know if a chunk of material was radioactive? Radiation is usually invisible. However, if the material is extremely radioactive, like radium, the enormous amount of energy being deposited in the material by the radiation can make it hotter than its surroundings. In extreme cases it may even glow. A less radioactive sample may leave evidence of its presence over a longer time. Pierre Curie developed a radioactive burn on his side because of his habit of carrying a piece of radium in his vest pocket.

We learn about the various types of radiation through their interactions with matter. In most cases, these interactions are outside the range of direct human observation. One exception is light. Although we don't see photons flying across the room, we do see the interaction of these photons with our retinas. Our eyes are especially tuned to a small range of photon energy. Other radiations are detected by extending our senses with instruments. For exam-

Everyday Physics | *Smoke Detectors*

Many of the uses of radioactive isotopes are not often apparent to most of us. For example, some smoke detectors (Figure A) commonly found in our homes use radioactive americium to detect the smoke particles. As shown in the schematic diagram of Figure B, the weak radioactive source ionizes the air, allowing a small current to flow between the electrodes. When smoke enters the detector, the ions are attracted by the smoke particles and stick to them. The larger mass of the smoke particles causes them to move slower than the ions, reducing the current in the circuit and setting off the alarm.

Americium-241 is an alpha emitter with a half-life of 433 years, so the level of ionization does not decrease significantly over the lifetime of the detector. This isotope is obtained as a by-product of the processing of fuel rods from nuclear reactors.

1. Explain how the air near a radioactive source may become ionized.

2. Explain how the smoke detector uses the ionized air to detect smoke particles.

© Radius Images RF/Jupiterimages

Figure A A household smoke detector.

Figure B Diagram of a smoke detector.

ple, the electromagnetic radiation from radio and television stations is not detected by our bodies but interacts with the electrons in antennas.

We also observe nuclear radiations through their interactions with matter. Alpha and beta particles interact with matter through their electric charge. For example, in Becquerel's discovery, the charged particles interacted with the photographic chemicals to expose the film. For the most part, these charged particles interact with the electrons in the material. As we saw in Rutherford's scattering experiment, they rarely collide with nuclei. As the alpha and beta particles pass through matter, they interact with atomic electrons, causing them to jump to higher atomic levels or to leave the atom entirely. This latter process is called **ionization.**

The relatively large mass of alpha particles compared with electrons means that the alpha particles travel through the material in essentially straight lines. This is like a battleship passing through a flotilla of canoes. Because beta particles are electrons, they are more like canoes hitting canoes and have paths that are more ragged. In both cases the colliding particles deposit energy in the material in a more or less continuous fashion. Therefore, the distance they travel in a material is a measure of their initial energy. We can determine this energy experimentally by measuring the distance the particle travels—that is, its range. Some sample ranges are given in Table 25-3.

The interaction of gamma rays with matter is quite different. Gamma rays do not lose their energy in bits and pieces. (There is no such thing as half a photon.) Whenever a photon interacts with matter, it is completely annihilated in one of three possible ways. The photon's energy can ionize an atom, be transferred to a free electron with the creation of a new photon, or be converted into a pair of particles according to Einstein's famous equation $E = mc^2$. If we count the number of gamma rays surviving at various distances in a material and make a graph, we would obtain a curve like the one in Figure 25-6. This curve has the same shape as the one for radioactive decay in Figure 25-5, which means that gamma rays do not have a definite range but are removed from the beam with a characteristic half-distance. That is, one-half of the original gamma rays are removed in a certain length, half of the remaining are removed in the next region of this length, and so on. From Table 25-4 you can see that gamma rays are much more penetrating than alpha or beta particles of the same energy.

© Photodisc/Getty Images

The electromagnetic radiation from the local radio station is not detected by our bodies, but interacts with the radio's antenna.

Figure 25-6 A graph of the number of gamma rays surviving at various distances through a material. Notice the similarity with the graph of the radioactivity in Figure 25-5.

Q: How far must a beam of 10-million-electron-volt gamma rays travel in aluminum before it is all gone?

A: Theoretically, this never happens; some of the gamma rays will travel very large distances. However, by the time the beam has traveled 7 half-distances, less than 1% of the gamma rays remain.

Are You On the Bus?

Table 25-3 *Ranges of Alpha Particles, Protons, and Electrons in Air and Aluminum*

Energy (MeV)*	Range in Air (cm)			Range in Aluminum (cm)		
	α	p	e^-	α	p	e^-
1	0.5	2.3	314	0.0003	0.0014	0.15
5	3.5	34	2000	0.0025	0.019	0.96
10	10.7	117	4100	0.0064	0.063	1.96

*MeV (million electron volts) is an energy unit equal to 1.6×10^{-13} joule.

© Royalty-free/Corbis/Jupiterimages

The 200,000 individual electronic components in this small crystalline chip are especially sensitive to radiation damage.

Table 25-4 *Half-Distances for the Absorption of Gamma Rays by Aluminum*	
Energy (MeV)*	Half-Distance (cm)
1	4.2
5	9.1
10	11.1

* MeV (million electron volts) is an energy unit equal to 1.6×10^{-13} joule.

Any radiation passing through matter deposits energy along its path, resulting in temperature increases, atomic excitations, and ionization. The ionization caused by radiation can damage a substance if its properties depend strongly on the detailed molecular or atomic structure. For example, radiation can damage the complex molecules in living tissues and cause cancer or genetic defects. Damage can also occur in nonliving objects, such as transistors and integrated circuits. These modern electronic devices are made of crystals in which the atoms have definite geometric arrangements. Integrated circuits, such as those found in pocket calculators and computers, contain thousands of individual electronic components in small crystalline chips. A disruption of these atomic structures can change the electronic properties of these components. If the radiation damage is severe, the device could fail. This hazard is obviously serious if the device is a control component for a nuclear reactor or part of the guidance system for a missile. However, there is not enough radiation around the university computer for you to expect all of your grades to change into A's.

WORKING IT OUT *Gamma Ray Shield*

Gamma rays are highly penetrating and produce serious damage when absorbed by living tissues. Consequently, those working near such dangerous radiation must be protected by a shield of absorbing materials. If you are working near a source of 5-MeV gamma rays, and you are protected by an aluminum wall that is 27.3 cm thick, what fraction of the gamma rays are getting past your shield?

Using Table 25-4, we see that the half-distance for 5-MeV gamma rays is 9.1 cm of aluminum. Only half of the original gamma rays survive after passing through the first 9.1 cm of the wall. Because the aluminum wall is three times as thick as this half-distance, we find that $\frac{1}{2} \times \frac{1}{2} \times \frac{1}{2} = \frac{1}{8}$ of the gamma rays are getting through the shield and reaching you. Perhaps you should have built the shield wall thicker, or used a more absorbing material, such as lead.

Biological Effects of Radiation

The ionization caused by radiation passing through living tissue can destroy organic molecules if the electrons are involved in molecular binding. If too many molecules are destroyed in this fashion or if DNA molecules are destroyed, cells may die or become cancerous.

The effects of radiation on our health depend on the amount of radiation absorbed by living tissue and the biological effects associated with this absorption. A **rad** (the acronym from *r*adiation *a*bsorbed *d*ose) is the unit used to designate the amount of energy deposited in a material. One rad of radiation deposits 0.01 joule per kilogram of material. Another unit, the **rem** (*r*oentgen *e*quivalent in *m*ammals), was developed to reflect the biological effects caused by the radiation. Different radiations have different effects on our bodies.

Alpha particles, for example, deposit more energy per centimeter of path than electrons with the same energy; therefore, the alpha particles cause more biological damage. One rad of beta particles can result in 1–1.7 rem of exposure, 1 rad of fast neutrons or protons may result in 10 rem, and 1 rad of alpha particles produces between 10 and 20 rem. The rad and the rem are nearly equal for photons. Because most human exposure involves photons and electrons, the two units are roughly interchangeable when talking about typical doses.

Knowing the amount of energy deposited or even the potential damage to cells from exposure to radiation is still not a prediction of the future health of an individual. The biological effects vary considerably among individuals depending on their age, health, the length of time over which the exposure occurs, and the parts of the body exposed. Large doses, such as those resulting from the nuclear bombs exploded over Japan or the accident at the nuclear reactor at Chernobyl in the former Soviet Union (Figure 25-7), cause radiation sickness, which can result in vomiting, diarrhea, internal bleeding, loss of hair, and even death. It is virtually impossible to survive a dose of more than 600 rem to the whole body over a period of days. An exposure of 100 rem in the same period causes radiation sickness, but most people recover. For comparison, the maximum recommended occupational dose is 5 rem per year, the maximum rate recommended for the general public by the U.S. Environmental Protection Agency is 0.5 rem per year, and the natural background radiation is about 0.3 rem per year.

Obtaining data on the health effects of exposures to low levels of radiation is difficult, primarily because the effects are masked by effects not related to radiation and the effects do not show up for a long time. Although cancers such as leukemia may show up in two to four years, the more prevalent tumorous cancers take much longer. Imagine conducting a free-fall experiment with an apple in which the apple doesn't begin to move noticeably for three years. Even after you spot its motion, you are compelled to ask whether it occurred because of your actions or something else. When cause and effect are so widely separated in time, making connections is difficult. For example, because there is already a high incidence of cancer, it is hard to determine whether any increase is due to an increase in radiation or to some other factor. On the other hand, large doses of radiation do cause significant increases in the incidence of cancer. Marie Curie died of leukemia, most likely the result of her long-term exposure to radiation.

Genetic effects due to low-level radiation are also difficult to determine. These effects can be passed on to future generations through mutation of the reproductive cells. Although mutations are important in the evolution of species, most of them are recessive. It is generally agreed that an increase in the rate of mutations is detrimental.

Any additional nuclear radiation hitting our bodies is detrimental. However, we live in a virtual sea of radiation during our entire lives. Exposure to some of this radiation is beyond our control, but other exposure is within our control. Indeed, many of the "nuclear age" debates center on this issue of assessing the effects of *additional* radiation on the health of the population and evaluating its risks and benefits. As an example, the increased radiation exposure during air travel due to the reduced shielding of the atmosphere is about 0.5 millirem per hour. This increase is not much for an occasional traveler, but may result in an additional 0.5 rem per year for an airplane pilot, which equals the maximum recommended dose.

The average dosage rates for a variety of sources are given in Table 25-5. A word of caution is appropriate here. Although the concept of averaging helps make some comparisons, in some situations the value of the average is quite meaningless. Consider, for example, the amount of radiation our population receives from the nuclear power industry. The bulk of this radiation exposure

Figure 25-7 One of the world's major nuclear reactor accidents happened in April 1986 at Chernobyl in the former Soviet Union. This photograph shows the structure built to contain the leaking radiation.

Air travelers implicitly accept the risk of the increased radiation for the benefits of quicker travel.

Table 25-5 *Average Annual Radiation Doses**

Source	Dose (mrem/yr)
Cosmic rays	30
Surroundings	30
Internal	40
Radon	200
Naturally occurring	300
Medical	50
Consumer products	5
Nuclear power	5
Human-made	60
Total	360

*Frank E. Gallagher III, California Campus Radiation Safety Officers Conference, September 1993.

occurs for the workers in the industry. Most of the rest of us get very little radiation from this source. This situation is similar to putting one foot in ice water, the other foot in boiling water, and claiming that on the average you are doing fine.

Radiation around Us

It is impossible to get away from all radiation. The radiation that is beyond our control comes from the cosmic rays arriving from outer space, from reactions in our atmosphere initiated by these cosmic rays, from naturally occurring radioactive decays in the material around us, and from the decay of radioactive isotopes (mostly potassium-40) within our bodies. In fact, even table salts that substitute potassium for sodium are radioactive. Radiation is part of the world in which we live. The radiations within our control include testing of nuclear weapons, nuclear power stations, and medical radiation used for diagnosing and treating diseases.

Medical sources of radiation *on average* contribute less to our yearly dose than the background radiation. The risks of diagnostic X rays have decreased over the years with new improvements in technology. Modern machines and more sensitive photographic films have greatly reduced the exposure necessary to get the needed information. The exposure from a typical chest X ray is 0.01 rem.

We saw in Chapter 23 that X rays can be used to examine internal organs and other structures for abnormalities. During this process, however, some cells are damaged or killed. We need to ask, is the possible damage worth the benefits of a better treatment based on the information gained? In most cases the answer is yes. In other cases the answer ranges from a vague response to a definite no. If your physician suspects that your ankle is broken, the benefits of having an X ray are definitely worth the risks. On the other hand, having an X ray of your lungs as part of an annual physical exam is not felt to be worth the risks. However, if there is good reason to suspect lesions or a tumor, the odds are in favor of having the X ray.

The largest doses of medically oriented radiation occur during radiation treatment of cancer. In these cases radiation is used to deliberately kill some cells to benefit the patient. Because a beam of radiation causes damage along its entire path, the beam is rotated around the body to minimize damage to normal cells. This procedure is shown in Figure 25-8. Only the cells in the overlap region (the cancerous tumor) get the maximum dosage. Alternative techniques use beams of charged particles. Because charged particles deposit energy at the highest rate near the end of their range, the damage to normal cells along the entrance path is minimized.

Sometimes radioactive materials are ingested for diagnostic or treatment purposes. Our bodies do not have the ability to differentiate between various isotopes of a particular chemical element. If a tumor or organ is known to accumulate a certain chemical element, a radioactive isotope of that element can be put into the body to treat the tumor or to produce a picture of the tumor or organ. For instance, because iodine collects in the thyroid gland, iodine-131, a beta emitter, can be used to kill cells in the thyroid with little effect on more distant cells.

With each source of radiation the debate returns to the question of risk versus benefit: does the potential benefit of the exposure exceed the potential risk?

Figure 25-8 During cancer treatment, the direction of the radiation beam is changed to minimize the damage to normal cells surrounding a tumor.

Radiation Detectors

Many devices have been developed to detect nuclear radiation. The earliest was the fluorescent screen Rutherford used in his alpha-particle experiments. Alpha particles striking the screen ionized some of the atoms on the screen. When these atomic electrons returned to their ground states, visible photons were emitted. Rutherford and his students used a microscope to count the individual flashes.

The fluorescent screen is just one example of a *scintillation detector,* a detector that emits visible light on being struck by radiation. The most familiar device of this type is the television screen. Scintillation detectors can be combined with electronic circuitry to automatically count the radiation. When a photon enters the device, it strikes a photosensitive surface and ejects an electron via the photoelectric effect (Chapter 23). A voltage then accelerates this electron so that it strikes a surface with sufficient energy to free two or more electrons. These electrons are accelerated and release additional electrons when they strike the next electrode. A typical photomultiplier, like the one in Figure 25-9, may have 10 stages, with a million electrons released at the last stage. This signal can then be counted electronically. Furthermore, the signal is proportional to the energy deposited in the scintillation material, and hence it gives the energy of the incident radiation.

Everyday Physics | *Radon*

A source of radiation that is partly under our control is the radioactive gas radon. Much attention has been focused by the media on the presence of indoor radon, especially in the air of some household basements. This awareness came about in part because of energy conservation efforts. In attempts to reduce heating bills, many homeowners improved their insulation and sealed many air leaks. This resulted in less air circulation within homes and allowed the concentration of radon in the air to increase.

Radioactive Decay

$$^{238}_{92}\text{U} \rightarrow \quad ^{234}_{90}\text{Th} \quad + \alpha \quad (9.5 \times 10^9 \text{ years})$$
$$\downarrow$$
$$^{234}_{91}\text{Pa} \quad + \beta^- \quad (24 \text{ days})$$
$$\downarrow$$
$$^{234}_{92}\text{U} \quad + \beta^- \quad (6.7 \text{ hours})$$
$$\downarrow$$
$$^{230}_{90}\text{Th} \quad + \alpha \quad (2.5 \times 10^5 \text{ years})$$
$$\downarrow$$
$$^{226}_{88}\text{Ra} \quad + \alpha \quad (7.5 \times 10^4 \text{ years})$$
$$\downarrow$$
$$^{222}_{86}\text{Rn} \quad + \alpha \quad (1622 \text{ years})$$
$$\downarrow$$
$$^{218}_{84}\text{Po} \quad + \alpha \quad (3.85 \text{ days})$$
$$\downarrow$$
$$^{214}_{82}\text{Pb} \quad + \alpha \quad (3 \text{ minutes})$$
$$\downarrow$$
$$^{214}_{83}\text{Bi} \quad + \beta^- \quad (27 \text{ minutes})$$
$$\downarrow$$
$$^{214}_{84}\text{Po} \quad + \beta^- \quad (19.7 \text{ minutes})$$
$$\downarrow$$
$$^{210}_{82}\text{Pb} \quad + \alpha \quad (10^{-4} \text{ seconds})$$
$$\downarrow$$
$$^{210}_{83}\text{Bi} \quad + \beta^- \quad (19 \text{ years})$$
$$\downarrow$$
$$^{210}_{84}\text{Po} \quad + \beta^- \quad (5 \text{ days})$$
$$\downarrow$$
$$^{206}_{82}\text{Pb} \quad + \alpha \quad (138 \text{ days})$$

A commercially available kit for testing radon gas levels in homes.

The source of the radon is uranium ore that exists in the ground; the ore's quantities vary geographically. Through a long chain of decays shown in the table, the uranium eventually becomes a stable isotope of lead. All the radioactive daughters are solids except one, the noble gas radon (Rn). This gas can travel through the ground and enter basements of homes through small (or large) cracks in the basement floors or walls. Because radon is an inert gas, it doesn't deposit on walls and leave the air. Thus, it can be inhaled into a person's lungs and become trapped there. In this case, each of the eight subsequent decays causes the lungs to experience additional radiation and possible damage. The maximum "safe" level of radon has been established by the Environmental Protection Agency at 4×10^{-12} curie per liter. Kits to test for excessive concentrations are available through retail outlets. If high levels are found, the homeowner can take measures to ventilate the radon and to block its entry into the house by sealing cracks around the foundation and in the basement.

1. What is the source of radon gas found in the basements of some houses?

2. What measures should you take to protect your family from possible radiation poisoning from radon gas?

The Geiger counter, popularized in old-time movies about prospecting for uranium, uses the ionization of a gas to detect radiation. The essential parts of a Geiger counter are shown in Figure 25-10. The cylinder contains a gas and has a wire running along its length. As the radiation interacts with the gas, it ionizes gas atoms. The electrons gain kinetic energy from the electric field in the cylinder, which they then lose in collisions with other gas atoms. These collisions release more electrons, and the process continues. This quickly results in an avalanche of electrons that produces a short electric current pulse that is detected by the electronics in the Geiger counter. In the movies the counter

(a)

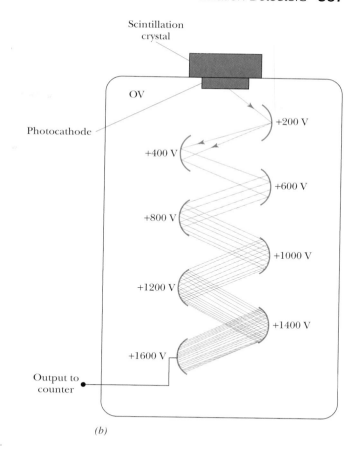

Scintillation crystal

OV

Photocathode

+200 V

+400 V

+600 V

+800 V

+1000 V

+1200 V

+1400 V

+1600 V

Output to counter

(b)

Figure 25-9 (a) Photograph and (b) schematic drawing of a photomultiplier tube used to detect radiation striking a scintillation counter.

(a)

Thin window

Wire electrode

(b)

Metal tube

To amplifier and counter

Figure 25-10 (a) A civil defense Geiger counter. (b) Schematic drawing of a Geiger counter.

produces audible clicks. Many clicks per second means that the count rate is high and our hero is close to fame and fortune.

Another type of detector produces continuous tracks showing the paths of charged particles. These tracks are like those left by wild animals in snow. The big difference, however, is that although we may eventually see a deer, we cannot ever hope to see the charged particles. One track chamber is a bubble chamber in which a liquid is pressurized so that its temperature is just below the boiling point. When the pressure is suddenly released, boiling does not begin unless there is some disturbance in the liquid. When the disturbance is caused by the ionization of a charged particle, tiny bubbles form along its path. These bubbles are allowed to grow so that they can be photographed, and then the liquid is compressed to remove them and get ready for the next picture.

The paths of the charged particles are bent by placing the bubble chamber in a magnetic field; positively charged particles bend one way, and negatively charged particles bend the other way. Examination and measurement of the tracks in bubble chamber photographs, such as the one in Figure 25-11, yield information about the charged particles and their interactions.

Modern particle detectors come in a variety of forms and sizes. In many of these, the particles cannot be seen; their locations and identities are all determined by large arrays of electronics connected to large computers like the one shown in Figure 25-12.

Summary

Radioactivity is a nuclear effect rather than an atomic one. It is virtually unaffected by outside physical conditions, such as strong electric and magnetic fields or extreme pressures and temperatures. Three types of radiation can occur: alpha (α) particles, or helium nuclei; beta (β) particles, which are electrons; and high-energy photons, or gamma (γ) rays. These different radiations have different penetrating properties, ranging from the gamma ray that can travel through many meters of air or even through thick walls to the alpha ray that won't penetrate a piece of paper.

Nuclei are composed of protons and neutrons. Nuclei of the same element have the same number of protons, but isotopes of this element have different numbers of neutrons. Isotopes of a particular element have their own nuclear properties but nearly identical chemical characteristics, because adding a neutron to a nucleus does not change the atom's electronic structure appreciably.

Radioactive decays are spontaneous, obey the known conservation laws (mass–energy, linear momentum, angular momentum, electric charge, and nucleon number), and most often change the chemical nature of their atoms. There are five main naturally occurring nuclear processes: a nucleus (1) emits an alpha particle, producing a nucleus two elements lower in the periodic chart; (2) beta decays, emitting an electron and producing the next higher

Figure 25-11 Photograph of tracks in a hydrogen bubble chamber shows the production of some strange particles that we will discuss in Chapter 27. The spiral track near the top of the photograph was produced by an electron.

Fermi National Accelerator Laboratory

Figure 25-12 This very large particle detector is used at the Fermi National Accelerator Laboratory to study collisions of particles at very high energies.

Fermi National Accelerator Laboratory

element; (3) inverse beta decays via electron capture, producing a daughter nucleus one lower in the periodic table; (4) beta decays with a positron, an antiparticle of the electron, decreasing the number of protons by one and increasing the number of neutrons by one; or (5) emits gamma rays, leaving its identity unchanged.

Equal amounts of different radioactive materials do not give off radiation at the same rate. The unit of activity, a curie, has a value of 3.7×10^{10} decays per second. Two factors determine the activity of a sample of material—the number of radioactive atoms in the sample and the characteristic half-life of the isotope. The quantum-mechanical nature of nucleons results in a wide range of decay rates and prohibits us from predicting the behavior of individual nuclei. Radioactive samples are good clocks for dating events in the past.

Radiation interacts with matter, often outside the range of direct human observation. Alpha and beta particles interact with matter through their electric charge, whereas gamma rays ionize atoms or convert into a pair of particles. All radiation interactions with matter deposit energy, resulting in temperature increases, atomic excitations, and ionization. The ionization caused by radiation passing through living tissue can kill or damage cells. It is impossible to get completely away from radiation.

CHAPTER 25 | *Revisited*

If the artifact is made from material that was once living, its organic material had the same ratio of carbon-14 to carbon-12 as the atmosphere at the time the artifact was made. Over time, the ratio of these carbon isotopes decreases because the carbon-14 is radioactive while the carbon-12 is stable. The change in this isotopic ratio serves as a clock, dating the artifact.

Key Terms

activity   The number of radioactive decays that take place in a unit of time. Activity is measured in curies.

alpha particle   The nucleus of helium, consisting of two protons and two neutrons.

alpha (α) radiation   The type of radioactive decay in which nuclei emit alpha particles (helium nuclei).

beta particle   An electron emitted by a radioactive nucleus.

beta (β) radiation   The type of radioactive decay in which nuclei emit electrons or positrons (antielectrons).

curie   A unit of radioactivity; 3.7×10^{10} decays per second.

daughter nucleus   The nucleus resulting from the radioactive decay of a parent nucleus.

electron capture   A decay process in which an inner atomic electron is captured by the nucleus. The daughter nucleus has the same number of nucleons as the parent but one less proton.

gamma (γ) radiation   The type of radioactive decay in which nuclei emit high-energy photons. The daughter nucleus is the same as the parent.

half-life   The time during which one-half of a sample of a radioactive substance decays.

ionization   The removal of one or more electrons from an atom.

isotope   An element containing a specific number of neutrons in its nuclei. Examples are $^{12}_{6}C$ and $^{14}_{6}C$, carbon atoms with six and eight neutrons, respectively.

neutron   The neutral nucleon; one of the constituents of nuclei.

nucleon   Either a proton or a neutron.

nucleus   The central part of an atom that contains the protons and neutrons.

parent nucleus   A nucleus that decays into a daughter nucleus.

positron   The antiparticle of the electron.

proton   The positively charged nucleon; one of the constituents of nuclei.

rad   The acronym for *radiation absorbed dose*; a rad of radiation deposits 0.01 joule per kilogram of material.

rem   The acronym for *roentgen equivalent in mammals*, a measure of the biological effects caused by radiation.

Questions and exercises are paired so that most odd-numbered are followed by a similar even-numbered.

Blue-numbered questions and exercises are answered in Appendix B.

ʃ indicates more challenging questions and exercises.

WebAssign   Many Conceptual Questions and Exercises for this chapter may be assigned online at WebAssign.

Conceptual Questions

1. What was the biggest difference between Becquerel's radiation and Roentgen's X rays?

2. What makes modeling the behavior of neutrons and protons inside the atomic nucleus more difficult than modeling the behavior of electrons in an atom?

3. Why can't we use visible light to look at nuclei?

4. What first led scientists to conclude that radioactivity is a nuclear phenomenon?

5. Which of the three types of radiation will interact with an electric field?

6. Why do beta rays and alpha rays deflect in opposite directions when moving through a magnetic field?

7. Which of the following names do *not* refer to the same thing: beta particles, cathode rays, alpha particles, and electrons? Explain.

8. Why are X rays not a fourth category of radiation?

9. Is the chemical identity of an atom determined by the number of neutrons, protons, or nucleons in its nucleus? Why?

10. How would the activity of 1 gram of uranium oxide compare with that of 1 gram of pure uranium?

11. What is a nucleon?

12. What observation told scientists that nuclei had to contain more than just protons?

13. What is the name of the element represented by the X in each of the following?
    a. $^{89}_{39}$X
    b. $^{140}_{58}$X
    c. $^{81}_{35}$X

14. What is the name of the element represented by the X in each of the following?
    a. $^{197}_{79}$X
    b. $^{34}_{16}$X
    c. $^{118}_{50}$X

15. How many neutrons, protons, and electrons are in each of the following atoms?
    a. $^{24}_{12}$Mg
    b. $^{59}_{27}$Co
    c. $^{208}_{82}$Pb

16. How many neutrons, protons, and electrons are in each of the following atoms?
    a. $^{19}_{9}$F
    b. $^{63}_{29}$Cu
    c. $^{221}_{88}$Ra

17. What is the approximate mass of $^{90}_{40}$Zr in atomic mass units?

18. What is the approximate mass of $^{69}_{31}$Ga in atomic mass units?

19. The three naturally occurring isotopes of neon are $^{20}_{10}$Ne, $^{21}_{10}$Ne, and $^{22}_{10}$Ne. Given that the atomic mass of natural neon is 20.18 atomic mass units, which of these three isotopes must be the most common?

20. The two naturally occurring isotopes of chlorine are $^{35}_{17}$Cl and $^{37}_{17}$Cl. Which of these isotopes must be the most common?

21. What happens to the charge of the nucleus when it decays via electron capture?

22. What happens to the charge of the nucleus when it decays via beta plus decay?

23. Each of the following isotopes decays by alpha decay. What is the daughter isotope in each case?
    a. $^{224}_{92}$U
    b. $^{197}_{83}$Bi

24. Each of the following isotopes can be produced by alpha decay. What is the parent isotope in each case?
    a. $^{238}_{96}$Cm
    b. $^{240}_{94}$Pu

25. What daughter is formed when each of the following undergoes beta minus decay?
    a. $^{18}_{7}$N
    b. $^{90}_{38}$Sr

26. Each of the following isotopes can be produced by beta minus decay. What is the parent isotope in each case?
    a. $^{47}_{18}$Ar
    b. $^{128}_{49}$In

27. Each of the following isotopes decays by electron capture. What is the daughter isotope in each case?
    a. $^{181}_{77}$Ir
    b. $^{237}_{94}$Pu

28. Each of the following isotopes can be produced by electron capture. What is the parent isotope in each case?
    a. $^{65}_{29}$Cu
    b. $^{253}_{99}$Es

29. Each of the following isotopes decays by beta plus decay. What is the daughter isotope in each case?
    a. $^{28}_{15}$P
    b. $^{22}_{11}$Na

30. Each of the following isotopes can be produced by beta plus decay. What is the parent isotope in each case?
    a. $^{50}_{24}$Cr
    b. $^{52}_{25}$Mn

31. The isotope $^{40}_{15}$P can beta decay to the isotope $^{39}_{16}$S. What else must have happened during this reaction?

32. Beta plus decay and electron capture both result in one additional neutron and one less proton in the nucleus. Why does beta minus decay not have a "mirror" reaction that produces the same change in the nucleus?

33. The isotope $^{64}_{29}$Cu can decay via beta plus and via beta minus decay. What are the daughters in each case?

34. The isotope $^{153}_{66}$Dy decays by alpha decay or by electron capture. What are the daughters in each case?

35. What type of decay process is involved in each of the following?
    a. $^{228}_{92}$U $\rightarrow$ $^{228}_{91}$Pa + (?)
    b. $^{254}_{100}$Fm $\rightarrow$ $^{250}_{98}$Cf + (?)

36. What does the (?) stand for in each of the following decays?
    a. $^{233}_{90}$Th $\rightarrow$ $^{233}_{91}$Pa + (?)
    b. $^{18}_{9}$F $\rightarrow$ $^{17}_{8}$O + (?)

37. Uranium-238 decays to a stable isotope of lead through a series of alpha and beta minus decays. Which of the following is a possible daughter of ^{238}U: ^{206}Pb, ^{207}Pb, ^{208}Pb, or ^{209}Pb?

38. What isotopes of lead (element 82) could be the decay products of $^{234}_{90}$Th if the decays are all alpha and beta minus decays?

39. A proton strikes a nucleus of $^{20}_{10}$Ne. Assuming an alpha particle comes out, what isotope is produced?

40. What isotope is produced when a neutron strikes a nucleus of $^{10}_{5}$B and an alpha particle is emitted?

41. What changes in the numbers of neutrons and protons occur when a nucleus is bombarded with a deuteron (the nucleus of deuterium containing one neutron and one proton) and an alpha particle is emitted?

42. A nucleus of $^{238}_{92}$U captures a neutron to form an unstable nucleus that undergoes two successive beta minus decays. What is the resulting nucleus?

43. You place a chunk of radioactive material on a scale and find that it has a mass of 4 kilograms. The half-life of the material is 10 days. What will the scale read after 10 days?

44. You place a chunk of radioactive material on a scale and find that it has a mass of 10 kilograms. The half-life of the material is 20 days. How long will you have to wait for the scale to read 5 kg?

45. A radioactive material has a half-life of 50 days. How long would you have to watch a particular nucleus before you would see it decay?

46. The isotope $^{239}_{93}$Np has a half-life of about $2\frac{1}{2}$ days. Is it possible for a nucleus of this isotope to last for more than one year? Explain.

47. How do radioactive clocks determine the age of things that were once living?

❢48. Assume that the atmospheric ratio of ^{14}C to ^{12}C has been increasing by a small amount every 1000 years. Would determinations of age by radiocarbon dating be too short or too long? Why?

49. Can carbon-14 dating be used to date the stone pillars at Stonehenge? Why?

50. Why can carbon-14 dating not be used to determine the age of old-growth forests?

51. In decreasing order, rank the following in terms of range: a 10-million-electron-volt alpha particle, a 10-million-electron-volt electron, and a 10-million-electron-volt gamma ray.

52. A beam containing electrons, protons, and neutrons is aimed at a concrete wall. If all particles have the same kinetic energy, which particle will penetrate the wall the least?

53. Why do we often not worry about the distinction between rads and rems?

54. Which of the following causes the most biological damage if they all have the same energy: photons, electrons, neutrons, protons, or alpha particles?

55. Which of the following sources of radiation contributes the least to the average yearly dose received by humans: surroundings, medical, internal, cosmic rays, or nuclear power?

56. Which external source would most likely cause the most damage to internal organs, X rays or alpha particles? Why?

57. Why is it difficult to determine the biological effects of low levels of exposure to radiation?

58. What are the methods for using radiation to treat cancerous tumors?

59. Why are radioactive substances that emit alpha particles more dangerous inside the body than outside? Explain.

60. Would you expect more radiation damage to occur with a rad of X rays or a rad of alpha particles? Explain.

Exercises

61. Your younger brother plans to build a model of the gold atom for his science fair project. He plans to use a baseball (radius = 6 cm) for the nucleus. How far away should he put his outermost electrons?

62. The radius of the Sun is 7×10^8 m, and the average distance to Pluto is 6×10^{12} m. Are these proportions approximately correct for a scaled-up model of an atom?

63. What is the mass of the neutral carbon-12 atom in kilograms?

64. The atomic mass of neutral nickel-60 is 59.93 amu. What is its mass in kilograms?

65. The neutral atoms of an unknown sample have an average atomic mass of 3.19×10^{-25} kg. What is the element?

66. What element has an average atomic mass of 4.48×10^{-26} kg?

67. A hydrogen atom consists of a proton and an electron. What is the atomic mass of the hydrogen atom?

❢68. A deuterium atom consists of a single electron orbiting a nucleus consisting of a proton and a neutron bound together. Compare the sum of the masses of three constituents of the deuterium atom to its measured atomic mass of 2.014 101 8 amu. How do you account for any difference?

69. The activity of a radioactive sample is initially 32 μCi (microcuries). What will the activity be after 4 half-lives have elapsed?

70. If the activity of a particular radioactive sample is 60 Ci and its half-life is 200 years, what would you expect for its activity after 800 years?

71. A radioactive material has a half-life of 10 min. If you begin with 512 trillion radioactive atoms, approximately how many would you expect to have after 30 min?

72. You have a sample of material whose activity is 64 Ci. After 60 minutes the material's activity has decreased to 4 Ci. What is the material's half-life?

73. If the ratio of carbon-14 to carbon-12 in a piece of bone is only one-quarter of the atmospheric ratio, how old is the bone?

74. If the ratio of carbon-14 to carbon-12 in a piece of parchment is only one-sixteenth of the atmospheric ratio, how old is the parchment?

75. A beam of 1-MeV (million electron volts) gamma rays is incident on a block of aluminum that is 12.6 cm thick. What fraction of the gamma rays penetrates the block?

76. How thick an aluminum plate would be required to reduce the intensity of a 5-MeV beam of gamma rays by a factor of 4?

77. How many chest X rays are required to equal the maximum recommended radiation dose for the general public?

78. Cosmic rays result in an average annual radiation dose of 41 mrem at sea level and 160 mrem in Leadville, Colorado (elevation 10,500 ft). How many chest X rays would a person living at sea level have to have each year to make up for this difference in natural exposure?

Nobel Laureates in Physics

The Nobel Prizes are awarded under the will of Alfred Nobel (1833–1896), the Swedish chemist and engineer who invented dynamite and other explosives. The annual distribution of prizes began on December 10, 1901, the anniversary of Nobel's death. No Nobel Prizes were awarded in 1916, 1931, 1934, 1940, 1941, or 1942. You can learn a lot more about the Nobel Prizes in all categories at the Nobel Prize Internet Archive: http://nobelprizes.com.

1901   **Wilhelm Roentgen*** (Germany), for the discovery of X rays.

1902   **Hendrik Lorentz** and **Pieter Zeeman** (both Netherlands), for investigation of the influence of magnetism on radiation.

1903   **Henri Becquerel*** (France), for the discovery of radioactivity, and **Pierre*** and **Marie Curie***† (France), for the study of radioactivity.

1904   **Lord Rayleigh** (Great Britain), for the discovery of argon.

1905   **Philipp Lenard** (Germany), for research on cathode rays.

1906   **Sir Joseph Thomson*** (Great Britain), for research on the electrical conductivity of gases.

1907   **Albert A. Michelson*** (United States), for spectroscopic and metrologic investigations.

1908   **Gabriel Lippmann** (France), for photographic reproduction of colors.

1909   **Guglielmo Marconi** (Italy) and **Karl Braun** (Germany), for the development of wireless telegraphy.

1910   **Johannes van der Waals** (Netherlands), for research concerning the equation of the state of gases and liquids.

1911   **Wilhelm Wien** (Germany), for laws governing the radiation of heat.

1912   **Gustaf Dalén** (Sweden), for the invention of automatic regulators used in lighting lighthouses and light buoys.

1913   **Heike Kamerlingh Onnes*** (Netherlands), for investigations into the properties of matter at low temperatures and the production of liquid helium.

1914   **Max von Laue** (Germany), for the discovery of the diffraction of X rays by crystals.

1915   **Sir William Bragg** and **Sir Lawrence Bragg** (both Great Britain), for the analysis of crystal structure using X rays.

1916   No award.

1917   **Charles Barkla** (Great Britain), for the discovery of the characteristic X-rays of the elements.

1918   **Max Planck*** (Germany), for the discovery of the quantum theory of energy.

1919   **Johannes Stark** (Germany), for the discovery of the Doppler effect in canal rays and the splitting of spectral lines by electric fields.

1920   **Charles Guillaume** (Switzerland), for the discovery of anomalies in nickel–steel alloys.

1921   **Albert Einstein*** (Germany), for the explanation of the photoelectric effect.

1922   **Niels Bohr*** (Denmark), for the investigation of atomic structure and radiation.

1923   **Robert A. Millikan*** (United States), for work on the elementary electric charge and the photoelectric effect.

1924   **Karl Siegbahn** (Sweden), for investigations in X-ray spectroscopy.

1925   **James Franck** and **Gustav Hertz** (both Germany), for the discovery of laws governing the impact of electrons on atoms.

1926   **Jean B. Perrin** (France), for work on the discontinuous structure of matter and the discovery of equilibrium in sedimentation.

1927   **Arthur H. Compton** (United States), for the discovery of the Compton effect, and **Charles Wilson** (Great Britain), for a method of making the paths of electrically charged particles visible by vapor condensation (cloud chamber).

1928   **Sir Owen Richardson** (Great Britain), for the discovery of the Richardson law of thermionic emission.

1929   **Prince Louis de Broglie*** (France), for the discovery of the wave nature of electrons.

1930   **Sir Chandrasekhara Raman** (India), for work on light diffusion and discovery of the Raman effect.

1931   No award.

1932   **Werner Heisenberg*** (Germany), for the development of quantum mechanics.

1933   **Erwin Schrödinger*** (Austria) and **Paul A.M. Dirac*** (Great Britain), for the discovery of new forms of atomic theory.

1934   No award.

1935   **Sir James Chadwick*** (Great Britain), for the discovery of the neutron.

1936   **Victor Hess** (Austria), for the discovery of cosmic radiation, and **Carl D. Anderson** (United States), for the discovery of the positron.

1937   **Clinton J. Davisson*** (United States) and **Sir George P. Thomson** (Great Britain), for the discovery of the diffraction of electrons by crystals.

1938   **Enrico Fermi*** (Italy), for identification of new radioactive elements and the discovery of nuclear reactions effected by slow neutrons.

*These winners are mentioned in the text.

†Marie Curie also received a Nobel Prize in chemistry in 1911.

1939 **Ernest Lawrence** (United States), for development of the cyclotron.

1940– No awards.
1942

1943 **Otto Stern** (United States), for the discovery of the magnetic moment of the proton.

1944 **Isidor I. Rabi** (United States), for work on nuclear magnetic resonance.

1945 **Wolfgang Pauli*** (Austria), for discovery of the Pauli exclusion principle.

1946 **Percy Bridgman** (United States), for studies and inventions in high-pressure physics.

1947 **Sir Edward Appleton** (Great Britain), for discovery of the Appleton layer in the ionosphere.

1948 **Lord Patrick Blackett** (Great Britain), for discoveries in nuclear physics and cosmic radiation using an improved Wilson cloud chamber.

1949 **Hideki Yukawa*** (Japan), for prediction of the existence of mesons.

1950 **Cecil Powell** (Great Britain), for the photographic method of studying nuclear process and discoveries about mesons.

1951 **Sir John Cockcroft** (Great Britain) and **Ernest Walton** (Ireland), for work on the transmutation of atomic nuclei.

1952 **Felix Bloch** and **Edward Purcell** (both United States), for the discovery of nuclear magnetic resonance in solids.

1953 **Frits Zernike** (Netherlands), for the development of the phase contrast microscope.

1954 **Max Born** (Great Britain), for work in quantum mechanics, and **Walter Bothe** (Germany), for work in cosmic radiation.

1955 **Polykarp Kusch** (United States), for measurement of the magnetic moment of the electron, and **Willis E. Lamb, Jr.** (United States), for discoveries concerning the hydrogen spectrum.

1956 **William Shockley, Walter Brattain,** and **John Bardeen** (all United States), for development of the transistor.

1957 **Tsung-Dao Lee** and **Chen Ning Yang** (both China), for discovering violations of the principle of parity.

1958 **Pavel Cerĕnkov, Ilya Frank,** and **Igor Tamm** (all U.S.S.R.), for the discovery and interpretation of the Cerĕnkov effect.

1959 **Emilio Segrè** and **Owen Chamberlain** (both United States), for confirmation of the existence of the antiproton.

1960 **Donald Glaser** (United States), for the invention of the bubble chamber.

1961 **Robert Hofstadter** (United States), for determination of the size and shape of nuclei, and **Rudolf Mössbauer** (Germany), for the discovery of the Mössbauer effect of gamma-ray absorption.

1962 **Lev D. Landau** (U.S.S.R.), for theories about condensed matter (superfluidity in liquid helium).

1963 **Eugene Wigner, Maria Goeppert-Mayer*** (both United States), and **J. Hans D. Jensen*** (Germany), for research on the structure of nuclei.

1964 **Charles Townes** (United States), **Nikolai Basov,** and **Alexsandr Prokhorov** (both U.S.S.R.), for work in quantum electronics leading to the construction of instruments based on maser-laser principles.

1965 **Richard Feynman,* Julian Schwinger*** (both United States), and **Shinichiro Tomonaga*** (Japan), for research in quantum electrodynamics.

1966 **Alfred Kastler** (France), for work on atomic energy levels.

1967 **Hans Bethe** (United States), for work on the energy production of stars.

1968 **Luis Alvarez** (United States), for the study of subatomic particles.

1969 **Murray Gell-Mann*** (United States), for the study of subatomic particles.

1970 **Hannes Alfvén** (Sweden), for theories in plasma physics, and **Louis Néel** (France), for discoveries in antiferromagnetism and ferrimagnetism.

1971 **Dennis Gabor*** (Great Britain), for the invention of the hologram.

1972 **John Bardeen,* Leon Cooper,*** and **John Schrieffer*** (all United States), for the theory of superconductivity.

1973 **Ivar Giaever** (United States), **Leo Esaki** (Japan), and **Brian Josephson** (Great Britain), for theories and advances in the field of electronics.

1974 **Antony Hewish** (Great Britain), for the discovery of pulsars, and **Martin Ryle** (Great Britain), for radio-telescope probes of outer space.

1975 **James Rainwater** (United States), **Ben Mottelson,** and **Aage Bohr** (both Denmark), for the development of the theory of the structure of nuclei.

1976 **Burton Richter** and **Samuel Ting** (both United States), for discovery of the subatomic J/ψ particles.

1977 **Philip Anderson, John van Vleck** (both United States), and **Nevill Mott** (Great Britain), for work underlying computer memories and electronic devices.

1978 **Arno Penzias*** and **Robert Wilson*** (both United States), for the discovery of the cosmic microwave background radiation, and **Pyotr Kapitsa** (U.S.S.R.), for research in low-temperature physics.

1979 **Steven Weinberg,* Sheldon Glashow*** (both United States), and **Abdus Salam*** (Pakistan), for developing the theory that the electromagnetic force and the weak force are facets of the same phenomenon.

1980 **James Cronin** and **Val Fitch** (both United States), for work concerning the asymmetry of subatomic particles.

1981 **Nicolaas Bloembergen** and **Arthur Schawlow** (both United States), for contributions to the development of laser spectroscopy, and **Kai Siegbahn** (Sweden), for contributions to the development of high-resolution electron spectroscopy.

1982 **Kenneth Wilson** (United States), for the study of phase transitions in matter.

1983 **Subramanyan Chandrasekhar** (United States), for theoretical studies of the processes important in the evolution of stars, and **William Fowler** (United States), for studies of the formation of the chemical elements in the Universe.

1984 **Carlo Rubbia** (Italy) and **Simon van der Meer** (Netherlands), for work in the discovery of intermediate vector bosons.

1985 **Klaus von Klitzing** (Germany), for the discovery of the quantized Hall effect.

1986 **Ernst Ruska** (Germany), for the design of the electron microscope, and **Gerd Binning** (Germany) and **Heinrich Rohrer** (Switzerland), for the design of the scanning tunneling microscope.

1987 **K. Alex Müller** (Switzerland) and **J. Georg Bednorz** (Germany), for development of a "high-temperature" superconducting material.

1988 **Leon Lederman*, Melvin Schwartz,** and **Jack Steinberger** (all United States), for the development of a new tool for studying the weak nuclear force and the discovery of the muon neutrino.

1989 **Norman Ramsay** (United States), for various techniques in atomic physics, and **Hans Dehmelt** (United States) and **Wolfgang Paul** (Germany), for the development of techniques for trapping single charged particles.

1990 **Jerome Friedman, Henry Kendall** (both United States), and **Richard Taylor** (Canada), for experiments important to the development of the quark model.

1991 **Pierre de Gennes** (France), for discovering methods for studying order phenomena in complex forms of matter.

1992 **Georges Charpak** (France), for his invention and development of particle detectors.

1993 **Russell Hulse*** and **Joseph Taylor*** (both United States), for discovering evidence of gravity waves.

1994 **Bertram Brockhouse** (Canada) and **Clifford Shull** (United States), for pioneering contributions to the development of neutron-scattering techniques for studies of condensed matter.

1995 **Martin Perl** and **Frederick Reines*** (both United States), for pioneering experimental contributions to lepton physics.

1996 **David Lee, Douglas Osheroff,** and **Robert Richardson** (all United States), for their discovery of superfluidity in helium-3.

1997 **Steven Chu** (United States), **Claude Cohen-Tannoudji** (Algeria), and **William Phillips** (United States), for the development of methods to cool and trap atoms with laser light.

1998 **Robert Laughlin** (United States), **Horst Störmer** (Germany), and **Daniel Tsui** (China), for their discovery of a new form of quantum fluid with fractionally charged excitations.

1999 **Gerardus 't Hooft** and **Martinus Veltman** (both Netherlands), for elucidating the quantum structure of electroweak interactions.

2000 **Zhores I. Alferov** (Russia) and **Herbert Kroemer** (United States), for developing semiconductor heterostructures used in high-speed- and opto-electronics, and **Jack St. Clair Kilby** (United States), for his part in the invention of the integrated circuit.

2001 **Eric A. Cornell, Wolfgang Ketterle,** and **Carl E. Wieman** (all United States), for the achievement of Bose–Einstein condensation in dilute gases of alkali atoms and for early fundamental studies of the properties of the condensates.

2002 **Raymond Davis, Jr.,* Riccardo Giacconi** (both United States), and **Masatoshi Koshiba*** (Japan), for pioneering contributions to astrophysics.

2003 **Alexei A. Abrikosov, Vitaly L. Ginzburg** (both Russia), and **Anthony J. Leggett** (Great Britain and United States), for pioneering contributions to the theory of superconductors and superfluids.

2004 **David J. Gross, H. David Politzer,** and **Frank Wilczek** (all United States), for the discovery of asymptotic freedom in the theory of the strong interaction.

2005 **Roy J. Glauber** (United States), for his contribution to the quantum theory of optical coherence, and **John L. Hall** (United States) and **Theodor W. Hänsch** (Germany), for their contributions to the development of laser-based precision spectroscopy, including the optical frequency comb technique.

2006 **John C. Mather** and **George F. Smoot** (both United States), for their discovery of the blackbody form and anisotropy of the cosmic microwave background radiation.

2007 **Albert Fert** (France) and **Peter Grünberg** (Germany), for the discovery of Giant Magnetoresistance.

2008 **Yoichiro Nambu** (United States), for the discovery of the mechanism of spontaneous broken symmetry in subatomic physics; and **Makkoto Kobayashi** and **Toshihide Maskawa** (both Japan), for the discovery of the origin of the broken symmetry which predicts the existence of at least three families of quarks in nature.

Appendix B

Answers to Most Odd-Numbered Questions and Exercises

CHAPTER 1

Questions
1. A physics world view incorporates data from outside the range of human sensations.
3. It does not have any scientific basis.
5. It must (a) account for known data, (b) make testable predictions, and (c) have a scientific basis.
7. The more prestigious the scientist who proposes the theory, the more likely the scientific community will commit resources to test the theory.
9. United States
11. About 170 cm
13. About 2.5 m
15. About 85 kg
17. 10^3 miles (3000 miles) or 10^3 kilometers (4800 km) or 10^6 meters

Exercises
19. 86,400 s
21. 109 yd
23. 39.4 in.
25. (a) 8.976×104 in. (b) 7.07×10^{-13} g
27. (a) 4300 g (b) 0.000 081 2 m
29. (a) 1.56×102 (b) 3.4×108
31. 10^4 times

Chapter 2

Questions
1. The puck speeds up and then slows down.
3. In the middle
5. ●●●● ● ● ● ● ● ● ● ● ● ●●
7. ●● ● ● ● ● ● ● ● ● ● ●●
9. The truck
11. Less than 45 mph
13. They had the same average speed, but Chris had higher instantaneous speeds.
15. The average speed is greater than the instantaneous speed at C and less than the instantaneous speed at D.
17. Determine the time it takes to travel between mile markers.
19. No, the average speed doesn't tell us any instantaneous speeds.
21. A stopwatch and an odometer give average speed; a speedometer gives instantaneous speed.
23. Velocity has a direction.
25. They both have zero acceleration.
27. Anything that changes the speed (or direction) is an accelerator.
29. The bicycle has the greatest acceleration.
31. The motorcycle
33. Carlos could be going 60 mph and slowing while Andrea is going 5 mph and speeding up.
35. They are falling in a vacuum.
37. 40 m/s
39. It stays the same.
41. It will be beside you.
43. They hit at the same time.
45. Galileo concluded that the object falls with a constant acceleration when air resistance is ignored. Aristotle hypothesized that the object quickly reaches a constant speed.
47. The heavier ball hits first.
49. The marble has the greater acceleration.
51. The accelerations are the same.
53. The increasing upward force due to the air resistance causes the downward acceleration to continually decrease.
55. Greater than

Exercises
57. 3530 km/h
59. 66 mph
61. 6.05 mph
63. 480 miles
65. 2.4 mph
67. 4 s, compared to 10 s for humans
69. 25 h; no
71. 12.5 mph/s
73. 5 mph/s
75. 27 m/s
77. 13 m/s
79. 20 m
81.

Time (s)	Height (m)	Velocity (m/s)
0	80	0
1	75	10
2	60	20
3	35	30
4	0	40

83. 30 m/s; 3 s
85. 1.3 m/s

Chapter 3

Questions
1. The snacks appear to fall straight down.
3. An unbalanced force (friction) opposes the motion of the car.
5. The net force on each car is zero.
7. Because of its inertia
9. Because of your inertia
11. The inertia of the anvil keeps it from moving.
13. No, inertia refers to how hard it is to speed up or slow down.
15. No, the net force will not normally point in the direction of any of the individual forces.
17. 50 N; 130 N
19. The wagon now accelerates (speeds up indefinitely) because the net force is not zero.
21. East; still east
23. The net force is up.
25. It doubles.
27. 4 m/s^2
29. Her mass doesn't change.
31. It triples.
33. For a skier slowing down and moving to the right:

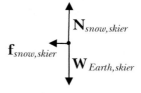

35. If there is no air resistance
37. Upward force of resistance balances downward force of gravity, so acceleration is zero.
39. The acceleration of each is zero, so the net force on each must be zero.
41. Newton's first law is always valid. There must be other forces opposing the friction so that the net force is zero.
43. The frictional force is equal to 400 N because the net force is zero.
45. 250 N
47. They are equal and opposite by Newton's third law.
49. They are equal and opposite by Newton's third law.
51. Zero
53. According to Newton's third law, there is a reaction force on the cannon.
55. The frictional force of the floor on your feet
57. 40 N
59. The forces act on different objects. The frictional force of the ground on the horse's hooves allows the horse to move the cart.
61. By Newton's second law because the net force is zero

Exercises
63. 14 N; 2 N; 10 N
65. 250 N backward
67. 3 m/s²
69. 900,000 m/s²
71. 120 N
73. 1.67 m/s²
75. 75 kg
77. 5 N down
79. 4 m/s²
81. 162 N
83. 10 N
85. 3 m/s²

Chapter 4

Questions
1. At each point the instantaneous velocity is in the direction of motion (tangent to the circle), and the net force is toward the center of the circle.
3. Gravity provides the required centripetal force.
5. The velocity is tangent to the path; all the others point radially inward.
7. They point in opposite directions; they have the same length.
9. The speed changes.
11. Southwest
13. (a)   (b)

15. (a) Speeding up in a straight line (b) Speeding up and turning left
17. It must point southward and eastward. The angle would be greater than 90°.
19. (a) Any example that is moving down and speeding up would work, such as a falling rock. (b) Any example that is moving up and slowing down would work, such as an elevator slowing as it reaches the tenth floor from the lobby.
21. It is down while you are speeding up and up while you are slowing down.
23. The friction force exerted by the ground on the bicycle
25. The tension force must be greater than the monkey's weight.
27. At the bottom
29. It must be nearly uniform.
31. Doubling speed would quadruple acceleration; reducing radius to half would only double acceleration.
33. Only the force of gravity

35. The ball's velocity is horizontal toward home plate; the net force and acceleration are vertically downward.
37. The accelerations would be the same. Acceleration is independent of horizontal velocity.
39. They hit at the same time. They start with the same vertical velocity.
41. The demonstration works.
43. Kicks the ball at a larger angle
45. 5 N, by Newton's third law
47. The diagram contains only the gravitational force exerted by the Sun.
49. Point C is most likely nearest the object's center of mass.

Exercises
51. (a) 5 m/s west (b) 5 m/s east (c) 15 m/s east
53. 50 m/s at 37° north of west
55. 5 m/s² (53° north of east)
57. (a) 2 m/s² (b) 240 N
59. (a) 3.16×10^7 s (b) 9.42×10^{11} m (c) $5.96 \times 10^{(3}$ m/s²
    (d) 3.56×10^{22} N
61. 45 m/s; 8 m/s
63. 125 m; 200 m
65. 2 s; 60 m
67. 9.83 m/s²

Chapter 5

Questions
1. None
3. Same, by Newton's third law
5. Same
7. The surface area quadruples for cube and sphere.
9. 200 N
11. Because the Moon's acceleration does not depend on its mass
13. Twice as great
15. Because it is also moving sideways fast enough to match Earth's curvature.
17. He is in free fall.
19. They are in free fall.
21. No
23. The gravitational forces are very small.
25. Atmospheric friction, though weak, still acts.
27. Send a satellite to orbit Venus.
29. Larger
31. A slow decrease in size of the planetary orbits
33. No, Paris is not on the equator.
35. The satellite would appear to move westward.
37. Because the mass of Earth is so much larger than that of the Moon, it has a much smaller orbit about their common center of mass.
39. Nearer low tide
41. Half the rotational period, 4 h 55 min
43. d; the effects of the Moon and Sun act together.
45. The effects reinforce, producing a larger bulge.
47. The extra tidal force is minuscule.
49. Decreases with the square of the distance
51. Not possible

Exercises
53. 0.006 m/s²
55. $g/9 = 1.11$ m/s²
57. 60 N
59. $\frac{1}{2}F_{old}$; no
61. 0.25 $\mathbf{F}_{Earth}$
63. 12,660 km; 6290 km
65. 1.07×10^{-7} N compared to 200 N
67. 200 N

69. 3.8 m/s²
71. 264,000 km; 3060 m/s
73. 1.93
75. 40 N

Chapter 6

Questions
1. Its numerical value does not change.
3. Toby is correct.
5. They have large inertia.
7. The net force on an object is equal to the change in its momentum divided by the time required to make the change.
9. They lengthen the time for your leg to stop.
11. Same
13. The impulses are equal.
15. 8 kg · m/s directed up
17. Jeff feels the greater force because his flowerpot experienced the greater change in momentum.
19. They lengthen the interaction time and decrease the forces.
21. At each interaction there are equal and opposite momentum changes. The total momentum is conserved at all times.
23. 4 N acting for 4 s
25. Conservation of linear momentum requires the bullet to go one way and the rifle the other.
27. The net force exerted by the exhaust gases on the rocket–plate system is nearly zero.
29. Earth's large mass allows it to acquire an equal and opposite momentum to the ball without a measurable velocity.
31. At rest
33. Moving to the left
35. 250 kg · m/s
37. Zero
39. They give the rowboat momentum away from the dock.
41. It is conserved.
43. Not without outside help unless the astronaut has something to throw
45. Path B
47. Path D

Exercises
49. 36,000 kg·m/s
51. 62 m/s
53. 5625 N
55. 35,000 kg·m/s (= 35,000 N·s)
57. Impulse = 45,000 N · s; F_{av} = 5625 N
59. 180 N upward
61. 900 N
63. 1.5 m/s
65. Zero
67. 66,800 kg · m/s (north)
69. 2.5 m/s

Chapter 7

Questions
1. The initial momentum of the system is zero.
3. Conservation of momentum
5. Yes
7. Motorboat pulling a water skier
9. Minivan
11. Same kinetic energies
13. Two balls leave the other side with the same speed as the incoming balls.
15. How high she lifted the ball
17. Decrease
19. Kinetic energies are the same; momentum is greater for the heavier sled.
21. Frictional forces and air resistance do an equal amount of negative work.
23. Less than
25. Both cars stop in the same distance.
27. No, the kinetic energy lost by one object during a collision could be converted to many different forms of energy.
29. 4 N through 7 m do more work.
31. The choice for the zero of potential energy is arbitrary. Only differences in potential energy are important.
33. The gain in the kinetic energy equals the loss in the gravitational potential energy, keeping the mechanical energy the same.
35. Potential energy is a maximum at either end, and kinetic energy is a maximum at the bottom.
37. It will decrease.
39. As the satellite moves closer to Earth, its gravitational potential energy decreases, and its kinetic energy increases. The reverse happens as the satellite moves farther from Earth.
45. The work done in plowing through the dirt or sand reduces the kinetic energy of the truck.
47. The chemical potential energy of the battery is converted to thermal energy in the socks.
49. We cannot recover the energy lost because of frictional effects.
51. If the winch will run for more than 1 s, it can do more than 600 J of work.
53. A unit of energy
55. A kilowatt-hour is a unit of energy.

Exercises
57. 630,000 J
59. Momentum is not conserved; therefore, the collision could not have taken place.
61. 10 J
63. 1.6 J
65. 4 N
67. Negative for the next 6 months; zero for the year
69. 396 kJ
71. (a) −8.7 J (b) 0 (c) 0
73. 30 J
75. 144 kJ
77. 2.73 horsepower
79. 120 Wh = 0.12 kWh

Chapter 8

Questions
1. Rotational speed is the same for all points on a rotating body.
3. The same
5. Toward the North Star
7. Rotational inertia
9. None
11. F_2 because it acts farther from the pivot
13. The force on the nail is larger because it acts through a smaller radius than that of the force applied to the handle.
15. Your torque about the base of the ladder is balanced by the torque exerted by the window—the higher you climb, the greater your torque.
17. Because Sam has to exert the greater force, Sam must be closer to the center of mass.
19. The flywheel of larger radius has the greater rotational inertia and is harder to stop.
21. Increases; decreases
23. Sphere
25. There is no appreciable external torque acting on Earth.

27. The front–back location can be determined by separately weighing the front and back tires on a truck scale. The ratio of those two readings will give the ratio of the distances of the center of mass from the two axles. The same procedure works for the left-right position.
29. Below the foot of the figure
31. One diagram should show that the force acting at the center of mass lies inside the base for the upside-down cone even when it is tipped.
33. Stable equilibrium
35. The wall prevents you from rocking backward, to keep your center of mass over your feet.
37. The jumper's body is curved as it passes over the bar, causing the center of mass to lie outside the jumper's body below the bar.
39. The cylinder wins! Both have the same kinetic energy, but the hoop has more of this energy in the form of rotational kinetic energy because of its larger rotational inertia.
41. To counteract the torque of the main rotor acting on the helicopter. This torque is a reaction to the torque applied to the main rotor.
43. Tuck
45. The lower rotational inertia requires a larger rotational speed to conserve angular momentum.
47. No, one part of its body has angular momentum in one direction while the rest of its body has an equal angular momentum in the opposite direction.
49. The grooves give the bullet angular momentum so that it doesn't tumble in flight.
51. The push exerted by the exhaust gases produces a torque on the merry-go-round.
53. In the same (horizontal) direction
55. The car would not go around corners because of the large angular momentum of the flywheel. Use two flywheels rotating in opposite directions.
57. Clockwise
59. Polaris appears stationary because Earth's spin axis is aligned with it.
61. (a) Toward the ground (b) Toward the sky (c) Toward the ground (d) Zero

Exercises

63. 200 rpm, or 3.33 rev/s
65. 2.78×10^{-4} rev/s
67. −2500 rpm/s
69. 210 N · m
71. 1800 N · m
73. 2 m
75. Left-hand side will fall because it has larger torque.
77. $N_C = 793$ N $\Rightarrow N_W = 910$ N $- 793$ N $= 117$ N
79. 300 kg · m²/s
81. $L_M = 0.132 L_E$; Earth has the larger angular momentum.

Chapter 9

Questions

1. No, Newton's first law is true only in inertial reference frames.
3. Alice would not see any motion of the jar relative to her. Someone sitting on a shelf would see the jar fall freely.
5. No experiment can distinguish between the two.
7. The ball lands on the white spot.
9. Both would agree that there are no horizontal forces.
11. The woman would claim that the ball began with zero kinetic energy, whereas the observer on the ground would calculate a nonzero initial kinetic energy. Both observers would agree on the change in kinetic energy.

13. They would not agree on the value of the kinetic energy at an instant, but they would both agree that the change in kinetic energy is zero.
15. Forward because of its inertia
17. If the velocity is to the right, it is speeding up; if the velocity is to the left, it is slowing down.
19. The ball will land farther from her feet than the white spot. The inertial force will be in the direction opposite to the acceleration, independent of the velocity.
21. It would initially be zero and increase in the direction away from her.
23. The woman would claim that there is a constant horizontal force directed away from her. The observer on the ground would claim that there is no horizontal force.
25. Greater, because the inertial force acts in the same direction as the gravitational force
27. Less, because the inertial force acts in the opposite direction to the gravitational force
29. Same
31. Inertial forces act backward in the noninertial reference system of the accelerating plane.
33. The same, because you are comparing masses. The weights increase by the same amount during the acceleration.
35. If the train's velocity is to the west, it must be speeding up; if its velocity is to the east, it must be slowing down.
37. (a) Up (b) Up (c) Down
39. Up is not defined.
41. (a) Constant velocity (b) Slowing down
43. The "up" direction is toward the ground.
45. Your figure should have the plants leaning toward the center.
47. The centrifugal force pulls the mud off.
49. Parallax would be easier to observe.
51. No, because there is no Coriolis force on the equator
53. There is no Coriolis force on the equator.

Exercises

55. 40 m/s; 10 m/s
57. (a) 70 mph (b) 30 mph backward
59. (a) 4 m/s² downward (b) 6 m/s² upward
61. 0; 25 J; 25 J
63. 1500 N
65. 390 N
67. 100 N
69. 240 N
71. 10 m/s²

Chapter 10

Questions

1. Because the physical laws are the same in all inertial systems, it is impossible to determine your speed.
3. No, the speed of light is an absolute quantity.
5. 70 mph
7. He will obtain a speed equal to *c*.
9. No, the special theory of relativity does not allow one to travel backward in time.
11. No, the supernova in the Whirlpool galaxy happened first.
13. The flash at the front of the train occurred first.
15. The one at the back of the skateboard. The third person must have a speed to the right that is larger than that of the skateboard.
17. A could not have triggered B, so the order could be reversed.
19. Less than 3 min
21. Same note
23. Shorter
25. Greater

27. Peter will be the younger.
29. It is not possible to travel backward in time.
31. The observers would agree because the clocks move toward or away from the light by the same amount.
33. Earth's reference system
35. Less than 100 m
37. Less than 400 m
39. Same
41. (a) The caboose entered first. (b) The tunnel is longer. (c) Yes.
43. Decrease
45. Matter and energy can be converted to each other.
47. The energy radiated away causes the mass to decrease.
49. Special relativity does not include noninertial reference systems, which are included in the general theory.
51. No
53. Blake is comparing the gravitational masses. Jordan is comparing the inertial masses.
55. The penny
57. Passengers in spaceship A
59. Over laboratory distances, the bending of light is less than the diameter of an atom.
61. (a) Slower because of the stronger gravitational field
    (b) Slower because of the centripetal acceleration

Exercises
63. 500s = 8 min 20 s; 1.28 s
65. 300 s = 5 min
67. 1.09
69. 0.379 ns
71. 28.4 h
73. 69 m
75. 200 m
77. 1.31×10^{13} m
79. 672 N
81. 0.866 c
83. 5 m/s² backward
85. 11 m/s² downward
87. 5 m
89. $3 \times 180° = 540°$. It is almost a great circle, but it has slight bends in three places; for instance, at one pole and both crossings of the equator.

Chapter 11

Questions
1. Lack of predictive power
3. Seal off the machine, thus taking away essentials such as food and water for a period of time.
5. The next brown can may sink. A model can never be proven true.
7. The atomistic nature of matter is not important for most day-to-day activities.
9. Water, salt, and granite are not elements.
11. It is a compound because the mercury combines with something in the air.
13. Compounds are new substances with their own properties that result from substances combining according to the law of definite proportions.
15. 14 amu
17. 1 g of hydrogen
19. They have the same number of molecules.
21. 32 g
23. The ideal gas model does not apply to liquids, but the particles must be much closer together in the liquid.
25. Ball

27. Increased pressure at lower elevation results in decreased volume.
29. More molecules will be hitting the walls per unit time.
31. The perfume molecules travel very crooked paths.
33. To amplify the rise in the narrow tube
35. Atmospheric pressure and the purity of the water must be maintained.
37. 39°C
39. 39°C
41. 273 K
43. Average kinetic energy of the molecules
45. The average kinetic energies are equal because they are at the same temperature.
47. The particles have more kinetic energy. Therefore, they are moving faster and strike the walls more frequently and with more momentum.
49. Volume
51. Temperature drops to one-fourth on the Kelvin scale.
53. The particles strike the walls more frequently, producing a larger average impulse with the walls of the container.
55. The more energetic molecules leave via evaporation, lowering the average kinetic energy and therefore the temperature of the remaining water.

Exercises
57. 3 g
59. 18 g
61. 20 sandwiches; 1200 g; 800 g of ham
63. 5.02×10^{22} atoms
65. 3.35×10^{25} molecules
67. 1 nitrogen and 3 hydrogens
69. 10^{16}
71. 382 kPa ≈ 3.7 atm
73. $\frac{1}{3}$ L
75. 240 balloons
77. 3.2 atm

Chapter 12

Questions
1. Solid, liquid, gas, and plasma
3. They have the same densities.
5. Magnesium
7. The gold atoms must be closer together.
9. They are not the same because the crystals have different shapes.
11. Diamond has strong bonds in all directions; graphite has stronger bonds between atoms in two-dimensional layers.
13. The interatomic forces are stronger.
15. Spherical
17. The surface tension allows the surface of the water to rise without overflowing.
19. The water molecules are more strongly attracted to the glass than to each other.
21. Gas particles are neutral, whereas the particles in a plasma are electrons and charged ions.
23. The reduced surface area in the case of the gravel requires greater pressure.
25. The volume of liquid water is less than that of vapor, resulting in lower pressure in the can.
27. Denver has a lower atmospheric pressure, so fewer horses would be needed.
29. Denver would always be listed as a low-pressure region.
31. Your ears would hurt the same in both.
33. Same
35. Greater for the cold water

37. It is the atmospheric pressure that determines the maximum height.
39. With the pump at the bottom, you can apply much higher pressures.
41. The purple liquid has the larger density.
43. Higher in salt water
45. The buoyant forces are the same.
47. The reduction in volume reduces the buoyant force and causes you to sink.
49. Lead will float and gold will sink.
51. The volume does not change. Expelling water reduces weight.
53. The buoyant force is the same because they both displace the same volume.
55. It stays the same because the volume below the surface is the same as the volume of the ice when melted.
57. The fast-moving air creates lower pressure above the dime, which results in a net upward force.
59. So the balls will drop over the net
61. The wind blowing by the base reduces the outside pressure.

Exercises
63. 2.7 g/cm³; aluminum
65. 0.3 g/cm³
67. 1200 kg
69. 0.07 m³
71. 1 cm³
73. 100,000 N; 100,000 Pa; same
75. 7.5 in.
77. 15 lb/in.²
79. 750 kg/m³; float
81. 5 N
83. 0.9 g/cm³
85. 79,300 N

Chapter 13

Questions
1. The gravitational potential energy is converted to thermal energy.
3. The heat seemed to be produced endlessly, implying that it was not a substance.
5. Both change the internal energy of a system and are measured in joules. Heat operates at a microscopic level, whereas work is a macroscopic concept.
7. If the thermometer is initially at a different temperature, heat must flow.
9. Yes, if they start with different temperatures.
11. Temperature is not a physical quantity and therefore cannot flow between objects.
13. Heat is a flow of thermal energy, whereas temperature is a measure of the average kinetic energy of the atoms and molecules. Notice that temperature is not a form of energy.
15. Neither student is correct.
17. Under all conditions
19. Water
21. No
23. More
25. It takes a tremendous amount of thermal energy to change the temperature of the water in the oceans.
27. It will freeze if heat is removed and melt if heat is added.
29. It requires a lot of thermal energy to melt the ice without changing its temperature. Also, ice is not a good conductor of thermal energy.
31. Less than 40°C
33. More internal energy in steam

35. The internal energy must increase by the first law. However, if the increase in internal energy causes a change of state, the system can remain at a fixed temperature.
37. The molecules of the rod exchange kinetic energy via collisions.
39. The order from best to worst is polyurethane foam, static air, glass, and concrete.
41. It will not freeze.
43. Stainless steel has a thermal conductivity that is about one-sixth that of iron.
45. The ground beneath the road keeps the road warm for a while.
47. The air beneath the clouds is shielded from the Sun and is cooler. Being cooler, the air is denser and creates a downdraft.
49. Both cars are heated by radiation.
51. The vacuum reduces conduction and convection, and the silvering reduces radiation.
53. Expansion and contraction of the roof
55. The spacing would be smaller on the wide portion.

Exercises
57. 5000 cal
59. 800 J; 190 cal
61. 10 Cal
63. 16 J
65. 330 J
67. 0.278 cal/g·°C
69. 124 cal
71. −1°C; +4°C
73. (a) 1600 cal (b) 1600 cal (c) It stays the same.
75. 6771 kJ
77. 1.73 × 10⁶ J; 1188 kJ
79. 0.8 mm

Chapter 14

Questions
1. It converts heat to mechanical energy.
3. This would violate the second law of thermodynamics because heat will not naturally flow between two reservoirs at the same temperature.
5. The second law
7. Yes
9. Yes, provided the water that the tube accesses is at a different temperature.
11. You cannot get more energy out than you put in.
13. First: You cannot get more energy out than you put in.
    Second: You cannot convert all the heat to mechanical work.
15. About 20% of the thermal energy available in the food that is consumed shows up as mechanical work.
17. The efficiency increases.
19. Heat engine A is more efficient.
21. This would allow the hot region to operate at a higher temperature.
23. Because an efficiency of 1 or more means that the mechanical energy produced is as large as or larger than the thermal input.
25. In the shade
27. No
29. Temperature rises
31. A refrigerator will always need a motor.
33. The statement implies that work must be expended to make the thermal energy flow from a cooler region to a hotter region.
35. Greater than 1
37. Both strategies are equally invalid.
39. Sums 2 and 12 have highest order; sum 7 has the lowest order.
41. There are more ways of getting this sum than any other.

43. There is only one way of getting this arrangement.
45. The universe continually becomes more disordered.
47. The disorder increases as collisions cause all the particles to reach a common kinetic energy.
49. The entropy of the universe increases.
53. The order increases. Entropy must increase elsewhere.
55. It is highly unlikely for all the molecules to move in the same direction at the same time.
57. No, because atomic motions are symmetric with time. The direction of time in the release of air from a balloon, however, would be observable.
59. b
61. Electricity
63. If the oven heats the kitchen, the baseboard heaters will run less.

Exercises
65. 700 kJ
67. 5000 cal (each minute)
69. 0.25, or 25%
71. 2250 J
73. 9000 cal/min; 0.25
75. 0.0273, or 2.73%
77. 750 K = 477°C
79. 800 J
81. 800 J (each second)
83. 1600 J (each second)
85. HHHH, HHHT, HHTH, HTHH, THHH, HHTT, HTTH, TTHH, HTHT, THTH, THHT, HTTT, THTT, TTHT, TTTH, TTTT
87. $\frac{5}{36}$, or 13.9%
89. $\frac{6}{216} = \frac{1}{36}$, or 2.78%

Chapter 15

Questions
1. Its inertia carries it through the equilibrium point.
3. Down toward the equilibrium point in both cases
5. The period will increase, and the frequency will decrease.
7. A 3-kg block
9. The period stays the same.
11. The period is 60 s, and the frequency is 1/(60 s) = 0.017 Hz.
13. Move the mass down to lengthen the period.
15. The exhaust system has a natural frequency at 2000 rpm.
17. 1 Hz and 0.5 Hz
19. No
21. It is a longitudinal wave because the medium is a fluid.
23. No
25. Tension and mass per unit length
27. The reflected pulse will be on the opposite side and will cancel with the first one.
29.

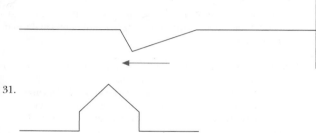

31.

33. Frequency and wavelength
35. The wave with the lower frequency
37. The wavelength is doubled.
39. The frequency stays the same, but the spacing decreases.
41.

43. Five times as high
45. Three antinodes
47.

49. One-half the rope's length
51. Twice the length of the rod
53. Antinode
55. The spacing decreases.
57. Increase
59. Decrease

Exercises
61. 3 s
63. 0.167 Hz
65. 3 cm
67. 0.628 s
69. 4
71. 4.45 s
73.

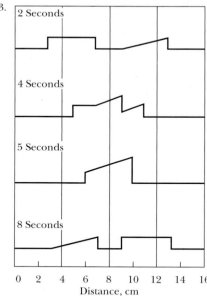

75. 4.17 m/s
77. 75 cm/s
79. 15 m
81. 343 Hz
83. 6 m; (6 m)/2 = 3 m; (6 m)/3 = 2 m; (6 m)/4 = 1.5 m; (6 m)/5
85. 1.5 Hz

Chapter 16

Questions

1. Interference and superposition
3. Temperature
5. Water
7. The speed of light is much faster than the speed of sound.
9. Sound waves of all frequencies have the same speed.
11. The intensity is reduced to one-hundredth.
13. Yes
15. Mainly frequency
17. Equal to
19. 400 Hz
21. Increasing the tension or decreasing its mass per unit length
23. Twice the length of the string
25. One quarter the distance from either end; no
27. No
29. The same
31. Nodes
33. Twice the length of the pipe
35. The wavelength doubles.
37. Open at both ends
39. Decreases
41. Equal to
43. The unique blend of higher harmonics is hard to reproduce electronically.
45. Increases
47. You cannot tell.
49. Increases
51. The frequency of the whistle decreases as the train passes by.
53. Speed
55. The frequency is decreasing, and the sound is getting louder.
57. Billy's has the higher frequency.
59. Both produce a sharp rise in pressure.
61. You hear nothing until the shock wave arrives. You will then hear the characteristic boom-boom followed by engine noise.

Exercises

63. 0.003 82 s
65. 0.275 m
67. 343 Hz
69. 17.2 m
71. 2.74 km
73. 600 m
75. 40 dB
77. 125 Hz, 250 Hz, 375 Hz, 500 Hz, . . .
79. 470 Hz
81. $f_1 = 71.5$ Hz, $f_3 = 215$ Hz, $f_5 = 358$ Hz, . . .
83. 17.4 cm
85. 75 Hz
87. 260 Hz or 264 Hz

Chapter 17

Questions

1. Unless there is dust in the air, none of the light gets scattered into the eyes until the light hits the wall.
3. No light scatters back toward your eyes.
5. Point source
7. Shrinking
9. There will be no umbra when the shadow is not long enough to reach the screen. This occurs when the Moon's shadow does not reach Earth.
11. The image gets brighter and fuzzier.
13. The image is upside down and left–right reversed.
15. C
17. A

19. 0.7 m
21. B and C
23. One; infinite
25. Half her height; no
27. The images are progressively farther from the observer. They are virtual.
29. To produce a beam of light
31. The concave mirror
33. Concave
35. The image is very close to the mirror on the back side and only slightly magnified. Moving away makes the image get larger and move away.
37. The light actually converges to form a real image; there is no light at the location of a virtual image.
39. No; virtual
41. At the focal point
43. The speed of light is much faster than the speed of sound.
45. The time delay to send a signal to Earth and get a response was on the order of 20 min.
47. Magenta
49. Red; green; black
51. The letter would blend into the background because both would appear red.
53. With light beams you are adding colors, whereas with pigments you are subtracting colors. Red paint absorbs all but red, and green paint then absorbs the red, leaving muddy brown.
55. Magenta
57. Cyan
59. The sky would appear green, and the Sun would appear magenta.

Exercises

61. 20-cm diameter
65. 5 m
67. 36.9°
69. 10 m
71. The image is real, inverted, and reduced in size.
73. The image is located 30 cm in front of the mirror and is 3 cm tall.
75. The image is located 30 cm behind the mirror.
77. The image is located 20 cm behind the mirror and is two-thirds as big.
79. The image is located 40 cm in front of the mirror.
81. 0.1 ms; it is 1/2000 as big.
83. 16 ms
85. 9.46×10^{15} m

Chapter 18

Questions

1. C
3. C
5. The bending of light at the surface
7. Greater at the water–air surface
9. The light rays hit the sides of the light pipe at angles greater than the critical angle.
11. 141 million miles
13. Decrease (remember that the Moon is shortened in the vertical direction)
15. The Sun appears to rise before it actually does. Therefore, sunrise is earlier.
17. They are shortened in the vertical direction, yielding oval shapes.
19. Assuming that there is no deflection of the spear as it enters the water, you would aim low, because the fish appears to be higher than it actually is.

21. No
23. The bend will not be a problem for the blue light, but red light will leak out.
25. Yellow and green
27. The angle formed by the line from the Sun to the raindrop and the line from your eye to the raindrop must have a fixed value for a given color.
29. In the west at about midmorning
31. Diverging lens
33. Send the two beams in parallel to the optic axis and find the point where they cross.
35. Send the two beams in parallel to the optic axis and trace their diverging paths. Extend the paths backward to find their crossing point.
37. Converging lens
39. Through the principal focal point
41. The image becomes dimmer.
43. The lens with the larger index of refraction
45. The light converges to form the image and then diverges again. Your eye must be to the right of the image to see this diverging light.
47. Real image
49. To regulate the amount of light entering the eye
51. Dispersion occurs when light is refracted, not reflected.
53. Converging
55. Converging
57. Larger
59. Consider the symmetry of the letters.

Exercises
61. 23°; 20°
63. From Figure 18-2, the fish is down 32° from normal, or 13° below image of the fish.
65. 39°–40°
67. 48°
71. The image is virtual, located at the other focal point, and the magnification is 2.
73. 20 cm on the near side of the lens
75. Farther than $2f$ from the lens
77. The diverging lens will spread the principal rays slightly, such that their convergence will be shifted to the right.
79. The image is virtual, erect, and reduced in size.
81. 2.5 diopters
83. 0.143 m
85. −0.5 m; diverging

Chapter 19

Questions
1. The particles must be moving very fast to not have noticeable deflections due to gravity.
3. Friction would change the component of the momentum parallel to the surface, and an inelastic collision would change the normal component. Either change would affect the angle of reflection.
5. Less than
7. The light particles experience forces directed into the material with the higher index of refraction.
9. Frequency
11. Red light
13. Amplitude
15. The wave equation $(v = \lambda f)$ predicts that if the wavelength decreases while the frequency remains constant, the speed decreases.
17. From a slow medium to a fast medium
19. Red light has a longer wavelength.

21. All colors emerge at the same angle.
23. Blue light
25. Decrease
27. Yellow light
29. Phase
31. Diffraction effects are more pronounced for sound because the wavelength is much longer.
33. Decrease
35. 300 nm
37. Molecules are smaller than the wavelength of visible light.
39. No
41. Thinner
43. Thicker
45. Yes
47. Rotate the lens of one pair 90° relative to the lens of another. If both lenses are polarized, no light will pass through.
49. Tilt your head by 90°.
51. Objects at different depths in the hologram move relative to one another when the hologram is viewed from different angles.
53. Monochromatic light with a constant phase relationship
55. A powerful laser can expose the film before the person can move less than the wavelength of the light.

Exercises
57. 1.88×10^8 m/s
59. 2.42
61. 1.5
63. 1.85
65. 211 m
67. 4.74×10^{14} Hz
69. 396 nm
71. 244 nm
73. 0.006 84°
75. 1.38×10^{-2} arcseconds
77. 100 nm
79. 221 nm

Chapter 20

Questions
1. Insulator
3. Metal rod
5. The moisture allows some of the charge to leave the balloon.
7. To prevent the buildup of charge that may cause sparks
9. Either the fur and plastic produced opposite charges on the rods or one rod is not charged.
11. Neutral objects have equal positive and negative charges.
13. The north pole of a magnet is attracted to positive objects and negative objects but not to neutral objects. It must also be electrically neutral.
15. Repel
17. Zero
19. The balloon induces a charge in the wall.
21. The induced charges on the near sides are always opposite the charged object, producing attraction.
23. Bring the charged rod next to the electroscope. If the leaves get closer together, the rod is negative. If the leaves get farther apart, the rod is positive.
25. Bring the rod near the electroscope and touch the electroscope with your hand.
27. The negative charges flow through the hand to the ground, neutralizing the electroscope.
29. Sphere A is negative, and sphere B is neutral.
31. Touching a charged sphere to a neutral one yields two spheres with one-half charge on each one, using two neutral spheres yields charges of one-third, and so on.

33. Diagram d indicates equal and opposite forces required by Newton's third law.
35. The force increases by a factor of 9.
37. The force increases by a factor of 9.
39. 3
41. The gravitational force is always attractive, the elementary electric charge is not proportional to inertial mass, electric charge comes in one size, and gravity is much weaker.
43. You have very little net electric charge, and the gravitational force is too weak.
45. Because they have different masses
47. 60 N directed to the left
49. The same force at any point between the plates
51. Equal in size, opposite in direction, and force on proton is upward.
53. b > a > c = d, and tangent to the field lines.
55. Greater
57. The speed at point B is less than v_o.
59. Work required to bring 1 coulomb from reference zero to point in question
61. Moves toward lower electric potential energy and lower electric potential

Exercises

63. -1.6×10^{-19} C
65. 1.47×10^{-17} C
67. 4.05×10^{10} N
69. 2.13×10^{-6} N (attractive)
71. 1.23×10^{36}
73. 400 N/C north; 8 N south
75. 4×10^{10} N/C toward the charge
77. 2.16×10^{11} N/C outward
79. 3.6×10^{10} N/C toward the 2-C charge
81. 4.8×10^{-16} N; 2.87×10^{11} m/s^2
83. 20 J
85. 24 mJ
87. 39,000 V

Chapter 21

Questions

1. Lifetime
3. 18 V
5. Differences are voltage, maximum current available, and maximum charge available; both supply direct current.
7. c; a
9. b; d
11. Turn the battery end for end.
13. No difference
15.

17. A volt is a measure of potential difference, and an ampere is a measure of current.
19. Water, volume per unit time, and pressure
21. All of them
23. Doubles
25. The small resistance yields large currents.
27. Wire 20 bulbs in series.
29. Equal to
31. Increase; decrease
33. Increased; decreased
35. Greater
37. Less than
39. Batteries in series with the bulbs in parallel
41. A is the brightest; B and C are equally dim.

43. Stay the same
45. They will all be shorted.
47. They are equally bright.
49. They all go out.
51. A is shorted out; the others get brighter.
55. Parallel
57. The power is cut in half.
59. The same
61. 120-W bulb

Exercises

63. 240 Ω
65. 12 A
67. 4 V
69. 1 A
71. 6 Ω; 2 A
73. 6 A; 2 Ω
75. 11.7 A
77. 0.2 A
79. 3 A; 36 W
81. 4800 W
83. 240 Ω
85. 0.16 kWh
87. 36¢

Chapter 22

Questions

1. Both ends of the unmagnetized rod will attract both ends of the other two rods.
3. The second experiment
5. No
7. 4
9. The north pole of a compass needle points in the direction of the field.
11. Clockwise
13. Opposite
15. Down at A; up at B
17. Into the page
19. The magnetic field from one wire cancels that from the other wire.
21. Out of the page
23. He could strike the needle while holding it in Earth's magnetic field.
25. Decrease
27. 4 N/m in the opposite direction, by Newton's third law
29. If two parallel wires each carry a current of 1 ampere, the force per unit length on each wire will be 2×10^{-7} newtons per meter.
31. Magnetic north pole
33. 15° east of north
35. The cosmic rays spiral along the magnetic field lines toward the South Pole.
37. No
39. They are bent in opposite directions. The electron will have the larger acceleration because of its smaller mass.
41. Up; counterclockwise
43. Toward; push against it
45. Equal to
47. Inserting the south end into the coil or removing the north end
49. Larger
51. To reverse the direction of the current in the loop, ensuring that the torque is in the same direction
53. The electric and magnetic fields are perpendicular to the direction of travel and to each other. They oscillate in phase and travel at the speed of light.

55. Sound
57. They have different frequencies and wavelengths.
59. By modulating the amplitude of the carrier wave
61. 102.1 MHz

Exercises
63. 4.5×10^5 G
65. 4.3×10^{-4} T
67. 1.2×10^{-12} N; 1.32×10^{18} m/s^2
69. 8×10^{-4} N; 0.267 m/s^2
71. 37.5 μC
73. 1.5×10^6 m/s
75. 0.12 T
77. 20 turns
79. 0.025 A; 20
81. 20 A
83. 1.28 s
85. 12.2 cm; about one-fifth the size of the oven
87. 1×10^{15} Hz
89. 200–545 m
91. 275 m

Chapter 23

Questions
1. The structure of the atom was not understood in Mendeleev's time.
3. Carbon
5. Each gas has its own unique collection of spectral lines.
7. Absorption
9. Element A and element B
11. There are many fewer lines in the absorption spectrum.
13. The wavelength of lines can be determined by position alone.
15. B
17. Their deflections by electric and magnetic fields; no
19. Excess
21. Because the ratio of the electron's charge to mass was already known, allowing the mass to be determined
23. They were electrically charged and had a mass less than that of the hydrogen atom.
25. Accelerating charges radiate away energy.
27. The intensity curves are the same for all materials.
29. Decreasing
31. Object B
33. A star's temperature determines the color it appears to be.
35. That the atomic oscillators have quantized energies that are whole-number multiples of a lowest energy, $E = hf$
37. The number emitted from the surface per unit of time
39. Different electrons require different amounts of energy to reach the surface.
41. Each photon in red light has an energy below the minimum.
43. (1) Only angular momenta equal to whole-number multiples of a smallest angular momentum are allowed. (2) Electrons do not radiate when they are in allowed orbits. (3) A single photon is emitted or absorbed when an electron changes orbits.
45. They are equal.
47. The spectrum is produced by a large number of hydrogen atoms.
49. The inner electrons shield part of the positive nucleus, resulting in one electron orbiting a hydrogen-like nucleus.
51. The first two shells are filled with two and eight electrons, and the outer shell has the remaining seven electrons.
53. It easily gives up its single outer electron and forms a bond with other atoms.
55. X-ray
57. The X-ray photon has higher energy, shorter wavelength, and higher frequency.

Exercises
59. 1.76×10^{11} C/kg
61. 2.7×10^3 kg/m^3; 2.7 times as much
63. 5×10^{-6} m
65. 3.32×10^{-19} J
67. 5.43×10^{14} Hz
69. 9.95×10^{-3} m
71. $h/2 = 1.06 \times 10^{-34}$ J·s
73. 8.48×10^{-10} m
75. 7.24×10^{14} Hz
77. 2.46×10^{15} Hz; above
79. 1:64
81. 8.29 keV

Chapter 24

Questions
1. *Successes:* Accounting for the stability of atoms, the numerical values for wavelengths of spectral lines in hydrogen and hydrogen-like atoms, and the general features of the periodic table. *Failures:* Could not account for why accelerating electrons didn't radiate; the spectral lines in nonhydrogen-like atoms; the splitting of spectral lines into two or more lines; the relative intensities of the spectral lines; details of the periodic table, including the capacity of each shell; and relativity.
3. 20 Hz
5. No, the electrons would still be accelerating.
7. Photoelectric effect
9. This implies that photons are massless.
11. Energy
13. Electrons and photons are neither particles nor waves; they simply exhibit behavior that we interpret as wavelike and particle-like.
15. No
17. The wavelengths associated with the electrons must form standing wave around the nucleus.
19. The wavelength is too small.
21. Both electrons and photons behave as waves when producing the pattern and as particles when detected.
23. Electrons are deflected by electric and magnetic fields, electrons have mass and charge, photons travel at the speed of light.
25. An interference pattern
27. Intensities
29. The probability of finding the "particle" at a particular location
31. In the middle of the box
33. You cannot meaningfully say that the electron was ever at one particular location in the box, and therefore it does not have to move.
35. Along the y axis, not too near the origin and not too far away
37. 6
39. $n = 1$, $\ell = 0$, $m_\ell = 0$, and $m_s = +\frac{1}{2}$; $n = 1$, $\ell = 0$, $m_\ell = 0$, and $m_s = -\frac{1}{2}$; $n = 2$, $\ell = 0$, $m_\ell = 0$, and $m_s = +\frac{1}{2}$; $n = 2$, $\ell = 0$, $m_\ell = 0$, and $m_s = -\frac{1}{2}$
41. The Heisenberg uncertainty principle places precise limits on our simultaneous knowledge of specific variables that are important for systems on the atomic scale. It should not be generalized to apply outside this domain.
43. Momentum and energy do not form a Heisenberg pair.
45. There must be uncertainty in the momentum and therefore in the wavelength. A range of wavelengths is used to describe a localized electron.
47. The electrons have precisely defined positions and velocities.
49. Einstein was troubled by the idea that the probability associated with quantum mechanics was all that can be known.
51. Energy

53. Nothing
55. The electrons return to their lower energy levels through a series of jumps, emitting visible light.
57. The electrons must remain in the excited energy levels for a relatively long time.
59. A photon induces an electron in an excited state to return to a lower state, producing a second, identical photon.

Exercises

61. 2.21×10^{-38} m
63. 2.88×10^{-11} m
65. 6.22×10^{-7} m
67. 7.28×10^{6} m/s
69. 1.66×10^{-25} kg·m/s
71. 1.77×10^{-33} kg·m/s
73. 1.33×10^{-23} kg·m/s
75. 3.32×10^{-36} m
77. 0.132 nm
79. 1.46×10^{7} m/s; 607 eV
81. 2.07×10^{-7} eV
83. 2.07×10^{-11} s

Chapter 25

Questions

1. Becquerel's radiation was found to occur naturally.
3. The wavelengths of visible light are much bigger than the size of nuclei.
5. Alpha and beta radiation
7. Alpha particles
9. Protons
11. A neutron or a proton
13. (a) Yttrium (b) Cerium (c) Bromine
15. (a) 12 neutrons, 12 protons, and 12 electrons (b) 32 neutrons, 27 protons, and 27 electrons (c) 126 neutrons, 82 protons, and 82 electrons
17. 90
19. $^{20}_{10}$Ne
21. Decreases by 1
23. (a) $^{220}_{90}$Th (b) $^{193}_{81}$Tl
25. (a) $^{18}_{8}$O (b) $^{90}_{39}$Y
27. (a) $^{181}_{76}$Os (b) $^{237}_{93}$Np
29. (a) $^{28}_{14}$Si (b) $^{64}_{30}$Zn
31. A neutron must leave the nucleus.
33. $^{64}_{28}$Ni or $^{64}_{30}$Zn
35. (a) Beta plus (b) Alpha
37. 206Pb
39. $^{17}_{9}$F
41. There would be a decrease of one proton and one neutron.
43. 4 kg
45. It could happen at any time because the process is random.
47. By determining the fraction of the radioactive ^{14}C that has decayed to ^{12}C
49. No
51. Gamma ray, electron, and alpha particle
53. Most natural radiation exposure comes from photons and electrons, for which there is little difference between rads and rems.
55. Nuclear power
57. It takes a long time for the effects to show up, and there are other causes.
59. Skin provides protection because alpha particles cannot penetrate very far

Exercises

61. 6 km
63. 1.99×10^{-26} kg

65. 192.2 amu; Ir
67. 1.007825 amu
69. 2 μCi
71. 64 trillion
73. 11,400 years
75. $\frac{1}{8}$
77. 50

Chapter 26

Questions

1. 5,000,000 V
3. They all acquire the same kinetic energy.
5. Alpha particle, because of its larger charge
7. The energy losses due to the radiation of the accelerating charges are less for massive particles.
9. Strong, electromagnetic, weak, gravitational
11. There must be a strong force to counteract the electric repulsive force and hold the nucleons together.
13. Both are fundamental forces that can be attractive or repulsive. The strong force is stronger, has a finite range, and changes from attractive to repulsive at short distances. The electromagnetic force has an infinite range and remains either attractive or repulsive.
15. A 2.2-MeV gamma ray
17. 96 protons and 138 neutrons
19. Carbon nucleus
21. Nitrogen-17
23. The process will continue until the nucleons are most tightly bound, which occurs in the region of the peak of the average-binding-energy curve at iron.
25. They are about equal because the energy spacing between proton states and neutron states is about the same.
27. It is energetically more favorable to add neutrons than it is to add protons.
29. It would decay via beta plus or electron capture, converting a proton into a neutron.
31. Beta minus decay
33. The splitting of a heavy nucleus into two or more lighter ones
35. Two
37. Because it releases more than one neutron on the average.
39. To initiate additional fission reactions
41. Nuclear bomb
43. To absorb enough neutrons to ensure that an average of one neutron from each fission process initiates another
45. Plutonium-239
47. New fuel is bred by converting an isotope that does not readily fission into one that does.
49. The combining of two or more light nuclei to form a heavier one.
51. The temperature is high, but the density is low, so there is little heat energy.
53. Fuel is much more available, and there is less risk.
55. Agrees with solar models and with the age of Earth, as determined by radioactive dating
57. Fusion is the combining of lighter elements to form heavier ones, whereas fission is the splitting of heavier elements to form lighter ones.

Exercises

59. 7.31×10^{-21} kg · m/s; 9.07×10^{-14} m
61. Use $E = \frac{1}{2}mv^2$ and solve for $v = 8.75 \times 10^{9}$ m/s $> c$.
63. Use $E = mc^2$ and convert units.
65. 105 MeV
67. 7.012 161 amu
69. 18.6 keV

71. 5.24 MeV
73. Approximately 200 MeV
75. Subtract the product and reactant masses and convert to million electron volts (MeV).
77. 121
79. 3.87×10^{26} W

Chapter 27

Questions

1. Electron, proton, neutron, and photon
3. Positron; same magnitude but opposite sign
5. No
7. It will mutually annihilate with its corresponding particle.
9. Because they are electrically neutral, they do not leave tracks.
11. The total linear momentum is zero. If a single photon were emitted, it would carry away momentum E/c, yielding nonzero total momentum.
13. Conservation of charge and nucleon number (or baryon number)
15. Weak
17. The uncertainty principle provides the mechanism to create exchange particles without violating conservation of energy.
19. The intermediate vector bosons are quite massive, and their discovery awaited more energetic particle accelerators.
21. Exchange photons with very little energy can exist for very long times and can therefore travel infinite distances.
23. Its infinite range
25. Mesons have spins that are whole numbers, whereas baryons have spins of $\frac{1}{2}, \frac{2}{3}, \ldots$
27. Hadrons are not elementary; they are composed of quarks. Hadrons also participate in the strong interaction.
29. Yes, as evidenced by its beta decay.
31. 10^{-10} s
33. It doesn't have enough rest-mass energy.
35. e^+
37. (a) Muon and electron lepton numbers (b) Spin and baryon number (c) Strangeness
39. (a) 1 (b) −1 (c) 0
41. Two photons
43. uud for the antiproton and udd for the antineutron
45. $\overline{u}s$; −1
47. uuu
49. dss
51. ds; no
53. No
55. It would be a neutral meson.
57. By assigning the color quantum number to the quarks

Chapter 28

Questions

1. Supernovae, orbiting neutron stars, and orbiting black holes
3. The energy radiated in the form of gravitational waves is very small.
5. Taylor and Hulse found a pair of neutron stars orbiting each other and measured the orbital period to be decreasing, as predicted by theory.
7. Not if we observe a large number of protons
9. Gravitational force
11. Tiny loops
13. 15 billion years
15. Gravitational
17. To escape cosmic radiation, which would swamp the tiny neutrino signals
19. Measurements of ripples in the cosmic-ray background
21. Any physical system, no matter how complex, may be understood in terms of its component parts.

Glossary

aberration A defect in a mirror or lens causing light rays from a single point to fail to focus at a single point in space.

absolute zero The lowest possible temperature: 0 K, −273°C, or −459°F.

absorption spectrum The collection of wavelengths missing from a continuous distribution of wavelengths because of the absorption of certain wavelengths by the atoms or molecules in a gas.

activity The rate at which a collection of radioactive nuclei decay. One curie corresponds to 3.7×10^{10} decays per second; also called radioactivity.

alloy A metal produced by mixing other metals.

alpha (α) radiation The type of radioactive decay in which nuclei emit alpha particles (helium nuclei).

alpha particle The nucleus of helium consisting of two protons and two neutrons.

ampere The SI unit of electric current; 1 coulomb per second. The current in each of two parallel wires when the magnetic force per unit length between them is 2×10^{-7} newton per meter.

amplitude The maximum distance from the equilibrium position that occurs in periodic motion.

angular momentum A vector quantity giving the rotational momentum. For an object orbiting a point, the magnitude of the angular momentum is the product of the linear momentum and the radius of the path, $L = mvr$. For a spinning object, it is the product of the rotational inertia and the rotational velocity, $L = I\omega$.

antinode One of the positions in a standing-wave or interference pattern where there is maximum movement; that is, the amplitude is a maximum.

antiparticle A subatomic particle with the same-size properties as those of the particle, although some may have the opposite sign. The positron is the antiparticle of the electron.

apparent weight The support force needed to maintain an object at rest relative to a reference system. For inertial systems, the apparent weight is equal in magnitude to the true weight, the gravitational force acting on the object.

Archimedes' principle The buoyant force is equal to the weight of the displaced fluid.

astigmatism An aberration, or defect, in a mirror or lens that causes the image of a point to spread out into a line.

atom The smallest unit of an element that has the chemical and physical properties of that element. An atom consists of a nucleus surrounded by an electron cloud.

atomic mass The mass of an atom in atomic mass units. Sometimes this refers to the atomic mass number—the number of neutrons and protons in the nucleus.

atomic mass unit One-twelfth the mass of a neutral carbon atom containing six protons and six neutrons.

atomic number The number of protons in the nucleus or the number of electrons in the neutral atom of an element. This number also gives the order of the elements in the periodic table.

average acceleration The change in velocity divided by the time it takes to make the change, $\bar{a} = \Delta v/\Delta t$; measured in units such as (meters per second) per second. An acceleration can result from a change in speed, a change in direction, or both.

average speed The distance traveled divided by the time taken, $\bar{s} = d/t$; measured in units such as meters per second or miles per hour.

average velocity The change in position—displacement—divided by the time taken, $\bar{v} = \Delta x/\Delta t$.

Avogadro's number The number of molecules in 1 mole of any substance. Equal to 6.02×10^{23} molecules.

baryon A type of hadron having a spin of $\frac{1}{2}, \frac{3}{2}, \frac{5}{2}, \ldots$ times the smallest unit. The most common baryons are the proton and neutron.

beats A variation in the amplitude resulting from the superposition of two waves that have nearly the same frequencies. The frequency of the variation is equal to the difference in the two frequencies.

Bernoulli's principle The pressure in a fluid decreases as its velocity increases.

beta (β) radiation The type of radioactive decay in which nuclei emit electrons or positrons (antielectrons).

beta particle An electron emitted by a radioactive nucleus.

binding energy The amount of energy required to take a nucleus apart. The analogous amount of energy for other bound systems.

black hole A massive star that has collapsed to such a small size that its gravitational force becomes so strong that not even light can escape from its "surface."

bottom The flavor of the fifth quark.

British thermal unit The amount of heat required to raise the temperature of 1 pound of water by 1°F.

buoyant force The upward force exerted by a fluid on a submerged or floating object. *See* Archimedes' principle.

calorie The amount of heat required to raise the temperature of 1 gram of water by 1°C.

camera obscura A room with a small hole in one wall, used by artists to produce images.

cathode ray An electron emitted from the negative electrode in an evacuated tube.

center of mass The balance point of an object. The location in an object that has the same translational motion as the object if it were shrunk to a point.

centi- A prefix meaning "one-hundredth.." A centimeter is $\frac{1}{1000}$ meter.

centrifugal An adjective meaning "center-seeking."

centrifugal force A fictitious force arising in a rotating reference system. It points away from the center, in the direction opposite to the centripetal acceleration.

centripetal An adjective meaning "center-fleeing."

centripetal acceleration The acceleration of an object directed toward the center of its circular path. For uniform circular motion, it has a magnitude v^2/r.

centripetal force The force on an object directed toward the center of its circular path. For uniform circular motion, it has a magnitude mv^2/r.

chain reaction A process in which the fissioning of one nucleus initiates the fissioning of others.

change of state The change in a substance between solid and liquid or between liquid and gas.

charge A property of an elementary particle that determines the strength of its electric force with other particles possessing charge. Measured in coulombs, or in integral multiples of the charge on the proton.

charged Possessing a net negative or positive charge.

charm The flavor of the fourth quark.

chromatic aberration A property of lenses that causes different colors (wavelengths) of light to have different focal lengths.

coherent A property of two or more sources of waves that have the same wavelength and maintain constant phase differences.

complementarity principle A complete description of an atomic entity, such as an electron or a photon, requires both a particle description and a wave description, but not at the same time.

complementary color For lights, two colors that combine to form white.

complete circuit A continuous conducting path from one end of a battery (or other source of electric potential) to the other end of the battery.

compound A combination of chemical elements that forms a new substance with its own properties.

conduction, thermal The transfer of thermal energy by collisions of the atoms or molecules within a substance.

conductor A material that allows the passage of electric charge or the easy transfer of thermal energy. Metals are good conductors.

conservation of angular momentum If the net external torque on a system is zero, the total angular momentum of the system does not change.

conservation of charge In an isolated system, the total charge is conserved.

conservation of energy The total energy of an isolated system does not change.

conservation of linear momentum The total linear momentum of a system does not change if there is no net external force.

conservation of mass The total mass in a closed system does not change even when physical and chemical changes occur.

conserved This term is used in physics to mean that a number associated with a physical property does not change; it is invariant.

convection The transfer of thermal energy in fluids by means of currents such as the rising of hot air and the sinking of cold air.

Coriolis force A fictitious force that occurs in rotating reference frames. It is responsible for the direction of the winds in hurricanes.

coulomb The SI unit of electric charge. The amount of charge passing a given point in a conductor carrying a current of 1 ampere. The charge of 6.25×10^{18} protons.

crest The peak of a wave disturbance.

critical Describes a chain reaction in which an average of one neutron from each fission reaction initiates another reaction.

critical angle The minimum angle of incidence for which total internal reflection occurs.

critical mass The minimum mass of a substance that will allow a chain reaction to continue without dying out.

crystal A material in which the atoms are arranged in a definite geometric pattern.

curie A unit of radioactivity; 3.7×10^{10} decays per second.

current A flow of electric charge; measured in amperes.

cycle One complete repetition of a periodic motion. It may start anywhere in the motion.

daughter nucleus The nucleus resulting from the radioactive decay of a parent nucleus.

definite proportions, law of When two or more elements combine to form a compound, the ratios of the masses of the combining elements have fixed values.

density A property of material equal to the mass of the material divided by its volume; measured in kilograms per cubic meter or grams per cubic centimeter.

diaphragm An opening that is used to limit the amount of light passing through a lens.

diffraction The spreading of waves passing through an opening or around a barrier.

diffuse reflection The reflection of rays from a rough surface. The reflected rays do not leave at fixed angles.

diopter A measure of the focal length of a mirror or lens, equal to the inverse of the focal length measured in meters.

disordered system A system with an arrangement equivalent to many other possible arrangements.

dispersion The spreading of light into a spectrum of color; the variation in the speed of a periodic wave due to its wavelength or frequency.

displacement A vector quantity giving the straight-line distance and direction from an initial position to a final position. In wave (or oscillatory) motion, the distance of the disturbance (or object) from its equilibrium position.

Doppler effect A change in the frequency of a periodic wave due to the motion of the observer, the source, or both.

efficiency The ratio of the work produced to the energy input. For an ideal heat engine, the Carnot efficiency is given by $1 - T_c / T_h$.

elastic A collision or interaction in which kinetic energy is conserved.

electric field The space surrounding a charged object where each location is assigned a vector equal to the force experienced by one unit of positive charge placed at that location; measured in newtons per coulomb.

electric field lines A representation of the electric field in a region of space. The electric field is tangent to the field line at any point, and its magnitude is proportional to the local density of field lines.

electric potential The electric potential energy divided by the object's charge; the work done in bringing a positive test charge of 1 coulomb from the zero reference location to a particular point in space; measured in joules per coulomb.

electric potential energy The work done in bringing a charged object from some zero reference location to a particular point in space.

electromagnet A magnet constructed by wrapping wire around an iron core. The electromagnet can be turned on and off by turning the current in the wire on and off.

electromagnetic wave A wave consisting of oscillating electric and magnetic fields. In a vacuum, electromagnetic waves travel at the speed of light.

electron A basic constituent of atoms; a lepton.

electron capture A decay process in which an inner atomic electron is captured by the nucleus. The daughter nucleus has the same number of nucleons as the parent but one less proton.

electron volt A unit of energy equal to the kinetic energy acquired by an electron or proton falling through an electric potential difference of 1 volt; equal to 1.6×10^{-19} joule.

element Any chemical species that cannot be broken up into other chemical species.

emission spectrum The collection of discrete wavelengths emitted by atoms that have been excited by heating or by electric currents.

entropy A measure of the order of a system. The second law of thermodynamics states that the entropy of an isolated system tends to increase.

equilibrium position A position where the net restoring force is zero.

equivalence principle Constant acceleration is completely equivalent to a uniform gravitational field.

ether The hypothesized medium through which light was believed to travel.

exclusion principle No two electrons can have the same set of quantum numbers. This statement also applies to protons, neutrons, and other baryons.

field A region of space where each location is assigned a value or a vector. *See* electric field, gravitational field, and magnetic field.

fission The splitting of a heavy nucleus into two or more lighter nuclei.

flavor The types of quarks: down, up, strange, charm, bottom, or top.

fluorescence The property of a material whereby it emits visible light when it is illuminated by ultraviolet light.

focal length The distance from a mirror or the center of a lens to its focal point.

focal point The location at which a mirror or a lens focuses rays parallel to the optic axis or from which such rays appear to diverge.

force A push or a pull, measured by the acceleration it produces on a standard, isolated object, $\mathbf{F}_{net} = m\mathbf{a}$; measured in newtons.

frequency The number of times a periodic motion repeats in a unit of time. It is equal to the inverse of the period, $f = 1/T$; measured in hertz.

fundamental frequency The lowest resonant frequency for an oscillating system.

fusion The combining of light nuclei to form a heavier nucleus.

gamma (γ) radiation The type of radioactive decay in which nuclei emit high-energy photons; the range of frequencies of the electromagnetic spectrum that lies beyond the X rays.

gas Matter with no definite shape or volume.

gauss A unit of magnetic field strength; 10^{-4} tesla.

geocentric model A model of the universe with Earth at its center.

gluon An exchange particle responsible for the force between quarks. The eight gluons differ only in their color quantum numbers.

gravitation, law of Universal $F = Gm_1 m_2/r^2$, where F is the force between any two objects, G is a universal constant, m_1 and m_2 are the masses of the two objects, and r is the distance between their centers.

gravitational field The space surrounding an object where each location is assigned a vector equal to the gravitational force experienced by a 1-kilogram mass placed at that location.

gravitational mass The property of a particle that determines the strength of its gravitational interaction with other particles.

gravitational potential energy The work done by the force of gravity when an object falls from a particular point in space to the location assigned the value of zero, $GPE = mgh$.

graviton The exchange particle responsible for the gravitational force.

gravity wave A wave disturbance caused by the acceleration of masses.

ground state The lowest energy state of a system allowed by quantum mechanics.

grounding Establishing an electric connection to Earth in order to neutralize an object.

hadrons The family of particles that participate in the strong interaction. Baryons and mesons are the two subfamilies.

half-life The time during which one-half of a sample of a radioactive substance decays.

halo A ring of light that appears around the Sun or Moon. It is produced by refraction in ice crystals.

harmonic A frequency that is a whole-number multiple of the fundamental frequency.

heat Energy flowing because of a difference in temperature.

heat capacity The amount of heat required to raise the temperature of an object by 1°C.

heat engine A device for converting heat into mechanical work.

heat pump A reversible heat engine that acts as a furnace in winter and an air conditioner in summer.

heliocentric model A model of the universe with the Sun at its center.

hologram A three-dimensional record of visual information.

holography The photographic process for producing three-dimensional images.

hyperopia Farsightedness. Images of distant objects are formed beyond the retina.

ideal gas An enormous number of tiny particles separated by relatively large distances. The particles have no internal structure, are indestructible, and do not interact with each other except when they collide; all collisions are elastic.

ideal gas law $PV = nRT$, where P is the pressure, V is the volume, T is the absolute temperature, n is the number of moles, and R is the gas constant.

impulse The product of the force and the time during which it acts, $\mathbf{F}\Delta t$. This vector quantity is equal to the change in momentum.

in phase Describes two or more waves with the same wavelength and frequency that have their crests lined up.

index of refraction An optical property of a substance that determines how much light bends on entering or leaving it. The index is equal to the ratio of the speed of light in a vacuum to that in the substance.

inelastic A collision or interaction in which kinetic energy is not conserved.

inertia An object's resistance to a change in its velocity. *See* inertial mass.

inertia, law of *See* motion, Newton's first law of.

inertial force A fictitious force that arises in accelerating (noninertial) reference systems. Examples are centrifugal and Coriolis forces.

inertial mass An object's resistance to a change in its velocity; measured in kilograms.

inertial reference system Any reference system in which the law of inertia (Newton's first law of motion) is valid.

instantaneous speed The limiting value of the average speed as the time interval becomes infinitesimally small. The magnitude of the instantaneous velocity.

insulator A material that does not allow the passage of electric charge or is a poor conductor of thermal energy. Ceramics are good electrical insulators; wood and stationary air are good thermal insulators.

interference The superposition of waves.

intermediate vector bosons The exchange particles of the weak nuclear interaction: the W^+, W^-, and Z^0 particles.

internal energy The total microscopic energy of an object, which includes its atomic and molecular translational and rotational kinetic energies, its vibrational energy, and the energy stored in the molecular bonds.

inversely proportional A relationship in which two quantities have a constant product. If one quantity increases by a certain factor, the other decreases by the same factor.

inverse-square Describes a relationship in which a quantity is related to the reciprocal of the square of a second quantity. Examples include the force laws for gravity and electricity; the force is proportional to the inverse-square of the distance.

ion An atom with missing or extra electrons.

ionization The removal of one or more electrons from an atom.

isotope An element containing a specific number of neutrons in its nuclei. Examples are and , carbon atoms with six and eight neutrons, respectively.

joule The SI unit of energy, equal to 1 newton acting through a distance of 1 meter.

kilo- A prefix meaning "one thousand." A kilometer is 1000 meters.

kilogram The SI unit of mass. A kilogram of material weighs about 2.2 pounds on Earth.

kilowatt-hour A unit of energy; 3,600,000 joules. One kilowatt-hour of energy is transformed to other forms when a machine runs at a power of 1000 watts for 1 hour.

kinetic energy The energy of motion, mv^2; measured in joules.

kinetic friction The frictional force between two surfaces in relative motion. This force does not depend much on the relative speed.

Kirchhoff's junction rule The sum of the currents entering any junction in a circuit must equal the sum of the currents leaving that junction.

Kirchhoff's loop rule Along any path from the positive terminal to the negative terminal of a battery, the voltage drops across the resistive elements encountered must add up to the battery voltage.

laser A device that uses stimulated emissions to produce a coherent beam of electromagnetic radiation. *Laser* is the acronym for *l*ight *a*mplification by *s*timulated *e*mission of *r*adiation.

latent heat The amount of heat required to melt (or vaporize) a unit mass of a substance. The same amount of heat is released when a unit mass of the same substance freezes (or condenses); measured in calories per gram or kilojoules per kilogram.

lepton A family of elementary particles that includes the electron, muon, tau, and their associated neutrinos.

light ray A line that represents the path of light in a given direction.

line of stability The locations of the stable nuclei on a graph of the number of neutrons versus the number of protons.

linear momentum A vector quantity equal to the product of an object's mass and its velocity, $p = mv$.

liquid Matter with a definite volume that takes the shape of its container.

liquid crystal A liquid that exhibits a rough geometrical ordering of its atoms.

longitudinal wave A wave in which the vibrations of the medium are parallel to the direction the wave is moving.

macroscopic Describes the bulk properties of a substance such as mass, size, pressure, and temperature.

magnetic field The space surrounding a magnetic object, where each location is assigned a value determined by the torque on a compass placed at that location. The direction of the field is in the direction of the north pole of the compass.

magnetic monopole A hypothetical, isolated magnetic pole.

magnetic pole One end of a magnet; analogous to an electric charge.

magnitude The size of a vector quantity. For example, speed is the magnitude of a velocity.

mass *See* center of mass, critical mass, gravitational mass, and inertial mass.

matter-wave amplitude The wave solution to Schrödinger's equation for atomic and subatomic particles. The square of the matter-wave amplitude gives the probability of finding the particle at a particular location.

mechanical energy The sum of the kinetic energy and various potential energies, which may include the gravitational and the elastic potential energies.

meson A type of hadron with whole-number units of spin. This family includes the pion, kaon, and eta.

meter The SI unit of length equal to 39.37 inches, or 1.094 yards.

microscopic Describes properties not visible to the naked eye such as atomic speeds or the masses and sizes of atoms.

milli- A prefix meaning "one-thousandth." A millimeter is $\frac{1}{1000}$ meter.

mirage An optical effect that produces an image that looks as if it has been reflected from the surface of a body of water.

moderator A material used to slow the neutrons in a nuclear reactor.

mole The amount of a substance that has a mass in grams numerically equal to the mass of its molecules in atomic mass units.

molecule A combination of two or more atoms.

momentum Usually refers to linear momentum. *See* angular momentum, conservation of angular momentum, conservation of linear momentum, and linear momentum.

motion, Newton's first law of The velocity of an object remains constant unless an unbalanced force acts on the object.

motion, Newton's second law of $\mathbf{F}_{net} = m\mathbf{a}$. The net force on an object is equal to its mass times its acceleration. The net force and the acceleration are vectors that always point in the same direction.

motion, Newton's third law of If an object exerts a force on a second object, the second object exerts an equal force on the first object.

muon A type of lepton; often called a heavy electron.

myopia Nearsightedness. Images of distant objects are formed in front of the retina.

neutrino A neutral lepton; one exists for each of the charged leptons (electron, muon, and tau).

neutron The neutral nucleon; a member of the baryon and hadron families of elementary particles.

newton The SI unit of force. A net force of 1 newton accelerates a mass of 1 kilogram at a rate of 1 (meter per second) per second.

node One of the positions in a standing-wave or interference pattern where there is no movement; that is, the amplitude is zero.

noninertial reference system Any reference system in which the law of inertia (Newton's first law of motion) is not valid. An accelerating reference system is noninertial.

normal A line perpendicular to a surface or curve.

nucleon Either a proton or a neutron.

nucleus The central part of an atom that contains the protons and neutrons.

ohm The SI unit of electrical resistance. A current of 1 ampere flows through a resistance of 1 ohm under 1 volt of potential difference.

Ohm's law The resistance of an object is equal to the voltage across it divided by the current through it, $R = V/I$.

optic axis A line passing through the center of a curved mirror and the center of the sphere from which the mirror is made; a line passing through a lens and both focal points.

order of magnitude The value of a quantity rounded off to the nearest power of 10.

ordered system A system with an arrangement belonging to a group with the smallest number (possibly one) of equivalent arrangements.

oscillation A vibration about an equilibrium position or shape.

pair production The conversion of energy into matter in which a particle and its antiparticle are produced. This usually refers to the production of an electron and a positron (antielectron).

parallel Two circuit elements are wired in parallel when the current can flow through one or the other but not both. Elements that are wired parallel to each other are directly connected to each other at both terminals.

parent nucleus A nucleus that decays into a daughter nucleus.

particle accelerator A device for accelerating charged particles to high velocities.

penumbra The transition region between the darkest shadow and full brightness. Only part of the light from the source reaches this region.

period The shortest length of time it takes a periodic motion to repeat. It is equal to the inverse of the frequency, $T = 1/f$.

periodic wave A wave in which all the pulses have the same size and shape. The wave pattern repeats itself over a distance of 1 wavelength and over a time of 1 period.

phosphorescence The property of a material whereby it continues to emit visible light after it has been illuminated by ultraviolet light.

photoelectric effect The ejection of electrons from metallic surfaces by illuminating light.

photon A particle of light. The energy of a photon is given by the relationship $E = hf$, where f is the frequency of the light and h is Planck's constant. It is the exchange particle for the electromagnetic interaction.

pion The least massive meson. The pion has three charge states: +1, 0, and −1.

plasma The fourth state of matter in which one or more electrons have been stripped from the atoms forming an ion gas.

polarized A property of a transverse wave when its vibrations are all in a single plane.

positron The antiparticle of the electron.

pound The unit of force in the U.S. customary system; the weight of 0.454 kilogram on Earth.

power The rate at which energy is converted from one form to another, $P = (E/\Delta t;$ measured in joules per second, or watts. In electric circuits the power is equal to the current times the voltage, $P = IV$.

powers-of-ten notation A method of writing numbers in which a number between 1 and 10 is multiplied or divided by 10 raised to a power.

pressure The force per unit area of surface; measured in newtons per square meter, or pascals.

projectile motion A type of motion that occurs near Earth's surface when the only force acting on the object is that of gravity.

proportional A relationship in which two quantities have a constant ratio. If one quantity increases by a certain factor, the other increases by the same factor.

proton The positively charged nucleon; a member of the baryon and hadron families of elementary particles.

quantum (pl., **quanta**) The smallest unit of a discrete property. For instance, the quantum of charge is the charge on the proton.

quantum mechanics The rules for the behavior of particles at the atomic and subatomic levels.

quantum number A number giving the value of a quantized quantity. For instance, a quantum number specifies the angular momentum of an electron in an atom.

quark A constituent of hadrons. Quarks come in six flavors of three colors each. Three quarks make up the baryons, whereas a quark and an antiquark make up the mesons.

rad The acronym for *r*adiation *a*bsorbed *d*ose. A rad of radiation deposits 0.01 joule per kilogram of material.

radiation The transport of energy via electromagnetic waves; particles emitted in radioactive decay.

real image An image formed by the convergence of light.

reference system A collection of objects not moving relative to each other that can be used to describe the motion of other objects. *See* inertial reference system and noninertial reference system.

reflecting telescope A type of telescope using a mirror as the objective.

reflection, law of The angle of reflection (measured relative to the normal to the surface) is equal to the angle of incidence. The incident ray, the reflected ray, and the normal all lie in the same plane.

refracting telescope A type of telescope using a lens as the objective.

refraction The bending of light that occurs at the interface between two transparent media. It occurs when the speed of light changes.

refrigerator A heat engine running backward.

relativity, Galilean principle of The laws of motion are the same in all inertial reference systems.

relativity, general theory of An extension of the special theory of relativity to include the concept of gravity.

relativity, special theory of A comprehensive theory of space and time that replaces Newtonian mechanics when velocities get very large.

rem The acronym for *r*oentgen *e*quivalent in *m*ammals, a measure of the biological effects caused by radiation.

resistance The impedance to the flow of electric current; measured in volts per ampere, or ohms. The resistance is equal to the voltage across the object divided by the current through it, $R = V/I$.

resonance A large increase in the amplitude of a vibration when a force is applied at a natural frequency of the medium or object.

rest-mass energy The energy associated with the mass of a particle; given by $E_o = mc^2$, where c is the speed of light.

retroreflectors Three flat mirrors at right angles to each other that reflect light back to its source.

rotational acceleration The change in rotational speed divided by the time it takes to make the change.

rotational inertia The property of an object that measures its resistance to a change in its rotational speed.

rotational kinetic energy Kinetic energy associated with the rotation of a body, Ω.

rotational speed The angle of rotation or revolution divided by the time taken, .

rotational velocity A vector quantity that includes the rotational speed and the direction of the axis of rotation.

series An arrangement of resistances (or batteries) on a single pathway so that the current flows through each element.

shell A collection of electrons in an atom that have approximately the same energy.

shock wave The characteristic cone-shaped wave front that is produced whenever an object travels faster than the speed of the waves in the surrounding medium.

short circuit A path in an electric circuit that has very little resistance.

solid Matter with a definite size and shape.

sonar Sound waves in water.

spacetime A combination of time and three-dimensional space that forms a four-dimensional geometry expressing the connections between space and time.

special relativity, first postulate of The laws of physics are the same for all inertial reference systems.

special relativity, second postulate of The speed of light in a vacuum is a constant regardless of the speed of the source or the speed of the observer.

specific heat The amount of heat required to raise the temperature of a unit mass of a substance by 1 degree; measured in calories per gram-degree Celsius or joules per kilogram-kelvin.

spherical aberration A property of lenses and mirrors caused by grinding the surface to a spherical rather than a parabolic shape.

spring constant The amount of force required to stretch a spring by 1 unit of length; measured in newtons per meter.

stable equilibrium An equilibrium position or orientation to which an object returns after being slightly displaced.

standing wave The interference pattern produced by two waves of equal amplitude and frequency traveling in opposite directions. The pattern is characterized by alternating nodal and antinodal regions.

static friction The frictional force between two surfaces at rest relative to each other. This force is equal and opposite to the net applied force if the force is not large enough to make the object accelerate.

stimulated emission The emission of a photon from an atom because of the presence of an incident photon. The emitted photon has the same energy, direction, and phase as the incident photon.

strange particle A particle with a nonzero value of strangeness. In the quark model, it is made up of one or more quarks carrying the quantum property of strangeness.

strangeness The flavor of the third quark.

strong force The force responsible for holding the nucleons together to form nuclei.

subcritical Describes a chain reaction that dies out because an average of less than one neutron from each fission reaction causes another fission reaction.

supercritical Describes a chain reaction that grows rapidly because an average of more than one neutron from each fission reaction causes another fission reaction. An extreme example of this is the explosion of a nuclear bomb.

superposition The combining of two or more waves at a location in space.

Système International d'Unités The French name for the metric system, or International System (SI), of units.

temperature, absolute The temperature scale with its zero point at absolute zero and degrees equal to those on the Celsius scale. Also called the Kelvin temperature scale.

temperature, Celsius The temperature scale with the values of 0°C and 100°C for the temperatures of freezing and boiling water, respectively. Its degree is that of the Fahrenheit degree.

temperature, Fahrenheit The temperature scale with values of 32°F and 212°F for the temperatures of freezing and boiling water, respectively.

temperature, Kelvin The temperature scale with its zero point at absolute zero and a degree equal to that on the Celsius scale. Also called the absolute temperature scale.

terminal speed The speed obtained in free fall when the upward force of air resistance is equal to the downward force of gravity.

tesla The SI unit of magnetic field.

thermal energy Internal energy.

thermal equilibrium A condition in which there is no net flow of thermal energy between two objects. This occurs when the two objects obtain the same temperature.

thermal expansion The increase in size of a material when heated.

thermodynamics The area of physics that deals with the connections between heat and other forms of energy.

thermodynamics, first law of The increase in internal energy of a system is equal to the heat added plus the work done on the system.

thermodynamics, second law of There are three equivalent forms: (1) It is impossible to build a heat engine to perform mechanical work that does not exhaust heat to the surroundings. (2) It is impossible to build a refrigerator that can transfer heat from a lower temperature region to a higher temperature region without expending mechanical work. (3) The entropy of an isolated system tends to increase.

thermodynamics, third law of Absolute zero may be approached experimentally but can never be reached.

thermodynamics, zeroth law of If objects A and B are each in thermodynamic equilibrium with object C, then A and B are in thermodynamic equilibrium with each other. All three objects are at the same temperature.

top The flavor of the sixth quark.

torque The rotational analog of force. It is equal to the radius multiplied by the force perpendicular to the radius, $\tau = rF$. A net torque produces a change in an object's angular momentum.

total internal reflection A phenomenon that occurs when the angle of incidence of light traveling from a material with a higher index of refraction into one with a lower index of refraction exceeds the critical angle.

translational motion Motion along a path.

transverse wave A wave in which the vibrations of the medium are perpendicular to the direction the wave is moving.

trough A valley of a wave disturbance.

umbra The darkest part of a shadow where no light from the source reaches.

uncertainty principle The product of the uncertainty in the position of a particle along a certain direction and the uncertainty in the momentum along this same direction must be greater than Planck's constant, or $\Delta p_y \, \Delta y > h$. A similar relationship applies to the uncertainties in energy and time.

unstable equilibrium An equilibrium position or orientation that an object leaves after being slightly displaced.

vector A quantity with a magnitude and a direction. Examples are displacement, velocity, acceleration, momentum, and force.

velocity A vector quantity that includes the speed and the direction of the object; the displacement divided by the time taken, $\vec{v}$.

vibration An oscillation about an equilibrium position or shape.

virtual image The image formed when light only appears to come from the location of the image.

viscosity A measure of the internal friction within a fluid.

volt The SI unit of electric potential, 1 joule per coulomb. One volt produces a current of 1 ampere through a resistance of 1 ohm.

watt The SI unit of power, 1 joule per second.

wave The movement of energy from one place to another without any accompanying matter.

wavelength The shortest repetition length for a periodic wave. For example, it is the distance from crest to crest or from trough to trough.

weak force The force responsible for beta decay. This force occurs through the exchange of the W and Z^0 particles. All leptons and hadrons interact via this force.

weight The force of gravitational attraction of Earth for an object. $W = m\mathbf{g}$.

work The product of the force along the direction of motion and the distance moved, $W = Fd$. Measured in energy units, joules.

X ray A high-energy photon, usually produced by cathode rays or emitted by electrons falling to lower energy states in atoms; the range of frequencies in the electronic spectrum lying between the ultraviolet and the gamma rays.

Index

e = equation
g = definition in glossary
t = table

Physical Constants

Atomic mass unit	$\text{amu} = 1.66 \times 10^{-27}$ kg
Avogadro's number	$N_A = 6.02 \times 10^{23}$ particles/mol
Bohr radius	$r_1 = 5.29 \times 10^{-11}$ m
Electron volt	$\text{eV} = 1.60 \times 10^{-19}$ J
Elementary charge	$e = 1.60 \times 10^{-19}$ C
Coulomb's constant	$k = 8.99 \times 10^9$ N $\cdot$ m^2/C^2
Gravitational constant	$G = 6.67 \times 10^{-11}$ N $\cdot$ m^2/kg^2
Mass of electron	$m_e = 9.11 \times 10^{-31}$ kg
Mass of neutron	$m_n = 1.675 \times 10^{-27}$ kg
Mass of proton	$m_p = 1.673 \times 10^{-27}$ kg
Planck's constant	$h = 6.63 \times 10^{-34}$ J $\cdot$ s
Speed of light	$c = 3.00 \times 10^8$ m/s

Physical Data

Acceleration due to gravity	$g = 9.80$ m/s^2
Density of water	$D_w = 1.00 \times 10^3$ kg/m^3
Earth–Moon distance	$R_{EM} = 3.84 \times 10^8$ m
Earth–Sun distance	$R_{ES} = 1.50 \times 10^{11}$ m
Mass of Earth	$M_E = 5.98 \times 10^{24}$ kg
Mass of Moon	$M_M = 7.36 \times 10^{22}$ kg
Mass of Sun	$M_S = 1.99 \times 10^{30}$ kg
Radius of Earth	$R_E = 6.37 \times 10^6$ m
Radius of Moon	$R_M = 1.74 \times 10^6$ m
Radius of Sun	$R_S = 6.96 \times 10^8$ m
Speed of sound (20°C, 1 atm)	$v_s = 343$ m/s
Standard atmospheric pressure	$P_{atm} = 1.01 \times 10^5$ Pa

Standard Abbreviations

A	ampere
amu	atomic mass unit
atm	atmosphere
Btu	British thermal unit
C	coulomb
°C	degree Celsius
cal	calorie
eV	electron volt
°F	degree Fahrenheit
ft	foot
g	gram
h	hour
hp	horsepower
Hz	Hertz
in.	inch
J	joule
K	kelvin
kg	kilogram
lb	pound
m	meter
min	minute
mph	mile per hour
N	newton
Pa	pascal
psi	pound per square inch
rev	revolution
s	second
T	tesla
V	volt
W	watt
Ω	ohm